D0022992

DIGITAL DESIGN

A PRAGMATIC APPROACH

DIGITAL DESIGN

A PRAGMATIC APPROACH

Everett L. Johnson
The Wichita State University

Mohammad A. Karim
The University of Dayton

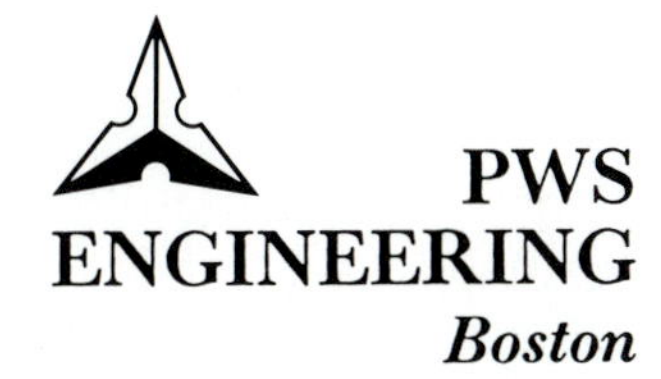

PWS
ENGINEERING
Boston

PWS PUBLISHERS

Prindle, Weber & Schmidt • Duxbury Press • PWS Engineering • Breton Publishers •
20 Park Plaza • Boston, Massachusetts 02116

PWS Publishers is a division of Wadsworth, Inc.

Library of Congress Cataloging in Publication Data

Johnson, Everett L.
Digital design.

Includes index.
1. Digital electronics. 2. Logic design.
I. Karim, Mohammad A. II. Title
TK7868.D5J634 1987 621.3815 86-93965
ISBN 0-534-06972-X

Printed in the United States of America

87 88 89 90 91—10 9 8 7 6 5 4 3 2 1

Sponsoring Editor: Bob Prior
Signing Representative: Robert Wolcott
Editorial Assistant: Kathleen Tibbetts
Production Coordinator: Jean Coulombre
Production: Technical Texts, Inc./Jean T. Peck
Interior and Cover Design: Jean Coulombre
Art Coordinator: Mary S. Mowrey
Interior Illustration: Oxford Illustrators, Ltd.
Typesetting: Alexander Typesetting, Inc.
Printing and Binding: Maple-Vail Book Manufacturing Group

PREFACE

Due to a phenomenal increase in the applications of digital circuits, the study of digital design techniques has assumed an important role in engineering and computer science curricula. The material in *Digital Design* supports that effort by encompassing both classical (Boolean algebra, combinational logic, and sequential logic) and modern (register transfer languages) perspectives. This text provides students with a wide but solid background for producing workable, reliable, and efficient digital systems. In an effort to be both clear and practical, the theory must be filtered for subjects that apply directly to design. We have pragmatically treated only the aspects necessary for acquiring expertise in digital design techniques with an integrated hardware–firmware approach.

It should be noted that this text is not tied to any commercial IC series, but has as its primary concern general design principles and the need to accommodate the rapidly changing technology. It is suggested that a manufacturer's specification book for a particular logic family or families be used in conjunction with this text to familiarize the student with the characteristics of the devices referred to throughout the text. In addition to standard classical subjects, this text also devotes considerable space to integrated circuits, Quine-McCluskey's tabular reduction method, bridging techniques, code conversions, error-correcting principles, complex arithmetic circuitry, memory, microprogramming, processor design, and input–output systems.

The material in *Digital Design* is divided into three areas: combinational logic design (Chapters One through Five); sequential logic design (Chapters Six through Ten); and complex digital systems design (Chapters Eleven through Fourteen). The first five chapters cover the core of combinational logic design. Chapter One introduces number systems, Boolean algebra, and discrete logic gates. Circuit simplification concepts using Karnaugh maps and Quine-McCluskey's tabular reduction technique are presented in Chapter Two. Chapter Three points out the practical limitations of various integrated circuits, and Chapter Four explores various design schemes to implement logic functions that compensate for IC limitations. The concepts of the earlier chapters are used in Chapter Five for designing various meaningful combinational logic circuits.

The next five chapters cover the design of sequential logic circuits. Chapter Six introduces different sequential devices and sequential designs. The next three chapters are devoted respectively to the design methods for clocked, pulse-mode, and fundamental-mode sequential circuits. Chapter Ten incorporates sequential design concepts in the design of counters and registers and introduces the concept of register transfer logic. The material pertaining to data and control units is presented here using register transfer sequences.

The last four chapters incorporate both combinational and sequential concepts for the design of robust digital systems. Chapter Eleven presents the algorithm and design for serial and parallel multi-bit arithmetic operations. These algorithms are effectively expressed in terms of register transfer sequences. Chapter Twelve examines binary data storage systems, both old and new. Chapter Thirteen discusses the important concepts of control and microprogramming that are utilized in the design of a digital processor. The final chapter considers the important topics of computer input–output and interfacing.

The basic concepts of each topic are developed in detail and then demonstrated with many worked-out examples. Problems at the end of each chapter allow the student to apply the techniques presented. Reading assignments from the bibliography included in each chapter should provide additional incentive to those who are considering a career in digital design. We have found that students who have successfully mastered the material in this text are ready to take additional courses in the design of digital systems and function well as digital designers in industry.

Preliminary versions of this text have been classroom-tested at The Wichita State University for several years with an enrollment of approximately 200 Electrical Engineering and Computer Science students per year. There are several course applications for this text, and the coupling between the chapters is loose enough to permit flexibility in adjusting the text to the needs of the students. Chapters One through Nine support a first course in digital logic design and switching theory. For a follow-up course in digital computer design, a review of Chapters Seven through Nine and complete coverage of Chapters Ten to Fourteen is suggested. This course can also make heavy use of manufacturers' specification books.

Acknowledgments

We wish to emphasize that the contributions of both authors have been equal and therefore our names appear alphabetically on this book. Successful creation of a textbook is a team effort with the authors only the most visible members. Among the most important members of this team are the reviewers, and we owe a debt of thanks to Aaron Collins, Tennessee Technological University; Eddie Fowler, Kansas State; James Herzog, Oregon State; John Maneely, University of Portland; Wm. Lee Pfefferman, University of the Pacific; Michael Shanblatt, Michigan State; and David Soldan, Oklahoma State. We express our thanks for the cooperation and encouragement of those persons responsible for the publishing of this book: Bob Prior and Jean Coulombre of PWS Publishers, and Jean Peck of Technical Texts, Inc.

Perhaps the most sustained support that we have received has come from Professor Roy H. Norris, Chairperson of the Electrical Engineering Department at The Wichita State University, who has provided us with continuous encouragement and good wishes. We would like to thank the many students who struggled through various versions of this text in manuscript form during their course work. We would particularly like to thank students Abdul Awwal, Abdul Basit, Michael Barfield, Mark Fiedler, Timothy Fisher, George Hassoun, Fayyaz Hussain, Meer Hussain, Richard Miller, Claudia Rollstin, Dinesh Sharma, Timothy Unruh, and Nancy Woodbridge for their help in examining the manuscript. Their suggestions, arguments, questions, and ideas were indispensable. Our colleagues at the Center for Energy Studies at WSU were extremely helpful in making us realize the full potential of word processing software. Finally, the help and encouragement of our families is sincerely appreciated. This project would have been abandoned long ago without their understanding.

E.L.J./M.A.K.

CONTENTS

CHAPTER ONE

Number Systems and Boolean Algebra

1.1 Introduction

A *digital system* usually consists of a combination of circuits connected in some specific way to perform a specified function where all variables involved are allowed only a finite number of discrete states. In particular, two-valued discrete systems are known as *binary systems*. The combinations of these two discrete states are used to represent numbers, characters, codes, and other information. There are many advantages that binary systems have over *analog (continuous) systems*. The solid-state devices used in digital circuits are extremely reliable and consistent in their functions as long as they are maintained in either of two states. In addition, because they are maintained in either of two states, they are less susceptible to variations of environment and have predetermined accuracy. The binary (two-valued) number system is the basis for the understanding of digital systems. In this chapter we shall examine the binary number system, its relationship to other specific number systems, and its application in the mathematical and logical operations that will allow us to study the design of digital systems.

This chapter also will present the rules and theorems that make up the fundamental concepts of a binary system, commonly known as *Boolean algebra*. The understanding of Boolean algebra is considered very vital because its applications directly lead to techniques that are essential in designing efficient digital systems. Boolean algebra serves as the basis for moving from a verbal description of the function of the desired digital device to an unambiguous mathematical description.

This first chapter will acquaint you with number systems and Boolean algebra and provide some facility in handling them. After studying this chapter, you should be able to:

- ○ Convert decimal numbers to nondecimal numbers;
- ○ Convert nondecimal numbers to decimal numbers;

- Find the sum and difference of two numbers given in any number system;
- Use various codes to represent numbers;
- Obtain truth tables from word descriptions;
- Obtain logic functions and logic networks from truth tables;
- Simplify Boolean functions.

1.2 Number Systems

The number system most familiar to us is the base-10 system, and the reason is obvious if we take a quick inventory of our fingers or toes. We have become so used to our number system that we seldom consider the fact that even though we only use the arabic digits 0, 1, 2, 3, 4, 5, 6, 7, 8, and 9, we still can represent numbers in excess of nine. We also can represent numbers that have an integer part and a fractional part set apart by a decimal point by assigning weights to the positions of the digits in a multi-digit number. For instance, 657.15 in base 10 can be written in polynomial notation as

$$(657.15)_{10} = (6 \times 10^2) + (5 \times 10^1) + (7 \times 10^0) + (1 \times 10^{-1}) + (5 \times 10^{-2})$$

This number is called a *fixed-point decimal* because the decimal point is in its proper place. The number $(657.15)_{10}$ can also be expressed as a *floating number* by changing the position of the decimal point and the value of the exponent. For example, the number can be written as $(0.65715 \times 10^3)_{10}$, or as $(6.5715 \times 10^2)_{10}$, or even as $(65715.0 \times 10^{-2})_{10}$. The correct location of the decimal point is indicated by the location of the floating point and by the exponent of 10.

In a fixed-point, base-R number system a number is written as

$$(N)_R = (\underbrace{a_m a_{m-1} a_{m-2} a_{m-3} \cdots a_0}_{\text{integer part}} \underset{\text{radix point}}{.} \underbrace{a_{-1} a_{-2} \cdots a_{-n}}_{\text{fraction part}}) \qquad [1.1]$$

where

$$0 \leq a_i \leq (R - 1)_R \qquad [1.2]$$

such that there are $m + 1$ integer digits and n fractional digits in the numbers. The base, R, is commonly known as the *radix* of the number system. Any decimal number, $(N)_{10}$, therefore, can be represented in another system with base R as follows:

$$(N)_{10} = a_m R^m + a_{m-1} R^{m-1} + \cdots + a_0 R^0 + a_{-1} R^{-1} + \cdots + a_{-n} R^{-n} \qquad [1.3]$$

where the part consisting of the coefficients a_m through a_0 is the integer portion, and the part consisting of the coefficients a_{-1} through a_{-n} is the fractional part.

TABLE 1.1 **Correspondence Table**

$(N)_{10}$	$(N)_2$	$(N)_8$	$(N)_{16}$
0	0	0	0
1	1	1	1
2	10	2	2
3	11	3	3
4	100	4	4
5	101	5	5
6	110	6	6
7	111	7	7
8	1000	10	8
9	1001	11	9
10	1010	12	A
11	1011	13	B
12	1100	14	C
13	1101	15	D
14	1110	16	E
15	1111	17	F
16	10000	20	10
242	11110010	362	F2

To meet the objective of this text, we shall examine numbers with bases 2 (binary), 8 (octal), 10 (decimal), and 16 (hexadecimal). Table 1.1 shows some corresponding numbers in these systems. An examination of this table should convince us that if we were to use the voltage output of an electronic device for representing a number, it would be easiest to do so in the binary system because this system requires only two values. The corresponding binary numbers, however, require a longer sequence of digits than either of the other three systems. Each binary digit is called a *bit*. The large number of bits in the binary representation presents no problem since in most digital systems operations occur in parallel.

The advantage of using the binary system is that there are many devices that display two stable states. One stable state can be assigned the value 1, the other 0. For example, a two-state device used in early computing machinery was the relay. If the relay contacts were together they represented a 1; if open they represented a 0. Relay-based systems were huge, slow, and consumed large amounts of energy. Computer systems progressed to vacuum tubes, then to transistors, and finally to the integrated circuits used today. Binary bits are now represented by voltages. We define a certain range of voltages as *logic 1* and another range of voltages as *logic 0*. Typically, the two ranges are separated by a forbidden range of voltages. The nominal values used for many technologies are 5 V for a 1 and 0 V for a 0.

Octal and hexadecimal numbers are used primarily as a convenient way to express binary numbers that have too many digits. Conversion between binary, octal, and hexadecimal is straightforward because the octal and hexadecimal bases are powers of 2.

1.3 Conversion between Different Bases

Initially, we shall restrict our discussion to conversion of positive numbers. The conversion of negative numbers is handled in a different way and will be discussed at a later stage. There are two different conversion processes that need to be considered: the conversion of a base-R number to its base-10 equivalent, and the conversion of a base-10 numbert to its base-R equivalent. Each of these conversion processes will involve two more processes: the conversion of the integer part and the conversion of the fraction part.

1.3.1 Conversion of Integers

The conversion of an m-digit, base-R integer to the equivalent base-10 integer is accomplished easily using Equation [1.3] and Table 1.1. Each of the base-R digits is converted to its base-10 equivalent and then multiplied by the respective R^j value, where j denotes the respective positional value of the integer between 0 and m. The products are then added to give the equivalent decimal integer. The following two examples illustrate the mechanics of conversion.

EXAMPLE 1.1

Find the decimal equivalent of the octal number 5672.

SOLUTION

The positional values of 5, 6, 7, and 2 are, respectively, 3, 2, 1, and 0. Therefore, the equivalent decimal is represented as follows:

$$\begin{aligned}(5 \times 8^3) + (6 \times 8^2) + (7 \times 8^1) + (2 \times 8^0) \\ &= [(5 \times 512) + (6 \times 64) + (7 \times 8) + (2 \times 1)]_{10} \\ &= (2560 + 0384 + 0056 + 0002)_{10} \\ &= (3002)_{10}\end{aligned}$$

EXAMPLE 1.2

Convert $(AF16B)_{16}$ to its base-10 equivalent number.

SOLUTION

We know from Table 1.1 that A = 10, F = 15, and B = 11 in the base-10 system. Therefore,

$$\begin{aligned}(AF16B)_{16} &= [(10 \times 16^4) + (15 \times 16^3) + (1 \times 16^2) + (6 \times 16^1) \\ &\quad + (11 \times 16^0)]_{10} \\ &= (655360 + 61440 + 256 + 96 + 11)_{10} \\ &= (717163)_{10}\end{aligned}$$

In order to convert a decimal integer to an equivalent base-R integer, each of the integer coefficients $(a_0, a_1, a_2, \ldots, a_m)$ of Equation [1.3] must be determined. In fact, the base-10 integer NI can be rewritten as

$$\begin{aligned}(NI)_{10} &= a_mR^m + a_{m-1}R^{m-1} + \cdots + a_1R^1 + a_0R^0 \\ &= (((\ldots(((a_mR + a_{m-1})R + a_{m-2})R + a_{m-3})\ldots)R + a_1) \\ &\quad R + a_0) \qquad [1.4]\end{aligned}$$

When both sides of this equation are divided by R, the least significant coefficient, a_0, is obtained as the remainder. Subsequent divisions by R would yield the remaining coefficients, a_1, a_2, a_3, and so on, until all of the coefficients have been determined. The conversion scheme is illustrated by the two examples that follow.

EXAMPLE 1.3

Convert $(624)_{10}$ to its base-2 equivalent.

SOLUTION

Successive division by 2 is carried out to obtain the remainders as follows:

Divisor	Quotient	Remainder	
2	624		
2	312	0	LSB
2	156	0	
2	78	0	
2	39	0	
2	19	1	
2	9	1	
2	4	1	
2	2	0	
2	1	0	
2	0	1	MSB

The remainder obtained at the last step is used as the *most significant bit* (*MSB*), while that obtained in the very beginning is used as the *least significant bit* (*LSB*). The rest of the remainders fall in between in the order of their appearances. Therefore,

$$(624)_{10} = (1001110000)_2$$

We can check the result by examining $\sum_n a_n R^n$. And indeed,

$$(1 \times 2^4) + (1 \times 2^5) + (1 \times 2^6) + (1 \times 2^9) = (624)_{10}$$

EXAMPLE 1.4

Find the hexadecimal equivalent of $(3875)_{10}$.

SOLUTION

Proceed as in the previous example:

Divisor	Quotient	Remainder	
16	3875		
16	242	3	LSB
16	15	2	
16	0	15	MSB

Note that the remainder, $(15)_{10}$, is expressed as F in hex-notation. Therefore,

$(3875)_{10} = (F23)_{16}$

Check: $(F \times 16^2) + (2 \times 16^1) + (3 \times 16^0) = (3875)_{10}$.

1.3.2 Conversion of Fractions

The technique for converting an n-digit, base-R fraction to the equivalent base-10 fraction is similar to the mechanism of conversion between integers. The base-10 equivalents of the base-R digits are multiplied by respective values of R^j, where j denotes the positional value of fractional bits and ranges between -1 and $-n$. The products are added together to obtain the equivalent decimal fraction.

EXAMPLE 1.5

Convert $(0.10111)_2$ to the base-10 equivalent.

SOLUTION

Each of the bits is multiplied by its corresponding positional multiplier, R^j:

$$\begin{aligned}(0.10111)_2 &= (1 \times 2^{-1}) + (0 \times 2^{-2}) + (1 \times 2^{-3}) + (1 \times 2^{-4}) \\ &\quad + (1 \times 2^{-5})_{10} \\ &= (0.5 + 0.125 + 0.0625 + 0.03125)_{10} \\ &= (0.71875)_{10}\end{aligned}$$

The last conversion process involves converting decimal fractions into equivalent base-R fractions. This requires the evaluation of the fraction coefficients $(a_{-1}, a_{-2}, \ldots, a_{-n})$ of Equation [1.3]. The base-10 fraction NF can be expressed as

$$(NF)_{10} = \sum_{k=1}^{n} \frac{a_{-k}}{R^k} \qquad [1.5]$$

If the base-10 fraction is multiplied by R, the integer part of the product yields the coefficient a_{-1}. If the remaining fractional part is again multiplied by R, the integer part of the product gives the next coefficient, a_{-2}. Continuing in this manner, we can obtain successive coefficients, a_{-3} through a_{-n}. Unlike the integer conversions, conversions between decimal and binary fractions do not always convert exactly and may require an infinite string of digits in the new base for complete conversion. For example, the binary equivalent of $(0.173)_{10}$ is $(0.00101100010010011011101\ldots)_2$. This problem is compounded by the fact that fractions like $(2/3)_{10}$ are themselves not exact. The decimal fraction 2/3, for example, yields $(0.666666666\ldots)_{10}$. Conversions between the binary, octal, and hexadecimal systems, however, may be accomplished exactly. The following examples illustrate the proper conversion technique.

EXAMPLE 1.6

Convert $(0.029296875)_{10}$ to the equivalent octal fraction.

SOLUTION

	Coefficients	
$0.029296875 \times 8 = 0.234375$	0	MSB
$0.234375 \times 8 = 1.875$	1	
$0.875 \times 8 = 7.000$	7	LSB

Therefore,

$$(0.029296875)_{10} = (0.017)_8$$

Check: $(0 \times 8^{-1}) + (1 \times 8^{-2}) + (7 \times 8^{-3}) = (0.029296875)_{10}$

EXAMPLE 1.7

Obtain the binary equivalent of $(110.375)_{10}$.

SOLUTION

The integer part of the binary equivalent number is determined as follows:

Divisor	Quotient	Remainder	
2	110		
2	55	0	LSB
2	27	1	
2	13	1	
2	6	1	
2	3	0	
2	1	1	
2	0	1	MSB

The fractional part of the binary equivalent is determined as follows:

	Coefficients	
$0.375 \times 2 = 0.75$	0	MSB
$0.750 \times 2 = 1.50$	1	
$0.50 \times 2 = 1.00$	1	LSB

The binary equivalent number is, therefore, $(1101110.011)_2$.

Check: $(1 \times 2^6) + (1 \times 2^5) + (1 \times 2^3) + (1 \times 2^2) + (1 \times 2^1) + (1 \times 2^{-2}) + (1 \times 2^{-3}) = (110.375)_{10}$

When a decimal fraction is converted to a binary fraction, or vice versa, the number of significant digits or bits must be determined to keep the same degree of accuracy. To convert the decimal fraction 0.357 to an equivalent binary fraction, for example, we must determine the number of bits required to distinguish between 0.356,

0.357, and 0.358. In other words, we need to find the smallest number of bits that will allow us to represent a value of 1/1000, or the smallest integer value of b in the following equation:

$$\left(\frac{1}{2}\right)^b \leq \frac{1}{1000} \qquad [1.6]$$

Equation [1.6] gives a minimum value for b of 10. To maintain the same accuracy of 1/1000 of the decimal number, we convert to a 10-bit binary value that allows an accuracy to 1/1024. In general, for any d-digit decimal fraction the relationship between the number of bits, b, and d is as follows:

$$\left(\frac{1}{2}\right)^b \leq \left(\frac{1}{10}\right)^d \qquad [1.7]$$

which when solved gives

$$b \geq 3.32d \qquad [1.8]$$

Since b has to be the minimum integer, the equation can be written as

$$b = I(3.32d + 1) \qquad [1.9]$$

where the function $I(x)$ is the integer value of x that is obtained by discarding the fractional part of the number. Similarly, by reversing the inequality in Equation [1.7] we can solve for the minimum number of decimal digits required to retain the accuracy in a b-bit binary number:

$$d = I(0.3b + 1) \qquad [1.10]$$

Equations [1.9] and [1.10] now can be used along with the techniques discussed earlier for adequate conversion of fractions. In actual cases the number of binary digits may be limited by the digital system, causing inherent inaccuracies in the system.

EXAMPLE 1.8

Convert $(416.712)_8$ to the decimal equivalent.

SOLUTION

Equation [1.10] is used to determine the value of d. Note that $b = 9$ since it takes nine binary bits to represent the three fractional octal bits. Therefore,

$$d = I[(0.3 \times 9) + 1] = 3$$

Consequently,

$$(416.712)_8 = [(4 \times 8^2) + (1 \times 8^1) + (6 \times 8^0) + (7 \times 8^{-1}) + (1 \times 8^{-2}) + (2 \times 8^{-3})]_{10}$$

$$= (270.89453)_{10}$$
$$= (270.894)_{10}$$

where d is limited to three.

EXAMPLE 1.9

Find the binary equivalent of $(416.712)_{10}$.

SOLUTION

Using $d = 3$ we can determine b from Equation [1.9] as follows:

$$b = I[(3.32 \times 3) + 1] = 10$$

Since $(416)_{10} = (110100000)_2$

and $(0.712)_{10} = (0.10110110010001\ldots)_2$

then $(416.712)_{10} = (110100000.1011011001)_2$

1.3.3 Shortcut Method of Conversion

The numbers 2, 8, and 16 can be obtained by raising 2 to the powers of 1, 3, and 4, respectively. This relationship allows easy conversion between the binary, octal, and hexadecimal number systems. In converting a number from octal to binary, for example, the octal number is first converted to the equivalent decimal number, and is subsequently converted to its binary equivalent. The process is relatively tedious. However, Table 1.2 can eliminate such complicated problems. Since $2^3 = 8$ and $2^4 = 16$, each octal number corresponds to three bits and each hexadecimal number corresponds to four bits. When the binary number is grouped into groups of four, starting from the binary point and grouping to the left and to the right, the equivalent hexadecimal number is obtained by replacing each group of four digits by its hexadecimal equivalent.

TABLE 1.2 **Relationship between Binary and Octal and between Binary and Hexadecimal**

Octal to Binary		Hexadecimal to Binary	
0	000	0	0000
1	001	1	0001
2	010	2	0010
3	011	3	0011
4	100	4	0100
5	101	5	0101
6	110	6	0110
7	111	7	0111
		8	1000
		9	1001
		A	1010
		B	1011
		C	1100
		D	1101
		E	1110
		F	1111

Similarly, if the bits to the left and right of the binary point are grouped into groups of three, the equivalent octal number is obtained. To convert from octal or hexadecimal to binary, write the binary equivalent of each octal or hexadecimal digit, maintaining the correct digit position.

EXAMPLE 1.10

Convert $(271.A)_{16}$ to the base-2 and base-8 equivalents.

SOLUTION

Use Table 1.2 to find the hexadecimal and binary equivalents:

$$
\begin{aligned}
(271.A)_{16} &= (0010\ 0111\ 0001.1010)_2 \\
& \qquad\ \ 2 \qquad 7 \qquad 1 \qquad A \\
&= (1001110001.101)_2
\end{aligned}
$$

To convert to octal, form groups of three bits, starting from the binary point, as follows:

$$
\begin{aligned}
(271.A)_{16} &= (001\ 001\ 110\ 001.101)_2 \\
&= (\ 1 \quad 1 \quad 6 \quad 1.\ \ 5)_8
\end{aligned}
$$

Therefore,

$$(271.A)_{16} = (1001110001.101)_2 = (1161.5)_8$$

Note that to make proper groupings you may have to add zeros to the extreme left of the integer part and to the extreme right of the fraction part.

Computers may have a set of switches as one means of input. If a set of 16 switches is to be used to input a 16-bit quantity, it is much easier to remember a number like $(AC6D)_{16}$ rather than $(1010110001101101)_2$. The switches may be color coded in groups of three or four to facilitate the use of octal and hexadecimal numbers. If input is made through a keyboard to an eight-bit storage cell in a digital system, only two key depressions would be required if hexadecimal coding were used, but three key depressions would be required for octal coding. In either case a circuit must be included that converts the key depressions to the corresponding binary code. The skills necessary to design such a circuit will be developed in the succeeding chapters.

1.4 Representation of Negative Numbers

Up until now we have considered positive numbers only. In this section we will present methods to incorporate negative numbers in our conversion process. There are three common ways to represent base-R negative fixed-point numbers:

sign-and-magnitude representation,

signed $(R - 1)$'s or diminished radix complement representation,

signed R's or true or radix complement representation.

Sign-and-magnitude representation is the method we use in our decimal system to indicate negative numbers. Conventionally, to indicate a negative number, a minus sign $(-)$ is placed before the magnitude of the number in question. However, such a convention would require the introdution of a separate symbol for the $-$ sign in our number system. This problem generally is avoided by introducing an additional bit, called the *sign bit*. The left-most bit of the number is reserved for the sign bit. In all three representations the sign bit of a positive number is indicated by a 0, and that of the negative number is represented by $R - 1$. This convention provides for an unambiguous representation of sign.

Sign-and-magnitude: In the negative representation the magnitude remains unchanged while the sign bit changes from 0 to $R - 1$.

Signed $(R - 1)$'s complement: The complement is obtained by subtracting each digit of the corresponding positive number from $R - 1$.

Signed R's complement: For negative representation a 1 is added to the least significant position of the corresponding $(R - 1)$'s complement.

Therefore, the binary positive number $(0\ 01011)_2$ when complemented yields $(1\ 01011)_2$ in sign-and-magnitude representation, $(1\ 10100)_2$ in signed 1's complement representation, and $(1\ 10101)_2$ in signed 2's complement representation. In all three representations the positive numbers are exactly alike, but the negative numbers are expressed differently.

For the binary number system the rules for finding the complement can be simplified and stated as follows:

Rule 1. To find 1's complement of a binary number, complement every digit; that is, change each 1 to a 0 and each 0 to a 1.

Rule 2. To find 2's complement of a binary number, either (a) add 1 to the least significant bit of the 1's complement and ignore the carry-out of the sign bit, or (b) while scanning the positive number from the right to the left, complement all bits that are encountered after the first 1.

It is interesting to note that the 2's complement technique is the only means by which both $+0$ and -0 have the same representation. Table 1.3 shows the corresponding values for $+0$ and -0 in all three of the complement representations.

***TABLE 1.3* Binary Number Representation of Zero**

	Sign-and-Magnitude	1's Complement	2's Complement
+0	0.00000	0.00000	0.00000
−0	1.00000	1.11111	0.00000

1.5 Binary Arithmetic Operations

It will be demonstrated that the only operations necessary to perform addition, subtraction, multiplication, and division in the binary system are addition and taking either 1's or 2's complement. The rules of decimal arithmetic apply equally well to binary arithmetic. Table 1.4 summarizes the addition operation of binary numbers.

The addition of two binary numbers involves only two variables, the *addend* and *augend*, at the least significant position. An additional variable, the *carry-in*, is required at the remaining bit positions. There are, however, two possible addition outputs, the *carry-out* and the *sum*. The sum output is a 1 when only one input is a 1 ($1 + 0 = 1$) or when all of the three inputs are equal to 1 ($1 + 1 + 1 = 1$ plus a carry-in to the next column). The carry-out is a 1 if two or more inputs are equal to 1. Example 1.11 illustrates the binary addition process.

***TABLE 1.4* Binary Addition**

Input			Output	
Augend	Addend	Carry-In	Carry-Out	Sum
0	0	0	0	0
0	0	1	0	1
0	1	0	0	1
0	1	1	1	0
1	0	0	0	1
1	0	1	1	0
1	1	0	1	0
1	1	1	1	1

EXAMPLE 1.11

Add the binary numbers representing 7_{10} and 15_{10}.

SOLUTION

	0 0 1 1 1 1	Carry-out
	0 1 1 1 1 0	Carry-in
$(7)_{10}$ =	$(0\ 0\ 0\ 1\ 1\ 1)_2$	Augend
$(15)_{10}$ =	$(0\ 0\ 1\ 1\ 1\ 1)_2$	Addend
	$(0\ 1\ 0\ 1\ 1\ 0)_2$	Sum

Therefore,

$$(7)_{10} + (15)_{10} = (010110)_2 = (22)_{10}$$

Note that in working out this example the values of carry-outs were placed in the column in which they were generated. The carry-out of one column appears as the carry-in of the next column.

Subtraction in binary may be thought of as the inverse of addition. The operations for binary subtraction follow directly from decimal subtraction as well, and have been organized in Table 1.5. The three input variables for subtraction are the *minuend*, the *subtrahend*, and the *borrow-in*, and the output variables are, respectively, the *borrow-out* and the *difference*. The example that follows illustrates the operation of binary subtraction.

TABLE 1.5 **Binary Subtraction**

Input			Output	
Minuend	Subtrahend	Borrow-In	Difference	Borrow-Out
0	0	0	0	0
0	0	1	1	1
0	1	0	1	1
0	1	1	0	1
1	0	0	1	0
1	0	1	0	0
1	1	0	0	0
1	1	1	1	1

EXAMPLE 1.12

Subtract $(21)_{10}$ from $(83)_{10}$ in binary.

SOLUTION

$$\begin{array}{rlr l}
 & \;\;0\,1\,1\,1\,1\,0\,0 & & \text{Borrow-out} \\
 & \;\;1\,1\,1\,1\,0\,0\,0 & & \text{Borrow-in} \\
(83)_{10} = & (1\,0\,1\,0\,0\,1\,1)_2 & & \text{Minuend} \\
(21)_{10} = & (0\,0\,1\,0\,1\,0\,1)_2 & & \text{Subtrahend} \\
\hline
 & (0\,1\,1\,1\,1\,1\,0)_2 & & \text{Difference}
\end{array}$$

Therefore,

$$(83)_{10} - (21)_{10} = (111110)_2 = (62)_{10}$$

Most students find subtraction to be a more difficult operation to follow. Fortunately it can be shown that subtraction of a number B from a number A is equivalent to the sum of A and the complement of B. Since multiplication uses repeated addition and division uses repeated subtraction, it becomes apparent that the addition process encompasses all possible arithmetic operations within itself. The rules for binary addition where negative numbers are represented in either 2's complement or 1's complement form are as follows:

Rule 1. When negative numbers are represented by 2's complement, all bits, including the sign bits, are added and the carry-out of the sign bit is discarded. If the resultant sign bit is a 1, the sum is negative and the sum is in 2's complement form. If the sign bit is 0, the sum is positive and the number appears as a positive sign-magnitude number. These conditions are true if the magnitudes of the number are such that no overflow occurs.

Rule 2. When negative numbers are represented in 1's complement form, all bits, including the sign bit, are added. The carry-out of the sign bit is then added to the result. If the sign bit is 0, the resultant sum appears as a sign-magnitude number. If the sign bit is 1, the sum is negative and the number is in 1's complement form.

When two numbers of n digits each are added and the sum occupies $n + 1$ bits, we say that an *overflow* has occurred. An overflow is indicated either by the presence of a 1 in the sign bit when the two numbers are positive, or by the presence of a 0 in the sign bit when the two numbers are negative. An overflow is a serious problem in a digital computer since the number of bits used to store each value is fixed. However, if the hardware can recognize the occurrence of an overflow, the user can be notified of the problem, and the values can be rescaled to eliminate the overflow.

EXAMPLE 1.13

Perform the addition of $(10)_{10}$ and $(-6)_{10}$ in the binary system.

SOLUTION

$(-6)_{10} = 1\ 111001_2$ in signed 1's complement

$(-6)_{10} = 1\ 111010_2$ in signed 2's complement

Therefore, working with signed 1's complement,

$$
\begin{array}{rrl}
(10)_{10} = & 0\ 001010_2 & \\
& + & \\
(-6)_{10} = & 1\ 111001_2 & \text{1's complement} \\
\hline
& 10\ 000011_2 & \\
& + & \\
& 1_2 & \text{carry-out is added (end-around carry)} \\
\hline
\end{array}
$$

$$(10)_{10} + (-6)_{10} = (0\ 000100)_2 = (4)_{10}$$

where the carry has been discarded.

Working with signed 2's complement,

$$
\begin{array}{rcll}
(10)_{10} & = & 0\ 001010_2 & \\
 & & + & \\
(-6)_{10} & = & \underline{1\ 111010_2} & \text{2's complement} \\
 & & 10\ 000100_2 &
\end{array}
$$

Therefore,

$$(10)_{10} + (-6)_{10} = (0\ 000100)_2 = (4)_{10}$$

where the carry has been discarded.

EXAMPLE 1.14

a. Subtract $(11)_{10}$ from $(8)_{10}$ in binary.

b. Subtract $(8)_{10}$ from $(11)_{10}$ in binary.

SOLUTION

a. Since $(11)_{10} = 0\ 01011_2$,

$(-11)_{10} = 1\ 10100_2$ in signed 1's complement

$(-11)_{10} = 1\ 10101_2$ in signed 2's complement

Therefore, working with signed 1's complement,

$$
\begin{array}{rcll}
(8)_{10} & = & 0\ 01000_2 & \\
 & & + & \\
(-11)_{10} & = & \underline{1\ 10100_2} & \\
 & & 1\ 11100_2 & \text{with no carry-out}
\end{array}
$$

$$(8)_{10} + (-11)_{10} = -(0\ 00011)_2 = -(3)_{10}$$

Working with signed 2's complement,

$$
\begin{array}{rcll}
(8)_{10} & = & 0\ 01000_2 & \\
 & & + & \\
(-11)_{10} & = & \underline{1\ 10101_2} & \\
 & & 1\ 11101_2 & \text{with no carry-out}
\end{array}
$$

$$(8)_{10} + (-11)_{10} = -(0\ 00011)_2 = -(3)_{10}$$

b. Since $(8)_{10} = 0\ 01000_2$,

$(-8)_{10} = 1\ 10111_2$ in signed 1's complement

$(-8)_{10} = 1\ 11000_2$ in signed 2's complement

Therefore, working with signed 1's complement,

$$\begin{array}{rr}
(11)_{10} = & 0\ 01011_2 \\
 & + \\
(-8)_{10} = & \underline{1\ 10111_2} \\
 & 10\ 00010_2 \\
 & + \\
 & \underline{\quad 1_2} \quad \text{end-around carry} \\
 & 0\ 00011
\end{array}$$

$$(11)_{10} + (-8)_{10} = 0\ 00011_2 = (3)_{10}$$

where the carry has been discarded.

Working with signed 2's complement,

$$\begin{array}{rr}
(11)_{10} = & 0\ 01011_2 \\
 & + \\
(-8)_{10} = & \underline{1\ 11000_2} \\
 & 10\ 00011_2
\end{array}$$

$$(11)_{10} + (-8)_{10} = (0\ 00011)_2 = (3)_{10}$$

where the carry-out has been discarded.

1.6 Binary Codes

Digital computers process only binary numbers, 0s and 1s. However, in addition to the normal base-2 number system, several other special binary codes have been developed over the years for performing specific functions. They have particular advantages and characteristics. One of the goals in coding is to standardize a set of universal codes that can be used by all. To code 2^n distinct quantities in binary, a minimum of n bits are required. Since there is no constraint on the maximum number of bits, if efficient codes are not a requirement, there could be many choices for binary coding. For example, the decimal numbers can be coded with ten bits, and each decimal number could be assigned a bit combination of a 1 and nine 0s. In this particular coding scheme, the digit 8 is coded as 0100000000 and, similarly, the digit 4 is coded as 0000010000.

Consider another coding scheme for the decimal number 13. We could write the equivalent 1101 in binary, or the individual decimal characters could be coded separately as 0001 0011. The latter means of expressing the number in binary form requires an additional four bits, but this method is useful because it provides ease of conversion to the decimal system. This method of coding each digit separately is called the *binary coded decimal* (*BCD*) code. BCD, also known as the 8-4-2-1 code (the weight of each bit), is the most common binary code. It makes use of the base-2 system by using the binary representation of each decimal digit 0–9. The bit combinations corresponding to the binary numbers greater than 1001 are not used, so BCD is an inefficient code.

EXAMPLE 1.15

Convert the number $(1462.35)_{10}$ to its binary coded decimal (BCD) form.

SOLUTION

Use the four-bit binary representation of each digit:

1	4	6	2 .	3	5	Decimal
0001	0100	0110	0010 .	0011	0101	BCD

Therefore,

$$1462.35_{10} = 0001010001100010.00110101_{BCD}$$

EXAMPLE 1.16

Convert the BCD code

1000011101100010.01100101

to the equivalent decimal form.

SOLUTION

Group bits in groups of four proceeding left and right from the binary point.

1000	0111	0110	0010 .	0110	0101	BCD
8	7	6	2 .	6	5	Decimal

Therefore,

$$1000011101100010.01100101_{BCD} = 8762.65_{10}$$

Some digital computers and many hand-held calculators store information in BCD form and also perform arithmetic operations using BCD arithmetic. Since binary adders and complement arithmetic are used, two problems arise:

1. The use of binary adders results in four-bit groups that do not represent allowable BCD codes without modification.
2. The 9's complement of BCD values is difficult to obtain from the BCD code.

These two problems can be solved by the introduction and use of a code called *Excess-3*, or *XS3* for short. In this code each decimal digit is represented by a binary number three *greater* than the decimal digit. The XS3 code has the advantage that when an ordinary binary addition is performed on two decimal digits in this code, their binary sum produces a carry bit from the most significant bit position under exactly similar circumstances as that of a decimal carry in a decimal addition.

EXAMPLE 1.17

a. Add the numbers 1983_{10} and 1824_{10} using the XS3 code.

b. Add the same numbers using the BCD codes.

SOLUTION

a. Represent each digit by the equivalent XS3 code and add:

	11	Carry	1	1			
Base 10	1983		0100	1100	1011	0110	XS3
	1824		0100	1011	0101	0111	
	3807		1001	1000	0000	1101	

b. Represent each digit by the equivalent BCD code and add:

Carry	1				
	0001	1001	1000	0011	
	0001	1000	0010	0100	BCD
	0011	0001	1010	0111	

Note that in Example 1.17 the carries in the addition of XS3 numbers occurred out of the same digits as in the decimal addition. In either case modification would have to be made to convert each binary coded decimal digit to its correct XS3 or BCD form.

The XS3 code can be easily complemented to allow storing information in negative form and performing subtraction by addition similar to the 2's complement arithmetic discussed earlier. The 9's complement of an XS3 coded decimal digit may be obtained by complementing each bit of the XS3 code.

EXAMPLE 1.18

Take the 9's complement of the decimal number 456 and its XS3 binary coded decimal equivalent.

SOLUTION

9's complement of 456 = 543

456 in XS3 code = 0111 1000 1001

543 in XS3 code = 1000 0111 0110

This demonstrates that the complementing technique for the XS3 code is accurate.

Table 1.6 shows codes that might be used to code each decimal digit to form a BCD number. Each of these codes, except for the XS3 code, is called a *weighted code* since each bit position is assigned a weight. In addition, 631(-1), 2421, 4221, 84(-2)(-1), and the

XS3 codes have a distinct advantage in that their logical complement is the same as their arithmetic complement. Consequently, these schemes also are known as *self-complementing BCD* codes. The 5043210 code is a two-of-seven code, meaning that only two bits are 1 in any code group. This code allows for error detection. It is very important to note that a string of binary digits in a decimal system may not always represent a binary number; the string of digits may represent some other information as specified by a particular binary coding scheme.

Sometimes it may be necessary for a designer to avoid a number system where two or more bits change value between successive numbers. The *Gray*, or *unit-distance*, code is generally chosen during such circumstances. The first 16 Gray code numbers are 0000, 0001, 0011, 0010, 0110, 0111, 0101, 0100, 1100, 1101, 1111, 1110, 1010, 1011, 1001, and 1000, respectively. To be specific, the Gray code is not a weighted code. It is unique because only one bit changes value in successive coded binary numbers.

The codes listed in Table 1.6 are just a sample of the many in use. Coding theory has become a separate discipline within the field of communications. Any codes needed in this text will be introduced and discussed when needed. Because of the many coding possibilities, digital systems usually are equipped with *translators* (called *encoders* and *decoders*) for translating one code to another. Some translators will be developed in a later chapter after we have mastered the skills needed to do so.

TABLE 1.6 **Examples of Binary Codes**

Decimal	BCD (8421)	XS3	4221	2421	5043210	84(−2)(−1)	631(−1)
0	0000	0011	0000	0000	0100001	0000	0011
1	0001	0100	0001	0001	0100010	0111	0010
2	0010	0101	0010	0010	0100100	0110	0101
3	0011	0110	0011	0011	0101000	0101	0111
4	0100	0111	1000	0100	0110000	0100	0110
5	0101	1000	0111	1011	1000001	1011	1001
6	0110	1001	1100	1100	1000010	1010	1000
7	0111	1010	1101	1101	1000100	1001	1010
8	1000	1011	1110	1110	1001000	1000	1101
9	1001	1100	1111	1111	1010000	1111	1100

1.7 Logic Operations

The logic functions defined in the following discussions are the allowed operations in Boolean algebra, which is covered in Section 1.9. An understanding of these operations is vital. These logic operations are important in translating a word statement to a digital logic network. These logic functions operate on variables that are allowed to assume only two values, 1 or 0. To define a complex logic function of several variables, only three logic functions are needed: AND, OR, and NOT. There are three other functions that can be derived from AND, OR, and NOT that will also be considered. The logic operations and their circuit symbols are defined in the following text. The electronic circuits corresponding to the logic functions are alternatively known as gates. Note that the usage of *gate* refers to an actual electronic *device*, and *function* or *operation* refers to a logic *operator*.

It is often easy to describe a function by means of a table, known commonly as a *truth table*. Such a table usually lists the output values for all possible combinations of the inputs. In the following subsections, therefore, the logic operations will be defined using truth tables.

1.7.1 AND Function

The *AND function*, defined by $f(A,B) = \mathrm{A} \cdot \mathrm{B}$, is a function that is a 1 if both *A and B* are 1, and is a 0 otherwise. The dot used for the AND operation is frequently omitted if the function is obvious. This definition may be extended to more than two input variables. The output is a 1 as long as *all* of the inputs are 1, and is a 0 otherwise. The AND logic operation resembles the multiplication process; however, it should not be confused with the arithmetic multiplication operation.

The AND function can be best illustrated by the analogy as shown in Figure 1.1[*a*]. *A* and *B* are two switches in series, and the output is indicated by the lamp *Z*. The open switch is analogous to logic 0, and the closed switch is analogous to logic 1. When all of the switches are closed the lamp illuminates, indicating an output of 1. For all other switch combinations the lamp does not illuminate. The truth table and the logic symbol for the AND function also are shown respectively in Figure 1.1.

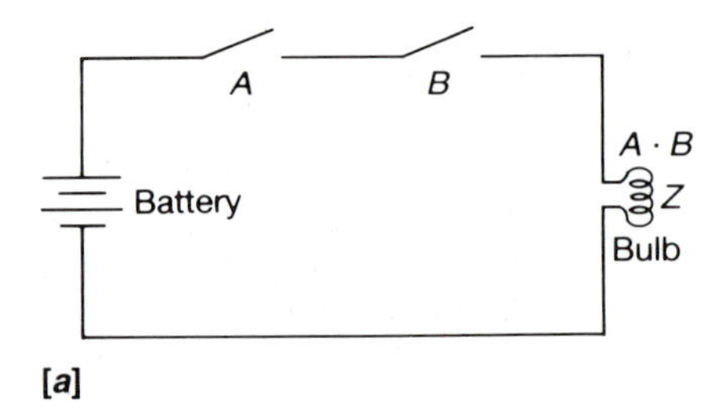

[*a*]

A	B	A · B
0	0	0
0	1	0
1	0	0
1	1	1

[*b*]

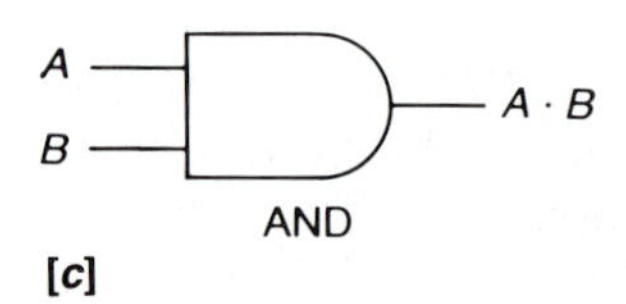

[*c*]

FIGURE 1.1 **Two-Input AND Function: [*a*] Analogy, [*b*] Truth Table, and [*c*] Logic Symbol.**

1.7.2 OR Function

The *OR function*, defined by $f(A,B) = A + B$, is a function that is a 1 when either A *or* B *or* both are 1. The + sign stands for the logical OR operation. To avoid confusion, this symbol will not be used in the rest of the text for the addition of numbers unless clearly indicated. The OR definition may be extended to more than two input variables. The output is a 0 as long as all of the inputs are 0, and is a 1 otherwise.

The OR function can be illustrated by the analogy of the switches and lamp as shown in Figure 1.2[*a*]. The two switches, A and B, are in parallel this time. The electric lamp will illuminate (logic 1) in this configuration as long as at least one of the switches is closed (logic 1), and it will remain turned off when all of the switches are open (logic 0). Figure 1.2 also shows the truth table and the logic symbol for the OR function.

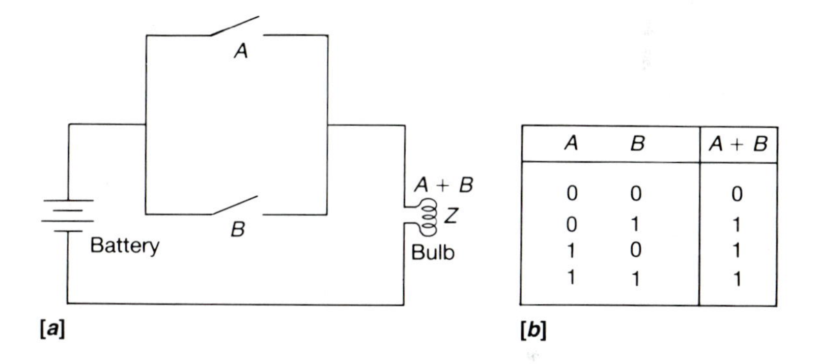

A	B	A + B
0	0	0
0	1	1
1	0	1
1	1	1

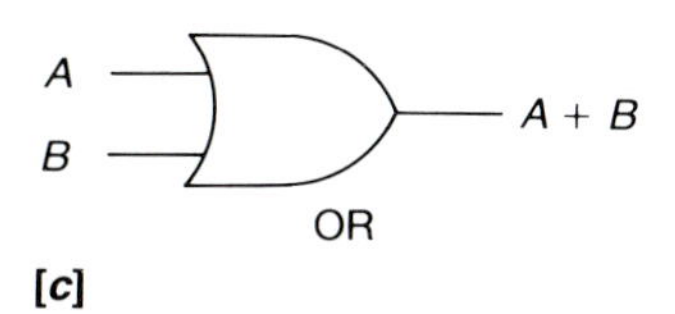

FIGURE 1.2 Two-Input OR Function: [*a*] Analogy, [*b*] Truth Table, and [*c*] Logic Symbol.

1.7.3 NOT Function

The *NOT function*, defined by $f(A) = \overline{A}$, is a function that is a 1 *only* when A is a 0. The bar indicates the operation and is called interchangeably the *not, complement*, or *inversion* of A. The bubble on the logic symbol illustrated in Figure 1.3[*b*] indicates an inversion. The NOT function is the only meaningful logic function that has a single input. Figure 1.3 shows the truth table and the logic symbol of the NOT function.

1.7.4 NAND Function

The *NAND function*, defined by $f(A,B) = \overline{A \cdot B}$, is a function that is identical to the AND function with the output inverted. When all of the inputs are 1, the output of the function is a 0, and is a 1 otherwise. In other words, a NAND function is a combination of an

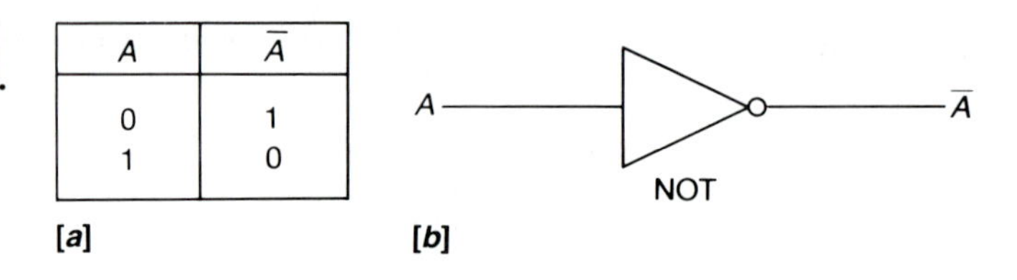

A	$\overline{A}$
0	1
1	0

FIGURE 1.3 **NOT Function: [*a*] Truth Table and [*b*] Logic Symbol.**

AND function followed by a NOT function. The bubble on the NAND logic symbol illustrated in Figure 1.4[*b*] indicates an inversion is involved. Figure 1.4 also shows the truth table and an equivalent circuit for the NAND function.

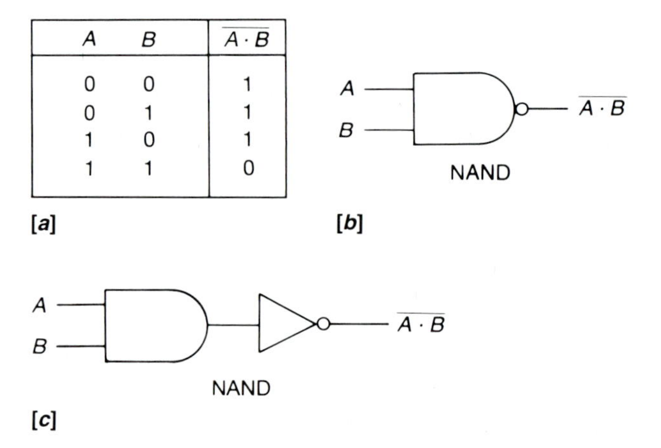

A	B	$\overline{A \cdot B}$
0	0	1
0	1	1
1	0	1
1	1	0

FIGURE 1.4 **Two-Input NAND Function: [*a*] Truth Table, [*b*] Logic Symbol, and [*c*] Equivalent Logic.**

1.7.5 NOR Function

The *NOR function*, defined by $f(A,B) = \overline{A+B}$, is a function identical to the OR function with the output inverted. When all of the inputs are 0, the output is a 1, and is a 0 otherwise. The NOR function is realizable by combining an OR gate and a NOT gate. Figure 1.5 shows the truth table, logic symbol, and equivalent logic for the two-input NOR function.

A	B	$\overline{A + B}$
0	0	1
0	1	0
1	0	0
1	1	0

[*a*]

A
B
$\overline{A + B}$
NOR

[*b*]

A
B
$\overline{A + B}$
NOR

[*c*]

FIGURE 1.5 **Two-Input NOR Function: [*a*] Truth Table, [*b*] Logic Symbol, and [*c*] Equivalent Logic.**

1.7.6 Exclusive-OR and Exclusive-NOR Functions

The *exclusive-OR function* (X-OR for short), defined by $f(A,B) = A \oplus B$, is a function that has an output of 1 if an odd number of its inputs are 1. The *exclusive-NOR* (X-NOR for short) is the inversion of the exclusive-OR and is defined by $f(A,B) = \overline{A \oplus B}$. Figure 1.6 shows the truth table and the logic symbols for both of these logic functions.

FIGURE 1.6 Two-Input X-OR and X-NOR Functions: [*a*] Truth Table and [*b*] Logic Symbols.

A	B	$A \oplus B$	$\overline{A \oplus B}$
0	0	0	1
0	1	1	0
1	0	1	0
1	1	0	1

[*a*]

A
B
$A \oplus B$
X-OR
$\overline{A \oplus B}$
X-NOR

[*b*]

Both of these functions can be derived by combinations of the other logic functions. In fact, either only NAND or only NOR functions can be successively combined to generate both X-OR and X-NOR functions. The X-OR and X-NOR functions occur so frequently in digital design that devices have been designed to perform these functions as well as the others just discussed.

Consider, for example, the circuit of Figure 1.7. We shall investigate the nature of the logic outputs available at points w, x, and y of the circuit, and eventually determine the output at z. In order to understand the working of this circuit, you must be familiar with

FIGURE 1.7 Circuit Consisting of Only NOR Logic.

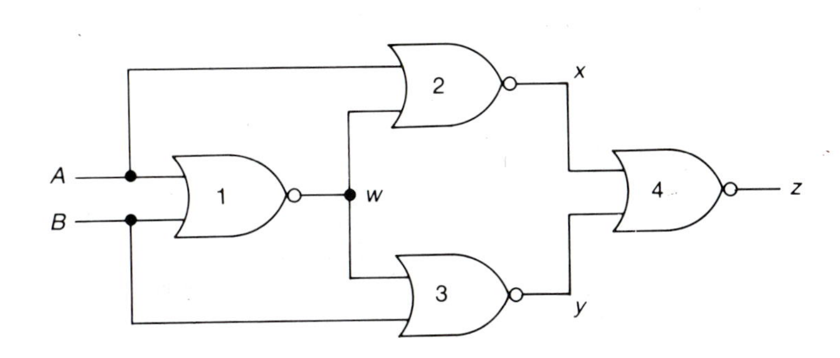

only NOR logic. Therefore, the output at w is determined as follows:

A	B	w
0	0	1
0	1	0
1	0	0
1	1	0

Note in the figure that w and A appear as inputs to the second NOR logic and, similarly, w and B appear as inputs to the third NOR logic. Consequently, x and y outputs could be determined as follows:

A	B	w	x	y
0	0	1	0	0
0	1	0	1	0
1	0	0	0	1
1	1	0	0	0

Finally, the x and y outputs appear as inputs to the fourth NOR logic. Accordingly, z can be determined by applying the NOR function to both x and y. The z output is obtained as follows:

A	B	w	x	y	z
0	0	1	0	0	1
0	1	0	1	0	0
1	0	0	0	1	0
1	1	0	0	0	1

Now we may compare the z output with the X-NOR output column of Figure 1.6[a]. The two columns are exactly alike for all combinations of the inputs. It can be concluded, therefore, that the circuit of Figure 1.7 functions exactly as an X-NOR logic. This example suggests that for most digital problems, there is more than one valid solution. It is the objective of the designers, therefore, to obtain an optimum solution. The techniques for determining the optimal design will be discussed in later sections.

1.8 Logic Functions from Truth Tables

Often a design problem is introduced to an engineer in the form of a verbal description. The descriptions are often vague and misleading in many respects. Consequently, the verbal description is translated into a truth table for the sake of clarity. A truth table is a tool that allows the designer to examine all possible inputs and the corresponding outputs; the design problem is completely specified by the truth table. In logic design the truth table is often overlooked, but it should be the first step in organizing the problem. We will introduce designing with truth tables by an application of them.

It will be advantageous to introduce at this time the concepts of minterm and maxterm that correspond to a set of inputs. A *minterm* is that particular AND combination of the variables that will yield a 1 if the variables are passed through an AND gate. For example, when $A = 0, B = 1$, and $C = 0$, then the corresponding minterm is $\overline{A}B\overline{C}$ since $\overline{A} \cdot B \cdot \overline{C}$ would yield a 1. Similarly, a *maxterm* is the OR combination of the variables that gives a 0 if the variables are passed through an OR gate. The maxterm corresponding to $A = 0$, $B = 1$, and $C = 0$ is, therefore, $A + \overline{B} + C$. In fact, $\overline{A}B\overline{C}$ (minterm) is the complement of $A + \overline{B} + C$ (maxterm) and can be verified easily once the next section has been covered.

Suppose we have been asked to design a voting machine that accepts three inputs: *A, B,* and *C*. The circuit is to output a 1 when a majority of the inputs are 1, except when *A* and *B* are 1 and *C* is a 0. The action of the circuit can described completely by considering each possible input combination. A truth table will be used to accomplish this. In general, for *N* variables the truth table contains a total of 2^N rows. The desired function is completely specified by the truth table in Table 1.7. The table lists both minterms and maxterms as well.

TABLE 1.7 **Truth Table for a Voting Machine**

Decimal Equivalents	A	B	C	Output	Minterms	Maxterms
0	0	0	0	0	$\overline{A} \cdot \overline{B} \cdot \overline{C}$	$A + B + C$
1	0	0	1	0	$\overline{A} \cdot \overline{B} \cdot C$	$A + B + \overline{C}$
2	0	1	0	0	$\overline{A} \cdot B \cdot \overline{C}$	$A + \overline{B} + C$
3	0	1	1	1	$\overline{A} \cdot B \cdot C$	$A + \overline{B} + \overline{C}$
4	1	0	0	0	$A \cdot \overline{B} \cdot \overline{C}$	$\overline{A} + B + C$
5	1	0	1	1	$A \cdot \overline{B} \cdot C$	$\overline{A} + B + \overline{C}$
6	1	1	0	0	$A \cdot B \cdot \overline{C}$	$\overline{A} + \overline{B} + C$
7	1	1	1	1	$A \cdot B \cdot C$	$\overline{A} + \overline{B} + \overline{C}$

The minterms are sometimes referred to by the decimal equivalents of the input variables that make them a 1. Minterm 5, for example, corresponds to $A\overline{B}C$. If the output is called $f(A,B,C)$, then we can write a logic equation of f describing when the output is a 1. This function is expressible in a form, called the *sum-of-product* (*SOP*) form, and can be obtained by ORing the minterms that cause the output to be a 1:

$$f(A,B,C) = \overline{A} \cdot B \cdot C + A \cdot \overline{B} \cdot C + A \cdot B \cdot C \qquad [1.11]$$

Equation [1.11], the SOP function, takes on the value 1 for three of the variable combinations: 011, 101, and 110. For every other input combination, the three ANDed terms would be 0, and consequently

Equation [1.11] would be 0. Alternatively, the SOP function can be concisely expressed as follows:

$$f(A,B,C) = \Sigma m(3,5,7) \qquad [1.12]$$

where 3, 5, and 7 are the three minterms that cause the function to be a 1. As seen from the truth table, there are three input combinations that cause the output function to be a 1, and they are 011, 101, and 111. Putting these combinations of variable values, one at a time, in Equation [1.12] will cause the function to be a 1.

Another form by which this same function f can be expressed is called the *product-of-sum* (*POS*) form. It is obtained by ANDing the maxterms corresponding to the 0 outputs:

$$f(A,B,C) = (A + B + C) \cdot (A + B + \overline{C}) \cdot (A + \overline{B} + C) \cdot (\overline{A} + B + C) \cdot (\overline{A} + \overline{B} + C) \qquad [1.13]$$

This POS form is also expressible in a concise form:

$$f(A,B,C) = \Pi M(0,1,2,4,6) \qquad [1.14]$$

where 0, 1, 2, 4, and 6 are the five maxterms forming the same voting function. The circuits that implement this voting function are shown in Figure 1.8.

The circuits just designed are more complex than necessary. The second circuit obviously is more complex than the first. To make the simplest possible circuit, a designer would look at the truth table to determine the number of ones in the output column. When the number of ones is less than or equal to the number of zeros, the SOP form would give the simpler circuit; otherwise the POS form would be used. However, the POS form of the function can be reduced further using Boolean algebraic techniques to

$$f(A,B,C) = (A + B) \cdot C \qquad [1.15]$$

Verify for yourself that the simplified function will be 1 for the same input values as the more complex function. Its circuit implementation is shown in Figure 1.9[*a*]. The truth table of Figure 1.9[*b*] clearly shows that the circuit of Figure 1.9[*a*] functions exactly as those of Figure 1.8. The circuit in Figure 1.9[*a*] obviously would please an employer more if he or she is in business for a profit. Assuming that most of the readers work for profit, we will spend a good portion of our time learning techniques that will help simplify our designs. The next section on Boolean algebra is the starting point.

FIGURE 1.8 **Realization of the Voting Machine: [*a*] SOP Form and [*b*] POS Form.**

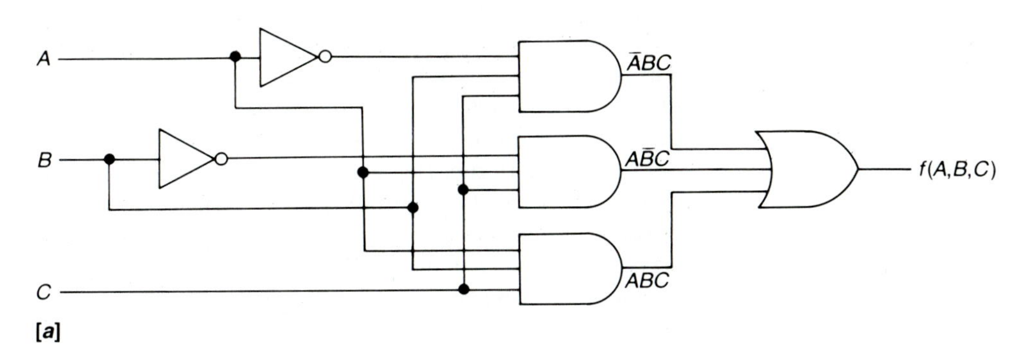

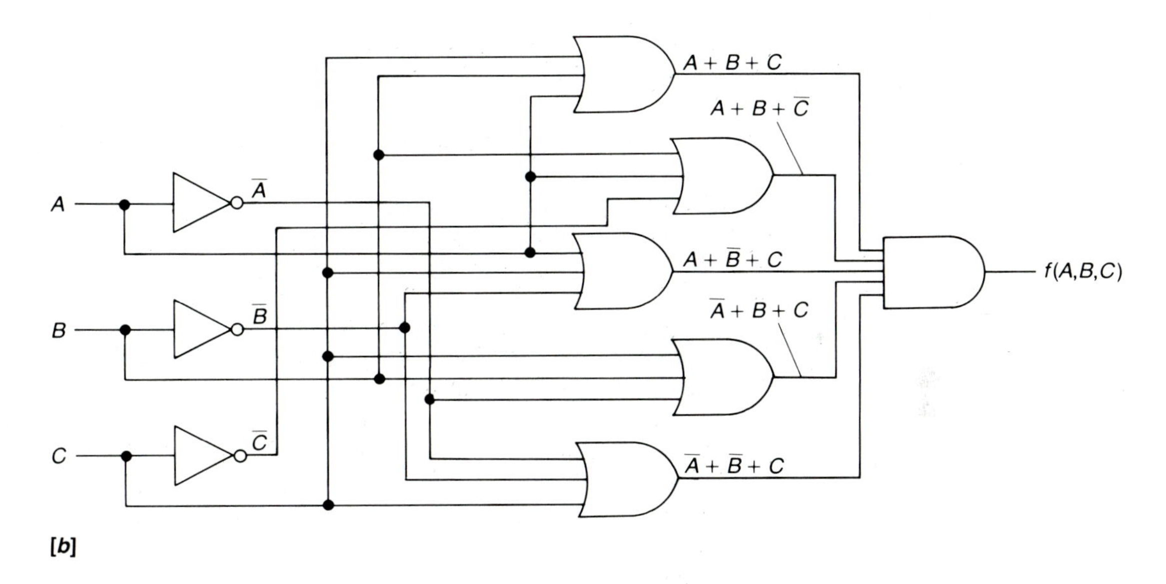

FIGURE 1.9 **Simplified Voting Machine: [*a*] Circuit and [*b*] Truth Table.**

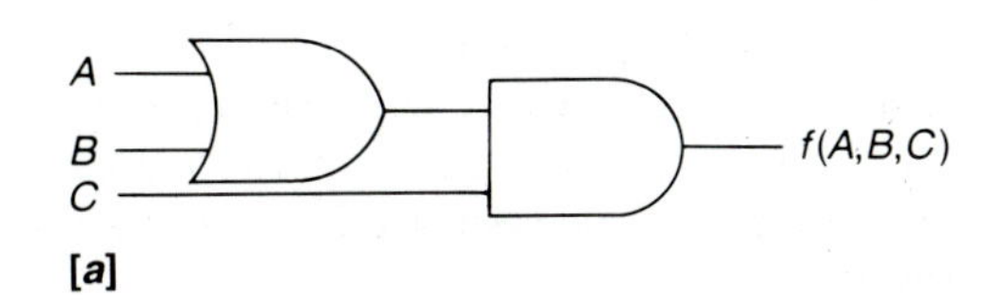

A	B	C	A + B	C · (A + B)
0	0	0	0	0
0	0	1	0	0
0	1	0	1	0
0	1	1	1	1
1	0	0	1	0
1	0	1	1	1
1	1	0	1	0
1	1	1	1	1

[b]

1.9 Boolean Algebra

As shown in the last section the initial POS form of the function determined by the use of only the truth table (Equation [1.13]) is correct, but it can be replaced usually by a less expensive and less complicated circuit (Equation [1.15]). This function complexity reduction is accomplished by means of Boolean algebra, which was first introduced by a mathematician named George Boole. Though not developed specifically for the purpose of function simplification, it is useful for manipulating logic expressions.

The formulation of Boolean algebra is based upon a set of axioms or postulates, known commonly as *Huntington's postulates*. As is true for other axiomatic systems, each Huntington's postulate has a corresponding dual. Briefly explained, the *principle of duality* states that if a given logic expression is valid, the dual of the same logic expression is also valid. The corresponding dual expression for Huntington's system is obtained by replacing all zeros with ones, all ones with zeros, all AND operations with ORs, and all OR operations with ANDs. For example, the dual expression of the voting machine equation, $\bar{A}BC + A\bar{B}C + ABC = (A + B)C$, is $(\bar{A} + B + C) \cdot (A + \bar{B} + C) \cdot (A + B + C) = A \cdot B + C$.

Boolean algebra is a closed system consisting of a finite set S of two or more elements, subject to an equivalence relation ($=$) and the three binary operators OR, AND, and NOT, such that for every element x and y in the set, the operations $x + y$, $x \cdot y$, $\bar{x}$, and $\bar{y}$ are also uniquely defined in the set, and the Huntington's postulates are satisfied. These postulates and their duals are listed as follows:

Postulate 1. For each operation there exists unique elements, 1 and 0, in set S such that for x in S,

(a) $x + 0 = x$ and $x \cdot 1 = x$

and

(b) $x \cdot 0 = 0$ and $x + 1 = 1$

Postulate 2. The operations are commutative for every x and y in set S such that

(a) $x + y = y + x$ and (b) $x \cdot y = y \cdot x$

Postulate 3. The operations are distributive. For all x, y, and z in set S,

(a) $x \cdot (y + z) = (x \cdot y) + (x \cdot z)$

and

(b) $x + (y \cdot z) = (x + y) \cdot (x + z)$

Postulate 4. The operations are associative. For every x, y, and z in set S,

(a) $x + (y + z) = (x + y) + z$

and

(b) $x \cdot (y \cdot z) = (x \cdot y) \cdot z$

Postulate 5. For every element x in the set S there exists an element $\bar{x}$ (called the *complement* of x) such that

(a) $x + \bar{x} = 1$ and (b) $x \cdot \bar{x} = 0$

Boolean theorems, whose primary application is in the minimization of logic circuits, can now be derived using the stated postulates. These theorems can be proven easily using the postulates, and the proofs are left as exercises. The theorems are as follows:

Theorem 1: The Law of Idempotency. For all x in set S,

(a) $x + x = x$ and (b) $x \cdot x = x$

Theorem 2: The Law of Absorption. For all x and y in set S,

(a) $x + (x \cdot y) = x$ and (b) $x \cdot (x + y) = x$

Theorem 3: The Law of Identity. For all x and y in set S, if

(a) $x + y = y$ and (b) $x \cdot y = y$ then $x = y$

Theorem 4: The Law of Complements. For all x and y in set S, if

(a) $x + y = 1$ and (b) $x \cdot y = 0$ then $x = \bar{y}$

Theorem 5: The Law of Involution. For all x in set S,

$x = \bar{\bar{x}}$

Theorem 6: DeMorgan's Law. For all $x, y, \ldots,$ *and* z in set S,

(a) $\overline{x \cdot y \cdot \cdots \cdot z} = \bar{x} + \bar{y} + \cdots + \bar{z}$

and

(b) $\overline{x + y + \cdots + z} = \bar{x} \cdot \bar{y} \cdot \cdots \cdot \bar{z}$

This theorem implies that any function can be complemented by changing the ORs to ANDs, ANDs to ORs, and complementing each of the variables.

Theorem 7: The Law of Elimination. For all x and y in set S,

(a) $x + \bar{x}y = x + y$ and (b) $x \cdot (\bar{x} + y) = x \cdot y$

Theorem 8: The Law of Consensus. For all $x, y,$ and z in set S,

(a) $xy + \bar{x}z + yz = xy + \bar{x}z$

and

(b) $(x + y) \cdot (\bar{x} + z) \cdot (y + z) = (x + y) \cdot (\bar{x} + z)$

Theorem 9: The Law of Interchange. For all x, y, and z in set S,

$$\text{(a)} \quad (x \cdot y) + (\bar{x} \cdot z) = (x + z) \cdot (\bar{x} + y)$$

and

$$\text{(b)} \quad (x + y) \cdot (\bar{x} + z) = (x \cdot z) + (\bar{x} \cdot y)$$

Theorem 10: The Generalized Functional Laws. The AND/OR operation of a variable X and a multi-variable composite function that is also a function of X is equivalent to similar AND/OR operation of X with the composite function whose X is replaced by 0:

$$\text{(a)} \quad X + f(X,Y,\ldots,Z) = X + f(0,Y,\ldots,Z)$$
$$\text{(b)} \quad X \cdot f(X,Y,\ldots,Z) = X \cdot f(0,Y,\ldots,Z)$$

For all X, Y, . . . , and Z in set S,

$$\text{(a)} \quad f(X,Y,\ldots,Z) = X \cdot f(1,Y,\ldots,Z) + \overline{X} \cdot f(0,Y,\ldots,Z)$$
$$\text{(b)} \quad f(X,Y,\ldots,Z) = [X + f(0,Y,\ldots,Z)] \cdot [\overline{X} + f(1,Y,\ldots,Z)]$$

The Generalized Functional Laws are very important in realizing a device called a *multiplexer*. They allow writing a function so that a selected variable and its complement appear only once. More will be said about this in Chapter 4. The following examples illustrate the use of many of these Boolean theorems.

EXAMPLE 1.19

Use Boolean algebra to show that the circuits of Figure 1.8[*a*] and Figure 1.9[*a*] are equivalent.

SOLUTION

The function corresponding to the circuit of Figure 1.8[*a*] is given by

$$\begin{aligned} f(A,B,C) &= \overline{A}BC + A\overline{B}C + ABC && [1.11] \\ &= \overline{A}BC + AC(\overline{B} + B) \\ &= \overline{A}BC + AC && [P.5(a)] \\ &= C(\overline{A}B + A) \\ &= C(A + B) && [Th.7(a)] \end{aligned}$$

This is the same as Equation [1.15] that was used for realizing the circuit of Figure 1.9[*a*]. Hence, the equivalency is proven.

EXAMPLE 1.20

Simplify

$$(x + y)(x + \bar{x}\bar{y})z + \bar{x}y + xyz + \overline{x(\bar{y} + \bar{z})}$$

SOLUTION

$$\begin{aligned} &(x + y)(x + \bar{x}\bar{y})z + \bar{x}y + xyz + \overline{x(\bar{y} + \bar{z})} \\ &= (x + y)(x + \bar{y})z + \bar{x}y + xyz + \overline{x(\bar{y} + \bar{z})} && [Th.7(a)] \\ &= (x + x\bar{y} + xy + 0)z + \bar{x}y + xyz + \overline{x(\bar{y} + \bar{z})} && [Th.1(b)] \end{aligned}$$

$$= x(1 + y + \bar{y})z + \bar{x}y + xyz + \overline{x(\bar{y} + \bar{z})}$$

$$= xz + \bar{x}y + xyz + \overline{x(\bar{y} + \bar{z})} \qquad [P.5(a)]$$

$$= xz(1 + y) + \bar{x}y + \overline{x(\bar{y} + \bar{z})}$$

$$= xz + \bar{x}y + \overline{x(\bar{y} + \bar{z})}$$

$$= xz + \bar{x}y + \bar{x} + \overline{\bar{y} + \bar{z}} \qquad [Th.6(a)]$$

$$= xz + \bar{x}y + \bar{x} + yz \qquad [Th.6(b)]$$

$$= xz + \bar{x}y + yz + \bar{x}$$

$$= xz + \bar{x}y + \bar{x} \qquad [Th.8(a)]$$

$$= xz + \bar{x}(y + 1)$$

$$= xz + \bar{x} \qquad [P.1(b)]$$

$$= \bar{x} + z \qquad [Th.7(a)]$$

EXAMPLE 1.21

Obtain the circuit organization within the two boxes, *m* and *n*, of the circuit shown in Figure 1.10. Assume that the function output $f = wx + \overline{yz} \cdot (\bar{w} + xy)$.

SOLUTION

From the generalized functional laws we may obtain

$$f(w,x,y,z) = [y + f(w,x,0,z)] \cdot [\bar{y} + f(w,x,1,z)]$$

This equation could be used because it meets the general characteristic of the given circuit. Since $f(w,x,y,z) = wx + \bar{y}\bar{z} \cdot (\bar{w} + xy)$, we may obtain

$$f(w,x,0,z) = wx + \bar{w}$$
$$= \bar{w} + x$$

and

$$f(w,x,1,z) = wx + \bar{z}(\bar{w} + x)$$
$$= wx + \bar{z}\bar{w} + \bar{z}x$$
$$= wx + \bar{z}\bar{w}$$
$$= wx + \overline{z + w}$$

FIGURE 1.10

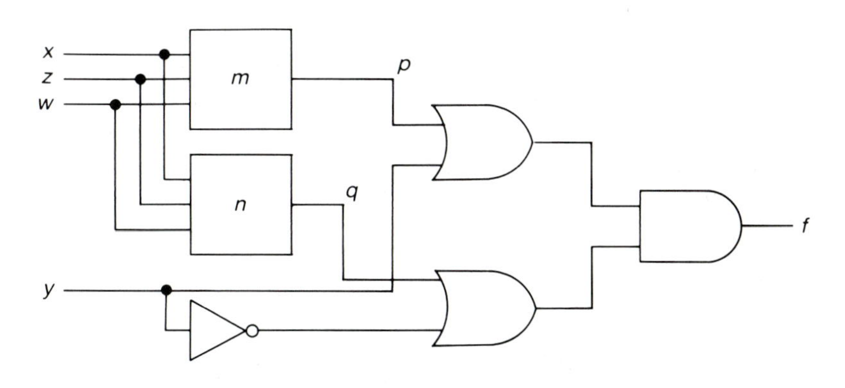

Consequently, the circuit within box m is obtained as shown in Figure 1.11[a] and that within unit n is given as shown in Figure 1.11[b].

FIGURE 1.11

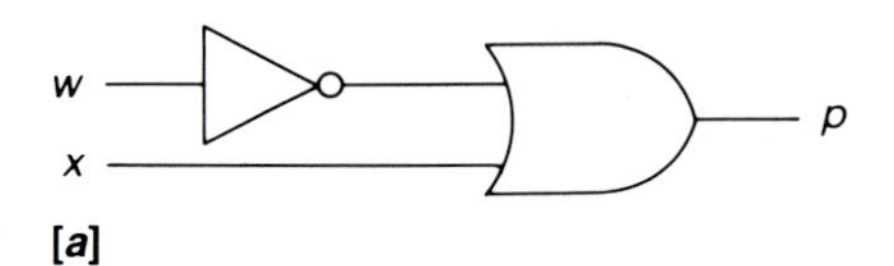

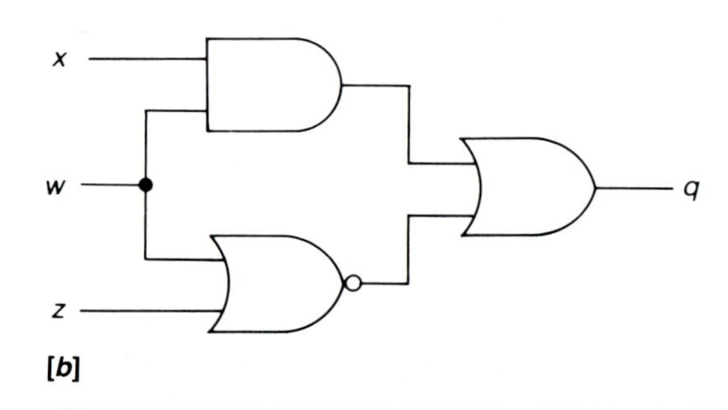

EXAMPLE 1.22

Show for a three-input logic circuit that

$\Sigma m(3,5,7) = \Pi M(0,1,2,4,6)$

SOLUTION

$$\begin{aligned}
\Pi M(0,1,2,4,6) &= (A + B + C)\cdot(A + B + \overline{C})\cdot(A + \overline{B} + C) \\
&\quad \cdot(\overline{A} + B + C)\cdot(\overline{A} + \overline{B} + C) \\
&= (A + AB + A\overline{C} + B + B\overline{C} + AC + BC + C\overline{C}) \\
&\quad \cdot(A + \overline{B} + C)\cdot(\overline{A} + B + C)\cdot(\overline{A} + \overline{B} + C) \\
&= [A(1 + B) + A(C + \overline{C}) + B(1 + \overline{C} + C)] \\
&\quad \cdot(A + \overline{B} + C)\cdot(\overline{A} + B + C)\cdot(\overline{A} + \overline{B} + C) \\
&= (A + B)(A + \overline{B} + C)(\overline{A} + B + C)(\overline{A} + \overline{B} + C) \\
&= (A + A\overline{B} + AC + AB + B\overline{B} + BC)(\overline{A} + B + C) \\
&\quad (\overline{A} + \overline{B} + C) \\
&= (A + BC)(\overline{A} + B + C)(\overline{A} + \overline{B} + C) \\
&= (A\overline{A} + AB + AC + \overline{A}BC + BC + BC)\,(\overline{A} \\
&\quad + \overline{B} + C) \\
&= (AB + AC + BC)(\overline{A} + \overline{B} + C) \\
&= A\overline{A}B + A\overline{B}B + ABC + A\overline{A}C + A\overline{B}C + AC \\
&\quad + \overline{A}BC + B\overline{B}C + BC \\
&= AC + A\overline{B}C + \overline{A}BC + BC \\
&= AC(B + \overline{B}) + A\overline{B}C + \overline{A}BC + (A + \overline{A})BC \\
&= ABC + A\overline{B}C + A\overline{B}C + \overline{A}BC + ABC + \overline{A}BC \\
&= ABC + A\overline{B}C + \overline{A}BC \\
&= \Sigma m(3,5,7)
\end{aligned}$$

This result justifies our earlier conclusions that the two circuits of Figure 1.8 are equivalent.

In logic expressions a variable x and its complement $\bar{x}$ are called *literals*. When we start connecting circuits we will see that each literal in a function has to be connected to an input. In actual circuits this would involve a wire connected to the pin at an integrated circuit. Every wire and every pin increase the cost of the circuit. The main objective of reducing or simplifying a function by the use of Boolean algebra is to achieve the reduction of the number of literals in the given function. The number of literals in a function of n variables can be at maximum $2n$. Fortunately, the reduction of literals can be achieved by more usable techniques. These techniques are the subject of the next chapter.

1.10 Summary

In this chapter the number systems were introduced. Schemes for converting numbers from one system to another were explained and demonstrated. The techniques for representing negative numbers were also discussed, and these representations were used for performing binary arithmetic.

The logic operations and gates were introduced and methods of producing the logic functions from truth tables were also presented. Finally, the techniques of manipulating and reducing Boolean expressions were also demonstrated.

Problems

1. Convert each following base-R integer to the equivalent base-10 integer:
   a. $(110101)_2$
   b. $(101010)_2$
   c. $(231230)_4$
   d. $(312312)_4$
   e. $(456703)_8$
   f. $(643200)_8$
   g. $(776764)_8$
   h. $(1ABF23)_{16}$
   i. $(F01120)_{16}$
2. Convert each following base-R fraction to the equivalent base-10 fraction:
   a. $(0.011011)_2$
   b. $(0.110111)_2$
   c. $(0.231232)_4$
   d. $(0.321231)_4$
   e. $(0.435671)_8$
   f. $(0.645317)_8$
   g. $(0.777666)_8$
   h. $(0.AACE42)_{16}$
   i. $(0.FF2310)_{16}$
3. Convert each following base-R number to the equivalent base-10 number:
   a. $(110110.11011)_2$
   b. $(101100.10011)_2$
   c. $(532.412)_8$
   d. $(3240.6724)_8$
   e. $(1A2325.5F1103)_{16}$
   f. $(5CD.31F)_{16}$

4. Convert each following base-10 number to its equivalent number in base 2, 3, 4, 8, and 16:
   a. $(12546.203)_{10}$ d. $(43002.001)_{10}$
   b. $(43562.6752)_{10}$ e. $(11909.0910)_{10}$
   c. $(432760.020)_{10}$ f. $(4321.00023)_{10}$
5. Derive the Equations [1.9] and [1.10] from Equation [1.6].
6. Perform the following conversions as indicated. Make sure that the converted number maintains an equal level of accuracy:
   a. $(1011.11011)_2$ to base 10 and base 5
   b. $(378.FC2)_{16}$ to base 10.
   c. $(893.2167)_{10}$ to base 8 and base 16
   d. $(546.3601)_8$ to base 9 and base 10
7. Perform addition of the following addend-augend pairs:
   a. $(1011011)_2 + (0011011)_2$
   b. $(54672.453)_8 + (11222.435)_8$
   c. $(A453.FF23)_{16} + (A43.22F)_{16}$
   d. $(1011.1101)_2 + (1000.0011)_2$
8. Perform the following subtractions using the radix complement method. The digits within parentheses represent magnitude only.
   a. $(1101.1111)_2 - (1110.0011)_2$
   b. $(1110.0011)_2 - (1001.110)_2$
   c. $(4567.7230)_8 - (564.2340)_8$
   d. $(564.2340)_8 - (4567.7230)_8$
   e. $(AA2F.34FE)_{16} - (9854.11FF)_{16}$
   f. $(9854.11FF)_{16} - (AA2F.34FE)_{16}$
9. Perform the subtractions of Problem 8 using the diminished radix method.
10. Perform the following operations using the radix complement for the subtraction part where $A = (C43.2)_{16}$, $B = (FFF.F)_{16}$, $C = (102.4)_{16}$, $D = -(382.9)_{16}$, and $E = -(F12.C)_{16}$.
    a. $A - B$ c. $A + B + E$
    b. $A + D$ d. $D - E + A$
11. Repeat Problem 10 using the diminished radix method for the subtraction operation.
12. Verify the Boolean theorems. Justify your reasonings.
13. Reduce the following functions to a minimum number of literals:
    a. $\bar{A} + \overline{AB}(A + \bar{C} + \bar{A}\bar{C})$ b. $\bar{A}(\bar{B} + \bar{C})(A + B + \bar{C})$
    c. $(\bar{A} + B + \bar{B}\bar{C})(A + \bar{B}C) + \bar{D}$
    d. $\overline{CD} + A + A + CD + AB$ e. $AB + \bar{A}B + AB\bar{C}D$
    f. $(A \oplus B) + BC(A + B)$ g. $(A \oplus B) + BC(A + B)$

14. Complement the Boolean expressions provided in Problem 13, and then reduce the complemented function to the minimum number of literals.

15. a. Obtain the truth table for converting a four-variable binary number into the corresponding four-variable Gray code. Obtain the Boolean equations necessary for this conversion and draw the optimum logic circuit.
    b. Obtain the truth table for converting a four-variable Gray code into the corresponding four-variable binary number. Obtain the Boolean equations necessary for this conversion and draw the optimum logic circuit.

16. Obtain a logic circuit for implementing the four-bit BCD to four-bit XS3 conversion.

17. There are four oscilloscopes, A, B, C, and D, in a laboratory that are equipped with individual indicator circuits. These indicator circuits give out a logic 1 any time the oscilloscope malfunctions. Design a circuit that would be able to indicate that two or more of the oscilloscopes are malfunctioning.

18. Realize the circuit of Figure 1.P1 using OR gates in the first stage and AND gates in the second stage.

FIGURE 1.P1

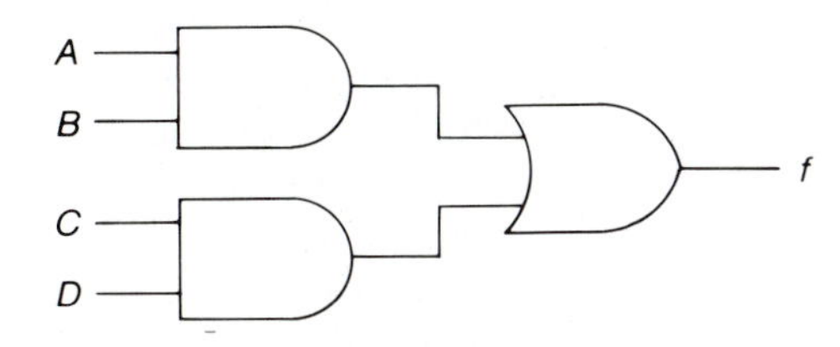

19. Find the complement of $f = \overline{BC}\,\overline{(A + \overline{CD})}$ and reduce it. Obtain the corresponding logic circuit.

20. Simplify the following Boolean expressions and obtain the corresponding logic circuit:
    a. $ABC + BCD + A\bar{B}\bar{D} + \bar{A}BC\bar{D} + \bar{A}\bar{B}\bar{D}$
    b. $(A + BC)(\bar{D} + BC)(\bar{A} + \bar{D})$
    c. $\overline{(\overline{ABC}\;\overline{BD}\;\overline{B})}E$

21. Find the output, f, of the circuit shown in Figure 1.P2 in terms of A and B.

FIGURE 1.P2

22. Show that $\overline{\overline{(A \oplus B)} \oplus \overline{(B \oplus C)}} = \overline{A \oplus C}$.
23. Show that $\overline{\overline{(A \oplus B)} \oplus C} = \overline{A \oplus \overline{(B \oplus C)}}$
24. Show that the following expressions are equivalent:
    a. $A \oplus B = \overline{A} \oplus \overline{B}$
    b. $A \oplus B \oplus AB = A + B$
    c. $\overline{A \oplus B} = A \oplus B \oplus 1$
25. Reduce the circuit of Figure 1.P3 and obtain an equivalent logic network.

FIGURE 1.P3

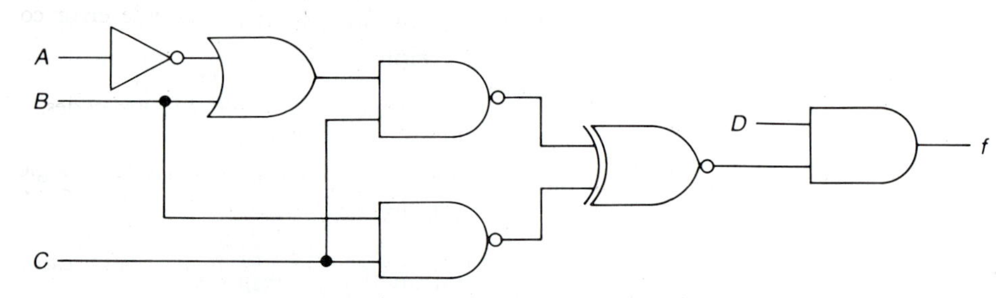

26. Organize the logic function $f = (XY\overline{Z} + \overline{X}\overline{Y}) + \overline{(Z + X)}$ in the following forms, where neither A nor B is a function of Z. Obtain the corresponding logic circuit:
    a. $f = ZA + \overline{Z}B$ b. $f = (Z + A)(\overline{Z} + B)$
27. An automobile is equipped with an alarm system to indicate each of the following conditions: (a) The key is in the ignition, the door is open, and the motor is not running; (b) the lights are on when the key is not in the ignition; (c) the seat belt is not fastened when the motor is running; or (d) the door is open and the motor is running. Obtain the alarm circuit when $K = 1$ if the key is in the ignition, $D = 1$ if the door is open, $L = 1$ if the lights are on, $M = 1$ if the motor is running, and $S = 1$ if the seat belt is not fastened.

Suggested Readings

Abraham, E.; Seaton, C. T.; and Smith, S. D. "The optical computer." *Sci. Am.* vol. 248 (1983): 85.

Agrawal, D. P. "Signed modified reflected binary code." *IEEE Trans. Comp.* vol. C-25 (1976): 549.

Asija, S. P. "Instant logic conversions." *IEEE Spectrum* vol. 5 (December 1968): 77.

Banerji, D. K. "A novel implementation method for addition and subtraction in residue number systems." *IEEE Trans. Comp.* vol. C-23 (1974): 106.

Bartee, T.C. "Computer design of multiple output logical networks." *IRE Trans. Elect. Comp.* vol. EC-10 (1961): 21.

Bhuyan, L. N., and Agrawal, D. P. "On the generalized binary system." *IEEE Trans. Comp.* vol. C-31 (1982): 335.

Cerny, E., and Marin, M. A. "An approach to unified methodology of com-

binational switching circuits." *IEEE Trans. Comp.* vol C-26 (1977): 745.
Clymer, B., and Collins, S. A., Jr. "Optical computer switching network." *Opt. Engn.* vol. 24 (1985): 74.
Cody, W. J., Jr. "Static and dynamic numerical characteristics of floating point arithmetic." *IEEE Trans. Comp.* vol. C-22 (1973): 598.
Culliney, J. H.; Young, M. H.; Nakagawa, T.; and Muroga, S. "Results of the synthesis of optimal networks of AND and OR gates for four-variable switching functions." *IEEE Trans. Comp.* vol C-27 (1979): 76.
Curtis, H. A. "Short-cut method of deriving nearly optimal arrays of NAND trees." *IEEE Trans. Comp.* vol. C-28 (1979): 521.
Da Rocha, V. C. "Multilevel double error correcting codes." *Elect. Lett.* vol. 17 (1981): 45.
El-Agamy, M. A., and Munday, E. "Probabilistic hard-decision table look-up decoding for binary block codes." *Elect. Lett.* vol. 20 (1984): 922.
Ellis, D. T. "A synthesis of combinational logic with NAND or NOR elements." *IEEE Trans. Elect. Comp.* vol. EC-14 (1965): 701.
Gaitanis, N. "Single error correcting and multiple unidirectional error detecting cyclic AN arithmetic codes." *Elect. Lett.* vol. 20 (1984): 638.
Garner, H. L. "A survey of some recent contributions to computer arithmetic." *IEEE Trans. Comp.* vol. C-25 (1976): 1277.
Garner, H. L. "Theory of computer addition and overflows." *IEEE Trans. Comp.* vol. C-27 (1978): 297.
Gaylord, T. K.; Mirsalehi, M. M.; and Guest, C. C. "Optical digital truth table look-up processing." *Opt. Engn.* vol. 24 (1985): 48.
Iliff, S. *Advanced computer design.* Englewood Cliffs, N. J.: Prentice-Hall, 1982.
Hong, S. J., and Ostapko, D. L. "On complementation of Boolean functions." *IEEE Trans. Comp.* vol. C-21 (1972): 1022.
Huntington, E. V. "Sets of independent postulates for the algebra of logic." *Trans. Am. Math. Soc.* vol. 5 (1904): 288.
Hwang, K. *Computer arithmetic.* New York: Wiley, 1979.
Ichioka, Y., and Tanida, J. "Optical parallel logic gates using a shadow-casting system for optical digital computing." *Proc. IEEE.* vol. 72 (1984): 787.
Mowle, F. J. *A systematic approach to digital logic design.* Reading, Mass.: Addison-Wesley, 1976.
Pal, A. "An iterative algorithm for testing two-assumability of Boolean functions." *Proc. IEEE.* vol. 69 (1981): 1164.
Papachristou, C. A. "An algorithm for optimal NAND cascade logic synthesis." *IEEE Trans. Comp.* vol. C-27 (1978): 1099.
Parhami, B., and Avizienis, A. "Detection of storage errors in mass memories using low-cost arithmetic error codes." *IEEE Trans. Comp.* vol. C-27 (1978): 302.
Peterson, W. W. "Error correcting codes." *Sci. Am.* vol. 215 (1962): 96.
Rajeraman, V., and Radhakrishnan, T. *Introduction to digital computer design.* Englewood Cliffs, N. J.: Prentice-Hall, 1983.
Shannon, C. E. "A symbolic analysis of relay and switching circuits." *Trans. AIEE.* vol. 57 (1938): 713.
Swartzlander, E. E. "Merged arithmetic." *IEEE Trans. Comp.* vol. C-29 (1980): 946.

Trivedi, K. S. "On the use of continued fractions for digital computer arithmetic." *IEEE Trans. Comp.* vol. C-26 (1977): 700.

Waser, S., and Flynn, M. J. *Introduction to arithmetic for digital systems designers.* New York: Holt, Rinehart & Winston, 1982.

Yuen, C. K. "A new representation for decimal numbers." *IEEE Trans. Comp.* vol. C-26 (1977): 1286.

CHAPTER TWO

Minimization of Logic Functions

2.1 Introduction

One of the objectives of the designer is to keep the number of gates and the number of literals to a minimum when implementing a logic function. Boolean functions can be simplified by algebraic manipulations, but the process is tedious and the designer cannot be certain that the manipulations have produced the minimum circuit. A much easier and faster method of minimization is presented in this chapter. This method involves plotting minterms on a two-dimensional map. Following the rules to be established in this chapter, algebraic manipulations will be traded for graphical operations. The graphical techniques, called *Karnaugh mapping* after their originator, allows minimum 2-level functions to be obtained with little effort.

It is appropriate to note that designers often use other methods for implementing complex logic circuits. For example, read-only-memory (see Chapter 4 for details) can be used to generate multiple functions of many variables without regard to any sort of minimization. Consequently, the traditional demand for absolute simplification has been diminished somewhat by such devices. However, there are still occasions when it is considered necessary to reduce complex logic functions.

Karnaugh mapping is not well suited for the manual solution of problems involving more than six variables and is not easily programmable for solution by digital computer. In the later sections of this chapter we will present a minimization algorithm, known as Quine-McCluskey's (Q-M) tabular reduction technique, that is reasonably straightforward but time-consuming. It is possible to come up with more than one valid minimization if Karnaugh maps are used because of the different ways the maps may be read. In contrast, the Q-M technique allows an optimum solution to be determined in a systematic way. The technique uses a systematic method to conduct an exhaustive search to determine all possible combinations of logically adjacent cell entries, and then directs the designer to the optimum solution. It is possible to write software programs using the Q-M method to allow computer-aided design of logic circuits. The Q-M technique also can be effectively used to optimize the logic functions for circuits that have common input variables but multiple output functions. After studying this chapter, you should be able to:

- Construct Karnaugh maps for minimizing Boolean functions;
- Obtain the minimal form (SOP or POS) for Boolean equations from Karnaugh maps;
- Eliminate hazard conditions from combinational circuits;
- Obtain the minimal form for single-output and multiple-output logic functions using Quine-McCluskey's tabular reduction scheme.

2.2 Binary Space and Karnaugh Maps

A binary function of one variable plotted in binary space has the possibility of being only positive and may be represented by one or several of the four corners of a unit square, because the values of the function and that of the independent variable can be either a 1 or a 0. Examples of single-variable plots are shown in Figure 2.1. Similarly, for functions of two variables the function values will all fall on the corners of a unit cube. The functions $f_1(A,B) = A\overline{B} + AB$ and $f_2(A,B) = \overline{A}\overline{B} + AB$ are shown plotted in Figure 2.2.

By using Boolean algebra, the function $f_1(A,B)$ can be reduced as follows:

$$f_1(A,B) = A\overline{B} + AB = A(\overline{B} + B) = A$$

It is sufficient to say that the function $f_1(A,B)$ is a 1 whenever A is a 1. In the corresponding binary space we can also see that the two points of the function are one unit distance apart.

FIGURE 2.1 **Representation of $f(A) = A$ and $f(A) = \overline{A}$ in Binary Space.**

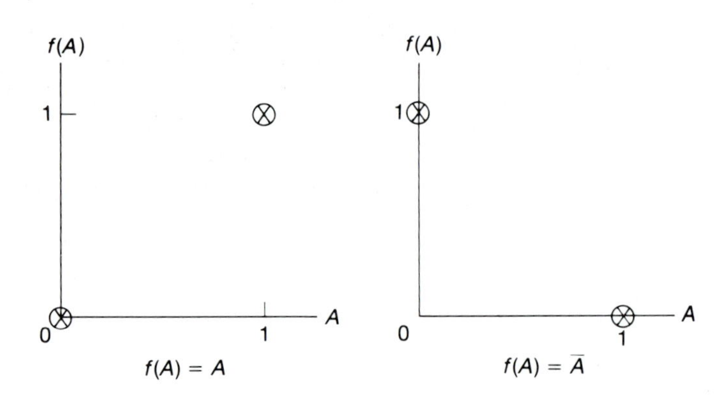

FIGURE 2.2 **Representation of Two-Variable Functions in Binary Space: [*a*] $f_1(A,B) = A\overline{B} + AB$ and [*b*] $f_2(A,B) = \overline{A}\overline{B} + AB$.**

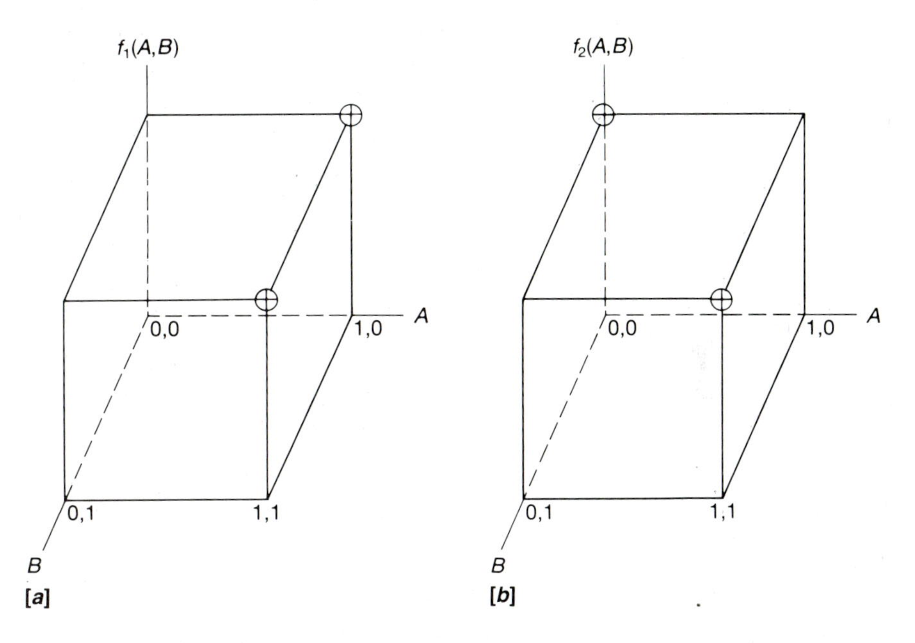

FIGURE 2.3 **Representation of $f_3(A,B) = AB + A\overline{B} + \overline{A}B$ in Binary Space.**

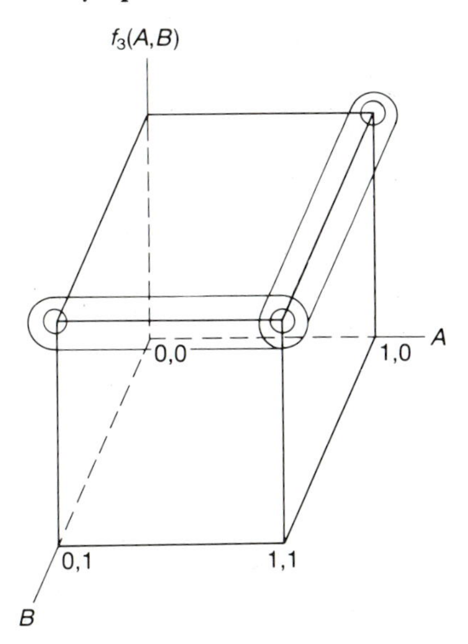

The function $f_2(A,B)$, however, is not reducible. The two points (0,0) and (1,1) graphically describe that the two minterms are two units apart and, therefore, are not adjacent to one another on the cube. Two minterms are considered to be *adjacent*, or neighbors, when they represent two successive Gray codes. It may be seen that two changes are necessary to convert 00 to 11 and vice versa. The two minterms in question are, therefore, said to be *nonadjacent*. This conclusion is the consequence of what is known as the *principle of adjacency*. We can now generalize and say that for a function of two variables, any two plotted points at one unit apart may be grouped and expressed as the literal that remains constant while moving from one point to the other.

A more complex example of a function, $f_3(A,B) = AB + A\overline{B} + \overline{A}B$, is shown in Figure 2.3. Grouping the adjacent pairs, the function can be simplified to $f_3(A,B) = A + B$. The function can also be reduced using Boolean algebra as follows:

$$
\begin{aligned}
f_3(A,B) &= AB + A\overline{B} + \overline{A}B \\
&= A(B + \overline{B}) + \overline{A}B \\
&= A + \overline{A}B \\
&= A + B
\end{aligned}
$$

It is apparent, therefore, that both binary space grouping of minterms and Boolean algebra manipulations give the same results.

Karnaugh devised a graphical scheme for representing minterms in a single plane. The Karnaugh map (K-map) is a set of 2^N squares arranged in an ordered two-dimensional array. Each of the squares on the map corresponds to one of the 2^N possible minterms (a point in binary space). The possible K-maps for functions of up to four variables are shown in Figure 2.4. Each of the squares on the map, called a *cell*, has an entry equal to the decimal value of the corresponding minterms. Along the top of the map, the variable values are listed in such a way that as one scans the values from left to right or from right to left from cell to cell, only one variable

FIGURE 2.4 **Karnaugh Maps for [*a*] *f*(*A*), [*b*] *f*(*A*,*B*,), [*c*] *f*(*A*,*B*, *C*), and [*d*] *f*(*A*,*B*,*C*,*D*).**

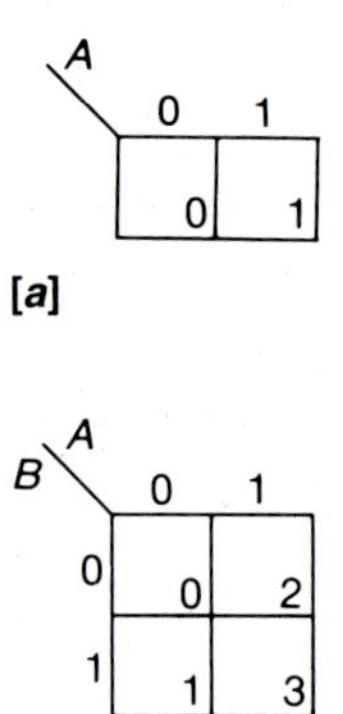

C \ AB	00	01	11	10
0	0	2	6	4
1	1	3	7	5

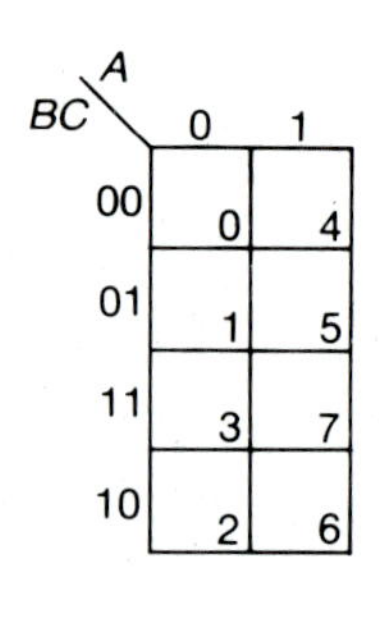

[*c*]

CD \ AB	00	01	11	10
00	0	4	12	8
01	1	5	13	9
11	3	7	15	11
10	2	6	14	10

[*d*]

changes. Each cell is thus adjacent to the next. In addition, the cells on the right end are only one unit distance from (adjacent to) the cells on the left end. The variables along the left edge are also labeled in the same way. Cells at the top of the map are adjacent to cells along the bottom of the map.

Simplification of a Boolean function plotted on a K-map is a process of correctly grouping the adjacent 1s. Entries of 1 are made in the cells corresponding to the minterms that make the function a 1 in SOP form. A 0 is entered in each of the remaining cells. The rules for forming groups are as follows:

Rule 1. If all entries in an N-variable K-map are 1s, the group size is 2^N and the corresponding minimized function is given by $f(A,B,\ldots) = 1$.

Rule 2. For N variables, the largest nontrivial group size is 2^{N-1}. All group sizes are powers of 2. For N variables, group sizes of 2^N, 2^{N-1}, $2^{N-2}, \ldots, 2^0$ are allowed.

Rule 3. In any group it must be possible to start at a cell and travel from cell to cell, with moves of one unit length, through each cell in the group without passing through any cell twice and return to the starting cell.

Using the rules for forming groups, we now apply the rules to arrive at a minimum SOP function as follows:

1. In the process of grouping, each 1 on the map must be included in at least one group.
2. For N variables, first look for 1s that cannot be grouped with any other 1 and circle these. These isolated 1s will appear as N literal minterms in the final function.
3. Look for 1s that can be grouped with only one other 1 and circle these groups as pairs.
4. Next look for 1s that can be grouped only in groups of four and circle these groups.
5. Continue the grouping process until group sizes of 2^{N-1} have been considered.
6. This grouping process should be stopped at any time that each 1 has been included in at least one group. Next, eliminate the smaller groups, if any, that are enclosed completely within a larger group.
7. For each group, determine which literals remain constant in each cell in the group. These literals when ANDed together form the product terms of the SOP function. For x groups, there will be x product terms. For an N-variable function, a group size of 2^k would contribute a product term of $N - k$ literals.

These instructions are best understood by considering the K-maps of Figure 2.5. There are only two minterms, 5 and 7, on the first three-variable map of Figure 2.5[*a*]. These minterms may be combined to form a single ANDed term using Boolean algebra as follows:

$$A\overline{B}C + ABC = AC(\overline{B} + B) = AC$$

The two minterms are adjacent, and the variables constant within this group-of-two are A and C. Similarly, the minterms, 1 and 5, of the next K-map would yield the following ANDed term:

$$\overline{A}\overline{B}C + A\overline{B}C = \overline{B}C(\overline{A} + A) = \overline{B}C$$

The possibility of Boolean simplification indicates that these two minterms can be grouped and, consequently, are neighbors. The K-maps are *rolled*, therefore, such that the left and right side edges are considered to be touching. Similar rolling demonstrates that (a) the top K-map of Figure 2.5[*b*] reduces to $\overline{B}$, (b) the bottom left K-map of Figure 2.5[*c*] reduces to $\overline{B}C$, and (c) the left K-map of Figure 2.5[*d*] reduces to $\overline{B}$.

The bottom right K-map of Figure 2.5[*c*], consisting of minterms 0, 2, 8, and 10, deserves extra attention because of its uniqueness. These minterms may be combined using Boolean algebra as follows:

$$\begin{aligned}\overline{A}\overline{B}\overline{C}\overline{D} + \overline{A}\overline{B}C\overline{D} + A\overline{B}\overline{C}\overline{D} + A\overline{B}C\overline{D} &= \overline{A}\overline{B}\overline{D}(\overline{C} + C) \\ &\quad + A\overline{B}\overline{D}(\overline{C} + C) \\ &= \overline{A}\overline{B}\overline{D} + A\overline{B}\overline{D} \\ &= \overline{B}\overline{D}(\overline{A} + A) = \overline{B}\overline{D}\end{aligned}$$

These four corner minterms are called a group-of-four. This is another rolling example. The maps are continuous in two directions so that both top and bottom edges and left and right edges are considered to be touching, forming a donut-shaped figure. The remaining K-maps of Figure 2.5 show other more obvious groupings. Example 2.1 illustrates the mapping scheme when there is more than one adjacent group.

EXAMPLE 2.1

Obtain the equation for a combinational circuit with four input variables, A, B, C, and D, that outputs 1 whenever $ABCD$ is a prime number.

SOLUTION

A four-variable problem requires a truth table of 2^4 different input combinations. The resulting truth table is shown in Figure 2.6. The function can also be expressed in the minterm form as follows:

$$f(A,B,C,D) = \Sigma m(1,2,3,5,7,11,13)$$

FIGURE 2.5 **K-maps: [*a*] Pair in Three-Variable Map, [*b*] Group-of-Four in Three-Variable Map, [*c*] Group-of-Four in Four-Variable Map, and [*d*] Group-of-Eight in Four-Variable Map.**

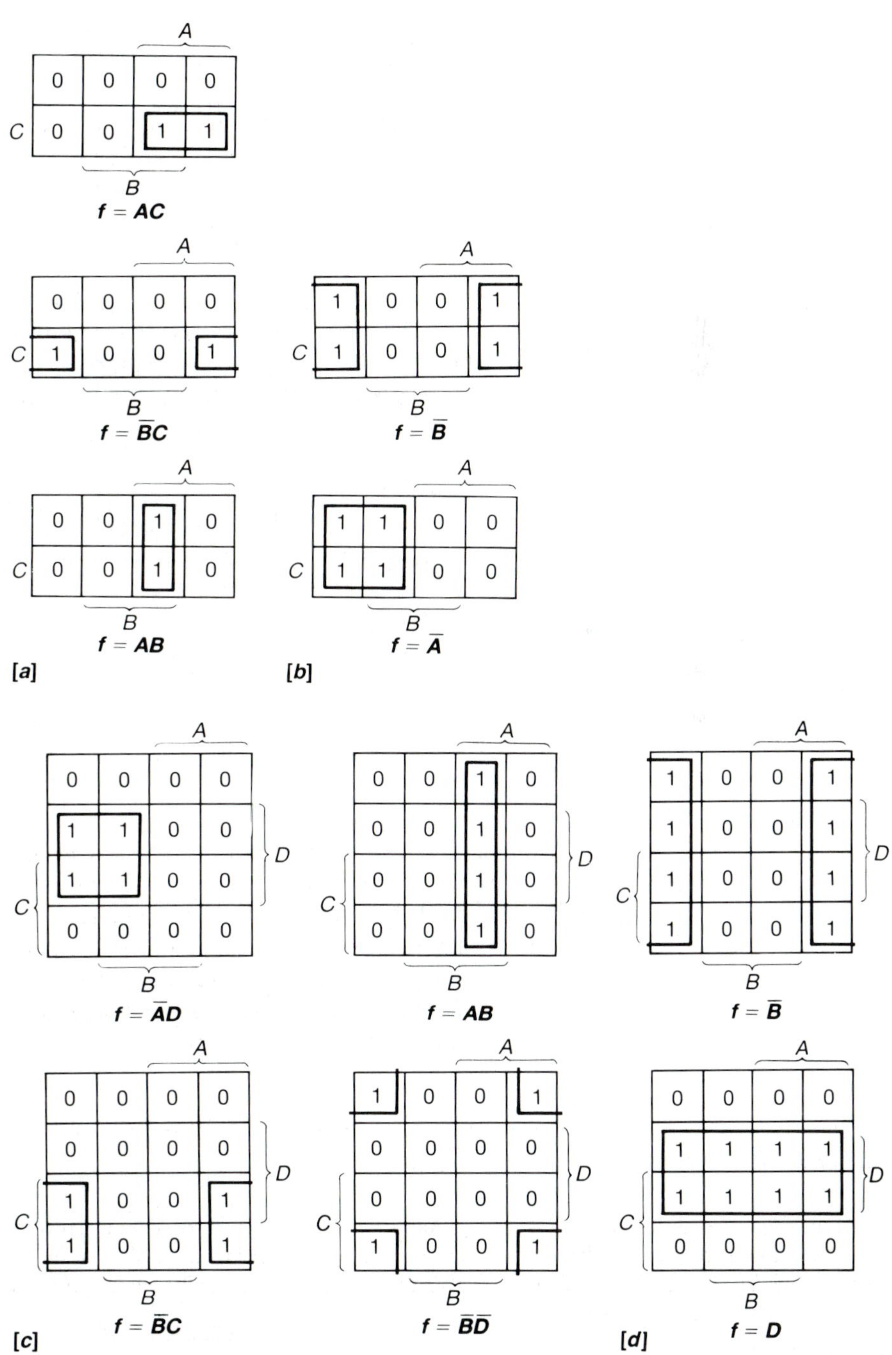

FIGURE 2.6

A	B	C	D	f
0	0	0	0	0
0	0	0	1	1
0	0	1	0	1
0	0	1	1	1
0	1	0	0	0
0	1	0	1	1
0	1	1	0	0
0	1	1	1	1
1	0	0	0	0
1	0	0	1	0
1	0	1	0	0
1	0	1	1	1
1	1	0	0	0
1	1	0	1	1
1	1	1	0	0
1	1	1	1	0

FIGURE 2.7

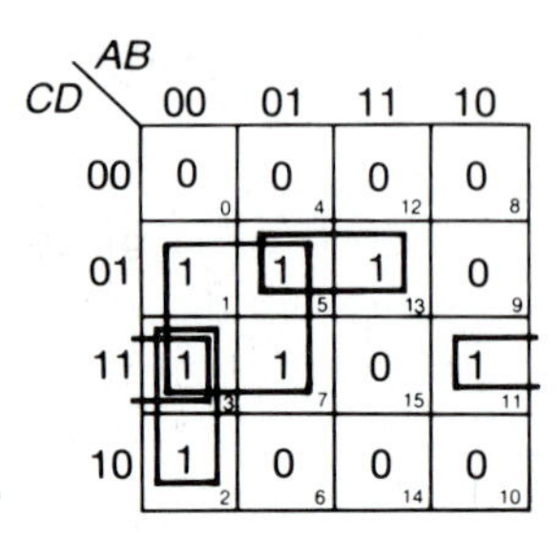

These minterms are entered as 1s in a four-variable K-map as shown in Figure 2.7. Using the rules to find a minimum function, we first look for 1s that cannot be grouped. There are none. Next, look for 1s that can be grouped with only one other 1. Minterm 11 can be grouped only with minterm 3, minterm 13 can be grouped only with minterm 5, and minterm 2 can be grouped only with minterm 3. Next look for groups-of-four. Minterms 1, 3, 5, and 7 form a valid group-of-four. At this point all 1s are included in a group, and the grouping of 1s is complete. No two chosen groups can be included in any other group. Note that minterm 5 is included in two groups and 3 in three groups. This is permitted. The important point is that all 1s must be included. The K-map groups are as follows:

(1,3,5,7), giving $\overline{A}D$
(2,3), giving $\overline{A}\overline{B}C$
(5,13), giving $B\overline{C}D$
(3,11), giving $\overline{B}CD$

Therefore, the minimized SOP function is given by

$$f(A,B,C,D) = \overline{A}D + \overline{A}\overline{B}C + B\overline{C}D + \overline{B}CD$$

In actual design most of the K-map applications will involve four or fewer variables. In the event a five- or six-variable problem is encountered, more than one four-variable map would be necessary for the minimization scheme. Note that the MSB of the first 16 five-variable minterms is always a 0, and that of the remaining 16 minterms is a 1. Similarly, the minterms of a six-variable problem could be split into four groups of 16. One group of 16 is distinguished from the other by their first two bits. Consequently, the five-variable K-map consists of two four-variable K-maps, and the six-variable K-map consists of four four-variable K-maps. The mapping schemes are shown in Figure 2.8.

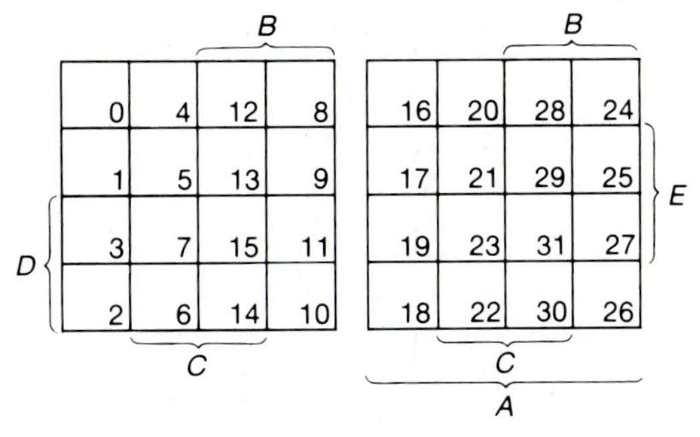

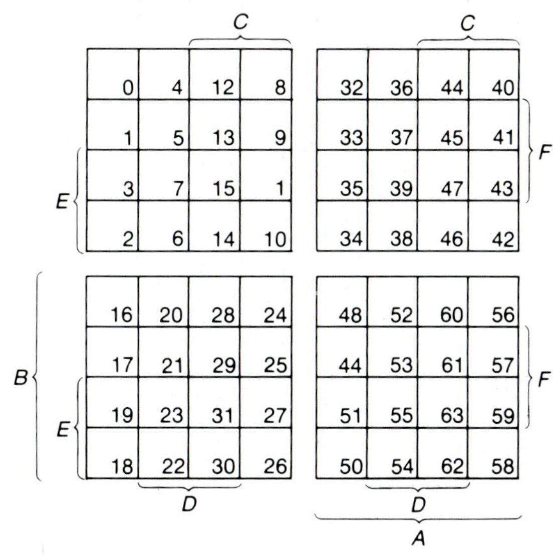

FIGURE 2.8 **[*a*] Five-Variable K-Map and [*b*] Six-Variable K-Map.**

In a five-variable map each minterm can have up to five possible neighbors. For example, in Figure 2.8[*a*] the minterms 1, 8, 11, 13, and 25 are all neighbors of the minterm 9. The minterms appearing in the same location on the two halves are logically adjacent. For the purpose of grouping, the two halves are assumed to be lying on top of each other. For a six-variable K-map, the right half and the left half and the top half and the bottom half are considered to be adjacent. Care must be taken in forming large groups. Examples of some of the valid groupings in a six-variable map are (0,16,32,48), (5,7,13,15,21,23,29,31,37,39,45,47,53,55,61,63), (0,1,2,3,16,17,18, 19, and (0,2,8,10,16,18,24,26,32,34,40,42,48,50,56,58).

EXAMPLE 2.2

Find the most reduced form of the function

$f(A,B,C,D,E) = \Sigma m(1,3,9,11\text{–}15,17\text{–}19,25,27,29,31)$

SOLUTION

The minterms are plotted and grouped in a five-variable K-map as shown in Figure 2.9. The groupings are as follows:

(1,3,9,11,17,19,25,27), yielding $\overline{C}E$
(9,11,13,15,25,27,29,31), yielding BE
(12,13,14,15), yielding $\overline{A}BC$
(18,19), yielding $A\overline{B}\overline{C}D$

Therefore, the reduced function is given by

$$f(A,B,C,D,E) = \overline{C}E + BE + \overline{A}BC + A\overline{B}\overline{C}D$$

FIGURE 2.9

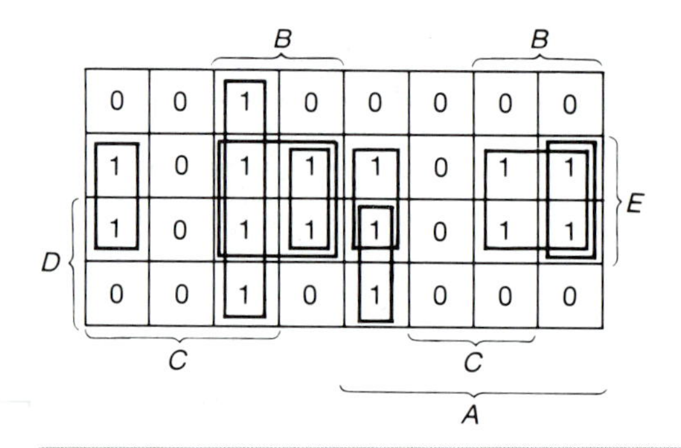

2.3 POS Form from K-Maps

Often it is easier for a designer to implement the POS form of a function rather than its regular SOP form. Such an instance occurs when it is determined to be more economical to implement the complement of a function than the function itself. However, it is not necessary to redraw the K-maps for finding the complement of a function. The standard K-map is used but only the 0s are grouped to obtain the complement function. The grouping of 0s follows the same rules that were used earlier in the grouping of 1s. When the resulting reduced SOP expression for the complement function is inverted, an expression logically equivalent to the original POS function is obtained.

As an example, consider finding the SOP and POS forms for the function $f(A,B,C,D) = \Sigma m(0,2,4,8,10,14)$. The function is plotted on two different K-maps, as shown in Figure 2.10. Map [*a*] is used for the SOP form and map [*b*] is used for the POS form. By grouping the 1s in map [*a*] we would obtain

$$f(A,B,C,D) = \overline{B}\overline{D} + AC\overline{D} + \overline{A}\overline{C}\overline{D}$$

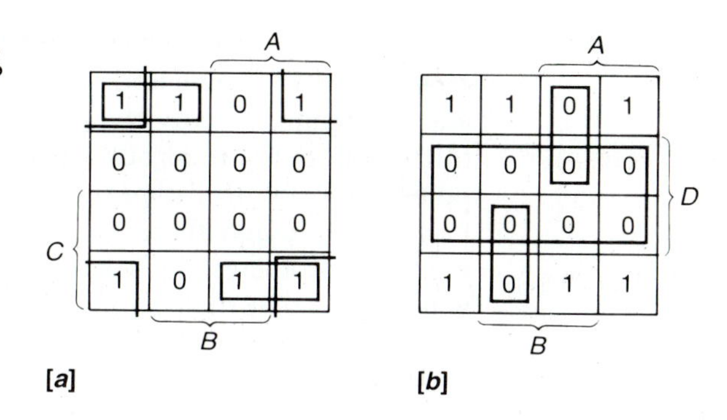

FIGURE 2.10 K-Maps for $f(A,B,C,D) = \Sigma m(0,2,4,8,10,14)$: [*a*] SOP Form and [*b*] POS Form.

and by grouping the 0s in map [*b*] we would obtain

$$\overline{f}(A,B,C,D) = D + \overline{A}BC + AB\overline{C}$$

The complement function now may be complemented to yield the original function:

$$f(A,B,C,D) = \overline{D}(A + \overline{B} + \overline{C})(\overline{A} + \overline{B} + C)$$

The circuits corresponding to these two forms are shown in Figure 2.11.

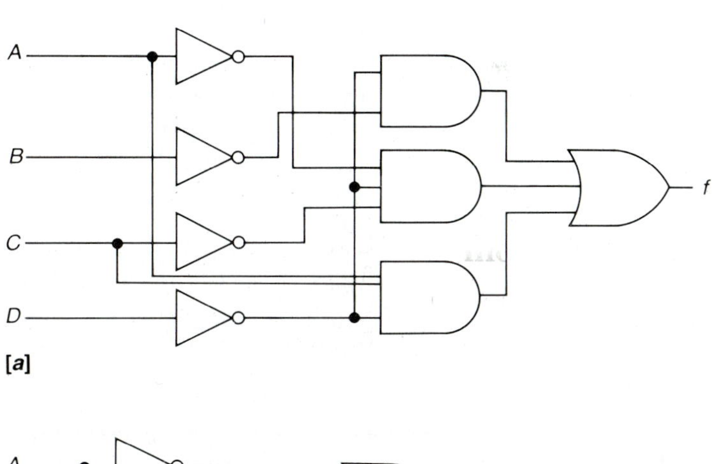

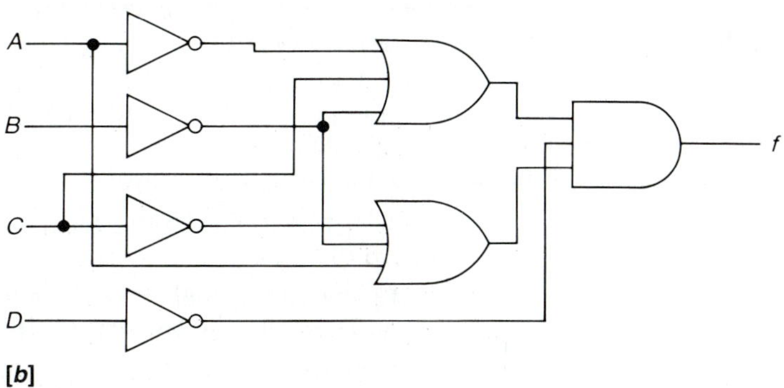

FIGURE 2.11 Logic Circuit for $f(A,B,C,D) = \Sigma m(0,2,4,8,10,14)$: [*a*] SOP Circuit and [*b*] POS Circuit.

You can see in Figure 2.11 that the SOP form requires a total of 8 gates and 15 inputs and the POS form requires a total of 7 gates and 13 inputs. Usually the cost of a circuit is directly proportional to the sum of the total number of inputs and discrete gates. At this point the POS form appears to be more economical than the SOP form even though there are fewer 1s than 0s. When we consider practical problems, such as the availability of certain gate types, you will see that the decision may not be so clear-cut.

2.4 Don't-Care Terms in K-Maps

Most logic circuit designs begin with a verbal statement of the problem. The next step in the process is making a truth table. Previously we have considered truth tables with output values for all possible input conditions specified. In this section we will consider logic functions that are incompletely specified. That is, practical constraints exclude a particular group of minterms from occurring, which makes their occurrence in the function strictly optional. We don't care if they are in the function since they wouldn't occur. These excluded minterms are called the *don't-care* minterms. For example, consider the BCD inputs that were discussed in Section 1.6. This four-bit code with weights 8-4-2-1 uses only 10 of a possible 16 combinations. The six combinations 1010, 1011, 1100, 1101, 1110, and 1111 are not used in BCD. Constraints of this type often become very helpful since the don't-care minterms may be included in the K-map as either a 0 or a 1, depending on which leads to the minimum logic function.

On the map the don't-care minterms are indicated with a – symbol. During the grouping procedure, those don't-cares that are grouped with other minterms to form larger groups are considered to be a 1 or 0, depending on which are being grouped. Groups of only don't-cares are never used. A minterm list for a function with don't-cares has the form $f = \Sigma m(\ldots) + d(\ldots)$ where the don't-cares are listed in $d(\ldots)$. Example 2.3 illustrates the benefit of having don't-cares in a logic function.

EXAMPLE 2.3

Design a combinational circuit that turns on a light-emitting diode (LED) for each even BCD input.

SOLUTION

It is assumed that an output of a 1 causes the LED to light up. The truth table and the corresponding K-map for such a circuit are shown in Figure 2.12. The function is written in concise form as follows:

$$f(A,B,C,D) = \Sigma m(0,2,4,6,8) + d(10\text{–}15)$$

In the K-map only those don't-cares are included that lead to larger grouping. Those don't-cares are 10, 12, and 14. The resulting reduced function is given by

$$f(A,B,C,D) = \overline{D}$$

FIGURE 2.12

A	B	C	D	f
0	0	0	0	1
0	0	0	1	0
0	0	1	0	1
0	0	1	1	0
0	1	0	0	1
0	1	0	1	0
0	1	1	0	1
0	1	1	1	0
1	0	0	0	1
1	0	0	1	0
1	0	1	0	–
1	0	1	1	–
1	1	0	0	–
1	1	0	1	–
1	1	1	0	–
1	1	1	1	–

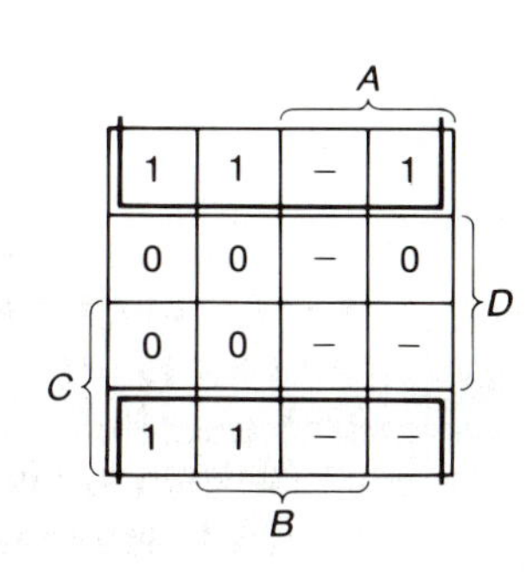

If we hadn't taken advantage of these don't-cares, the function would have reduced to

$$f(A,B,C,D) = \overline{A}\overline{D} + \overline{B}\overline{C}\overline{D}$$

It is important to understand that if the circuit providing the inputs to this circuit malfunctions and sends a 10, 12, or 14, your circuit will output a 1 and light the LED.

EXAMPLE 2.4

Obtain a six-input combinational circuit whose output, y_0, is given by

$$y_0 = \begin{cases} x_{-1} & \text{if } d = 1, s = 1, \text{ and } E = 1 \\ x_1 & \text{if } d = 0, s = 1, \text{ and } E = 1 \\ x_0 & \text{if } d = -, s = 0, \text{ and } E = 1 \\ 0 & \text{if } d = -, s = -, \text{ and } E = 0 \end{cases}$$

where d, s, E, x_0, x_1, and x_{-1} are the inputs.

SOLUTION

A six-variable truth table requires a total of 64 combinations of the inputs. But in this case we don't have to go through all of these. The outputs do not depend on all of the input variables simultaneously. The inputs not depended on can be treated as don't-cares. Consequently, the corresponding truth table is obtained as shown in Figure 2.13.

FIGURE 2.13

E	d	s	x_1	x_0	x_{-1}	y_0
1	1	1	–	–	0	0
1	1	1	–	–	1	1
1	0	1	0	–	–	0
1	0	1	1	–	–	1
1	–	0	–	0	–	0
1	–	0	–	1	–	1
0	–	–	–	–	–	0

FIGURE 2.14

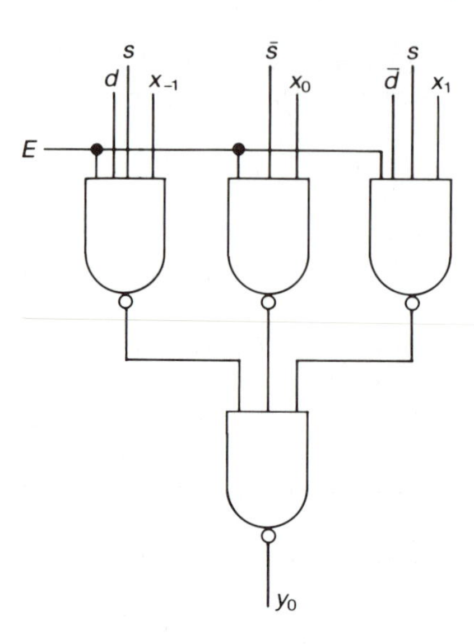

The equation for y_0 can be obtained readily from the truth table. Note that there is no need to go through any mapping scheme.

$$y_0 = (dsx_{-1} + \bar{d}sx_1 + \bar{s}x_0)E$$

The resulting circuit is obtained as shown in Figure 2.14. This circuit is commonly referred to as a *shifter* since it can be used for either shift-left, shift-right, or no-shift operations. Note, however, that the shifter produces 0 when $E = 0$.

2.5 Combinational Hazards

When the inputs are altered in a combinational circuit, some or all of the outputs may change in response. Any output that changes more than once for a single input variable change has actually responded to a circuit hazard. These spurious output changes are referred to as *glitches*. Such undesired circuit behavior results mainly from the fact that all of the practical logic gates have inherent propagation delays, that is, their outputs change some time after the inputs.

The hazard-related output perturbations, however, eventually cease to exist in a purely combinational network and do not pose a serious problem to the operation of the network. In later chapters it will be shown that the combinational output is used to drive circuits sensitive to 1→0 and 0→1 transitions. The presence of combinational hazards becomes extremely critical in these circuits.

An introduction to the concept of timing diagrams is vital in understanding hazards and other time-related phenomena. Often all the inputs are changing with considerable frequency, and it becomes difficult to determine the output that should occur. *Timing diagrams* generally are used to describe the time behavior of digital systems. They typically include such characteristics as delays and rise and fall times. Conventionally, a timing diagram consists of the input and output waveforms where each signal occupies a different horizontal stripe. Time is assumed to increase along the horizontal axis from left to right. The vertical axis corresponds to the values of the signals. In most cases only logical values are used because our interest primarily lies in the functional behavior of the system.

Two types of hazards are of interest: static hazards and dynamic hazards. A *static hazard* is an error condition resulting in a momentary change of output in response to an input change that should not cause output to change. In contrast, a *dynamic hazard* is an error condition that occurs when, in response to an input change, the output changes several times before settling down to its new steady-state value. These conditions are illustrated in the output waveforms of Figure 2.15.

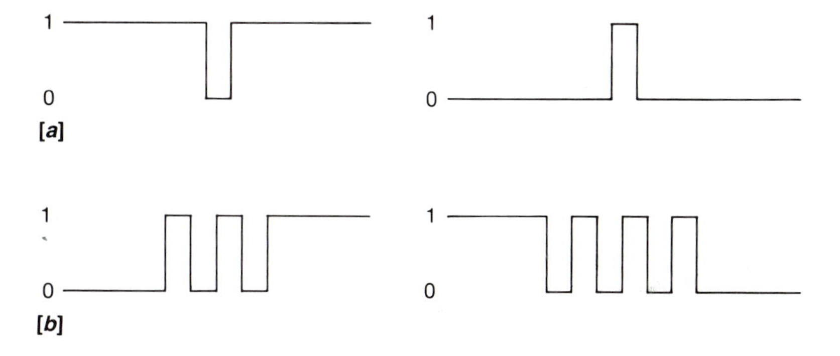

FIGURE 2.15 **Hazards: [*a*] Outputs with Static Hazard and [*b*] Outputs with Dynamic Hazards.**

As indicated previously, the classification of hazards as static or dynamic is based on the behavior of the circuit outputs. However, there are two sources that contribute to these hazards. A *function-induced hazard* (or, simply, function hazard) is a hazard caused by the function itself. A *logic-induced hazard* (logic hazard) is caused by the particular logic network implementing the function. A function hazard usually is eliminated by not allowing more than one input to change at a time. Logic hazards can exist even for single-input changes. Combinational circuits have the possibility of static hazards. Dynamic hazards will be considered in the chapters on sequential circuit design.

An example of a circuit with a static hazard induced by a logic hazard is shown in Figure 2.16. The performance of this circuit is illustrated by the associated timing diagram. This hazard is a result of the nonideal operation of physical devices. Due to the presence of capacitance in the electronics performing the logic functions, a small delay of a few nanoseconds exists between input changes and corresponding output response.

FIGURE 2.16 **Example of a Static Hazard: [*a*] K-Map, [*b*] Logic Circuit, and [*c*] Timing Diagram.**

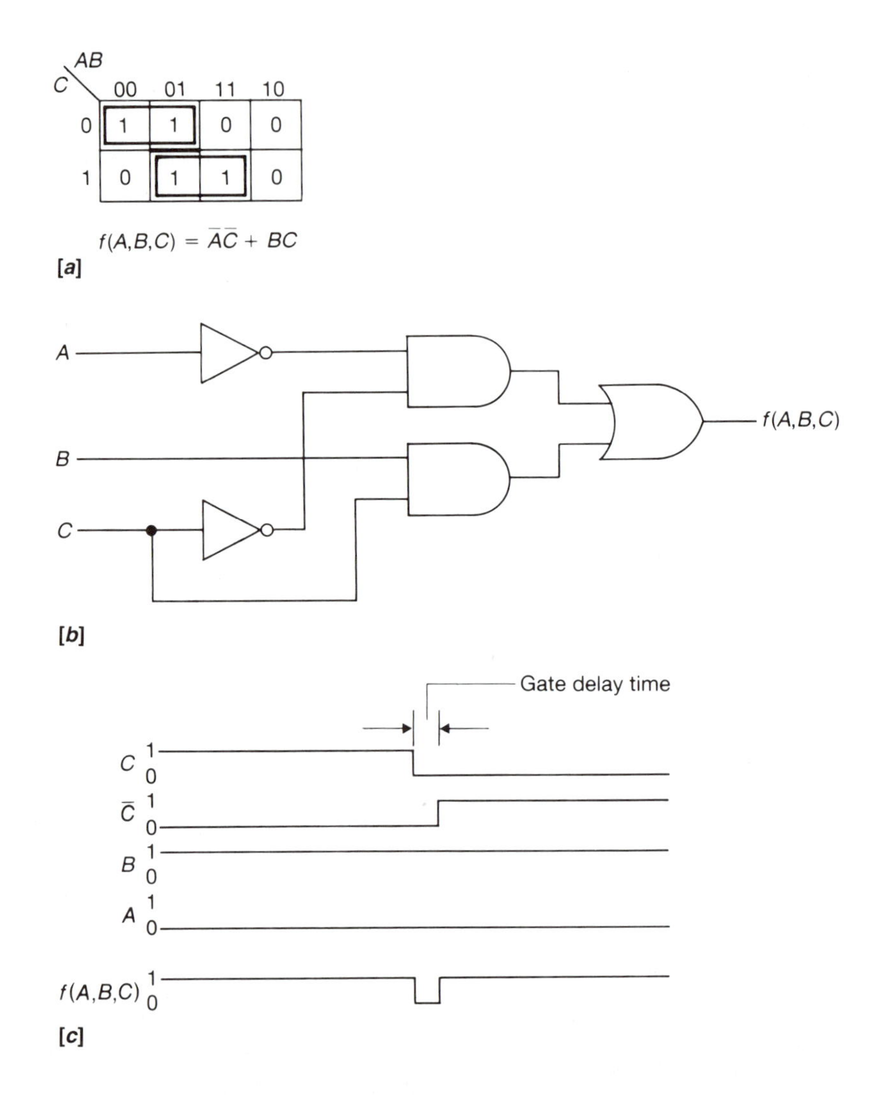

Figure 2.16[*a*] shows a K-map corresponding to the function $f(A,B,C,D) = \Sigma m(0,2,3,7)$. The output should be a 1 as long as the minterms corresponding to the current input conditions remain in the cells numbered 0, 2, 3, and 7. The circuit of Figure 2.16[*b*] has been obtained by grouping the 1s of the corresponding K-map. The minterms (0,2) and (3,7) are grouped to form two groups-of-two minterms, resulting in the equivalent logic function $f(A,B,C,) = \overline{A}\overline{C} + BC$. The corresponding timing diagram in Figure 2.16[*c*] represents the input condition changing from $ABC = 011$ to 010, that is, C changing from 1 to 0. During this change the circuit moves from the group represented by BC to the group represented by $\overline{A}\overline{C}$, that is, from cell 3 to cell 2. Due to the practical limitation of the inverter, C and $\overline{C}$ are both momentarily 0, which forces the output to be 0 for a time equal to the gate delay time. This momentary

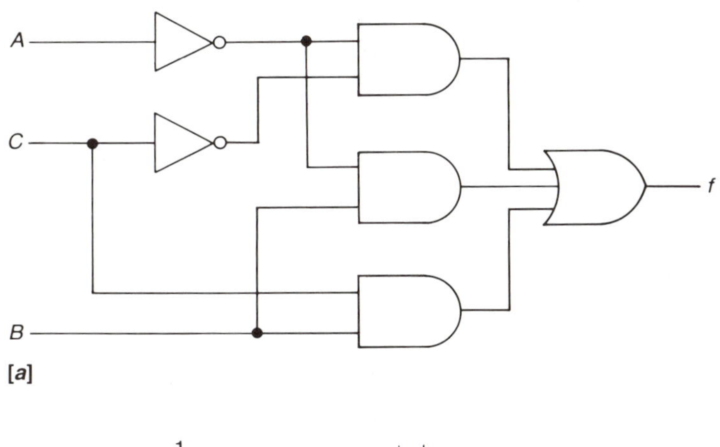

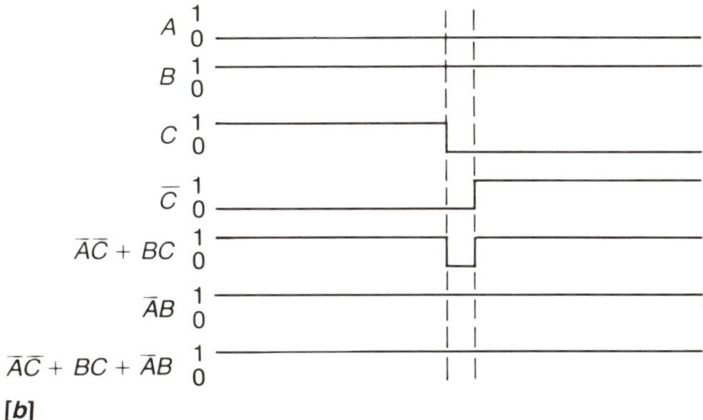

FIGURE 2.17 **[*a*] Modified Circuit with No Hazard and [*b*] Its Timing Diagram.**

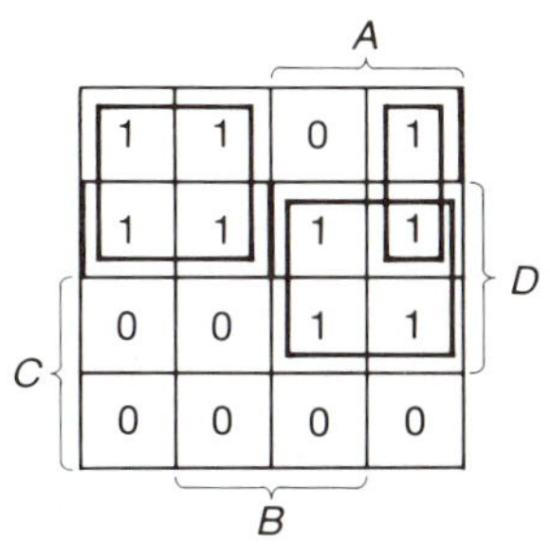

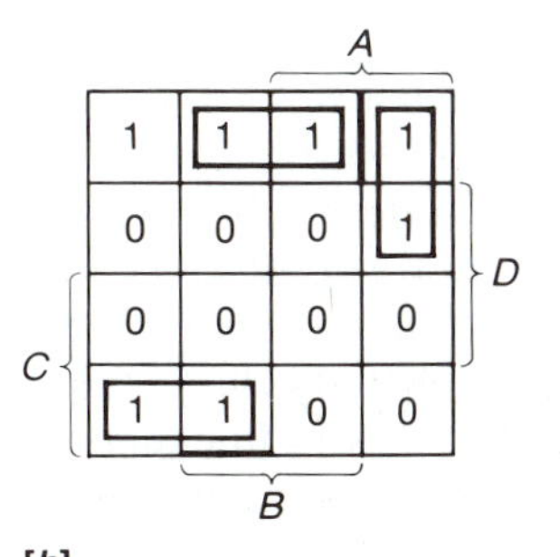

FIGURE 2.18 **Examples of K-Map Grouping That Results in Hazards.**

drop in the output is referred to as a static hazard since the initial and final outputs of the logic device are the same.

Consider the circuit in Figure 2.17 that entertains an additional pair of minterms, formed by grouping cells 2 and 3. The inclusion of this additional pair removes the hazard condition that was discussed earlier. In the corrected logic, the term $\overline{A}B$ holds the output at 1 when the circuit moves from cell 3 to cell 2.

Additional examples of logic groupings that result in hazards are shown in Figure 2.18. They can be recognized as contributing to static hazards by noting the map positions where one group is exited and another is entered with no other group covering the cells being traversed. As shown in our example, this type of hazard is eliminated by adding an additional term that covers transition from one group to another. The hazards in Figure 2.18[*a*] can be eliminated by replacing the pair (8,9) with the group (0,1,8,9) and including the group (1,5,9,13). Similarly, the static hazards in Figure 2.18[*b*] can be removed by including two additional pairs, (8,12) and (4,6).

There are occasions when a circuit must have these hazard conditions removed and other occasions when the hazards will not adversely affect the circuit operation. For example, in certain sequential circuits (as it will be shown in a later chapter) the hazard conditions may be extremely critical. However, if the variable change causing the hazard doesn't ever occur, or if it is determined that the spurious outputs do not affect the operations, no corrective circuitry is necessary.

2.6 Cellular Arithmetic

A reduced function in the SOP form can always be expanded to obtain the original function containing each minterm for which the function is a 1. This is called the *expanded sum-of-product* (*ESP*) form. Each of the ESP terms is a minterm and corresponds to an entry in the K-map of the original function.

EXAMPLE 2.5

Obtain the ESP form of the function

$f(A,B,C,D) = ABC + A\overline{B}\overline{C} + \overline{C}D$

SOLUTION

$$\begin{aligned} f(A,B,C,D) &= ABC(D + \overline{D}) + A\overline{B}\overline{C}(D + \overline{D}) + \overline{C}D(B + \overline{B}) \\ &= ABCD + ABC\overline{D} + A\overline{B}\overline{C}D + A\overline{B}\overline{C}\overline{D} + B\overline{C}D + \overline{B}\overline{C}D \\ &= ABCD + ABC\overline{D} + A\overline{B}\overline{C}D + A\overline{B}\overline{C}\overline{D} + B\overline{C}D(A + \overline{A}) \\ &\quad + \overline{B}\overline{C}D(A + \overline{A}) \\ &= ABCD + ABC\overline{D} + A\overline{B}\overline{C}D + A\overline{B}\overline{C}\overline{D} + AB\overline{C}D \\ &\quad + \overline{A}B\overline{C}D + A\overline{B}\overline{C}D + \overline{A}\overline{B}\overline{C}D \\ &= \overline{A}\overline{B}\overline{C}D + \overline{A}B\overline{C}D + A\overline{B}\overline{C}D + A\overline{B}\overline{C}\overline{D} + AB\overline{C}D \\ &\quad + ABC\overline{D} + ABCD \end{aligned}$$

which is the desired ESP form of the original function. In concise form this is also expressed as follows:

$$f(A,B,C,D) = \Sigma m(1,5,8,9,13,14,15)$$

In cellular notation each of the terms of an already reduced function may be represented by the Boolean cell, M, which is expressed as

$$M = (a_i a_{i-1} a_{i-2} a_{i-3} \ldots a_1)$$

where the cell component a_j is a member of the set [—,0,1]. Note that a_i is the MSB and a_1 is the LSB. A variable in the uncomplemented form is represented by 1, a variable in the complemented form is represented by a 0, and the absence of a variable is designated by the don't-care symbol —. The order of the cell is determined from the number of don't-cares present in that cell.

We are already familiar with several different ways to express a function. For example, the function shown in the K-map of Figure 2.19 may be expressed as

FIGURE 2.19 K-Map for $f(A,B,C,D) = \Sigma m(2,3,5,7,8,9,13,15)$.

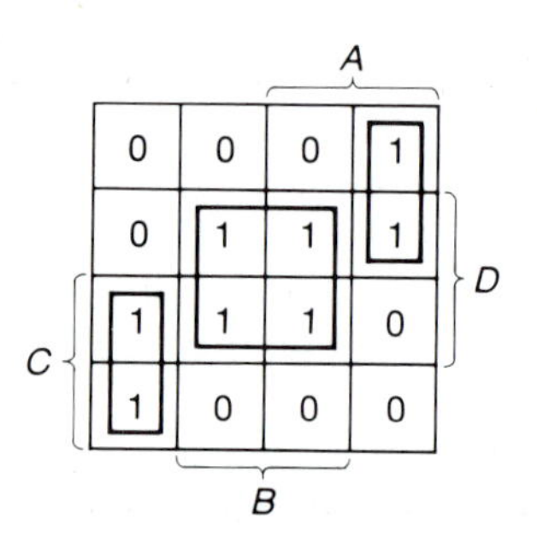

$$
\begin{aligned}
f(A,B,C,D) &= \Sigma m(2,3,5,7,8,9,13,15) && \text{ESP form} \\
&= A\bar{B}\bar{C} + BD + \bar{A}\bar{B}C && \text{SOP form}
\end{aligned}
$$

where no hazard term has been considered for the sake of simplicity. Note that minterms 2 and 3 may be grouped as follows:

$$
\begin{aligned}
&\bar{A}\bar{B}C\bar{D} + \bar{A}\bar{B}CD = \bar{A}\bar{B}C(\bar{D} + D) = \bar{A}\bar{B}C && \text{Algebraic form} \\
&(0010) \quad (0011) \qquad\qquad\qquad\qquad (001-) && \text{Cellular form}
\end{aligned}
$$

The binary representations of the minterms 2 and 3 differ in only one bit position, which has a weight of 1. We may introduce the notation 2,3(1) to indicate that the two minterms, 2 and 3, may be combined to form a pair by eliminating the variable corresponding to a weight of 1. Similarly, the minterms 5, 7, 13, and 15 can be simplified to give BD in the algebraic form or (—1—1) in the cellular form. Consequently, 5,7,13,15(2,8) indicates a group-of-four obtained by eliminating the variables corresponding to the weights of 2 and 8.

We now have another way to express a reduced function in terms of the minterms that are included in each of the groups. The K-map function of Figure 2.19, therefore, may be written as

$$f(A,B,C,D) = 2,3(1) + 5,7,13,15(2,8) + 8,9(1)$$

The function in algebraic form may be recovered from this representation by expressing any one of the minterms of each group in variable form and striking out those variables whose weights are listed within the corresponding parentheses. This technique is shown in Example 2.6.

EXAMPLE 2.6

Convert

$f(A,B,C,D) =$
$2,3(1) + 5,7,13,15(2,8) + 8,9(1)$

to the SOP algebraic form.

SOLUTION

Only one minterm from each group is necessary, but to demonstrate that they all give the same result, we will use them all:

2,3(1)	5,7,13,15(2,8)	8,9(1)
$\bar{A}\cdot\bar{B}\cdot C\cdot\cancel{\bar{D}}$	$\cancel{\bar{A}}\cdot B\cdot\cancel{\bar{C}}\cdot D$	$A\cdot\bar{B}\cdot\bar{C}\cdot\cancel{\bar{D}}$
$\bar{A}\cdot\bar{B}\cdot C\cdot\cancel{D}$	$\cancel{\bar{A}}\cdot B\cdot\cancel{C}\cdot D$	$A\cdot\bar{B}\cdot\bar{C}\cdot\cancel{D}$
	$\cancel{A}\cdot B\cdot\cancel{\bar{C}}\cdot D$	
	$\cancel{A}\cdot B\cdot\cancel{C}\cdot D$	

Striking out the literals in the positions with the weights as shown in parentheses, we obtain

$$f = \bar{A}\bar{B}C + BD + A\bar{B}\bar{C}$$

We are now in a position to conclude the following cellular grouping rules:

Rule 1. Any adjacent minterms grouped together will differ by a power of 2, that is, 1, 2, 4, 8, and so on.

Rule 2. Two minterms can be grouped only if the number of ones in their binary representation differs by one. The decimal value of a minterm with k ones in its binary representation is subtracted from each minterm with $k + 1$ ones. The positive difference, if it is a power of 2, indicates the weight of the variable that can be eliminated. If the resultant difference is negative, then no grouping can be made even if the difference is a power of 2.

These grouping concepts will be explored further in the following few sections.

2.7 Single-Output Quine-McCluskey's Tabular Reduction

The *Quine-McCluskey's* (*Q-M*) tabular reduction technique for a single-output function is summarized by the following list of steps. Example 2.7 follows with necessary explanations for each of the steps. It is suggested that these steps be read once and then each step reread as the worked-out example progresses.

Step 1. Convert the minterms to binary form and then group the minterms in accordance with the total number of 1s in their binary representations. The minterms will be referred to as *zero-cubes.*

Step 2. Make all possible groupings of two *zero-cubes*. These groups are known as *one-cubes* (they have one don't-care in their Boolean cell representation). This step must be followed by possible groupings of pairs that will be referred to as *two-cubes* (two don't-cares in their Boolean cell representation). This process is continued to produce higher-order cubes until no higher cubes can be formed.

Step 3. Identify those cubes that couldn't be used in the formation of higher-order cubes. These cubes are called *prime implicants* (*PIs*). The PIs represent all possible groups with no smaller group being part of a larger group.

Step 4. Identify those PIs that are necessary to provide a minimum covering of all the minterms, that is, each minterm of the original function must be in at least one group. This identification is accomplished by constructing an implication table that, in turn, is used for finding the desired minimum cover. The rules for constructing the implication table are summarized as follows:

1. Use all minterms of the original function as the column headings.

2. Use the PIs as the row labels. It is better to separate the PIs by order. If there are two PIs that cover a minterm, the higher-order PI is preferred.
3. A check mark, x, is entered in each matrix position that corresponds to a minterm covered by a particular PI.
4. Examine the table entries and locate the essential PIs. An essential PI is one that is the only cover for one or more minterms. All minterms that are covered by an essential PI are then removed from active consideration.
5. The secondary essential PIs are determined from the remaining PIs by making judicious choices. A PI is preferred over another if (a) it covers more of the remaining minterms, or (b) it is higher in order than the other. This process is continued until all of the minterms have been considered. There are some choices difficult to resolve for which a technique will be demonstrated later.

Example 2.7 illustrates the algorithm in detail.

EXAMPLE 2.7

Using the Q-M tabular procedure, find a minimum SOP expression for the Boolean function given by

$$f(A,B,C,D) = \Sigma m(0,2,3,5,7,8,10,13,15)$$

SOLUTION

Step 1. The minterms are classified as shown in Figure 2.20.

Step 2. Next we locate the one-cubes. Two minterms may be considered for combination only if the numbers of 1s present in their binary representation differs by one. (This was discussed in Section 2.6.) Moreover, the minterms must differ by a power of 2 to form a valid group.

Starting with the top section, we can match 0 and 2 because they differ by 2 − 0 = 2 to form 0,2(2), which is equivalent to the cell (00—0). This is listed in the top section of the one-cube column, as shown in Figure 2.21. And for bookkeeping purposes an x, is placed on the left side of 0 and also on the right side of 2. Continuing in like manner, 0 and 8 form another one-cube, 0,8(8), which

FIGURE 2.20

Minterms	Binary Representation	Number of 1s
0	0000	0
2	0010	1
8	1000	1
3	0011	2
5	0101	2
10	1010	2
7	0111	3
13	1101	3
15	1111	4

FIGURE 2.21

Zero-Cube		
xx	0	
xx	2	x
x	8	x
x	3	x
xx	5	
	10	xx
x	7	xx
x	13	x
	15	xx

One-Cube
0,2(2)
0,8(8)
2,3(1)
2,10(8)
8,10(2)
3,7(4)
5,7(2)
5,13(8)
7,15(8)
13,15(2)

is equivalent to the cell (—000). This completes matching the top two sections of the zero-cube column. A line is then drawn below the one-cubes just completed to separate the one-cubes with no 1s from those with one 1. We next match zero-cubes of the second section with those of the third section.

Between the second and third sections of the zero-cube column, a number of one-cubes are possible:

2 and 3, giving 2,3(1)

2 and 10, giving 2,10(8)

8 and 10, giving 8,10(2)

In all of these cases the difference between the minterms is always a positive power of 2. In addition, the minterm component from the upper section must always be smaller than the one from the lower section. Any time a group is formed, a check mark, x, is placed on the left side of the upper section minterm, and another mark is placed on the right of the lower section minterm. This is continued until no more one-cubes can be formed. The number of x's on the left side of a column should equal the number of x's on the right side of the same column. If not, recheck your work.

Next we proceed to determine the two-cubes from the already obtained one-cubes, as shown in Figure 2.22. A proper two-cube may be located by combining one-cubes from the neighboring sections only if (a) the entries within the parentheses are the same and (b) the minterms of one section differ from the corresponding minterms of the other section by a power of 2.

FIGURE 2.22

One-Cube		
x	0,2(2)	
x	0,8(8)	
	2,3(1)	
	2,10(8)	x
	8,10(2)	x
	3,7(4)	
x	5,7(2)	
x	5,13(8)	
	7,15(8)	x
	13,15(2)	x

Two-Cube	
0,2,8,10(2,8)	Two-cubes with zero 1s
	An empty two-cube group
5,7,13,15,(2,8)	Two-cubes with two 1s

Note that 0,2(2) and 8,10(2) may be combined since $8 - 0 = 10 - 2 = 2^3$ and both have 2 within their brackets. The resulting group-of-four is 0,2,8, 10(2,8). Note that 0,8(8) and 2,10(8) also produce the same group-of-four. For every two-cube, there will always be two pairs of one-cubes that will produce the same two-cube. Both of these pairs, therefore, are checked off. Next a line is drawn to separate two-cubes with no 1s from two-cubes with one 1.

Next we compare the second and the third sections under the one-cube column. There is no suitable match. We draw another line to indicate an empty two-cube group. We continue to the third and the fourth sections. Note that 5, 7(2) can be combined with 13,15(2), or, equivalently, 5,13(8) can be combined with 7,15(8) to give 5,7,13,15,(2,8). In both cases the same number is contained within the brackets, and $13 - 5 = 15 - 7 = 8$, and $7 - 5 = 15 - 13 = 2$. This completes the one-cube matching.

The two-cubes are compared in similar fashion. In this example no comparisons can be made since the comparisons must be made between adjacent two-cube groups, and the two-cube groups shown are separated by an empty group.

Step 3. All cubes without X's are called PIs. The PI list for this function is obtained as follows:

a 0,2,8,10(2,8)

b 5,7,13,15(2,8)

c 2,3(1)

d 3,7(4)

Step 4. Once the PIs have been determined, a table to determine which and how many of the PIs are necessary to assure that all of the minterms are covered is constructed. A PI table for this example is shown in Figure 2.23.

FIGURE 2.23

PI	0	2	3	5	7	8	10	13	15
**a* 0,2,8,10(2,8)	X	X				X	X		
**b* 5,7,13,15(2,8)				X	X			X	X
c 2,3(1)		X	X						
d 3,7(4)			X		X				
Covering PI	*a*	*a* or *c*	*c* or *d*	*b*	*b* or *d*	*a*	*a*	*b*	*b*

**Essential PIs.*

In Figure 2.23 both of the two-cubes are the essential PIs (indicated by *) since *a* is the only PI covering minterms 0, 8, and 10, and *b* is the only PI covering minterms 5,13, and 15. To complete the cover, either *c* or *d* may be chosen as the secondary essential PI. There is no particular disadvantage in choosing *c* over *d*. The reduced form of the function, therefore, is obtained as follows:

$$\begin{aligned} f(A,B,C,D) &= a + b + c \\ &= (-0-0) + (-1-1) + (001-) \\ &= \overline{B}\overline{D} + BD + \overline{A}\overline{B}C \end{aligned}$$

or, equivalently,

$$\begin{aligned} f(A,B,C,D) &= a + b + d \\ &= (-0-0) + (-1-1) + (0-11) \\ &= \overline{B}\overline{D} + BD + \overline{A}CD \end{aligned}$$

There are several ways to simplify a prime implication table:

1. If a minterm is covered by only one PI, then that column and all other columns covered by that PI may be eliminated. The corresponding PI is an essential PI and must appear in the function expression.
2. If two columns have identical entries, then one of the columns may be removed.
3. If two PIs of equal order cover the same minterms, but one covers an additional minterm, the one covering the fewer is eliminated.

Even after simplifying, it is sometimes difficult to pick the proper minimum cover from the remaining table entries. A technique that will always give the optimum cover is to pose a logic function that describes the possible choices. In Figure 2.24 the bottom row of a simplified PI table is shown. Note that some of its minterms are covered by more than one PI. Such conflicting situations are handled by defining a cover function f that is a 1 when a proper cover for the minterms in the reduced table is obtained.

FIGURE 2.24 **PI Checks of a Possible Implication Table.**

a or b	g or h	k or d

The function f, as described by the implication table of Figure 2.24, will be a 1 provided a or b is chosen and g or h is chosen and k or d is chosen. This could be expressed in the form of a logic equation as follows:

$$
\begin{aligned}
f &= (a + b) \cdot (g + h) \cdot (k + d) \\
&= agk + ahk + bgk + bhk + adg + adh + bdg + bdh
\end{aligned}
$$

This equation implies that there are eight possible choices of PIs that will cover the minterms. Each of these possible covers should be examined individually to determine which choice would require the minimum number of gates and gate inputs. This technique may be used without simplifying the implication table, but it is best to make all obvious PI choices prior to employing the technique to minimize algebraic manipulation.

2.8 Don't-Cares in Q-M Tabular Reduction

In Section 2.4 it was shown that the presence of don't-cares in a truth table provides an opportunity for reducing a function. The same opportunity exists when the Q-M technique is used for the simplification of functions. If the function to be reduced includes don't-cares, the tabular reduction proceeds exactly as before with one important difference. The don't-cares are included in all of the steps except in the implication table where they are not used as column headings. Example 2.8 illustrates the procedure.

EXAMPLE 2.8

Reduce the function

$f(A,B,C,D) = \Sigma m(0,1,2,7,8,13,15) + d(10,11)$

by the Q-M tabular reduction method.

SOLUTION

First form the cubes, treating 10 and 11 as regular minterms. This step is shown in Figure 2.25. Note that under the one-cube column there is a PI that covers minterms 10 and 11, both of which are don't-care terms. This particular PI, therefore, is not included in the implication table. However, the PI involving 11 and 15 is considered because at least one of these two minterms is not a don't-care. The resulting PI table, as shown in Figure 2.26, is formed without the don't-cares since a successful function cover does not have to include them. The reduced function is then

$$f(A,B,C,D) = a + b + c + e$$
$$= (-0-0) + (000-) + (-111) + (11-1)$$
$$= \overline{B}\overline{D} + \overline{A}\overline{B}\overline{C} + BCD + ABD$$

FIGURE 2.25

Number of 1s	Zero-Cube	One-Cube	Two-Cube
0	xxx 0	0,1(1) b	0,2,8,10(2,8) a
1	1 x	x 0,2(2)	
	x 2 x	x 0,8(8)	
	x 8 x	2,10(8) x	
2	x 10 xx	8,10(2) x	
3	x 7	10,11(1)	
	x 11 x	7,15(8) c	
	x 13	11,15(4) d	
4	15 xxx	13,15(2) e	

FIGURE 2.26

PI	0	1	2	7	8	13	15
*a	x		x		x		
*b	x	x					
*c				x			x
d							x
*e						x	x
Covering PI	a or b	b	a	c	a	e	c or e

EXAMPLE 2.9

Obtain the PIs of

$f(A,B,C,D,E,F,G,H) = \sum m(10,20,22,23,30,31,68,70,76,138,146,148,150,151,156,158,166,167,182,183,190,191,218,228–231,238,254) + d(0,16–19,21,24–29,71,74,77,78,95,118,119,122,127,128,144,152–155,157,159,173,202,255)$

SOLUTION

The symbol * is associated only with prime implicants. The remaining cubes are used in producing cubes of still higher order.

Zero-Cubes

a. Having not a single 1: 0

b. Having a single 1: 16 and 128

c. Having two 1s: 10, 17, 18, 20, 24, 68, and 144

d. Having three 1s: 19, 21, 22, 25, 26, 28, 70, 74, 76, 138, 146, 148, and 152

e. Having four 1s: 23, 27, 29, 30, 71, 77*, 78, 150, 153, 154, 156, 166, 202, and 228

f. Having five 1s: 31, 118, 122*, 151, 155, 157, 158, 167, 173*, 182, 218, 229, and 230

g. Having six 1s: 95, 119, 159, 183, 190, 231, 238, and 246

h. Having seven 1s: 127, 191, 247, and 254

i. Having eight 1s: 255

One-Cubes

a. 0,16(16) and 0,128(128)

b. 16,17(1); 16,18(2); 16,20(4); 16,24(8); 16,144(128); and 128, 144(16)

c. 10,74(64); 10,26(16); 10,138(128); 17,19(2); 17,21(4); 17,25(8); 18,19(1); 18,22(4); 18,26(8); 18,146(128); 20,21(1); 20,22(2); 20,28(8); 20,148(128); 24,25(1); 24,26(2); 24,28(4); 24, 152(128); 68,70(2); 68,76(8); 144,146(2); 144,148(4); and 144, 152(8)

d. 19,23(4); 19,27(8); 21,23(2); 21,29(8); 22,23(1); 22,30(8); 22, 150(128); 25,27(2); 25,29(4); 25,153(128); 26,27(1); 26,30(4); 26,154(128); 28,29(1); 28,30(2); 28,156(128); 70,71(1)*; 70, 78(8); 74,78(4)*; 74,202(128); 76,78(2); 138,154(16); 138, 202(64); 146,150(4); 146,154(8); 148,150(2); 148,156(8); 152, 153(1); 152,154(2); and 152,156(4)

e. 23,31(8); 23,151(128); 27,31(4); 27,155(128); 29,31(2); 29, 157(128); 30,31(1); 30,158(128); 150,151(1); 150,158(8); 150, 182(32); 153,155(2); 153,157(4); 154,155(1); 154,158(4); 154, 218(64); 156,157(1); 156,158(2); 166,167(1); 166,182(16); 166, 230(64); 202,218(16); 228,229(1); and 228,230(2)

f. 31,95(64)*; 31,159(128); 118,119(1); 118,246(128); 151,159(8); 151,183(32); 155,159(8); 157,159(2); 158,159(1); 158,190(32); 167,183(16); 167,231(64); 182,183(1); 182,190(8); 182,246(64); 229,231(2); 230,231(1); 230,238(8); and 230,246(16)

g. 95,127(32); 119,127(8); 119,247(128); 159,191(32); 183,191(8); 183,247(64); 190,191(1); 190,254(64); 231,247(16); 238,254(16); 246,247(1); and 246,254(8)

h. 127,255(128); 191,255(64); 247,255(8); and 254,255(1)

Two-Cubes

a. 0,16,128,144(16,128)*

b. 16,17,18,19(1,2); 16,17,20,21(1,4); 16,17,24,25(1,8); 16,18,20,22 (2,4); 16,18,24,26 (2,8); 16,18,144,146 (2,128); 16,20,144,148 (4, 128); 16,24,144,152 (8,128); and 16,20,20,28 (4,8)

c. 10,74,138,202(64,128)*; 68,70,76,78(2,8)*; 144,146,148,150 (2, 4); 144,146,152,154(2,8); 144,148,152,156(4,8); 17,19,21,23(2, 4); 17,19,25,27(2,8); 17,21,25,29(4,8); 18,19,22,23(1,4); 18,19, 26,27(1,8); 18,22,26,30(4,8); 18,22,146,150(4,128); 18,26,146, 154(8,128); 20,21,22,23(1,2); 20,21,28,29(1,8); 20,22,28,30(2,8); 20,22,148,150(2,128); 20,28,148,156(8,128); 24,25,26,27(1,2); 24,25,28,29 (1,4); 24,25,152,153(1,128); 24,26,28,30(2,4); 24,26, 152,154(2,128); 24,28,152,156(4,128); and 10,26,138,154(16, 128)*

d. 22,23,30,31(1,8); 22,23,150,151(1,128); 22,30,150,158(8,128); 138,154,202,218(16,64)*; 146,150,154,158(4,8); 148,150,156, 158(2,8); 152,153,154,155(1,2); 152,153,156,157(1,4); 152,154, 156,158(2,4); 19,23,27,31(4,8); 21,23,29,31(2,8); 25,27,29,31(2, 4); 25,27,153,155(2,128); 25,29,153,157(4,128); 26,27,30,31(1, 4); 26,27,154,155(1,128); 26,30,154,158(4,128); 28,29,30,31(1, 2); 28,29,156,157(1,128); and 28,30,156,158(2,128)

e. 23,31,151,159(8,128); 30,31,158,159(1,128); 150,151,158,159(1, 8); 150,151,182,183(1,32); 150,158,182,190(8,32); 153,155,157, 159(2,4); 154,155,158,159(1,4); 156,157,158,159(1,2); 166,167, 182,183(1,16); 166,167,230,231(1,64); 166,182,230,246(16,64); 228,229,230,231(1,2)*; 27,31,155,159(4,128); and 29,31,157, 159(2,128)

f. 118,119,246,247(1,128); 151,159,183,191(8,32); as elsewhere 158,159,190,191(1,32); 167,183,231,247(16,64); 182,183,190, 191(1,8); 182,183,246,247(1,64); 182,190,246,254(8,64); 230, 231,246,247(1,16); and 230,238,246,254(8,16)*

g. 119,127,247,255(8,128)*; as elsewhere 183,191,247,255(8,64); as elsewhere 190,191,254,255(1,64); and 246,247,254,255(1,8)

Three-Cubes

a. 16,17,18,19,20,21,22,23(1,2,4); 16,17,18,19,24,25,26,27(1,2,8); 16,17,20,21,24,25,28,29(1,4,8); 16,18,20,22,144,146,148,150(2,4, 128); 16,18,20,22,24,26,28,30(2,4,8); 16,18,24,26,144,146,152, 154(2,8,128); and 16,20,24,28,144,148,152,156(4,8,128)

b. 17,19,21,23,25,27,29,31(2,4,8); 18,19,22,23,26,27,30,31(1,4,8); 18,22,26,30,146,150,154,158(4,8,128); 20,21,22,23,28,29,30,31(1, 2,8); 20,22,28,30,148,150,156,158(2,8,128); 24,25,26,27,152,153, 154,155(1,2,128); 24,25,26,27,28,29,30,31(1,2,4); 24,25,28,29, 152,153,156,157(1,4,128); 24,26,28,30,152,154,156,158(2,4,128); and 144,146,148,150,152,154,156,158(2,4,8)

c. 22,23,30,31,150,151,158,159(1,8,128)*; 25,27,29,31,153,155,157,

159(2,4,128); 26,27,30,31,154,155,158,159(1,4,128); 28,29,30,31, 156,157,158,159(1,2,128); and 152,153,154,155,156,157,158, 159(1,2,4)

d. 150,151,158,159,182,183,190,191(1,8,32)* and 166,167,182,183, 230,231,246,247(1,6,64)*

e. 182,183,190,191,246,247,254,255(1,8,64)*

Four-Cubes

a. 16–31(1,2,4,8)* and 16,18,20,22,24,26,28,30,144,146,148,152, 154,152,154,156,158(2,4,8,128)*

b. 24–31,152–159(1,2,4,128)*

The search for higher cubes cannot be carried out beyond four-cubes. The process is completed at this point since there are no possible five-cubes. Note that some of the PIs are not to be accounted for since they are composed of only don't-care terms.

2.9 Multiple-Output Tabular Reduction

Occasionally it is necessary to generate multiple outputs in a combinational network that have all or some of their input variables in common. In such cases each of the functions may be generated separately, or the functions may be optimized so that they share a part of a circuit. It may be possible to design multiple-output circuits with common ports using K-maps, but it is much simpler to use the Q-M tabular reduction scheme. The rules for the multiple-output problem are similar to those of the single-output problem with the following exceptions:

1. Each minterm is labeled with symbols to indicate its association with one or more functions.
2. Cubes may be combined to form higher cubes only if both of them belong to one or more common outputs. The resulting higher-order cube carries the symbol or symbols common to both of the lower-order cubes.
3. A cube is checked off only if *all* of its associated symbols are also associated with the cube that combines with it.

An example of a multiple-output function derivation that illustrates the additions to the Q-M algorithm just described is given in Example 2.10.

EXAMPLE 2.10

Implement a multiple-output circuit for the following functions:

$$f_a(A,B,C,D) = \Sigma m(0,1,4,5,9,13,14,15)$$

$$f_b(A,B,C,D) = \Sigma m(0,4,5,7,9,13,14,15)$$

SOLUTION

If these two functions are reduced by the use of K-maps, we will require two separate K-maps as shown in Figure 2.27.

The functions reduce as follows:

$$f_a(A,B,C,D) = \overline{A}\overline{C} + \overline{C}D + ABC$$

FIGURE 2.27

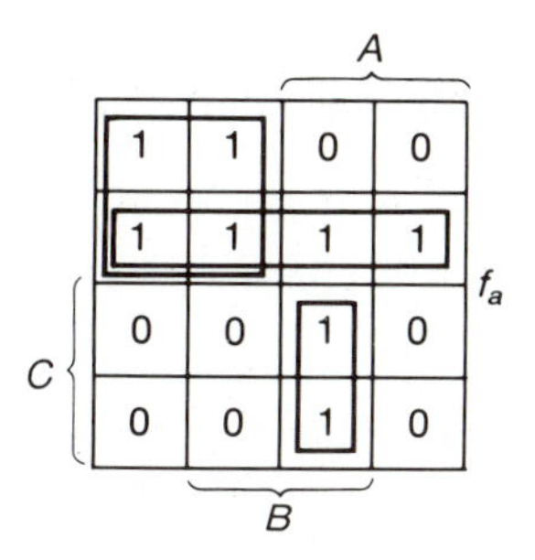

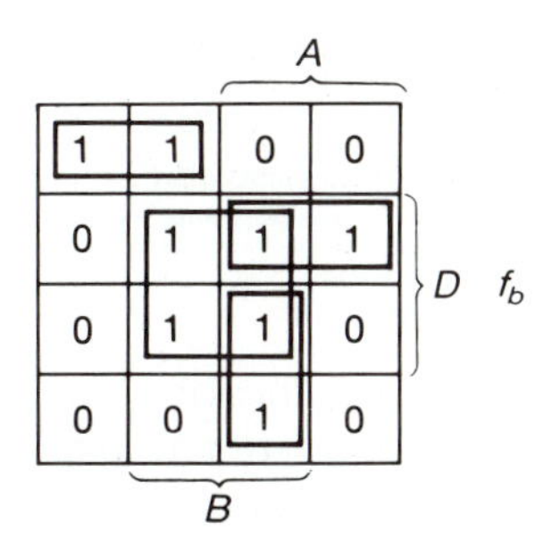

$$f_b(A,B,C,D) = \overline{A}\overline{C}\overline{D} + BD + A\overline{C}D + ABC$$

where the hazard has not been considered for simplicity. It can be seen that one term, ABC, is common to the two output functions. Using the Q-M technique, the minterms are listed, labeled with a and b to indicate their respective associations, and searched as shown in Figure 2.28.

FIGURE 2.28

Zero-Cube	One-Cube	Two-Cube
x 0 [ab]	x 0,1(1) [a]	0,1,4,5(1,4) [a] p
	0,4(4) [ab] s	1,5,9,13(4,8) [a] q
xx 1 [a] x		5,7,13,15(2,8) [b] r
x 4 [ab] x	x 1,5(4) [a] x	
x 5 [ab] x	4,5(1) [ab] t	
x 9 [ab]	x 1,9(8) [a]	
x 7 [b] x	x 5,7(2) [b]	
x 13 [ab] xx	5,13(8) [ab] u	
x 14 [ab]	9,13(4) [ab] v	
15 [ab] xx	14,15(1) [ab] w	
	7,15(8) [b] x	
	13,15(2) [ab] y	

PI	f_a								f_b							
	0	1	4	5	9	13	14	15	0	4	5	7	9	13	14	15
p	x	x	x	x												
q		x		x	x	x										
*r											x	x		x		x
*s	x		x						x	x						
t			x	x						x	x					
u				x		x					x			x		
*v					x	x							x	x		
*w							x	x							x	x
y						x		x						x		x
	s	q	s	q	v	v	w	w	s	s	r	r	v	v	w	r
		or		or	or	or										
		p		p	q	q										

The four PIs, r, s, v, and w, are found to be essential PIs. This leaves the minterms 1 and 5 in function a uncovered. Note that either p or q could be included to cover these two minterms. The inclusion of q eliminates the need of including v in the first function. Consequently, the reduced functions are, respectively,

$$f_a(A,B,C,D) = \overline{C}D + \overline{A}\overline{C}\overline{D} + ABC$$
$$f_b(A,B,C,D) = \overline{A}\overline{C}\overline{D} + BD + A\overline{C}D + ABC$$

Note that there are two terms that are common. The term $\overline{A}\overline{C}$ in the K-map solution of f_a is replaced by $\overline{A}\overline{C}\overline{D}$ in the corresponding Q-M solution. This technique is better than using gates to form both $\overline{A}\overline{C}\overline{D}$ and $\overline{A}\overline{C}$. This simplification may have been obvious without going through the Q-M technique, but with more variables and more outputs this wouldn't be the case. The Q-M multiple-output algorithm just illustrated increases the possibility of having a shared circuit and is, therefore, more desirable.

2.10 Summary

This chapter covered the fundamentals of logic simplification and how such simplification relates to the design of combinational logic circuits. This was done by introducing Karnaugh's map scheme, which was shown suitable for minimizing functions of up to six variables. The concepts of combinational hazards were then introduced and techniques were shown for eliminating such problems. Finally, the Q-M tabular reduction method was used not only to handle more than six variables, but also to reduce multiple-output logic networks.

Problems

1. Use a Karnaugh map to obtain the minimized SOP form of each of the following functions, ignoring the hazard conditions:
   a. $f(A,B,C) = \Sigma m(0,2,3,5,6)$
   b. $f(A,B,C) = \Sigma m(0,2,6) + d(5,7)$
   c. $f(A,B,C,D) = \Sigma m(1,3,5,8,9,11,15)$
   d. $f(A,B,C,D) = \Sigma m(1,3,5,8,11,15) + d(2,9)$
   e. $f(A,B,C,D) = \Sigma m(0,4\text{–}13,15)$
   f. $f(A,B,C,D,E) = \Sigma m(1,8\text{–}12,21,22,29)$
   g. $f(A,B,C,D,E) = \Sigma m(1,8\text{–}11,22,26,27) + d(3,23,31)$
   h. $f(A,B,C,D,E,F) = \Sigma m(1,8\text{–}12,34,48\text{–}57,63)$
   i. $f(A,B,C,D,E,F) = \Sigma m(1,8\text{–}12,34,48\text{–}57,63) + d(14,20\text{–}25,39,58)$
2. Repeat Problem 1 without ignoring hazard conditions. Identify the hazard locations in the map and include terms to eliminate the corresponding combinational hazards.
3. Obtain the most minimum and hazard-free POS expression for the following functions:

a. $f(A,B,C,D) = \Sigma m(1,3,4,6,7,11\text{–}13)$
b. $f(A,B,C,D) = \Sigma m(1,3,5,8,9,) + d(2,4)$
c. $f(A,B,C,D,E) = \Sigma m(0,4\text{–}7,9,14\text{–}17,23) + d(12,29\text{–}31)$
d. $f(A,B,C,D,E) = \Sigma m(0\text{–}3,12\text{–}15,26,28,31) + d(6,8\text{–}10)$
e. $f(A,B,C,D,E,F) = \Sigma m(0\text{–}3,12\text{–}15,26,34\text{–}38,54,59,61) + d(18\text{–}23,60,63)$

4. In a certain computer two separate sections proceed independently through four phases of operation. For the purpose of control, it is necessary to know when the two sections are in phase. Both sections put out a signal in parallel in two lines. Design a circuit that will give out a signal whenever both sections are in the same phase.
5. Find the minimal equivalent of $f = (X + Y)(X + \overline{Y} + Z)(\overline{X} + Z)$ using the K-map.
6. A digital circuit is to have an output F and four inputs: A, B, C, and D. The output is a 1 whenever the decimal equivalent of $(ABCD)_2$ is divisible by either 3 or 5. Design the circuit.
7. Repeat Problem 6 when the number of inputs are (a) five and (b) six.
8. Obtain the minimized SOP form of the following functions using the Q-M tabular reduction method:
   a. $f(A,B,C,D) = \Sigma m(0,1,4\text{–}10,12,14)$
   b. $f(A,B,C,D,E,F) = \Sigma m(0,2,4,5,7,8,16,18,24,32,36,40,48,56)$
   c. $f(A,B,C,D,E) = \Sigma m(0,2,3,7,10,11,12,24)$
   d. $f(A,B,C,D,E,F,G) = \Sigma m(0\text{–}5,7,15\text{–}19,42\text{–}45,67\text{–}79,8,90,94,96,100\text{–}112)$
   e. $f(A,B,C,D) = \Sigma m(2,4,8,11,15)$
9. Obtain the minimized SOP form for the following functions using the Q-M technique:
   a. $f(A,B,C,D) = \Sigma m(2,4,8,11,15) + d(1,10,12,13)$
   b. $f(A,B,C,D,E) = \Sigma m(0,2,3,7,8,10,11,12,24) + d(5,16,18)$
   c. $f(A,B,C,D,E) = \Sigma m(0,4,8\text{–}15,23,27) + d(20\text{–}22,31)$
   d. $f(A,B,C,D,E,F) = \Sigma m(0,2,4,5,7,8,10\text{–}12,24,32,36,40,48,56) + d(14\text{–}17,34,46,61\text{–}63)$
10. Obtain the minimized POS form for the functions of Problem 8.
11. Obtain the minimized POS form for the functions of Problem 9.
12. Determine the minimal SOP form for the following multiple-output systems:
    a. $f_x(A,B,C,D) = \Sigma m(0,1,2,3,6,7)$
       $f_y(A,B,C,D) = \Sigma m(0,1,6,7,14,15)$
       $f_z(A,B,C,D) = \Sigma m(0,1,3,8,9)$
    b. $f_x(A,B,C,D) = \Sigma m(4,5,10,11,12) + d(6,9)$

$f_y(A,B,C,D) = \Sigma m(0,1,3,4,8,11)$
$f_z(A,B,C,D) = \Sigma m(0,4,10,12,14)$

c. $f_x(A,B,C,D,E) = \Sigma m(0,5\text{–}11,13,22) + d(16,24,27)$
$f_y(A,B,C,D,E) = \Sigma m(6\text{–}13,19,28) + d(2,3,16,27)$
$f_z(A,B,C,D,E) = \Sigma m(1,3,6,9,11\text{–}15)$

13. A six-bit binary number appears at the inputs of a combinational network. One of the outputs, $Z1$, indicates if the multi-bit input is divisible by 2 without any remainder, and the other output, $Z2$, indicates if the multi-bit input is divisible by 3 without any remainder. Obtain the minimal SOP circuit.
14. Repeat Problem 13 when the input is a seven-bit number.
15. Repeat Problem 13 so that the resulting circuit follows a POS configuration.

Suggested Readings

Gimpel, J. F. "A reduction technique for prime implicant tables." *IEEE Trans. Elect. Comp.* vol. EC-14 (1965): 535.

Halder, A. K. "Karnaugh map extended to six or more variables." *Elect. Lett.* vol. 18 (1982): 868.

Hong, S. J.; Cain, R. G.; and Ostapko, D. L. "MINI: A heuristic approach for logic minimization." *IBM J. Res. & Dev.* vol. 18 (1974): 443.

Karnaugh, M. "The map method for synthesis of combinational logic circuits." *Trans. AIEE.* vol. 72 (1953): 593.

Levine, R. I. "Logic minimization beyond the Karnaugh map." *Comp. Des.* vol. 6 (1967): 40.

Luccio, F. "A method for the selection of prime implicants." *IEEE Trans. Elect. Comp.* vol. EC-15 (1966): 205.

McCluskey, E. J. "Minimization of Boolean functions." *Bell Syst. Tech. J.* vol. 35 (1956): 1417.

Quine, W. V. "The problems of simplifying truth functions." *Am Math. Monthly.* vol. 59 (1952): 521.

Rhyne, V. T.; Noe, P. S.; McKinney, M. H.; and Pooch, U. W. "A new technique for the fast minimization of switching functions." *IEEE Trans. Comp.* vol. C-26 (1977): 757.

Shiva, S. G., and Nagle, H. T. "A fast algorithm for complete minimization of Boolean functions." Project THEMIS Tech. Report. No. AU-T-24. Auburn, Ala.: Auburn University, June 1972.

Su, S. Y. H., and Cheung, P. T. "Computer minimization of multivalued switching functions." *IEEE Trans. Comp.* vol. C-21 (1972): 995.

Taconet, B. "New prime implicant algorithm, based on monotonocity properties." *Elect. Lett.* vol. 20 (1984): 89.

Zisapel, Y.; Krieger, M.; and Kella, J. "Detection of hazards in combinational switching circuits." *IEEE Trans. Comp.* vol. C-28 (1979): 52.

Integrated Circuits

CHAPTER THREE

3.1 Introduction

In the first two chapters we discussed various logic devices while deliberately avoiding their electrical characteristics. Often many of the electrical characteristics may be ignored and successful designs still made. However, there are some attributes and limitations of devices that must be of concern to the designer.

In the first digital computers the circuits were constructed by using discrete electrical components such as resistors, capacitors, tubes, and relays. However, with the advent of the transistor and solid-state fabrication techniques, now it is possible to produce miniature circuits on the surface of a small piece (chip) of semiconductor material. Such circuits are called *integrated circuits* (*ICs*), or *chips*, because the circuit components such as transistors, diodes, and resistors are integrated together to form a circuit. The ICs are distinctly different from discrete circuits in which each component is a distinct entity and is combined with other components to make the circuit. The components in the IC cannot be disconnected and the components inside the IC are not accessible except through the external access pins.

Advantages derived from using ICs are a reduction of cost, circuit size, energy consumption, and wiring, and a significant improvement in circuit reliability and operating speed. In this chapter we will consider digital ICs that perform the logic functions we introduced in Chapters 1 and 2.

The ICs come in various levels of complexity and are categorized by number of gates. ICs with fewer than 10 gates on the same chip are referred to as *small-scale integration* (*SSI*) chips. *Medium-scale integration* (*MSI*) chips usually contain between 10 and 100 gates. Digital subsystems like adders, comparators, and multiplexers fall into this category. ICs with more than 100 gates but less than 5000 gates are classified as *large-scale integration* (*LSI*) circuits. Microprocessors, direct memory access controllers, and calculators are examples of LSI devices. There also are *very large-scale integration* (*VLSI*) devices that contain the equivalent of more than 5000 gates.

This chapter introduces characteristics of IC families that will be used in the implementation of logic circuits. The intent is not to cover the solid-state physics and electronics involved. These topics are beyond the scope of this text, but courses on solid-state physics and electronics should be considered by those interested. It is essential, however, that the designer be familiar with certain relevant properties of ICs. This chapter will introduce the reader to these relevant properties in order to help avoid pitfalls that might not be avoided otherwise.

Some familiarity with electrical circuits and transistors is assumed. For those with no electrical background, this chapter may be skipped without loss of continuity in the logic design sequence. After studying this chapter, you should be able to:

- Read a data book and extract correct and important specifications of any particular IC;
- Decide whether or not to use a particular family of logic for a particular logic network application;
- Identify the characteristics of transistor-transistor logic (TTL), emitter-coupled logic (ECL), and metal-oxide semiconductor (MOS) devices;
- Interface devices belonging to different families.

3.2 Logic Circuit Families

ICs are basically of two types: bipolar and unipolar. In the *bipolar* circuits the charge carriers of both positive and negative polarities are required for the operation of active elements. In the *unipolar* or the metal-oxide semiconductor (MOS) case, either holes or electrons are required for the operation of active elements. Bipolar is the technology of choice for both MSI and SSI because of its high speed. MOS is the primary technology for both LSI and VLSI because more devices can be packed on a given chip area.

The bipolar ICs can be classified into three broader groups in terms of the input conditions necessary to activate them: current sourcing, current sinking, and current mode. Circuits subjected to such switching modes are shown, respectively, in Figure 3.1[*a–c*]. In each of these cases the circuit consists of at least one transistor that works as a source and one transistor that works as a sink.

Consider the circuit of Figure 3.1[*a*] where the transistor T_2 is in a common-emitter configuration. For those who have studied electronics, the circuit is easily identified as an inverter, that is, a NOT gate. In order to turn on the transistor T_2, a source like that of T_1 is needed, which is already turned off. Again, when T_2 is turned off, it

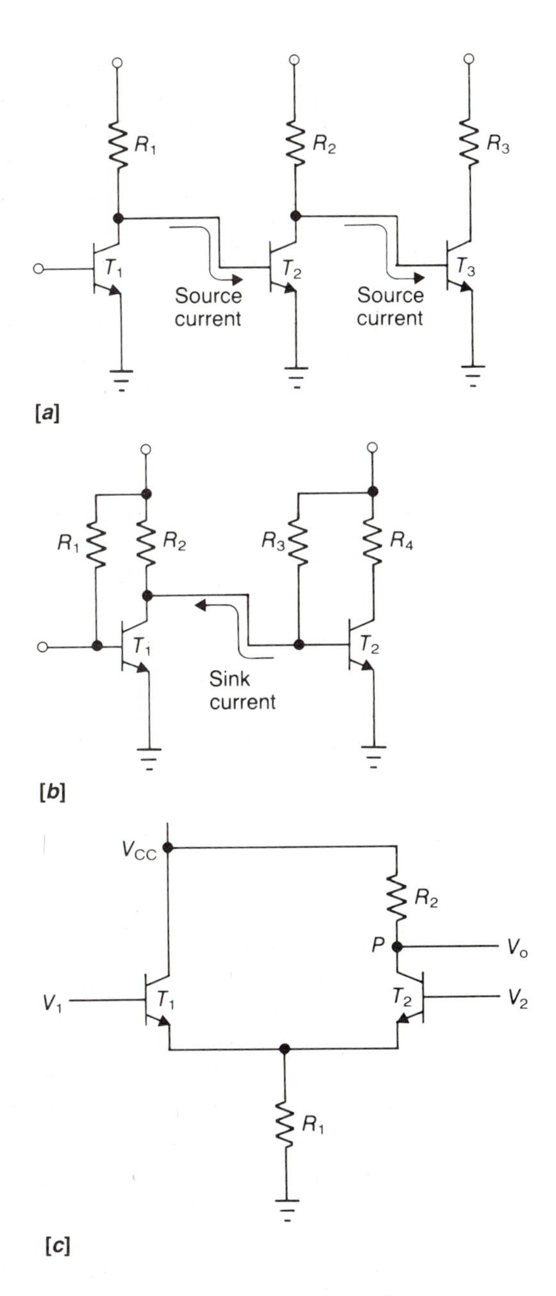

FIGURE 3.1 **Switching Modes: [*a*] Current Sourcing, [*b*] Current Sinking, and [*c*] Current Mode Switching.**

could be used to turn on T_3, and so on. This particular mode of switching is known as *current sourcing*.

There are many bipolar circuits similar to that of Figure 3.1[*b*] where the internal components are used to turn on the transistor. The transistor T_1 is turned on even when there is no input voltage because there is a bias current supplied through the resistor R_1. Consequently, the output of T_1 becomes low. This low output leads to a virtual short at the base of T_2, sinking the base current of T_2. This switching mode is called *current sinking*.

The third switching mode is illustrated in Figure 3.1[*c*] where the current direction is switched depending on the unequal voltage values of V_1 and V_2. When V_1 is greater than V_2, T_1 is turned on and T_2 is turned off, resulting in a high output at point *P*. If V_1 is less than V_2, T_1 is turned off and T_2 is turned on, causing a low voltage output. In either case the emitter current is constant and only the direction has been switched. This particular means of switching is called *current mode*.

Many different logic families of digital ICs have been introduced. The ICs within a family are compatible and easily interconnected. Some of these families have lost their importance, others are still popular, and there are some whose impact is still to be determined. The following is a current list of IC families:

RTL, resistor-transistor logic;
HTL, high-threshold logic;
DTL, diode-transistor logic;
TTL, transistor-transistor logic;
ECL, emitter-coupled logic;
PMOS, p-channel MOS;
NMOS, n-channel MOS;
CMOS, complementary MOS;
IIL, integrated injection logic;
LVIL, low-voltage inverter logic.

The RTL, DTL, and HTL families are relatively obsolete and have only historic significance. TTL, the most popular family, uses transistors almost exclusively and has an extensive list of digital functions. ECL is the fastest logic family and is used in systems that require high-speed operations. The PMOS and NMOS families are made up of LSI and VLSI devices because of their high-component density, and the CMOS family provides very low power consumption. The IIL and LVIL are very recent additions. They are important because of their high speed and low power consumptions. The *semiconductor quantum interference device* (*SQUID*) is the newest of all, much faster than the TTL and ECL, and it requires almost no power, but it is restricted to operation at an extremely low temperature.

The different IC families are usually ranked by examining the circuit of the basic gate in each family. The basic circuit in each family generally is either a NAND or a NOR gate. The most significant characteristics that are evaluated and compared are listed as follows:

Fan-out: Fan-out indicates the total number of gate inputs to which an output can be connected without losing its noise immunity or ceasing to perform its function. Fan-out normally is a function of the amount of current the gate can supply or absorb when it is turned on.

Noise margin: Noise margin is the minimum noise voltage that can be superimposed on the logic voltage without the logic sense being misinterpreted.

Power dissipation: Power dissipation is the average power required by the gate for active operation.

Gate delay: Gate delay is the average time taken by a signal change at the input to cause a change at the output.

In the sections that follow we shall limit our discussion to the most often used logic families: TTL, ECL, CMOS, and PMOS. Table 3.1 can be used as a general guideline of their characteristics. Relative values have been provided for only a few parameters. Manufacturers' data books should be consulted for numerical values. The authors are not endorsing a particular family. We will concentrate on TTL after a brief look at ECL and MOS in the next section since TTL is the most popular family. Once students are familiar with TTL, they can select other logic families as the application dictates.

TABLE 3.1 **IC Family Characteristics**

Parameter	TTL	ECL	CMOS	PMOS
Speed	Good	Very good	Fair	Poor
Noise immunity	Very good	Good	Very good	Fair
Cost	Low	High	Medium	Medium
Fan-out	Fair	Fair	Very good	Very good
Power*	Medium	High	Low	Low
Basic gate	NAND	NOR/OR	NOR or NAND	NAND

* Except for the CMOS, device power is fairly independent of input frequency. At higher frequency, power dissipation of CMOS increases. Consequently, CMOS requires little power in battery-driven circuits that switch slowly.

3.2.1 Transistor-Transistor Logic (TTL)

This particular family of logic devices was introduced in the mid-sixties and has become the most widely used of all bipolar ICs. The basic circuit for the TTL family is a NAND gate. The circuit of Figure 3.2 shows a three-input TTL NAND gate. A multiple-emitter input transistor is typical of all TTL gates and circuits. Each emitter acts like a diode and, consequently, the combination of transistor T_1 and the 4 kΩ resistor acts like a three-input AND gate. The remaining part of the circuit complements the signal, giving at the output the NAND of the three inputs.

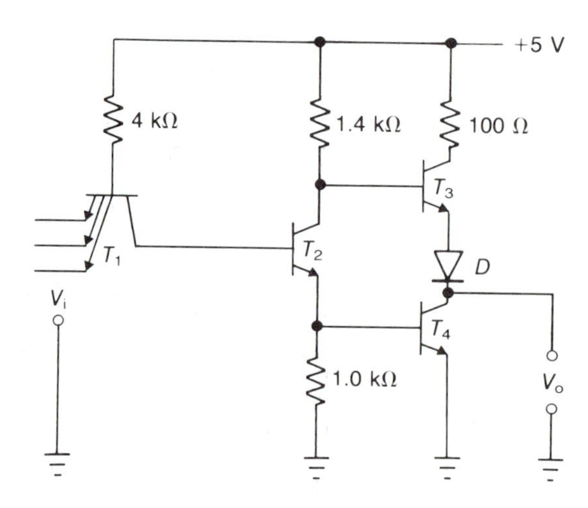

FIGURE 3.2 **Three-Input Standard TTL NAND Gate.**

The combination of the selected transistors T_3 and T_4 is known as a *totem-pole connection*, which is also typical of most TTL devices. One of these transistors is always on. When T_4 is on, the output V_o is low, and when T_3 is on, the output is high. This totem-pole configuration increases the switching speed of the device.

If one of the three input voltages is low, the transistor T_1 is forced to saturate, which in turn decreases the base voltage of T_2 to the cut-off of both T_2 and T_4. The net effect is that the transistor T_3 acts like an emitter-follower (i.e., its output is the same as its input), which causes the output voltage to be high. Again, when all of the three inputs are high, the T_1 collector conducts in the forward direction, forcing T_2 and T_4 to saturation and, consequently, the output voltage becomes low. The diode D maintains the base-emitter of T_3 reverse-biased so that T_4 is able to conduct freely when the output is low.

3.2.2 Emitter-Coupled Logic (ECL)

ECL is basically a current-mode logic. The transistors are not allowed to saturate, so the storage time is eliminated altogether and the operating speed is increased significantly. This particular logic family has the lowest propagation delay and is used in high-speed applications.

The basic circuit of the ECL family is a NOR gate, though many ECL ICs provide an alternate OR output. The main characteristics of an ECL device are as follows:

1. It usually has two outputs that are complementary.
2. Its transistors are never driven to saturation and as a result the storage time is eliminated.
3. It has high input impedance and very low output impedance. This makes high fan-out possible.

4. It is sensitive to electrical noise because the logic swing is small (only about 1 V).
5. It draws relatively constant current.
6. The speed-power product is reasonable even though the power dissipation is relatively high.

The outputs of two or more ECL gates can be combined externally with or without a resistor to form a *wire-OR* configuration. This capability allows the formation of other logic functions by combining several gate outputs. As an example, the wire-OR combination of three two-input ECL NOR gates results in an OR-AND-INVERT logic as shown in Figure 3.3.

FIGURE 3.3 **ECL NOR Gates in Wire-OR Configuration.**

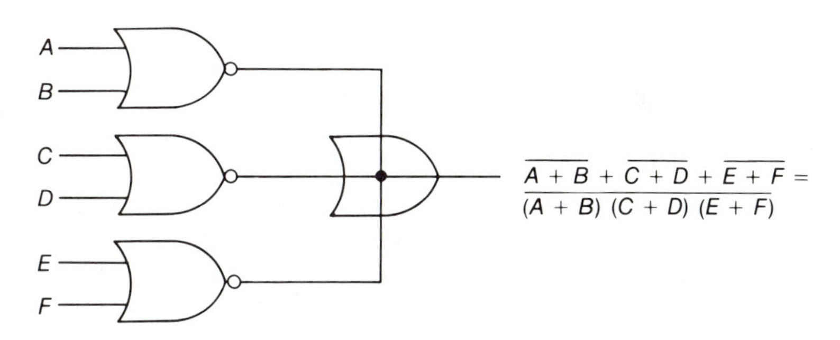

The basic ECL gate is shown in Figure 3.4, where both OR and NOR outputs are available. V_{CC1} and V_{CC2} are both grounded to eliminate the possible cross-talk resulting from high-speed noise. The supply voltage, V_{EE}, is -5.2 V, and logic 1 and 0 are equivalent to about -0.9 V and -1.75 V, respectively.

FIGURE 3.4 **Basic ECL Gate and Its Characteristics.**

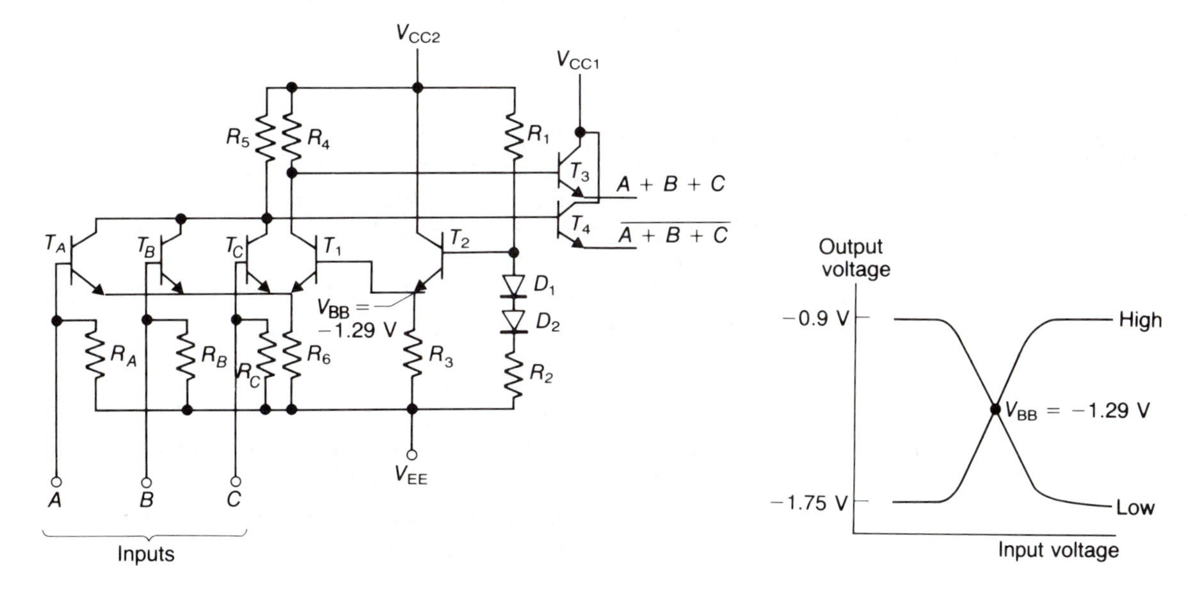

The input transistors, T_A, T_B, and T_C, operate as a difference amplifier in close conjunction with transistor T_1. At all times either one or more of the input transistors is on or T_1 is on. The resistors R_1 and R_2 and the diodes D_1 and D_2 clamp the base of the transistor T_2 so that its emitter and T_1 base are set to -1.29 V.

When all inputs are low, the transistor T_1 is turned on, which in turn clamps the emitters of the input transistors to approximately -2 V and turns off the input transistors. This results in a low OR output (equivalent to the sum of the voltage drop across R_4 and the base-to-emitter voltage) and a high NOR output (since the turned-off input transistors result in an insignificant voltage drop across R_5).

If any one of the inputs is high, the corresponding transistor will be turned on since its base voltage becomes higher than that of the T_1 base. This subsequently turns off T_1. Consequently, T_3 base will be close to 0 and the OR output becomes high. Further, if at least one of the input transistors is on, the voltage at T_4's base reaches approximately -1 V, which in turn causes the NOR output to be low.

3.2.3 Unipolar Logic (MOS ICs)

Unipolar logic is usually formed using MOS technology. The MOS transistors are field-effect transistors (FETs), which are used as resistors because of their high impedances. There are several different types determined by their manufacturing process. A p-channel MOS, referred to as PMOS, requires negative voltage for its operation, while an n-channel MOS, called NMOS, requires positive voltage for its operation. There is another version called the complementary MOS, referred to as CMOS, which consists of one PMOS and one NMOS combined in a complementary fashion. CMOS requires positive voltage for its operation. There are CMOS devices that are pin-for-pin compatible with many TTL devices.

The advantages of MOS over bipolar logic are that its high packing density allows fabrication of more circuits in a given chip area and that it consumes less power than the bipolar gates. Consequently, this technology is finding increased application in the fabrication of read-only memories and microprocessors.

FIGURE 3.5 **MOS Symbols: [*a*] NMOS and [*b*] PMOS.**

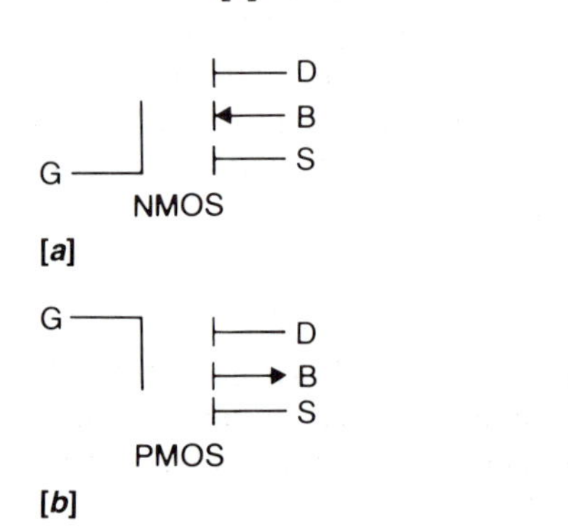

The symbols for MOS gates are shown in Figure 3.5, where G, D, B, and S, respectively, represent the gate, the drain, the active substrate, and the source. Enhancement-mode gates are used more commonly in logic circuits. The gate is separated from the active bulk or substrate by an oxide insulator that causes the MOS input impedance to be very high. Table 3.2 illustrates the differences between the characteristics and operations of the PMOS and NMOS devices.

Figure 3.6 shows a basic CMOS NOT gate. The NMOS and PMOS sections are arranged in such a way that they have their

TABLE 3.2 **MOS Characteristics**

Object	NMOS	PMOS
Substrate	p-type	n-type
Source	n-type	p-type
Drain	n-type	p-type
Gate effects (enhancement)	Current flows when gate is positive with respect to substrate.	Current flows when gate is negative with respect to substrate.
Conventional current	Drain to source	Source to drain

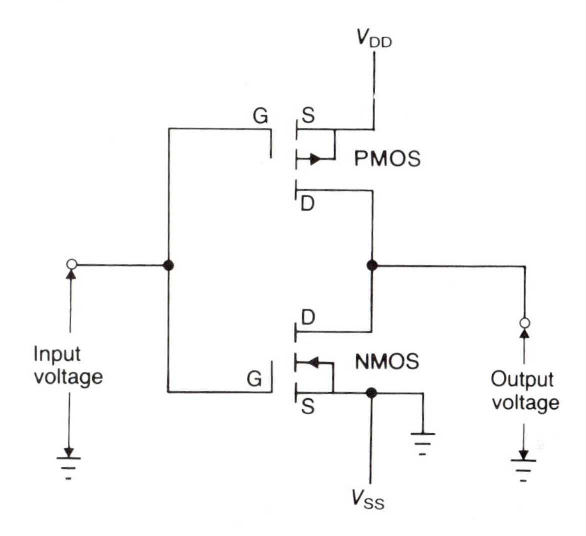

FIGURE 3.6 **CMOS Inverter Circuit.**

source and the substrate combinations tied, respectively, to ground and V_{DD} (3–15 V). The output voltage is high when it equals V_{DD} and is low when it equals 0 V. When the input is high, NMOS is enhanced but the PMOS is not. Because of the open PMOS the output displays a low voltage. It is interesting to note that there is no current and, therefore, no power is dissipated. When the input is low, PMOS is enhanced but NMOS is not. This time the output is shorted to V_{DD} and is high. The consumed power is still zero for all practical purposes because there is no current path. The only dissipation of power takes place when the states are switched. The power dissipation (which is still smaller than that of the TTL) is proportional to the switching frequency.

3.3 IC Specifications

Digital IC manufacturers usually provide the following numerical quantities as part of a document known as a *spec sheet.* These specifications play a significant role in the selection of devices for the system being designed.

Supply voltage, V_{CC}: This is the supply voltage.

High-level input voltage, V_{IH}: This is the minimum input voltage that is interpreted as high.

Low-level input voltage, V_{IL}: This is the maximum input voltage that is interpreted as low.

High-level output voltage, V_{OH}: This is the minimum output voltage that is interpreted as high.

Low-level output voltage, V_{OL}: This is the maximum output voltage that is interpreted as low.

Supply current, I_{CC}: This is the positive current into the V_{CC} supply terminal of the IC.

Noise margins: These are the voltage differences $V_{OH} - V_{IH}$ and $V_{IL} - V_{OL}$.

High-level input current, I_{IH}: This is the positive current into the input corresponding to V_{IH}.

Low-level input current, I_{IL}: This is the positive current into the input corresponding to V_{IL}.

High-level output current, I_{OH}: This is the positive current into the output corresponding to V_{OH}.

Low-level output current, I_{OL}: This is the positive current into the output corresponding to V_{OL}.

Maximum clock frequency, f_{max}: This is the highest frequency rate at which the clock input of a bistable circuit can be driven through the required sequence without ending up in unpredicted states.

Fan-out: This designates the number of similar gates that can be driven by the gate.

Propagation delay (high to low), t_{PHL}: This the time interval between the specified reference points on the input and output voltage, with the output changing from high to low.

Propagation delay (low to high), t_{PLH}: This is the time interval between the specified reference points on the input and output voltage, with the output changing from low to high.

There are many more specifications than those listed that help describe the device under consideration. It is the authors' opinion that the remaining specifications usually have a limited role to play in the design. However, students are urged to familiarize themselves with all the information on the spec sheets. Figure 3.7[*a*] shows the general characteristics of V_{OH}, V_{OL}, V_{IH}, V_{IL}, and noise margins. Figure 3.7[*b*] shows the respective propagation delays where the second waveform corresponds to the out-of-phase output.

There are two popular types of IC packages: the flatpack type

FIGURE 3.7 [*a*] IC Voltage Specifications and [*b*] IC Propagation Delays.

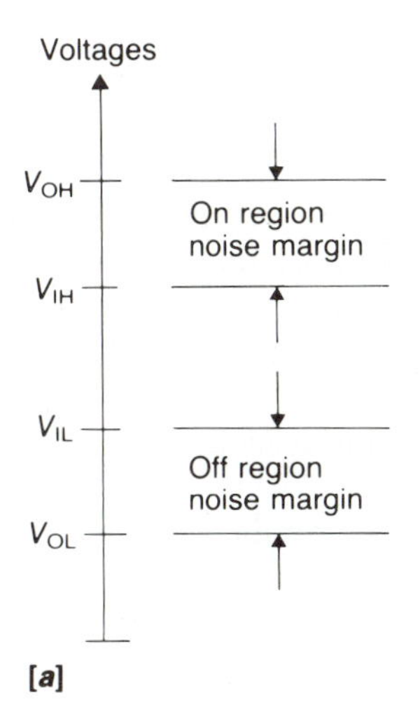

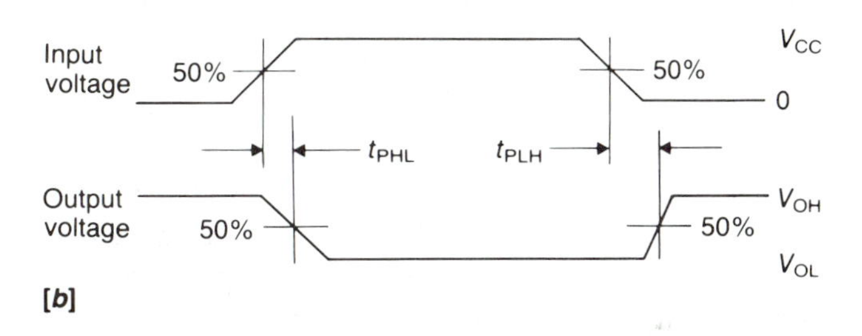

and the dual-in-line (DIP) type. The DIP type IC has the widest usage because of its easy installation. DIPs contain a single chip of silicon that is mounted in a plastic package with circuit leads brought out to pins that can be plugged into sockets or soldered to printed-circuit boards. A standard 14-pin DIP type IC is shown in Figure 3.8. Most of the ICs come in standard sizes and have between 14 and 64 terminals (referred to as *pins*). A new process has just appeared that allows surface mounting of ICs. Each of these hermetically sealed leadless chip carrier packages has a brazed seal for surface mounting. They are becoming popular particularly for VLSI devices with pin matrices as opposed to pins along the edges.

Each of the ICs has a numerical designation on its plastic or ceramic surface for the purpose of identification. For example, the TTL ICs are usually identified by their designation as part of the 5400 or the 7400 series. Some manufacturers provide TTL ICs with different designations as well. A version of the CMOS ICs are designated as the 54C00 and the 74C00 series, which are generally compatible with the TTL ICs. In a later section we shall discuss the TTL to CMOS and CMOS to TTL interfacing.

FIGURE 3.8 14-Pin Dual-in-Line Package (DIP).

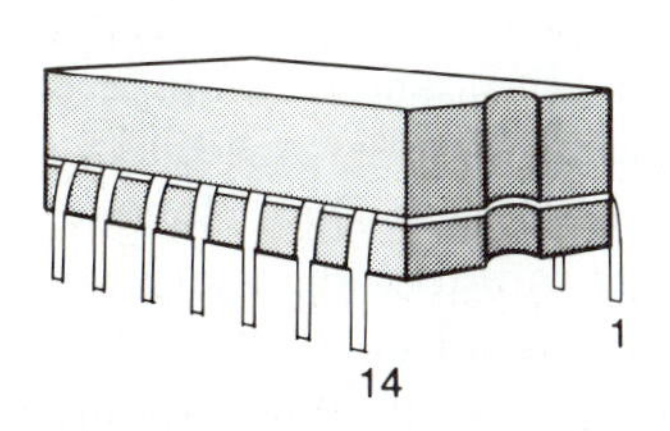

3.4 TTL Overview

3.4.1 Subfamilies

TTL devices, referred to as semiconductor networks (SN), have basically evolved into two series: 5400 and 7400. Devices in the 7400 series work over a relatively narrow temperature range of 0°C to 70°C and are adequate for commercial purposes only. The 5400 series devices, originally developed for military use, have exactly the same function as the corresponding 7400 devices except that they are operable over a wider range of temperatures, from −55°C to 125°C. The 5400 series devices are more expensive, so they are rarely used commercially except in severe environments.

The circuit of Figure 3.2 is a standard TTL three-input NAND gate. The propagation delays of this type of device are reduced by lowering the internal time constants. This can be accomplished by decreasing the resistances in the circuit of Figure 3.2. The resulting variation of TTL devices is called *high-speed TTL* and is numbered under the 54H00 and 74H00 series. High-speed TTL has a higher power dissipation because of the lower resistance values of the ICs. By increasing the internal resistances the manufacturer reduces the power dissipation of TTL gates but this contributes to slower operation. Devices with lower power dissipation are known as *low-power TTL* and are numbered under the 54L00 and 74L00 series.

Transistors in standard TTL, high-speed TTL, and low-power TTL saturate and cause flooding of extra carriers at the base. As a consequence there is a significant time delay, known as the *saturation delay time*, in switching a transistor from saturation to cutoff. To eliminate this problem a diode clamp is introduced between the base and the collector. Such an arrangement is shown in Figure 3.9.

FIGURE 3.9 **Transistor with a Schottky Diode Clamp.**

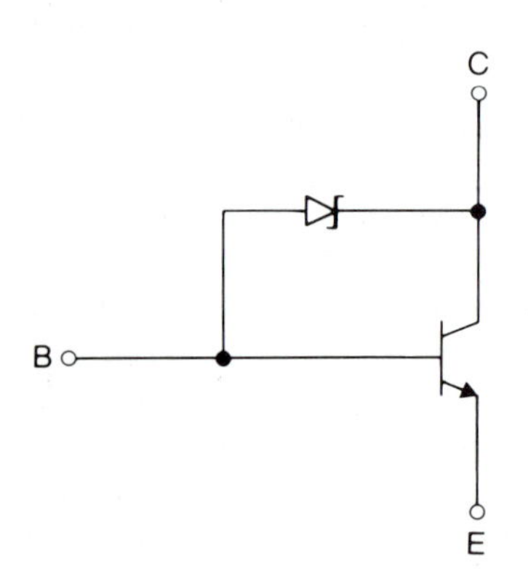

The transistor of Figure 3.9 uses a special diode known as the *Schottky diode* and is formed by the junction of a metal and a semiconductor. The Schottky diode has a forward voltage of 0.4 V and, therefore, prevents full saturation of the transistor. TTL devices of this type are called *Schottky TTLs* and are numbered under the 54S00 and 74S00 series. Unfortunately, Schottky TTLs have a higher power dissipation. A compromise is to increase the internal resistance of the Schottky TTL, which provides a device that has low power dissipation as well as good speed. This variation of TTL is known as the *low-power Schottky TTL* and is numbered under the 54LS00 and 74LS00 series. Presently both Schottky TTL and low-power Schottky TTL have undergone further innovation. One of this newer subfamily is *advanced Schottky TTL* and is numbered under the 54AS00 and 74AS00 series. This subfamily is twice as fast as the 54S/74S subfamily and dissipates very little power. However, the *advanced low-power Schottky* devices, numbered under the 54ALS00 and 74ALS00 series, dissipate even less power. Many times the speed-power product is used for rating one device against the other. In comparison, the 54ALS/74ALS subfamily has the lowest speed-power product value among all TTL subfamilies. The characteristics of these different TTL devices are listed in Table 3.3.

TABLE 3.3 **TTL Characteristics***

Type	Power Dissip. (mW)	Prop. Delay (ns)	Speed-Power Product (pJ)	V_{CC} (V)	V_{IH} (V)	V_{IL} (V)	V_{OH} (V)	V_{OL} (V)	I_{OL} (mA)	I_{IL} (mA)
54/74	10	10	100	5	2	0.8	2.4	0.4	16	−1.6
54H/74H	22	6	132	5	2	0.8	2.4	0.4	20	−2
54L/74L	1	33	33	5	2	0.7	2.4	0.4	3.6	−0.18
54S/74S	19	3	57	5	2	0.8	2.7	0.5	20	−2
54LS/74LS	2	9.5	19	5	2	0.8	2.7	0.5	8	−0.4
54AS/74AS	10	1.5	15	5	2	0.8	3.0	0.3	20	−0.5
54ALS/74ALS	1	4	4	5	2	0.8	3.0	0.25	12/24	−0.1

* The values of the current depend on the nominal value of the input pull-up resistor.

3.4.2 Characteristics

TTL inputs can be set either to high voltage (5 V) or to ground (0 V) but should never be left floating (unconnected). As shown in Figure 3.10, a positive emitter current, I_E, always exists corresponding to a low TTL input. Again when the input is high, the emitter diode cuts off. An input that is floating results in no emitter current. A floating TTL input is always treated as a high logic (equivalent to binary 1). This implies that a floating input at a TTL inverter results in a low output. Note that even though the floating inputs act like high logic, they appear low on an oscilloscope, which may be a reason for confusion during the debugging process.

In practice, all unused inputs should be tied to either a 0 or a 1 for improved noise immunity. The floating inputs in the case of MOS devices contribute to even more confusion since the corresponding outputs drift and are unpredictable. Unused MOS or CMOS inputs are *pulled up* by connecting them to V_{CC} through a resistor as in Figure 3.11[*a*] or grounding them. An alternative to this is achieved by *clamping* the floating inputs. The clamp circuit consisting of two diodes and resistors is shown in Figure 3.11[*b*]. This circuit is organized in such a way that the floating inputs are set at about 3.6 V to prevent IC input damage caused by possible voltage spikes that occur in the power supply. Another fairly popular technique involves tying used inputs to floating inputs as shown in Figure 3.11[*c*].

FIGURE 3.10 **TTL Transistor.**

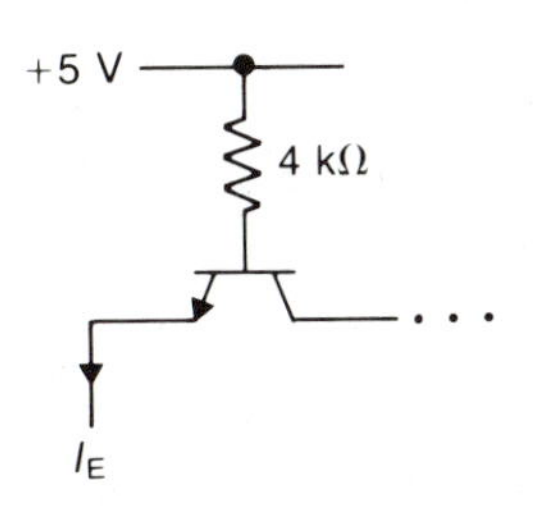

A standard TTL NOT gate, as shown in Figure 3.12, is now considered. When the input voltage, V_i, is grounded, the output voltage, V_o, is high. In this particular TTL NOT gate, the output is going to be high for all V_i values in the range of 0 V to 0.8 V. On

FIGURE 3.11 **[*a*] Pull-Up, [*b*] Clamping, and [*c*] Tying.**

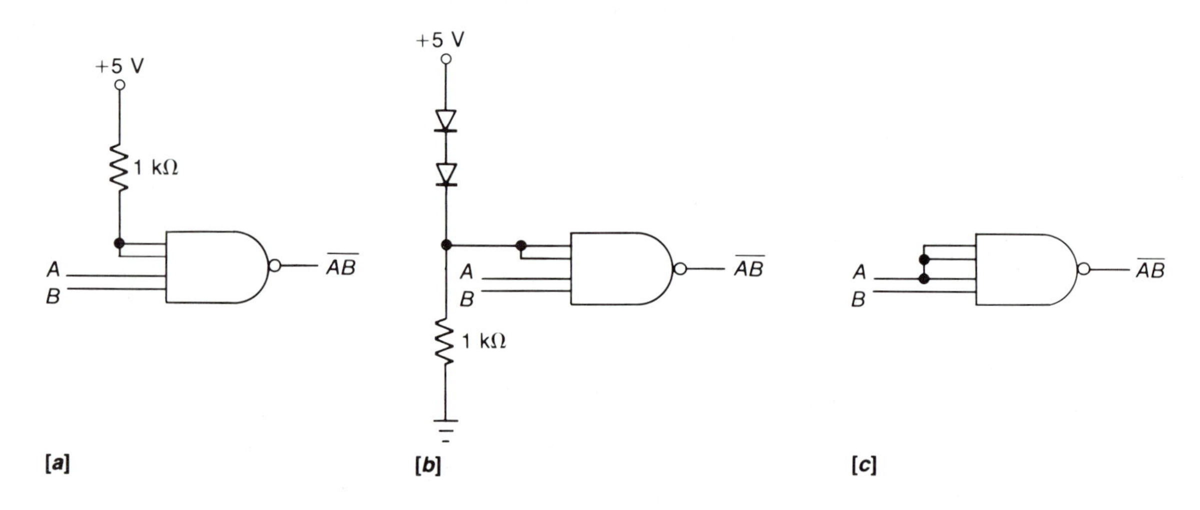

FIGURE 3.12 **Standard TTL Gate.**

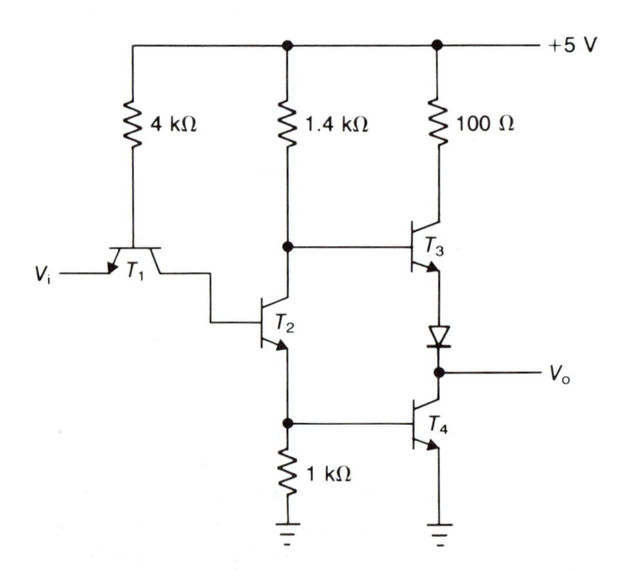

the other hand, for all input voltages greater than 2 V, the output is low. The worst-case output values are 0.4 V and 2.4 V for the low and the high, respectively. These worst-case values, as listed in Table 3.4, imply that the TTL devices are compatible and that the output of one TTL device can drive the input of another TTL device.

Figure 3.13[*a*] shows the example of a TTL low output connected to a second TTL device. The low output of the first device implies that its voltage could be anywhere between 0 V and 0.4 V.

TABLE 3.4 **Worst-Case TTL Voltages and Currents**

	Voltages		Currents	
Type	Input (V)	Output (V)	Input (mA)	Output (mA)
Low	0.8	0.4	−1.6	−16
High	2.0	2.4	0.04	+0.4

FIGURE 3.13 **TTL Device Interconnections: [*a*] Low Output to Second TTL and [*b*] High Output to Second TTL.**

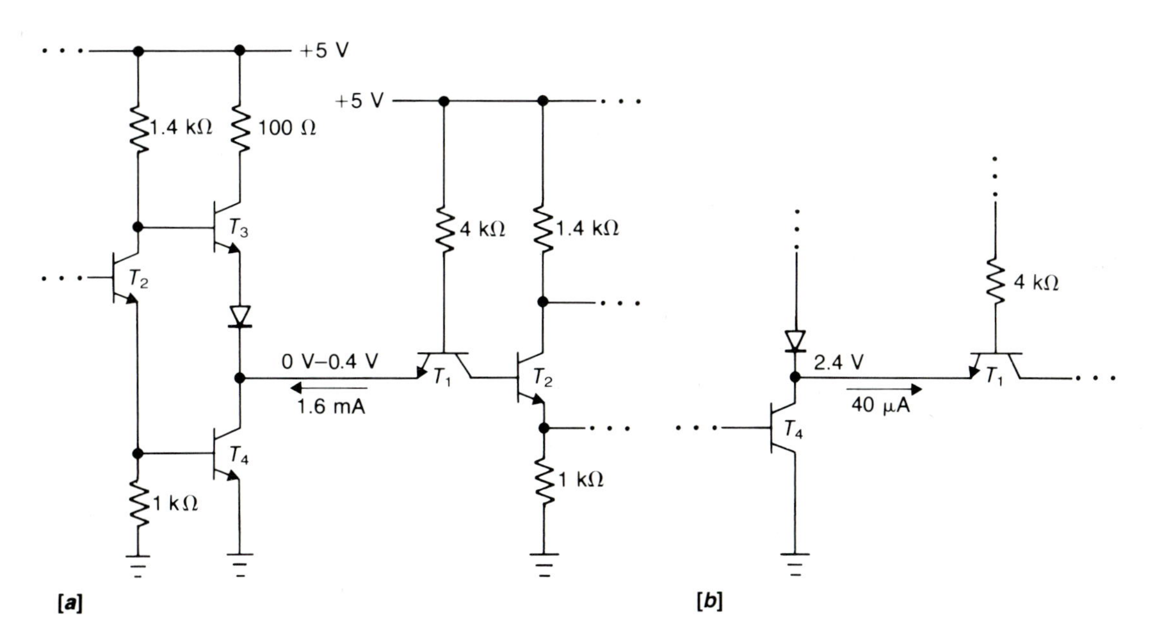

Therefore, it is very possible to drive the second device because any input voltage less than 0.8 V is treated as a low logic. Similarly, Figure 3.13[*b*] shows the other case where the high output of a device is connected to a second device. The minimum possible voltage at the output of the first device is 2.4 V and this is more than sufficient to drive the second device since it is well more than 2 V. We can see from Table 3.4 that the noise margin is about 0.4 V. As long as the induced stray noise voltages resulting from the device interconnection are less than this amount, the second TTL device will not have any erroneous outputs.

As apparent from both Figure 3.13 and Table 3.4, the T_4 transistor of the first TTL device acts as a current sink for the second TTL device when the output is low, and the T_3 transistor of the first TTL device acts as a current source for the second TTL device when the output is high. The first TTL device has the ability to sink current (low output) up to 16 mA and can source current (high output) up to 0.4 mA. A standard 74 series TTL device output can be connected to 10 different standard 74 series TTL emitters.

The maximum number of TTL gates that can be driven under the worst-case condition is called the *fan-out.* A standard TTL device has a fan-out of 10 as stated earlier. We can, of course, combine different TTL drivers and TTL loads to attain different fan-outs. Table 3.5 shows the fan-outs for different combinations of TTL subfamilies. Note that a Schottky TTL gate can drive up to 100 different advanced low-power Schottky TTL gates. At the other extreme, a low-power Schottky TTL gate can drive only four high-speed TTL gates or only four Schottky TTL gates. These drive considerations are important when mixing devices from different series. Note, however, that the electrical characteristics of specific devices within each subfamily may vary. It is advisable to look for the most recent parameter values from the manufacturer's data book.

TABLE 3.5 **Fan-Outs of TTL Combinations**

	Load					
Driver	54/74	54H/74H	54S/74S	54LS/74LS	54AS/74AS	54ALS/74ALS
54/74	10	8	8	40	80	80
54H/74H	12	10	10	50	100	100
54S/74S	12	10	10	50	100	100
54LS/74LS	5	4	4	20	40	40
54AS/74AS	12	10	10	50	100	100
54ALS/74ALS	5	4	4	20	40	40

3.4.3 Discrete Gates

The NAND gate is the basic internal gate of the TTL series. Figure 3.2 is an example of a three-input TTL NAND gate. By increasing the number of emitters at the transistor T_1 of the basic NAND gate, two-, three-, four-, and eight-input NAND gates have been obtained and are commercially available. NAND gates with more than eight inputs may be obtained by combining several NAND gates. A NAND gate with up to 64 inputs may be formed by NANDing eight inverted eight-input TTL NAND gates as shown in Figure 3.14[*a*]. Similarly, Figure 3.14[*b*] shows that a 32-input NAND gate may be obtained by NANDing eight four-input TTL AND gates. Commercially available TTL AND gates have a maximum of only four inputs.

The TTL NAND gate circuit is modified to obtain other discrete logic gates. Figure 3.15 shows a two-input NOR gate where the cir-

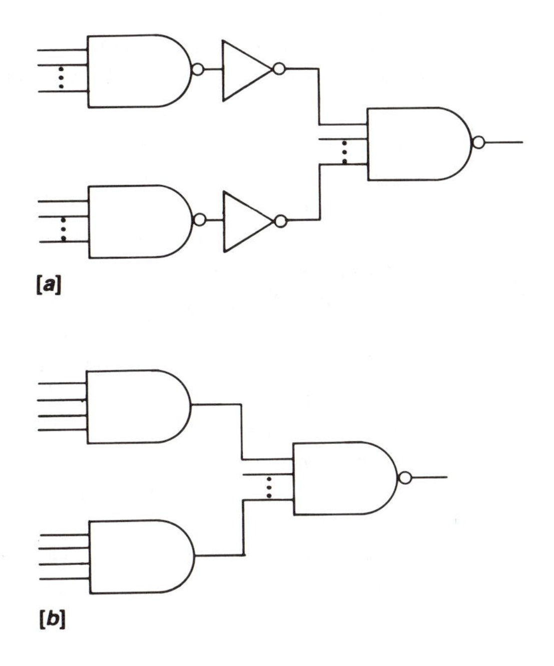

FIGURE 3.14 **Expanded TTL NAND Circuits: [a] 64 Possible Inputs and [b] 32 Possible Inputs.**

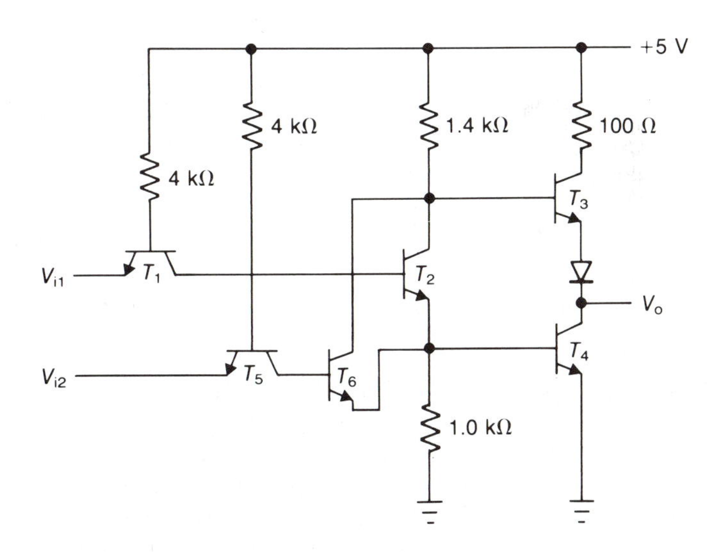

FIGURE 3.15 **TTL NOR Gate Circuit.**

cuit is similar to that of Figure 3.2, except for the two transistors, T_5 and T_6. When V_{i1} or V_{i2} or both are high, the two transistors in parallel, T_6 and T_2, are turned on since T_1 and T_5 remain turned off. This condition results in the saturation of T_4 that pulls down the output to a low. Again if both V_{i1} and V_{i2} are low, T_1 and T_5 saturate, cutting off both T_2 and T_6. As a consequence, T_3 works like an emitter-follower, resulting in a high output. Currently in the TTL family, two-input, three-input, and four-input NOR gates are available. NOR gates with higher numbers of inputs can be obtained by making use of a special TTL gate called AND-OR-INVERT.

In order to realize a TTL OR gate, a common-emitter stage is inserted before the totem-pole output of the circuit of Figure 3.15. This additional inversion converts the NOR to an OR gate. Similarly, addition of a common-emitter stage to the totem-pole output of Figure 3.2 would result in an AND gate. TTL AND gates are available with two, three, and four inputs. In fact, a 16-input AND gate can be realized by ANDing four four-input TTL AND gates.

The TTL family includes a particular device called an *AND-OR-INVERT* (*AOI*) gate that has proved useful to many designers. Figure 3.16[*a*] shows the circuit of a TTL AOI gate. Besides its basic NAND configuration, it has two additional transistors, T_5 and T_6. T_1 and T_5 function as a two-input AND gate, and the two transistors in parallel, T_6 and T_2, provide for the OR and NOT operations, respectively.

The largest AOI gate is equivalent to four two-input AND gates. Since the T_2-T_6 combination involves the OR operation, it is possible to expand AOI by connecting other gates to the collector and emitter points (shown by the broken line in Figure 3.16[*a*]) of T_6. Figure 3.16[*b*] shows the logic symbol for an expandable AOI gate. The gates that are actually connected to the expandable AOI are known as the expanders. *Expanders* are AND gates with a phase-splitter output transistor, as shown in Figure 3.17[*a*]. Figure 3.17[*b*] shows the logic symbol of a four-input expander, and Figure 3.17[*c*] shows the expandable AOI with two four-input expanders. Suitable

FIGURE 3.16 **[*a*] TTL AOI Schematic Diagram and [*b*] Logic Symbol for Expandable AOI.**

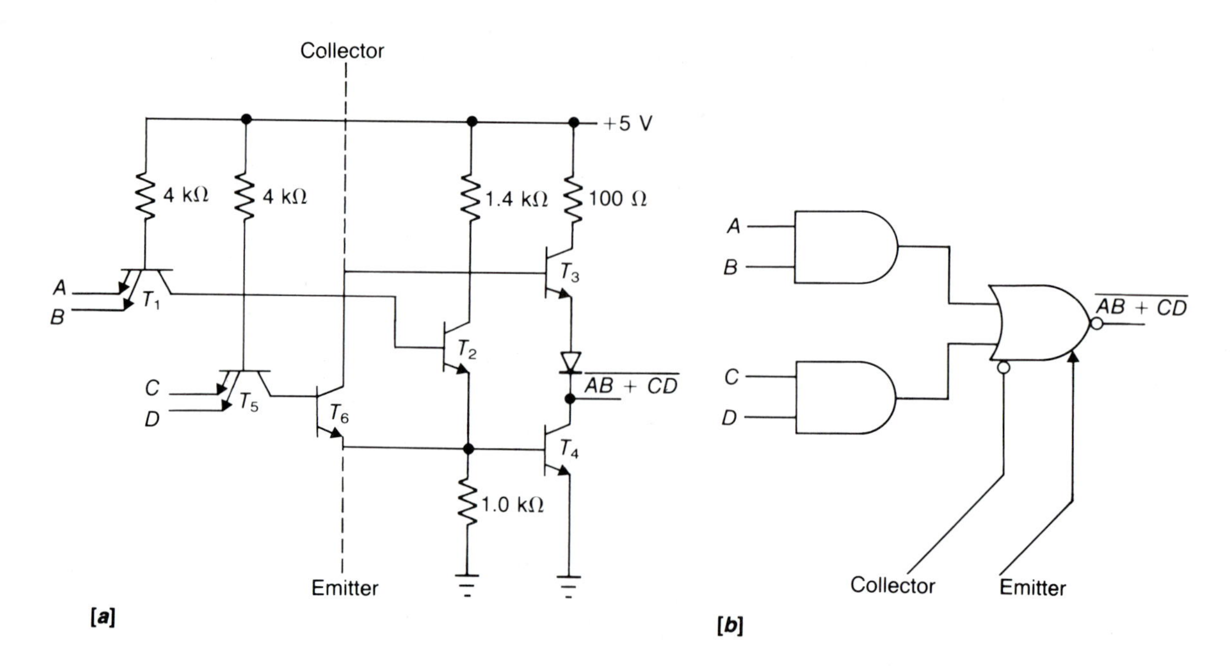

FIGURE 3.17 **[*a*] Expander Schematic, [*b*] Expander Symbol, and [*c*] Expandable AOI with Two Expanders.**

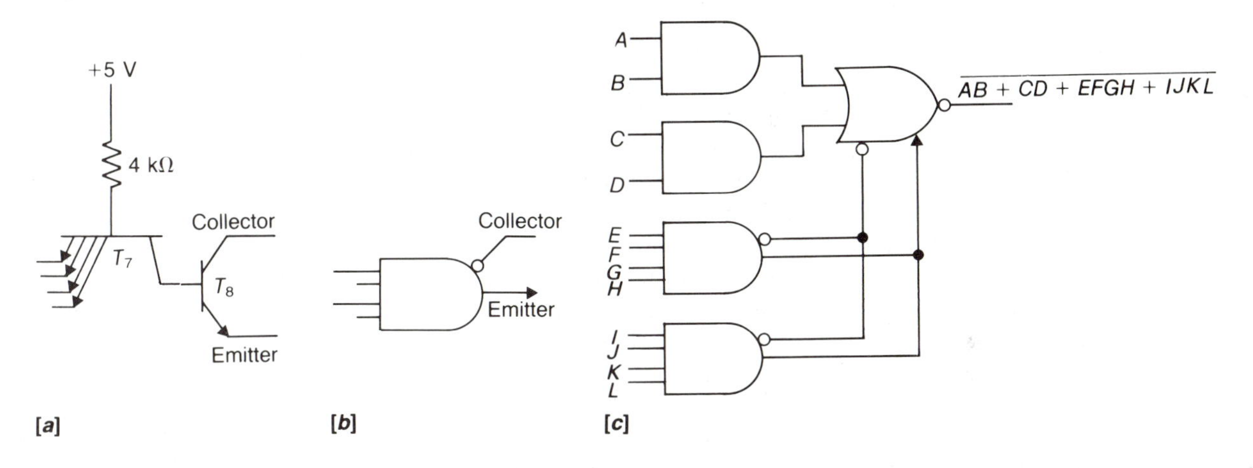

combinations of expander and expandable AOI lead to the formation of AND, OR, and NOR gates.

3.5 Open-Collector and Tri-State Outputs

DTL gates become low by saturating their transistors. Consequently, if two or more DTL outputs are tied together (wire-ANDed), as shown in Figure 3.18, an additional level of AND logic is obtained. In the configuration shown if any one of the transistors is saturated, it causes the output to be a low. Only when all of the transistors are cut off is the overall output a high.

It is not advisable to wire-AND the outputs of totem-pole TTL

FIGURE 3.18 **Wire-ANDing of Three DTL Gates.**

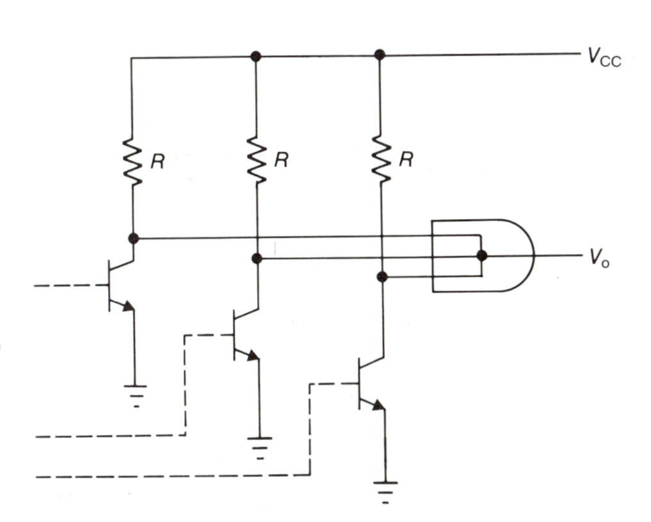

FIGURE 3.19 *Wire-ANDing of Two Totem-Pole TTL Gates.*

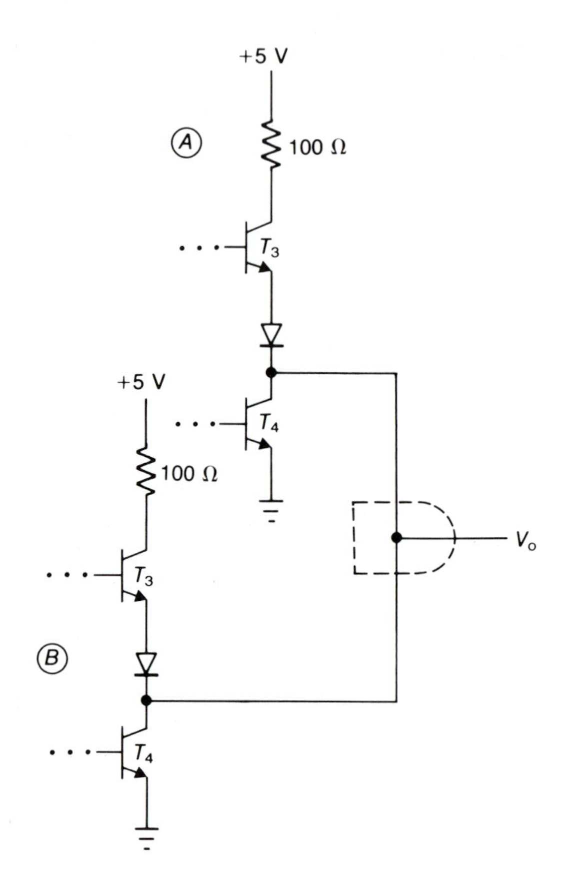

circuits, because it could lead to a short circuit, causing the gate to be destroyed. The wire-ANDing of the totem-pole TTL outputs is shown in Figure 3.19, where only the output stage of each circuit has been drawn for simplicity. If the output of gate A is high and if the output of gate B is low, the T_3 transistor of A and the T_4 transistor of B turns on. Since there is only a small resistance value in that path, the resulting current will be in excess of the device rating.

Open-collector TTL devices may be used for a so-called wired-OR. Figure 3.20[*a*] shows a four-input TTL NAND gate with an open-collector output. Notice in this configuration that the T_4 collector is open by the absence of transistor T_3. This circuit will not work without the use of an external load resistor, as shown in Figure 3.20[*b*]. Figure 3.20[*c*] shows the outputs of two open-collector circuits tied together with a common load resistor. This configuration is known as a wire-OR. In actuality the function performs a logic AND of the two gate outputs. The value of the pull-up resistor is determined by considering the fan-out of the tie and the number of devices in the tie.

Open-collector (OC) devices can be used to transmit data by means of a bus system. Figure 3.21 shows a possible scheme for a

FIGURE 3.20 **[*a*] Open-Collector TTL Circuit, [*b*] Open-Collector TTL with Pull-Up Resistor, and [*c*] Wire-OR.**

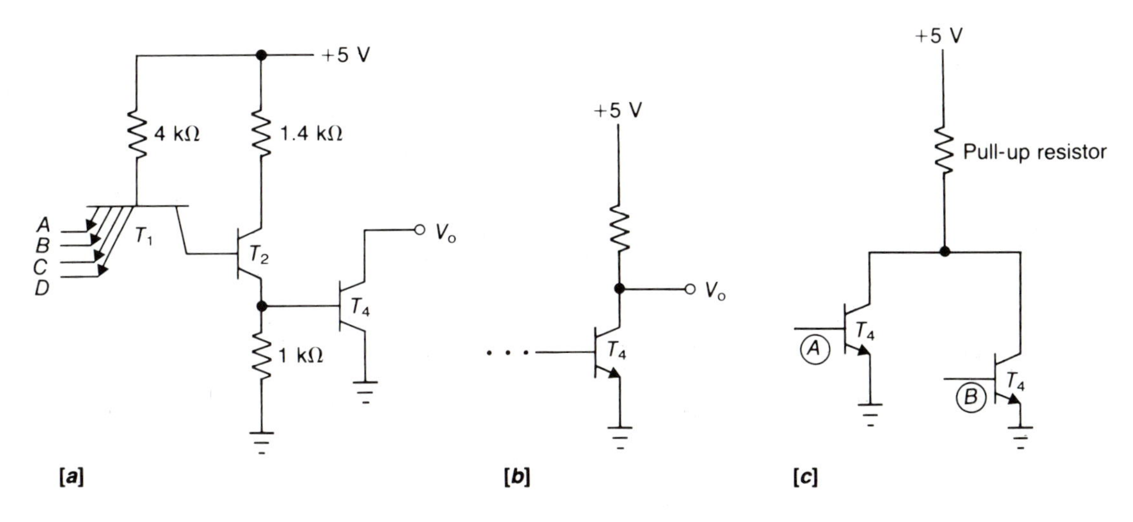

FIGURE 3.21 **Bus Scheme for Only One-Bit Location Using Open-Collector Devices.**

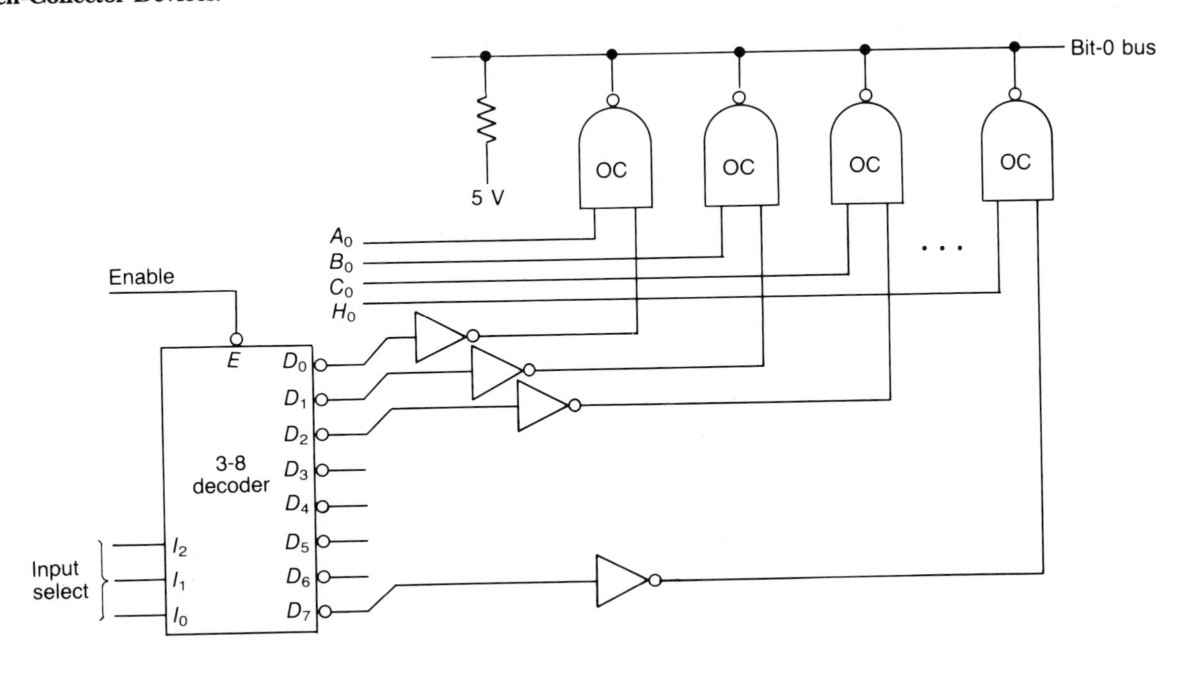

common *bus*. The 3-8 line decoder present in this figure is used to enable one of the eight outputs by means of three select inputs. For a system with n-bit words, n such circuits—one for each bit position—would be required. A bus like the one just described is often connected to a processor unit (see Chapter 13 for details) of a digital computer. A processor can select the input from which it wishes to obtain data. The selected data appear on the bus. The data are inverted by the NAND gate. The major disadvantage of the open-collector circuits, however, is their long gate delay.

Tri-state devices are now available in both TTL and CMOS and are eliminating many open-collector applications. Figure 3.22 is an example of a tri-state TTL device. Unlike normal logic circuits, the tri-state device has three distinct states. Two of the states are the normal low and high states; the third state is a high-impedance state.

FIGURE 3.22 **Tri-State TTL NOT Gate.**

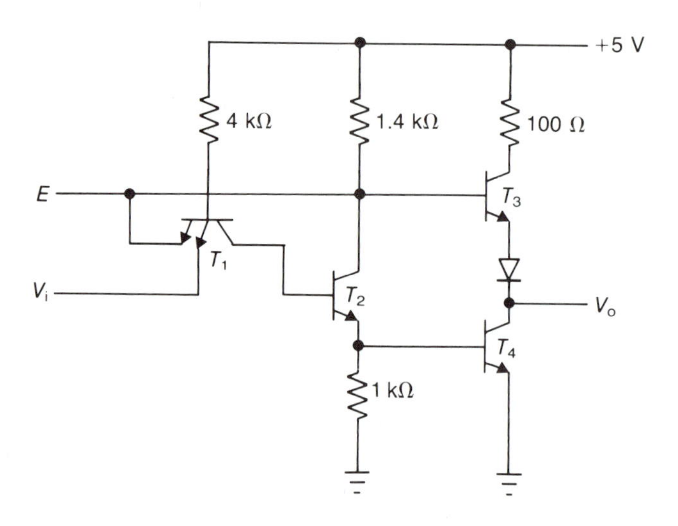

When the enable signal, E, is low, current flows through the base-emitter junction of T_1 and depletes T_2 of its base current, turning it off. The low value of E pulls the T_3 base down and cuts off both T_3 and T_4. This condition produces a high output impedance to both V_o and ground and results in the third state.

When E is high the circuit functions normally and the output depends strictly on the input condition. The truth table for the tri-state TTL inverter is shown in Figure 3.23. There is a difference between the wire-OR and a tri-state device. Tri-state devices tied together do not perform a Boolean function. Only the selected device places its input on the common output. Tri-state devices have the ability to multiplex many functions to a single bus. If several outputs are tied together and none of the gates are enabled, the collective output shows a high impedance and its voltage may be in the prohibitive region.

FIGURE 3.23 **Truth Table for a Tri-State Inverter.**

Input	Enable	Output
0	0	High impedance
0	1	1
1	0	High impedance
1	1	0

Figure 3.24[*a*] shows the symbol for a tri-state buffer. The two devices may be considered as a normal TTL device in series with a relay that is closed by the enable signal, which is illustrated in Figure 3.24[*b*]. The tri-state devices are used for tying several devices to a data bus. The selected device (only one device should be connected to the data bus at one time) uses the bus to transmit data. An example of their application is shown in Figure 3.25. By sending a 000 on the lines, the processor can connect the disk output to the data bus, a 001 would connect the tape unit, and so on. Tri-state devices are also used to realize a bidirectional data bus as shown in Figure 3.26. A low value at the control would transmit, while a high value at the control would receive; the incoming data. Several of these receiver-transmitter pairs may be tied onto the same bus without exceeding the drive capability of such devices.

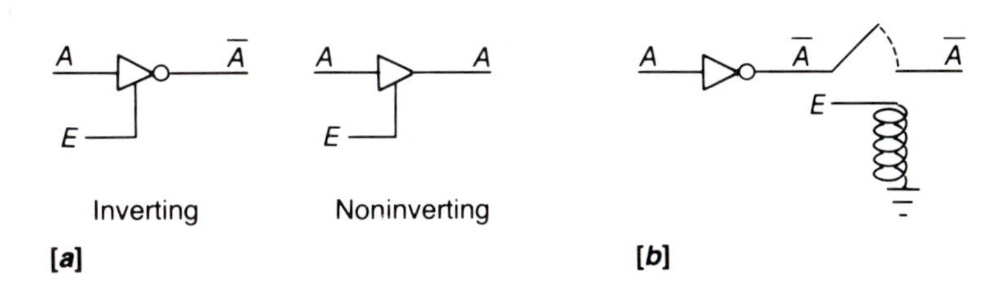

FIGURE 3.24 **[*a*] Tri-State Buffer Symbols and [*b*] Relay Analog of the Tri-State Buffer.**

FIGURE 3.25 **Data Bus Input Control.**

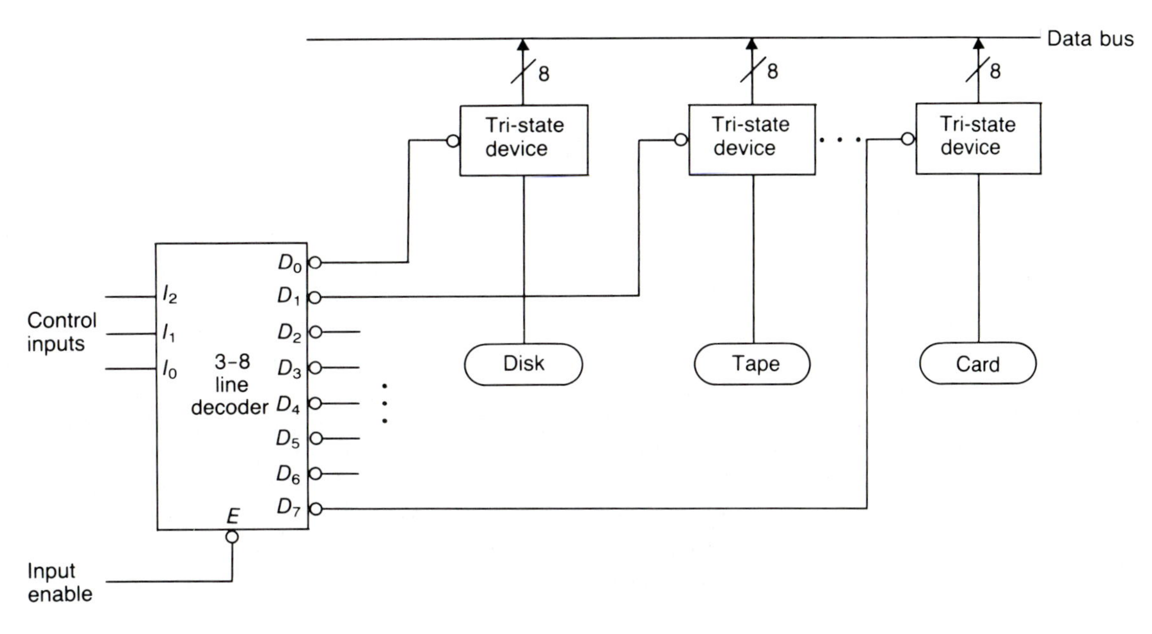

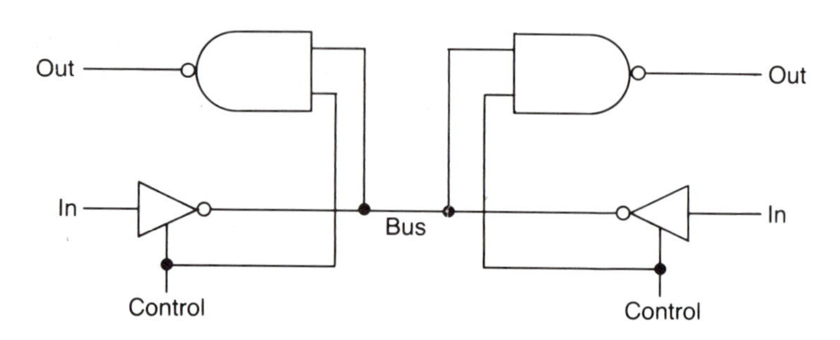

FIGURE 3.26 **Bidirectional Data Bus.**

3.6 Interfacing of Different Families

As long as the designer uses devices of the same family for making a complete system, no problems arise in their interconnection. When devices belonging to two or more families are used, however, the interfacing between them deserves special attention and often requires the inclusion of discrete bipolar transistors at the junctions. Interconnections between the TTL subfamilies require attention only to fan-out and fan-in.

Consider interfacing a TTL and a CMOS circuit when the TTL gate is driving *n* CMOS gates. From the compatibility standpoint a TTL can be interfaced with little trouble to 5 V CMOS. There is no problem at all when the TTL output is low. However, for a high TTL output the voltage is only 3.5 V. A pull-up resistor is required at the TTL output to raise the voltage to 5 V for reliable operation. CMOS devices source very little current, so the resulting fan-out from TTL to CMOS is not a problem.

The proposed arrangement of a TTL device driving *n* number of CMOS gates requires the following constraints:

$$V_{OL(TTL)} \leq V_{IL(CMOS)}$$
$$I_{OL(TTL)} \leq nI_{IL(CMOS)}$$
$$V_{OH(TTL)} > V_{IH(CMOS)}$$
$$I_{OH(TTL)} > nI_{IH(CMOS)}$$

Similar requirements must be satisfied whenever a device of one family is driving one or several devices of another family. Manufacturers' data books often provide information that is helpful in interfacing devices of different families.

3.7 Summary

In this chapter integrated circuits were introduced and several IC families were investigated. As a particular example, the TTL family and the characteristics of its various subfamilies were thoroughly studied. The three basic types of circuitry—namely, open-collector, totem-pole, and tri-state devices—were examined in detail. In addition open-inputs, wire-ANDing, and expanders were introduced and examples of their uses were discussed.

Problems

1. How does a TTL gate respond when the output of a disabled tri-state gate is tied to a TTL input? Explain.
2. What is the reason for not connecting an open-collector gate to a totem-pole gate?
3. Show how you would reduce fan-in for the following logic gates:
   a. AND and NAND
   b. OR and NOR
   c. X-OR and X-NOR
4. Show the scheme for increasing the fan-in of the following logic gates:
   a. AND and NAND
   b. X-OR and X-NOR
5. What is the switching mode of the TTL devices? Discuss their switching mode.
6. What is the switching mode of the ECL devices? Discuss their switching modes.
7. What will result if the OR outputs of ECL devices are wire-ANDed?

Suggested Readings

Chakrabarti, K., and Kolp, O. "Fan-in constrained tree networks of flexible cell." *IEEE Trans. Comp.* vol. C-23 (1974): 1238.

Danielsson, P. E. "A note on wired-OR gates." *IEEE Trans. Comp.* vol. C-19 (1970): 849.

Dao, T. T.; McCluskey, E. J.; and Russell, L. K. "Multivalued integrated injection logic." *IEEE Trans. Comp.* vol. C-26 (1977): 1233.

Fleming, D. "Enhancement of modular design capability by using tri-state logic." *Comp. Des.* vol. 10 (June 1971): 59.

Greenfield, J. W. *Practical Digital Design Using ICs.* 2d ed. New York: Wiley, 1983.

Holton, W. C. "The large-scale integration of microelectronic circuits." *Sci. Am.* vol. 237 (September 1977): 82.

Hurst, S. L. "Multiple-valued logic—its status and its future." *IEEE Trans. Comp.* vol. C-33 (1984): 1160.

Kodandapani, K. L., and Seth, S. C. "On combinational networks with restricted fan-out." *IEEE Trans. Comp.* vol. C-27 (1978): 256.

Liu, T. K. "Synthesis algorithms for two-level MOS networks." *IEEE Trans. Comp.* vol. C-24 (1975): 72.

McCluskey, E. J. "Logic design of multivalued I^2L logic circuits." *IEEE Trans. Comp.* vol. C-28 (1979): 546.

Meindl, J. D. "Microelectronic circuit elements." *Sci. Am.* vol. 237 (September 1977): 70.

Mori, R. D. "Suggestion for an IC fast parallel multiplier." *Elect. Lett.* vol. 5 (1969): 50.

Rosenberg, A. L. "Three-dimensional VLSI: a case study." *J. ACM.* vol. 30 (1983): 397.

Ruehli, A. E., and Diflow, G. S. "Circuit analysis, logic simulation, and design verification for VLSI." *Proc. IEEE.* vol. 71 (1983): 34.

Sheets, J. "Three-state switching brings wired-OR to TTL." *Electronics.* vol. 43 (September 1970): 78.

Taub, D. M. "Overcoming the effects of spurious pulses on wired-OR lines in computer bus systems." *Elect. Lett.* vol. 19 (1983): 340.

Taub, D. M. "Limitations of looped-line scheme for overcoming wired-OR glitch effects." *Elect. Lett.* vol. 19 (1983): 579.

CHAPTER FOUR

Implementation of Logic Functions

4.1 Introduction

The design mechanism of combinational logic circuits is usually a multi-step process. The realization, and the subsequent minimization, of the logic function is not the end of the design. We are already familiar with the various schemes for coming up with the reduced logic function either in the SOP or in the POS format. These forms can be translated easily into either a familiar AND-OR or OR-AND pattern of logic circuits. However, we have also seen in the last chapter that digital ICs have several practical limitations that may affect the implementation of circuits. These include the fan-in and fan-out limitations and the fact that ICs are more frequently available in the NAND and NOR form than in the AND and OR form. NAND and NOR gates are easier to realize with electronic components and are, therefore, the basic ingredients used in all of the logic families. Consequently, it is important for the designer to be familiar with the techniques for translating the reduced function so that either NAND gates or NOR gates may be used.

Combinational circuits may be realized using a standardized combinational unit called a multiplexer (MUX). In the MUX some of the input variables are used as input selectors for the unit and the remaining variables are entered as data inputs. Two other devices, read-only memories (ROMs) and programmable logic arrays (PLAs), are also frequently used to implement combinational networks. This chapter will explore the possibilities of using only one type of gate or one of the modules–MUX, ROM, PAL, or PLA–for the realization of combinational circuits. Such exploration is extremely useful because most designers usually choose to use only one type of basic gate unless there is a particular reason for doing otherwise. After studying this chapter, you should be able to:

- Design combinational circuits using only NAND gates;

- Design combinational circuits using only NOR gates;
- Design combinational circuits using single- or multi-level multiplexers (MUXs);
- Design combinational circuits using read-only memories (ROMs);
- Design combinational circuits using programmable logic arrays (PLAs);
- Construct a complex circuit using the outputs of a known functional unit.

4.2 Universal Logic Elements

In practice many logic circuits are built using only NAND and NOR gates because the basic gates in some of the logic families such as TTL and CMOS are NAND and NOR, respectively. NAND and NOR gates are considered universal logic elements since they both can be easily manipulated to obtain all possible logic functions. This simplification follows directly from Boolean theorems that we have discussed in the earlier chapters.

A close inspection of the truth table of these two functions, as described in Section 1.7, reveals that the NAND and NOR operators are duals of each other. Recall also from Chapter 1 that the dual of a Boolean expression is obtained by replacing every OR with AND, AND with OR, 0 with 1, and 1 with 0. Six of these dual properties are listed as follows:

NAND	NOR
1. $\overline{a \cdot 0} = 1$	$\overline{a + 1} = 0$
2. $\overline{a \cdot 1} = \bar{a}$	$\overline{a + 0} = \bar{a}$
3. $\overline{a \cdot a} = \bar{a}$	$\overline{a + a} = \bar{a}$
4. $\overline{a \cdot b} = \bar{a} + \bar{b}$	$\overline{a + b} = \bar{a} \cdot \bar{b}$
5. $\overline{\bar{a} \cdot \bar{b}} = a + b$	$\overline{\bar{a} + \bar{b}} = a \cdot b$
6. $\overline{\overline{a \cdot b}} = a \cdot b$	$\overline{\overline{a + b}} = a + b$

All of the logic functions may be generated using these properties of NAND and NOR logic. The corresponding NAND and NOR logic circuits for various functions are shown in Figure 4.1.

The circuits of Figure 4.1 show how NAND and NOR gates may be cascaded to form each of the logic functions NOT, AND, OR, and X-OR. Since either NANDs or NORs may be used to implement all of the logic operations, designers may prefer to use only NANDs or only NORs in order to decrease the inventory of spare parts. One of the methods by which to realize this is the *brute force*

FIGURE 4.1 **Logic Functions Using NANDs and NORs.**

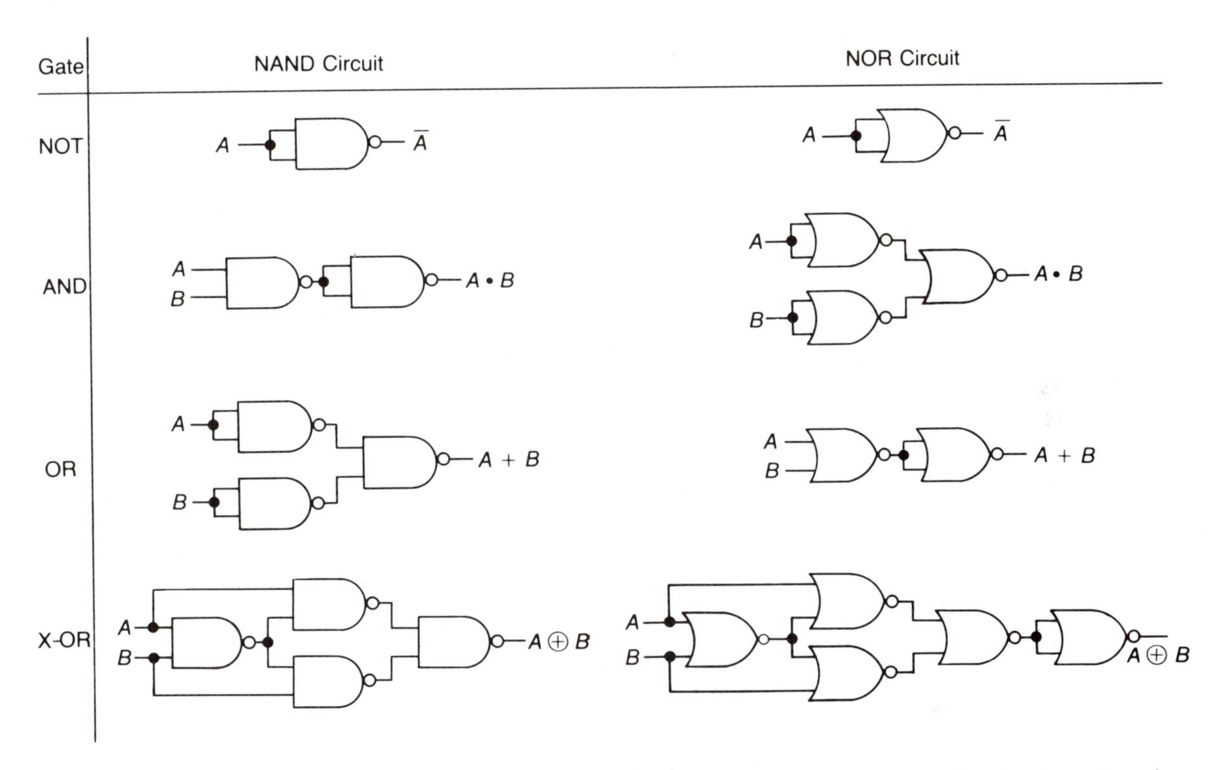

scheme, where each of the logic operations of the Boolean function is replaced by the corresponding NAND/NOR circuit. Note, however, that restricting the number of inputs to the gates will not cause any major problem if proper use of the involution and DeMorgan's laws are made.

EXAMPLE 4.1

Implement $\overline{A \oplus B}$ using only NAND gates.

SOLUTION

There are two different ways to implement this function: (a) using NAND equivalents of an X-OR gate and of a NOT gate, and (b) using the NAND equivalents of either the SOP or the POS terms.

a. The first possibility results in the circuit of Figure 4.2.

b. Otherwise, $\overline{A \oplus B}$ can be expressed in the SOP form as $AB + \overline{A}\overline{B}$. Consequently, the circuit appears as in Figure 4.3. The circuit of Figure 4.3 can be reduced further since $X = \overline{\overline{X}}$. Therefore, the circuit reduces to that of Figure 4.4. Note also that the function could be expressed in the POS form. Consequently, $\overline{A \oplus B} = (A + \overline{B})(\overline{A} + B)$, which leads to another variation of an X-NOR circuit as shown in Figure 4.5. The circuit of Figure 4.5

FIGURE 4.2

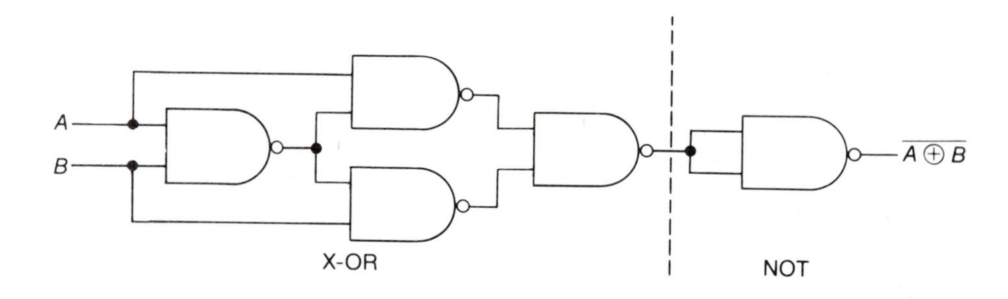

FIGURE 4.3

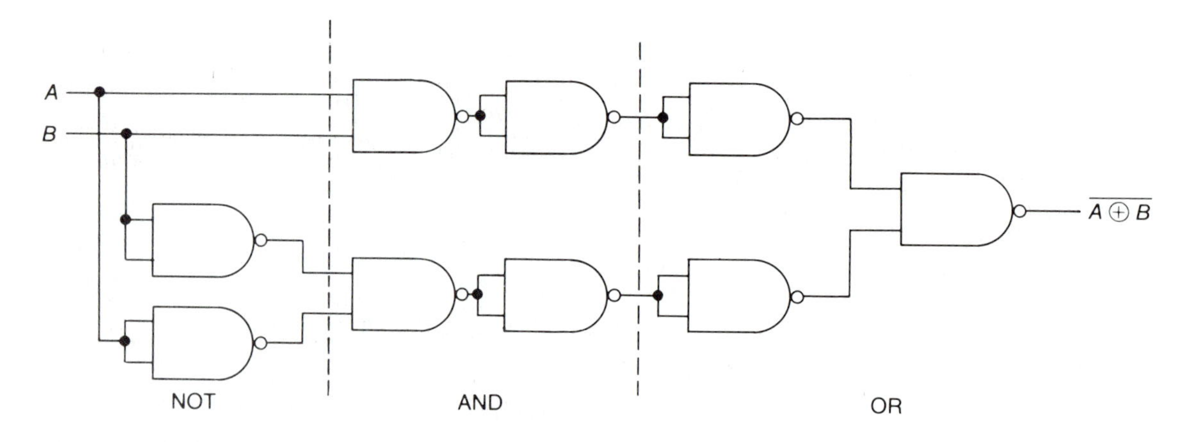

FIGURE 4.4

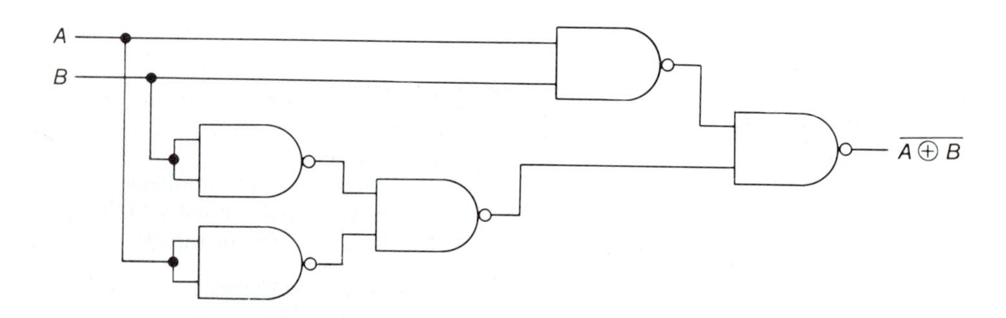

may be reduced further by making use of the law of involution. The resulting circuit is shown in Figure 4.6.

The first and the third forms, as shown in Figures 4.2 and 4.4, require five NAND gates each, and the fifth, as shown in Figure 4.6, requires a total of six NAND gates.

FIGURE 4.5

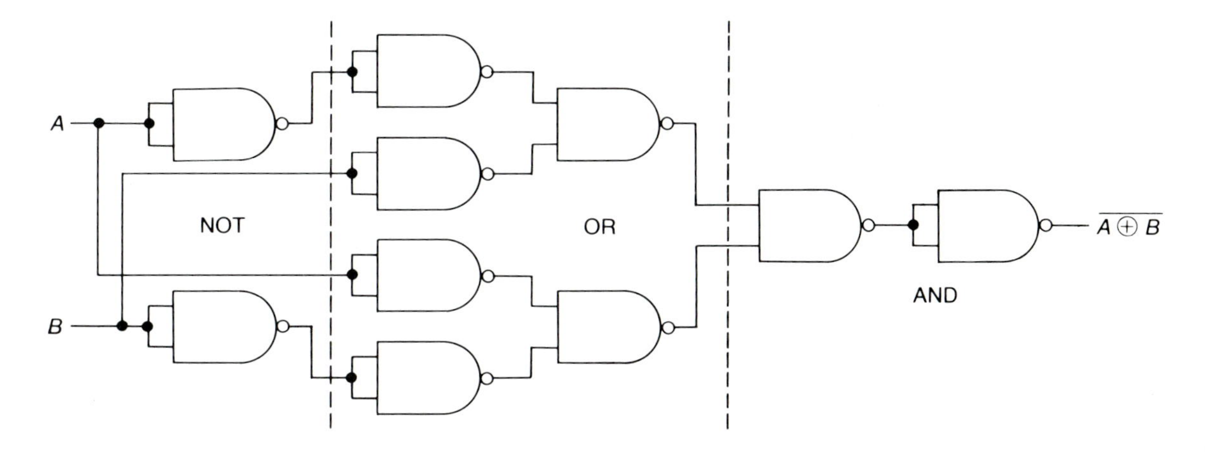

FIGURE 4.6

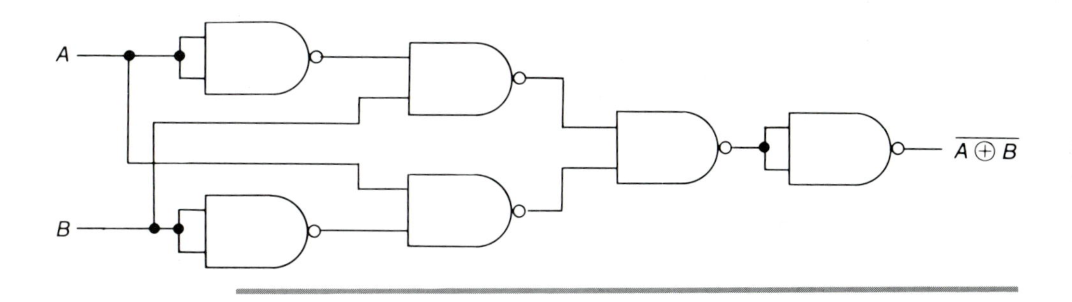

4.3 Function Implementation Using NANDs

It is quite easy to realize any SOP function using two levels of NAND gates. This method makes use of the fact that complementing a function twice returns the function to its original form. This result is achieved in two steps:

1. The function is complemented by complementing the ANDed terms and replacing the OR signs with AND signs.
2. The original function is then recovered by complementing the complement function.

It is not necessary to perform this operation each time a NAND realization is required. SOP forms always assume the same two-level NAND form.

The output of a NAND gate is also equivalent to the ORed output of the complements of the input. This statement follows directly from DeMorgan's theorem. Consequently, we are led to the following set of rules for obtaining the output function of a multi-level NAND circuit:

Rule 1. Consider the gate from which the output signal is derived as the first level, the preceding gate as the second level, and so on.

Rule 2. In odd-numbered levels the NAND gates perform OR operations. All ungated input variables entering the odd-level NAND gates will appear complemented in the final expression.

Rule 3. In even-numbered levels the NAND gates perform AND operations. All input variables entering the even-level NAND gates will appear uncomplemented in the final expression.

EXAMPLE 4.2

Using only NAND gates, implement the function given by

$$f(A,B,C,D) = ABC + A\overline{B}\overline{C} + \overline{C}D$$

SOLUTION

$$f = ABC + A\overline{B}\overline{C} + \overline{C}D$$

which yields

$$f = \overline{\overline{ABC} \cdot \overline{A\overline{B}\overline{C}} \cdot \overline{\overline{C}D}}$$

The final circuit, therefore, is obtained as shown in Figure 4.7. The circuit requires three three-input NAND gates and one two-input NAND gate provided B and C inputs are also available in the complemented form.

FIGURE 4.7

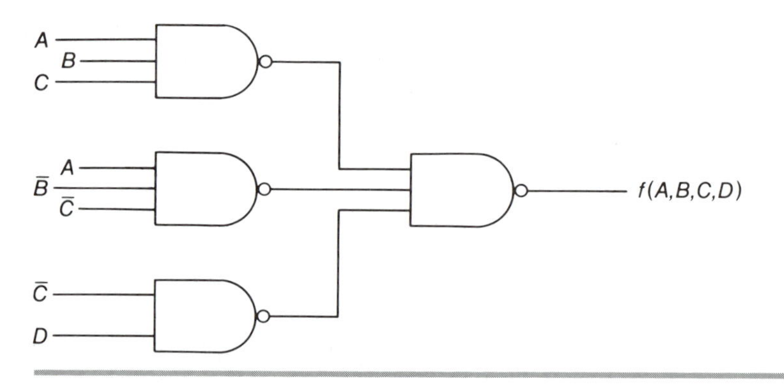

4.4 Function Implementation Using NORs

The implementation of an SOP function using only NOR gates is possible only if the function is first converted to the equivalent POS form. The process includes the following steps:

1. Plot the function on a K-map and obtain the complemented function by grouping all zeros.

2. Expand each of the ANDed terms by using DeMorgan's theorem.
3. Complement the whole Boolean expression.

NOR realizations of SOP functions always have the same two-level structure. Steps 1 through 3 should be followed until the designer is confident of the result.

DeMorgan's theorem may be used to interpret the NOR operation as well. The output of the NOR gate is equivalent to the ANDed output of the complements of the inputs. Rules for the interpretation of multi-level NOR circuits are listed as follows:

Rule 1. Consider the gate from which the output signal is derived as the first level, the preceding gate as the second level, and so on.

Rule 2. In odd-numbered levels the NOR gates perform AND operations. All ungated input variables entering the odd-level NOR gates will appear complemented in the final expression.

Rule 3. In even-numbered levels the NOR gates perform OR operations. Input variables entering the even-level NOR gates will appear uncomplemented in the final Boolean expression.

EXAMPLE 4.3

Using only NOR gates, implement the function given by

$$f(A,B,C,D) = \Sigma m(0,2,4,5,8,10,13)$$

SOLUTION

The minterms are plotted in a four-variable K-map and the corresponding zeros are grouped as shown in Figure 4.8. This gives

$$\begin{aligned}\bar{f} &= CD + BC + \bar{B}D + AB\bar{D} \\ &= \overline{\bar{C} + \bar{D}} + \overline{\bar{B} + \bar{C}} + \overline{B + \bar{D}} + \overline{\bar{A} + \bar{B} + D}\end{aligned}$$

Therefore,

$$f = \overline{\overline{\bar{C} + \bar{D}} + \overline{\bar{B} + \bar{C}} + \overline{B + \bar{D}} + \overline{\bar{A} + \bar{B} + D}}$$

FIGURE 4.8

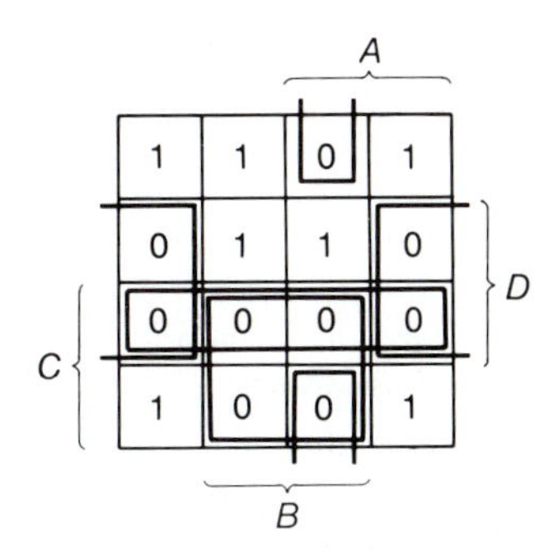

The equivalent NOR circuit, therefore, may be obtained as shown in Figure 4.9. It requires three two-input NOR gates, one three-input NOR gate, and one four-input NOR gate provided the inputs are also available in the complemented form.

FIGURE 4.9

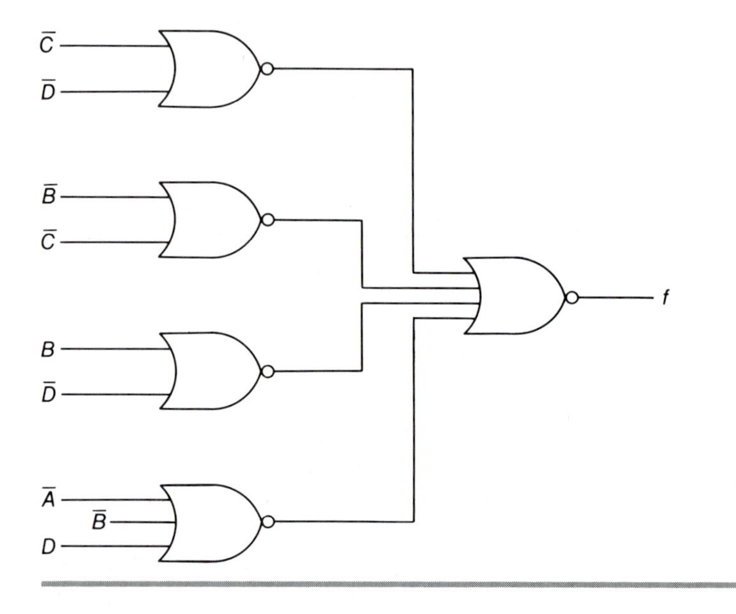

EXAMPLE 4.4

Write the output expression for the multi-level circuit shown in Figure 4.10.

SOLUTION

FIGURE 4.10

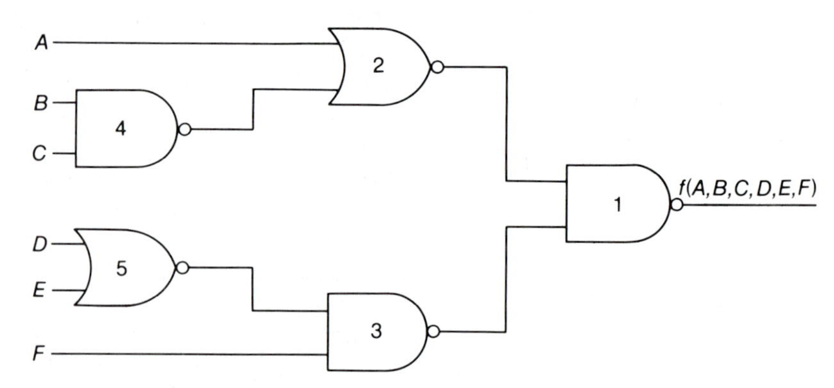

This is a three-level circuit. Therefore, the involved operations may be summarized as follows:

a. Gate 1 performs the OR operation;

b. Gate 2 performs the OR operation, and A must appear uncomplemented;

c. Gate 3 performs the AND operation, and F must appear uncomplemented;

d. Gate 4 performs the OR operation, and both B and C must appear complemented;

e. Gate 5 performs the AND operation, and both D and E must appear complemented.

Therefore,

$$f(A,B,C,D) = [(\bar{D} \cdot \bar{E}) \cdot F] + [A + (\bar{B} + \bar{C})] = \bar{D}\bar{E}F + A + \bar{B} + \bar{C}$$

Function output expressions may also be determined by starting at the inputs and making repeated use of DeMorgan's theorem. This latter technique is best when there is a mix of NOR and NAND gates in the circuit.

4.5 Function Implementation Using MUXs

A *multiplexer* (*MUX*), known also as a *data selector*, is a combinational network that has up to 2^n data inputs, n control inputs, and an output line. Commercial MUXs are limited to values of n of 1 through 4. An additional input is available that allows cascading of multiplexers to obtain higher-order devices. The MUX allows the selection of one of the 2^n data inputs as the device output. This selection is made by the control lines. A block diagram of a MUX with eight data input lines, D_0, D_1, D_2, D_3, D_4, D_5, D_6, and D_7, is shown in Figure 4.11[*a*]. Most MUXs are provided with at least two additional lines: $\bar{f}$ for the complemented output and E for enabling the device. The internal circuit configuration of the corresponding MUX is shown in Figure 4.11[*b*]. For every 2^n inputs the MUX has exactly n control lines. By applying appropriate signals to the control lines, any one of the data lines may be selected. For example, when $I_2I_1I_0 = 011$, the D_3 input is routed to the output provided $E = 0$. Note that whenever $E = 1$, the MUX is completely disabled; that is, regardless of the control variables or the data inputs, the output is 0. The enable input allows several of these devices to be cascaded together.

A MUX with 2^n input lines and n selection lines (such a device is also referred to as a 1-of-2^n MUX) may be wired to realize any Boolean function of $n + 1$ variables. This fact will be illustrated by implementing the function $f(A,B,C,D) = \Sigma m(0,2,4,5,6,8,10,13)$ using a 1-of-8 MUX. A 1-of-8 MUX has three control inputs where all but any one of the input variables may be entered. For example, we may consider A, B, and C inputs as the three control inputs, I_2, I_1, and I_0, respectively. The technique, therefore, consists of determining the function output in terms of the fourth input, D, for every possible combination of the control inputs. The values so obtained are then entered at the respective data inputs of the MUX. The truth table for the said function may then be reorganized as shown

FIGURE 4.11 **Eight-Input MUX: [*a*] Block Diagram and [*b*] Circuit.**

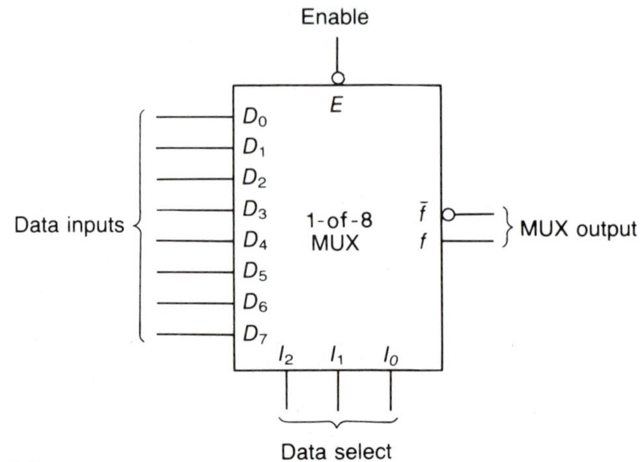

[*a*]

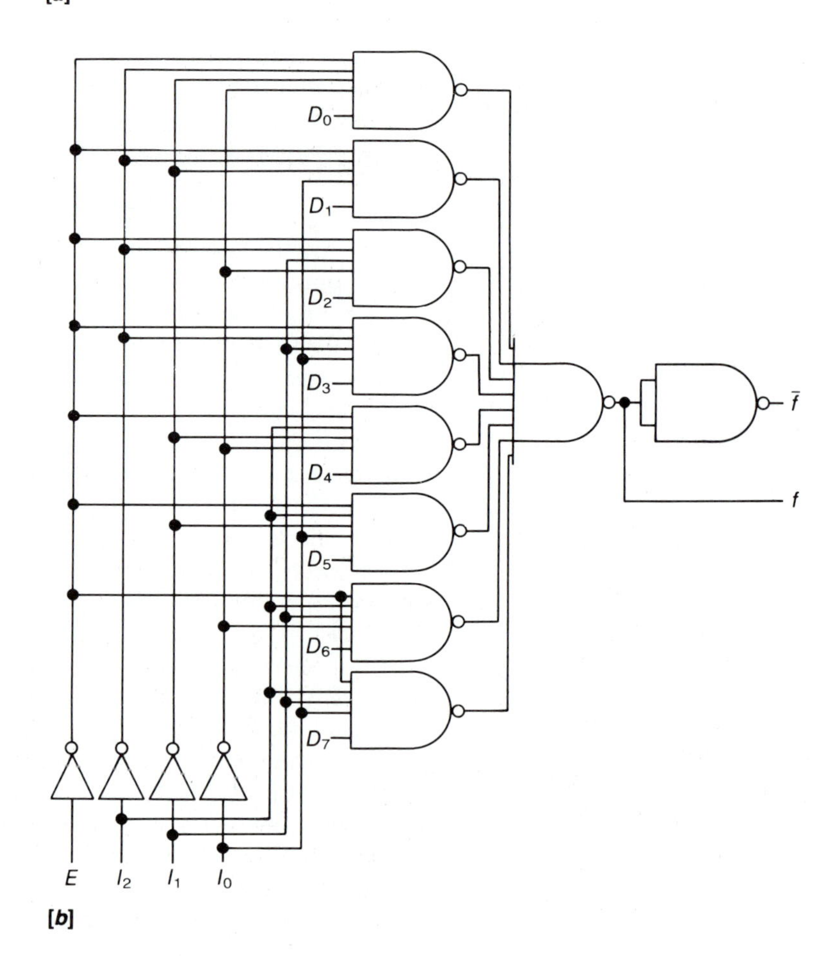

[*b*]

in Figure 4.12. The entries under the $f(D)$ column are determined by comparing the entries of the column D with that of the column f. For any combination of A, B, and C, the following are true:

1. If $f = 1$ irrespective of D, then $f(D) = 1$.
2. If $f = 0$ irrespective of D, then $f(D) = 0$.
3. If $f = x$ when $D = x$, then $f(D) = D$.
4. If $f = \bar{x}$ when $D = x$, then $f(D) = \bar{D}$.

FIGURE 4.12 Truth Table for $f(A,B,C,D) = \Sigma m(0,2,4,5,6,8,10,13)$.

A	B	C	D	f	f(D)
0	0	0	0	1	$\bar{D}$
			1	0	
0	0	1	0	1	$\bar{D}$
			1	0	
0	1	0	0	1	1
			1	1	
0	1	1	0	1	$\bar{D}$
			1	0	
1	0	0	0	1	$\bar{D}$
			1	0	
1	0	1	0	1	$\bar{D}$
			1	0	
1	1	0	0	0	D
			1	1	
1	1	1	0	0	0
			1	0	

The function, therefore, is implemented as shown in Figure 4.13[a]. The eight values of $f(D)$ are fed respectively into data inputs D_0 through D_7. Consequently, for any combination of the three control inputs, $f(D)$ would actually appear at the MUX output. The implementations of the same function for the other three combinations of the control inputs—B, C, and D; A, C, and D; and A, B, and D—can also be obtained in like manner. The corresponding MUX configurations are shown in Figures 4.13[b–d]. The designer might even change the order of the control inputs, resulting in a total of 24 different circuit configurations. In addition, if one of the control variables is available only in complemented form, the variable may be used without inverting and the inputs rearranged accordingly (see Problem 3b).

It is quite obvious that the use of MUXs provides the designer with numerous choices. Consequently, it is necessary to consider each of the solutions to determine which is the optimum. In Figure

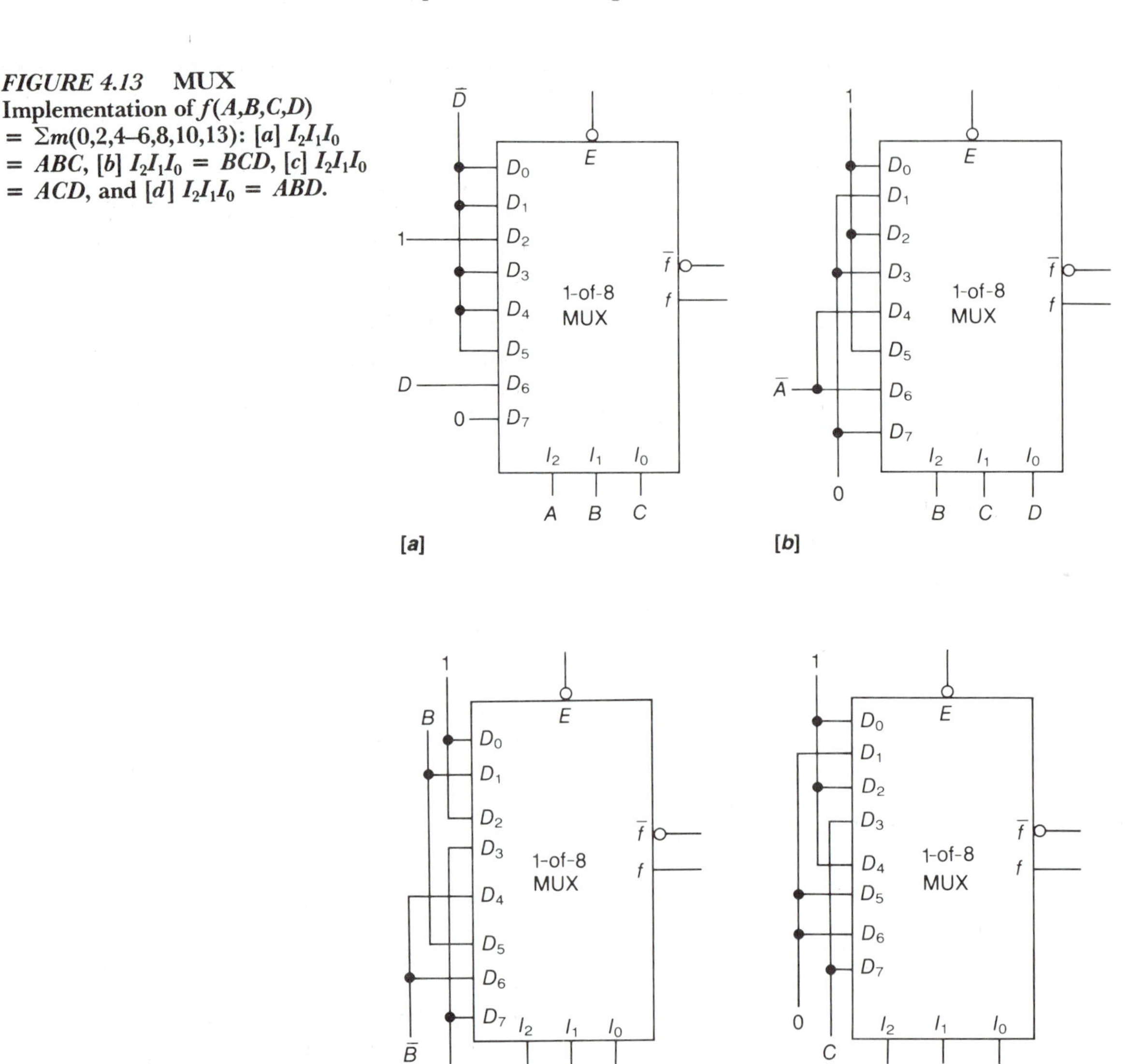

FIGURE 4.13 **MUX Implementation of** $f(A,B,C,D) = \Sigma m(0,2,4\text{–}6,8,10,13)$**:** **[*a*]** $I_2I_1I_0 = ABC$, **[*b*]** $I_2I_1I_0 = BCD$, **[*c*]** $I_2I_1I_0 = ACD$, **and [*d*]** $I_2I_1I_0 = ABD$.

4.13[*a*], $\bar{D}$ is seen to be tied to five of the data inputs. The gate that provides $\bar{D}$ must then have a fan-out of at least five. If we limit ourselves to the four choices of Figure 4.13, it is apparent that Figure 4.13[*d*] provides the most preferable circuit, because the C variable needs to be fed directly to only two of the data inputs. Example 4.5 illustrates the mechanism of obtaining a multi-level multiplexer circuit.

EXAMPLE 4.5

Implement the function

$f(A,B,C,D,) = \sum m(3,4,8\text{–}10,13\text{–}15)$

a. by using a 1-of-4 MUX and a few assorted gates,

b. by using 1-of-2 and 1-of-4 MUXs in two levels.

SOLUTION

a. The function may be expressed in the SOP form and then regrouped as

$$\begin{aligned} f(A,B,C,D) &= \bar{A}\bar{B}CD + \bar{A}B\bar{C}\bar{D} + A\bar{B}\bar{C}\bar{D} + A\bar{B}\bar{C}D + A\bar{B}C\bar{D} \\ &\quad + AB\bar{C}D + ABC\bar{D} + ABCD \\ &= \bar{A}\bar{B}(CD) + \bar{A}B(\bar{C}\bar{D}) + A\bar{B}(\bar{C}\bar{D} + \bar{C}D + C\bar{D}) \\ &\quad + AB(\bar{C}D + C\bar{D} + CD) \\ &= \bar{A}\bar{B}(CD) + \bar{A}B(\bar{C}\bar{D}) + A\bar{B}(\bar{C} + \bar{D}) + AB(C + D) \end{aligned}$$

Accordingly, the resultant circuit is obtained as shown in Figure 4.14.

FIGURE 4.14

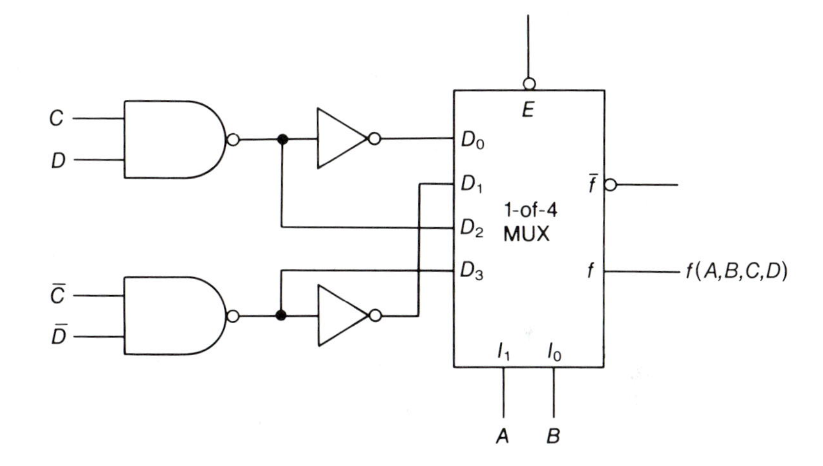

b. Also,

$$\begin{aligned} f(A,B,C,D) &= \bar{A}\bar{B}[\bar{C}(0) + C(D)] + \bar{A}B[\bar{C}(\bar{D}) + C(0)] \\ &\quad + A\bar{B}[\bar{C}(1) + C(\bar{D})] + AB[\bar{C}(D) + C(1)] \end{aligned}$$

This implies that a two-level MUX circuit would be able to generate this function. The first level of a 1-of-2 MUX essentially eliminates the need for discrete gates. The resultant circuit is obtained as shown in Figure 4.15. However it can be shown that

$$\overline{\bar{C}(0) + C(D)} = \bar{C}(1) + C(\bar{D})$$

and

$$\overline{\bar{C}(\bar{D}) + C(0)} = \bar{C}(D) + C(1)$$

Note also that the MUXs usually are provided with an additional output for providing the complemented result. Consequently, two of the first-level 1-of-2 MUXs may be removed. The resulting reduced multi-level MUX circuit is obtained as shown in Figure 4.16.

FIGURE 4.15

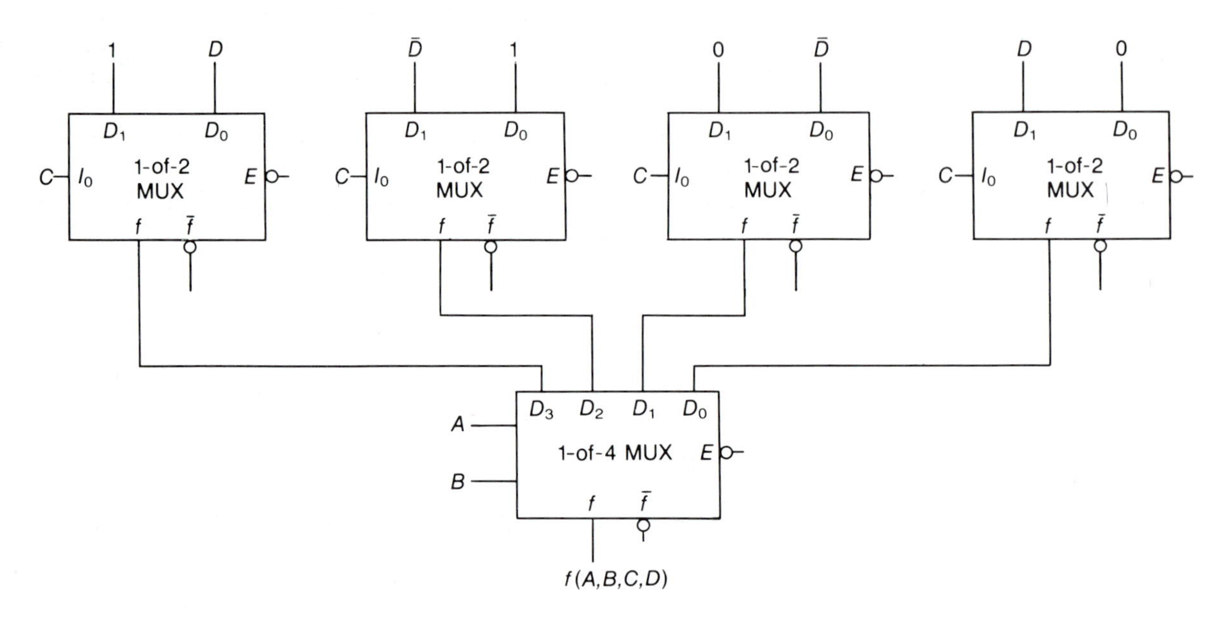

FIGURE 4.16

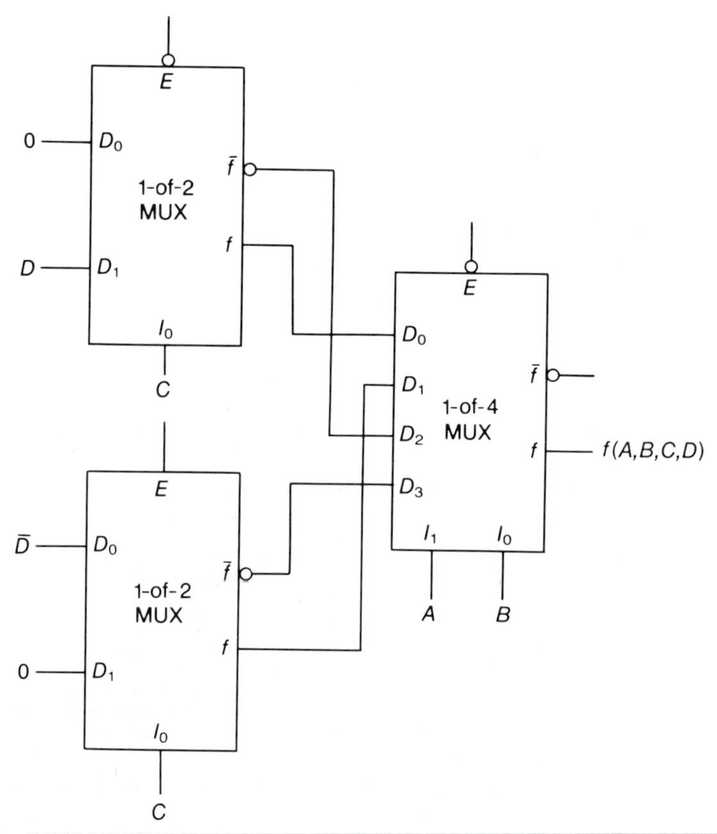

MUXs can also be applied in a more brute force manner. This technique involves 1-of-2^n MUXs for functions of n variables, while Example 4.5 used 1-of-2^{n-1} MUXs to implement a function. The function values from a function's truth table are transformed directly to the inputs of the MUX. The n variables are treated as control inputs to the MUX. Figure 4.17 shows the truth table for a three-variable function and the corresponding 1-of-8 MUX implementation. For two functions of the same variables, two MUXs may be used as shown in Figure 4.18.

FIGURE 4.17 **Realization of $f(A,B,C) = \Sigma m(1\text{–}3,5,6)$ Using 1-of-8 MUX.**

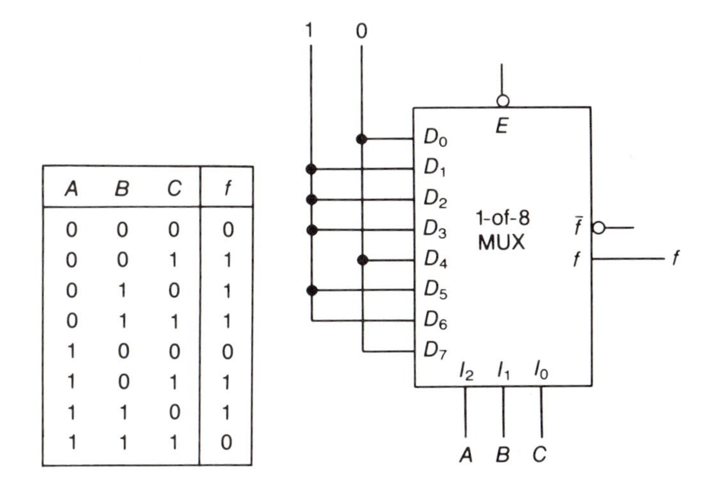

A	B	C	f
0	0	0	0
0	0	1	1
0	1	0	1
0	1	1	1
1	0	0	0
1	0	1	1
1	1	0	1
1	1	1	0

FIGURE 4.18 **Realization of $f_1(A,B,C) = \Sigma m(1,3,6,7)$ and $f_0(A,B,C) = \Sigma m(0,3\text{–}5)$ Using Two 1-of-8 MUXs.**

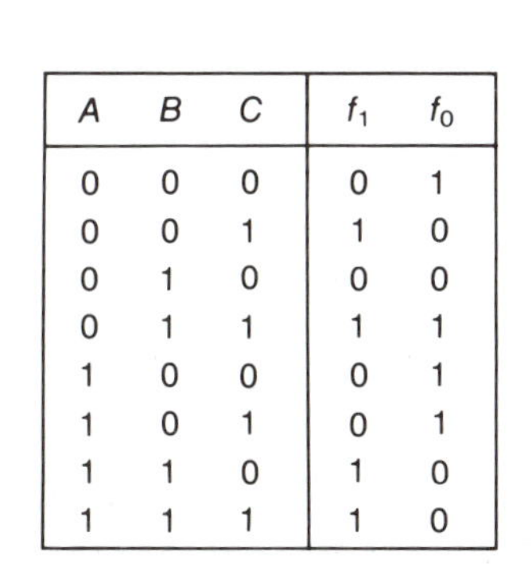

A	B	C	f_1	f_0
0	0	0	0	1
0	0	1	1	0
0	1	0	0	0
0	1	1	1	1
1	0	0	0	1
1	0	1	0	1
1	1	0	1	0
1	1	1	1	0

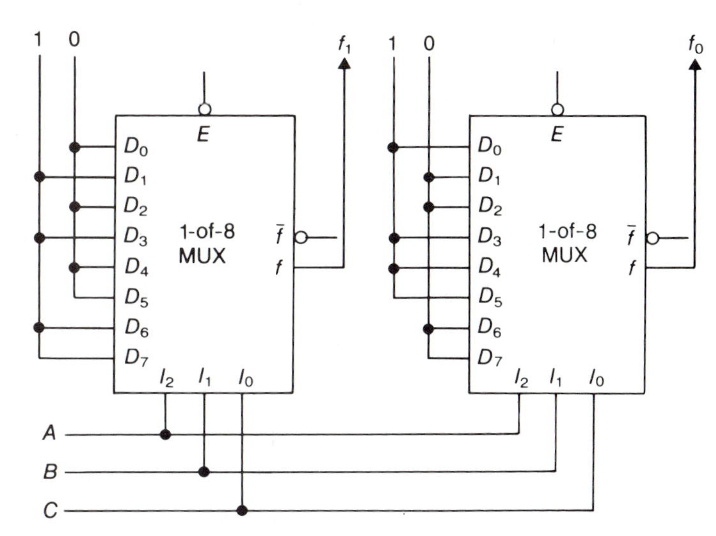

The technique just discussed is not as efficient as the one before in terms of MUX size, but it is certainly easy to implement and is a good way to introduce another device that may be used to implement combinational logic functions. If we put the circuit implementation of Figure 4.18 in a box, we could show the circuit symbolically as illustrated in Figure 4.19. Such a representation is called a *black-box* representation since the user is not required to know about the details of the internal logic. Not knowing what is in the box in Figure 4.19, we might explain the circuit action by simply saying that ABC forms an *address*, and f_1 and f_0 are what is stored there. This is precisely the explanation of the operation of a device with which we can implement multiple functions of a set of variables. This device is called a *read-only memory* (*ROM*). It can be considered as a set of storage cells with every cell having a value for each of the multiple functions. These ROMs can be programmed (1s and 0s applied to the inputs of each of the function's MUX inputs) by the manufacturer or by the designer if he or she has available blank ROM chips and an appropriate programmer. The information *stored* in the ROM remains there permanently. The ROM equivalent of the logic of Figure 4.18 is shown by the block diagram of Figure 4.19. When power is applied, f_1 and f_0 have the same values as before power was removed.

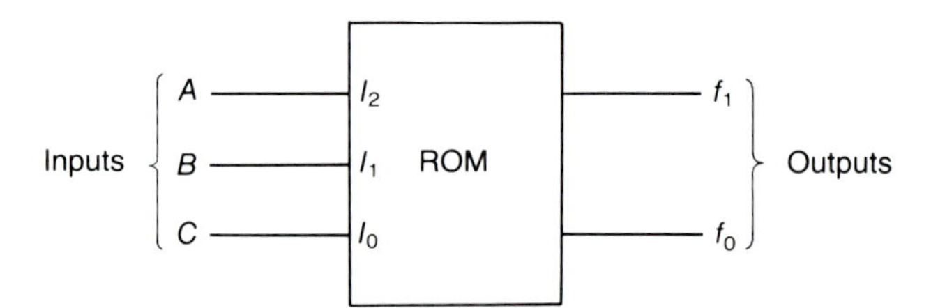

FIGURE 4.19 **Black-Box Representation of the Circuit of Figure 4.18.**

ROMs come in many sizes. The sizes are determined by the number of storage cells (corresponding to the number of MUX inputs) and the number of bits stored in each cell (the number of functions that can be implemented). In Figures 4.17 and 4.18 there are three variables that allow addressing 2^3 storage cells in our MUX-implemented ROM. In general, n variables would require 2^n storage cells and correspond to n address lines. Each storage cell would have one bit (1 or 0) for each function of the n variables.

Commercially available ROMs come in many sizes. They may be listed as 2K × 1 or 2K × 8, meaning 2048 storage locations one bit wide in the first case and 2048 storage locations eight bits wide in the second case. The first would allow implementing a function of up to 11 variables, the second up to eight functions of 11 variables. The designer selects a ROM of adequate size to implement the desired function.

The following section discusses ROMs in more detail. For most function generation applications, ROMs can be visualized as a set of MUXs—one for each function of the input variables (addresses).

EXAMPLE 4.6

Obtain a multi-bit shifter (see Example 2.4 for definition) that has an $(n + 2)$-bit input, x, an n-bit output, y, and three control inputs: s, d, and E. You may use only MUXs for the design.

SOLUTION

The characteristics of the shifter could be summarized as follows:

$$y_i = \begin{cases} x_{i-1} & \text{if } d = 1, s = 1, \text{ and } E = 1 \text{ (left-shift)} \\ x_{i+1} & \text{if } d = 0, s = 1, \text{ and } E = 1 \text{ (right-shift)} \\ x_i & \text{if } s = 0 \text{ and } E = 1 \text{ (no-shift)} \\ 0 & \text{if } E = 0 \end{cases}$$

where $0 \leq i \leq n - 1$.

A 1-of 4 MUX could be used corresponding to each bit of y. The variables s and d can be fed as its selectors, and E as the enable input. The inputs x_i, x_{i-1}, and x_{i+1} could be introduced at the MUX data inputs, and they could be suitably selected as the shifter output. The multi-bit shifter is obtained accordingly as shown in Figure 4.20. Whenever $E = 0$, the MUXs are disabled and the output becomes zero.

FIGURE 4.20

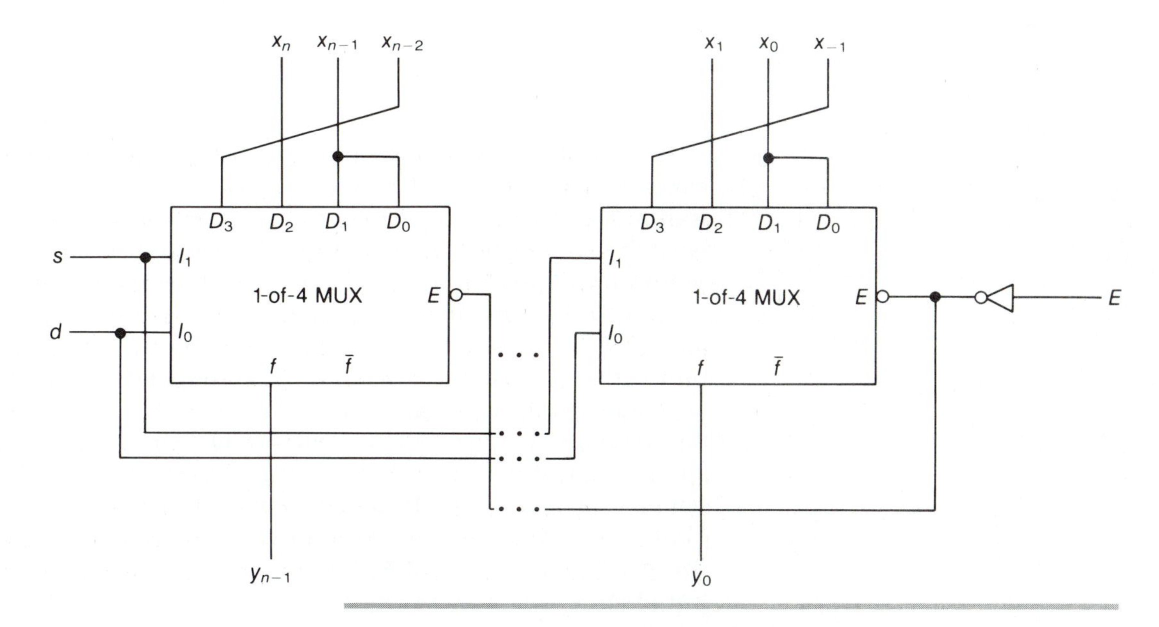

4.6 Function Implementation Using ROMs

A *read-only memory* (*ROM*), as the name implies, is intended to hold fixed information that can only be read, not altered. The primary use of the ROM is to provide a means for storing binary information. The storing is done during the fabrication of the ROM and may not be altered without undergoing a significantly involved process. The same is true for a combinational network that has been designed, fabricated, tested, and encapsulated with only the inputs and the outputs available. The ROM has become an important part of many digital systems because of the ease with which complex functions such as code conversion, program storage, and character generation can be implemented. The chip count of circuits, for which the access time of the ROM is not a restriction, may be greatly reduced by using ROMs.

A $2^m \times n$ ROM is an array of memory cells organized into 2^m words of n bits each, as shown by the block diagram of Figure 4.21. Such a ROM is accessed by means of m address lines and the stored information is retrieved via a total of n data-out lines, one for each bit of the word. The ROM corresponds to a combinational network with n outputs, where each of the outputs is associated with up to 2^m different minterms. A ROM may be provided with one or more chip-select lines to permit cascading smaller ROMs to form a ROM with more words (allowing implementation of functions of more variables).

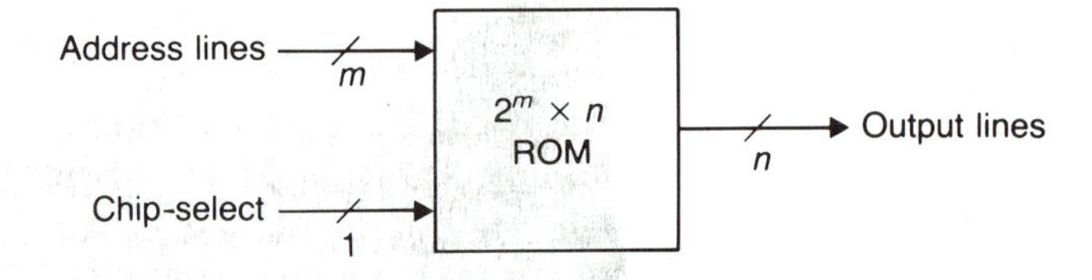

FIGURE 4.21 **Block Diagram of a ROM.**

As indicated in the previous section, a ROM is a combinational circuit. A ROM can be implemented by using only diodes, bipolar transistors, or MOS transistors. Although the diode matrix ROM no longer represents the current ROM technology, it serves as a simple model to show the basic concept. Figure 4.22 shows a simple diode ROM, where the row and column lines are interconnected via diodes placed at the respective intersections. The absence or presence of a diode indicates that the corresponding row and column intersection is programmed with a 1 or a 0. If the output is buffered with inverters, the converse is true. The output for the ith address depends on the ORing diodes connected to that line. For example, if $A_2A_1A_0 = 100$, the output will be $D_8D_7 \ldots D_1 = 01000001$. For $A_2A_1A_0 = 100$ input, a low is produced on the decoder output numbered 4. This low and the pull-up resistor forward biases each of the diodes connected to this row and pulls down

FIGURE 4.22 **8 × 8 Diode ROM.**

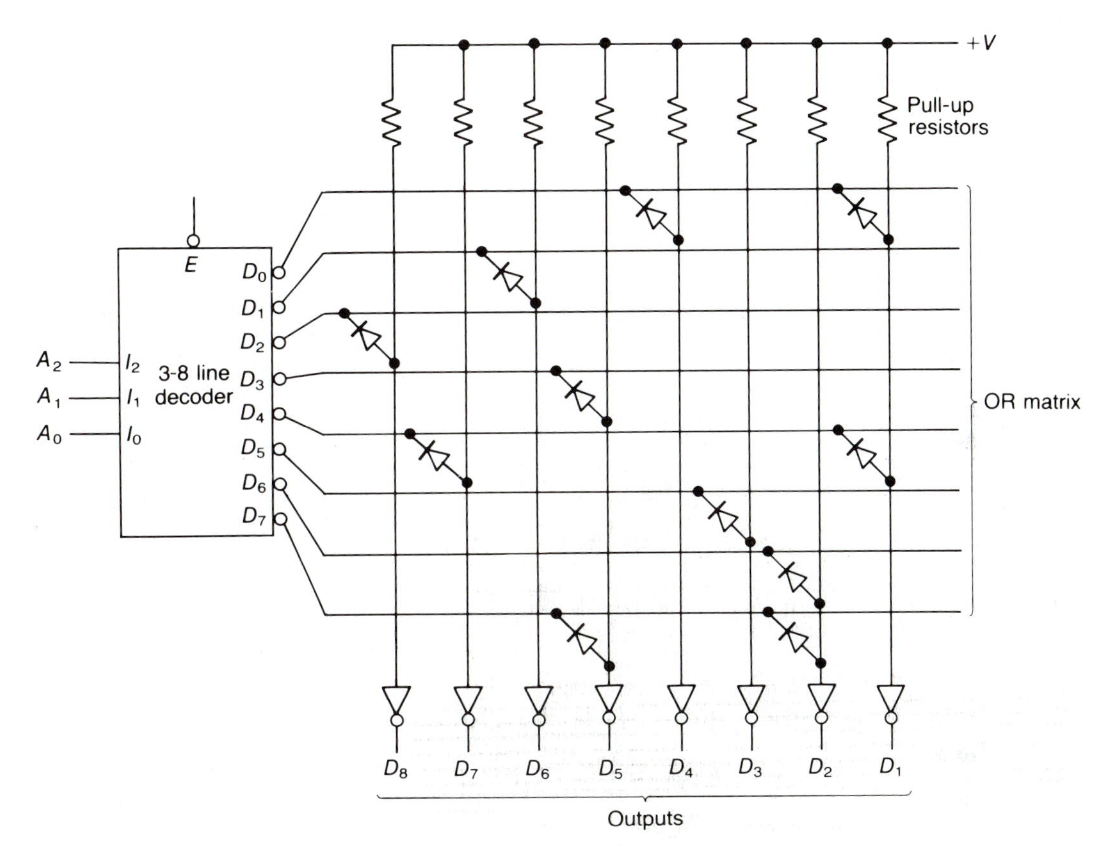

the corresponding column outputs to a low. In the absence of a diode, a high is maintained at the output due to the pull-up resistors. The capacity of a ROM usually is quoted as the number of possible intersections in the matrix, for example, $8 \times 8 = 64$ bits $= (1/16)$K bits in this case. A total of 1024 intersections is usually referred to as 1K bits. Examples 4.7 and 4.8 illustrate the use of diode ROMs in combinational problems.

EXAMPLE 4.7

Use ROM to realize the implementation of the integer function

$f(x) = x^3 \quad \text{for } 0 \leq x \leq 3$

SOLUTION

The function truth table is obtained as shown in Figure 4.23. A 2-4 line decoder would be sufficient to decode the numbers 0, 1, 2, and 3, and it is apparent that a maximum of five output lines are needed to represent the cube of the largest number. The two select lines, A_0 and A_1, can be used to select any one of these four outputs.

FIGURE 4.23

x	f(x)	f(x) in Binary
0	0	00000
1	1	00001
2	8	01000
3	27	11011

The resultant diode matrix implementation of x^3 ROM, therefore, is obtained as shown in Figure 4.24.

FIGURE 4.24

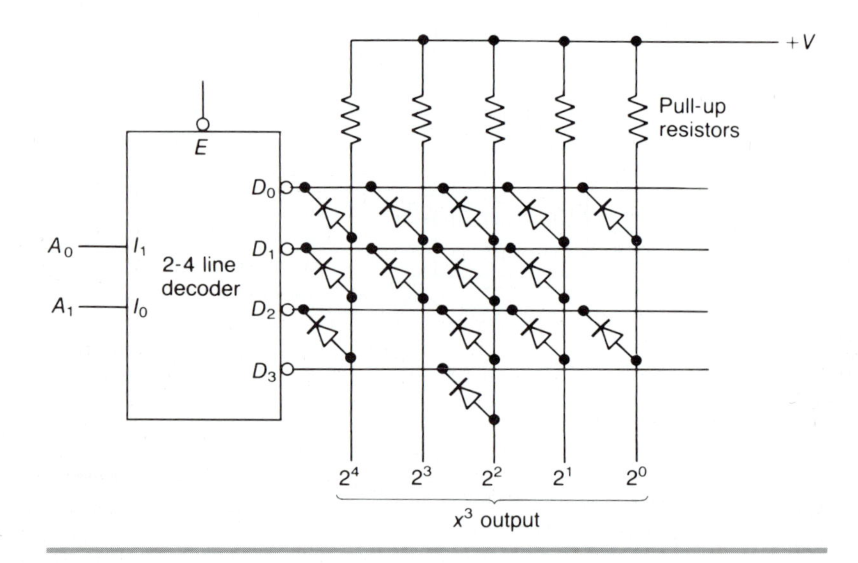

EXAMPLE 4.8

Using a minimal ROM implementation, design a seven-segment–to–BCD code converter. The seven-segment display device consists of seven LEDs, as shown in Figure 4.25, arranged in such a way that they could be used for displaying data.

SOLUTION

FIGURE 4.25

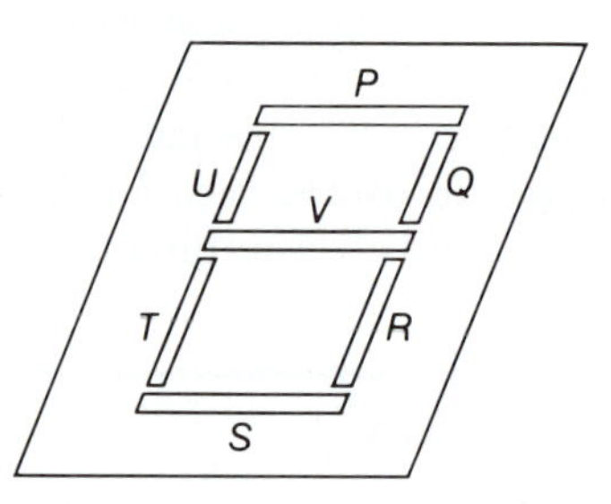

The seven-segment–to–BCD truth table is obtained as shown in Figure 4.26. The input entries are selected in such a way that the output would be displayed only if the corresponding LED segments are lighted. For example, when all of the seven segments are turned on, the display will be an 8.

FIGURE 4.26

Inputs							Outputs			
P	*Q*	*R*	*S*	*T*	*U*	*V*	*D*	*C*	*B*	*A*
1	1	1	1	1	1	0	0	0	0	0
0	1	1	0	0	0	0	0	0	0	1
1	1	0	1	1	0	1	0	0	1	0
1	1	1	1	0	0	1	0	0	1	1
0	1	1	0	0	1	1	0	1	0	0
1	0	1	1	0	1	1	0	1	0	1
0	0	1	1	1	1	1	0	1	1	0
1	1	1	0	0	0	0	0	1	1	1
1	1	1	1	1	1	1	1	0	0	0
1	1	1	0	0	1	1	1	0	0	1

Note, however, that there could be up to two choices for displaying a 1: either *U* and *T* or *Q* and *R*.

A direct implementation of this table requires a $2^7 \times 4 = 512$-bit ROM. Note that the partitioned table (with inputs *P, Q, R, S,* and *T*), as shown in Figure 4.27, would require only seven of the possible $2^5 = 32$ combinations. This situation suggests that a reasonable improvement is possible if the designer is willing to cascade at least two smaller ROMs. One of these ROMs should have at the least three outputs—*Z, Y,* and *X*—since $2^3 > 7$. The output of the first ROM—*Z, Y,* and *X*—can be fed along with the other two inputs—*U* and *V*—into the second ROM. However, care must be taken in organizing the second ROM so that its output becomes equivalent to that of Figure 4.26. Accordingly, the compressed truth table of Figure 4.28 is obtained such that it incorporates the same logic as that of Figure 4.26. The seven inputs have now been replaced by only five inputs, where the first three are functions of *P, Q, R, S,* and *T* and the other two are *U* and *V* themselves.

FIGURE 4.27

P	*Q*	*R*	*S*	*T*	*Z*	*Y*	*X*
1	1	1	1	1	0	0	0
0	1	1	0	0	0	0	1
1	1	0	1	1	0	1	0
1	1	1	1	0	0	1	1
1	0	1	1	0	1	0	0
0	0	1	1	1	1	0	1
1	1	1	0	0	1	1	0

Outputs *Z, Y,* and *X* are realizable using a 96-bit (3×2^5) ROM. A second ROM can be used where *Z, Y, X, U,* and *V* are the inputs. The second ROM size is $4 \times 2^5 = 128$ bits. Therefore, the total ROM size requirement is reduced to only $96 + 128 = 224$ bits. This solution reduces

FIGURE 4.28

Z	*Y*	*X*	*U*	*V*	*D*	*C*	*B*	*A*
0	0	0	1	0	0	0	0	0
0	0	1	0	0	0	0	0	1
0	1	0	0	1	0	0	1	0
0	1	1	0	1	0	0	1	1
0	0	1	1	1	0	1	0	0
1	0	0	1	1	0	1	0	1
1	0	1	1	1	0	1	1	0
1	1	0	0	0	0	1	1	1
0	0	0	1	1	1	0	0	0
1	1	0	1	1	1	0	0	1

the ROM size to about half of the original. The ROM implementation of the circuit, therefore, is given by the multi-level circuit of Figure 4.29.

FIGURE 4.29

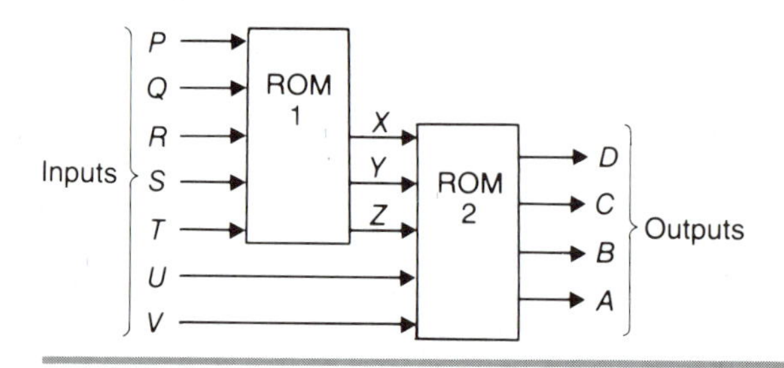

Although the diode matrix serves to demonstrate the ROM concept, ROMs are presently manufactured using bipolar and MOS transistors, as shown in Figure 4.30. The presence of a connection from a row line to either a transistor base or a MOSFET gate represents a logic 0, and the absence of such a connection represents a logic 1. For economical reasons MOS ROM is preferred to bipolar ROM for large numbers of bits. Access time for bipolar ROM, however, is much less than that of MOS ROM.

FIGURE 4.30 **[*a*] Bipolar $m \times n$ ROM and [*b*] MOS $m \times n$ ROM.**

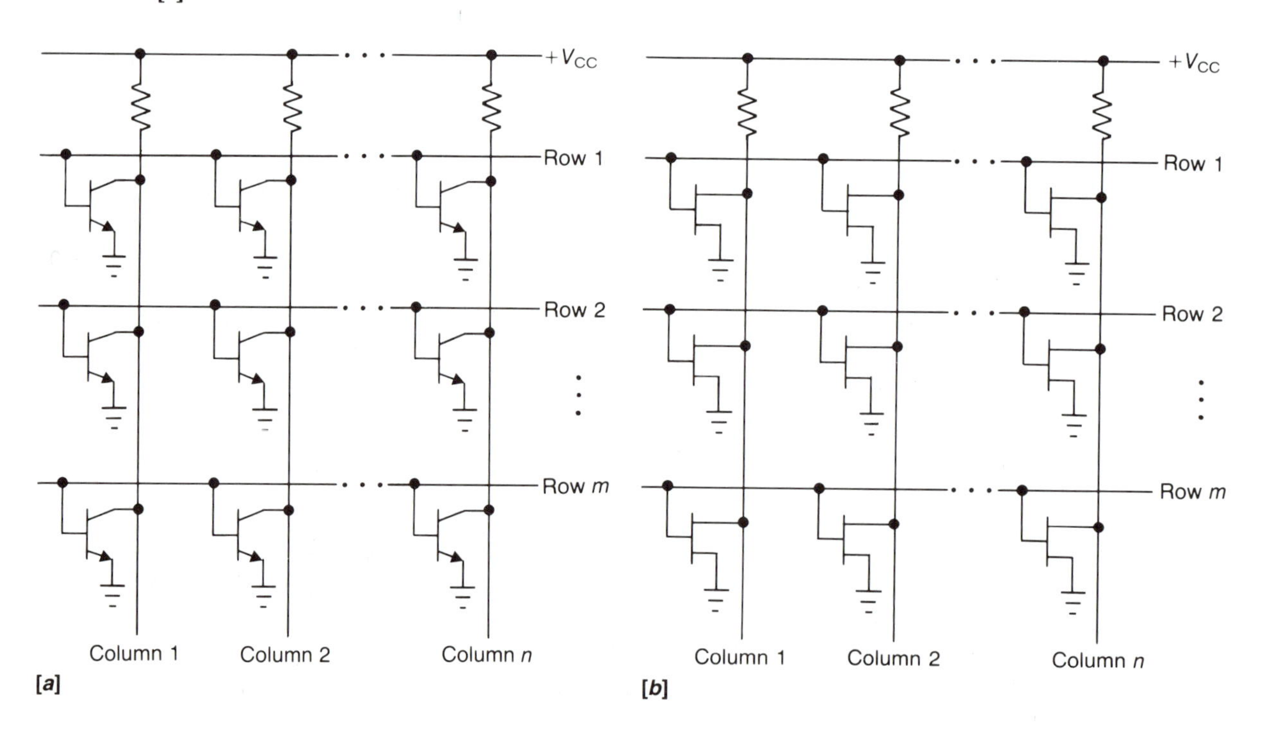

4.7 Function Implementation Using PLAs and PALs

Examples 4.7 and 4.8 illustrated how a ROM may be used to implement SOP logic expressions. There is another aspect of the ROM that deserves attention, however. A ROM consists of a level of AND gates, which constitute the decoder part, followed by a second level of OR gates (made up of diodes or transistors), which constitute the encoder section. A ROM may be thought of as a programmable array of logic gates. With this array of AND and OR gates, every combination of minterms of the input variables (addresses) can be formed. This flexibility is costly in that, when implementing complex functions, not all minterms are necessary to realize a given expression. For example, a ROM that processes 12 variables requires a total of 4K byte (eight bits are called a *byte*) memory. For example, such an arrangement is needed for the Hollerith code conversion circuit that has up to 12 input variables, but has only 96 eight-bit output combinations of these variables. This situation implies that 4000 out of 4096 bytes will remain unused. Such waste can be eliminated by the use of a *programmable logic array* (*PLA*).

A PLA consists of an array of AND-OR logic along with *inverters* that may be programmed to realize the desired output. In essence a PLA may be regarded as being made up of two separate ROMs: an AND ROM and an OR ROM. A typical PLA configuration is shown in Figure 4.31 in a 12 × 32 × 8 format. The circuit consists

FIGURE 4.31 **12 × 32 × 8 PLA Using Diodes.**

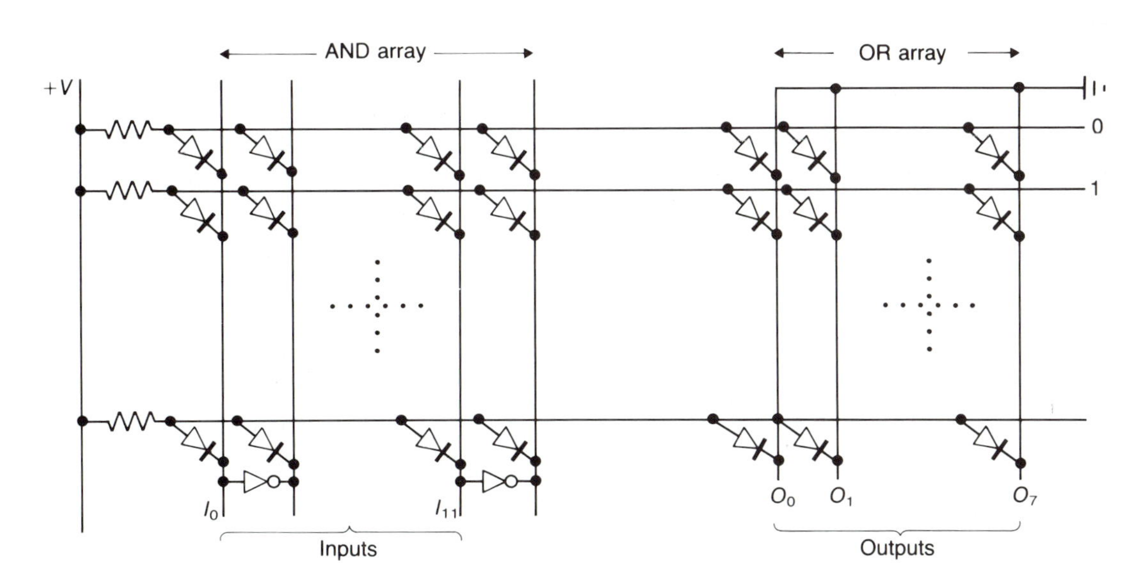

of an initial AND array, which can implement any one or more of the 32 product terms of up to 12 variables. The inclusion of an inverter with each input variable allows any minterm to be formed.

A PLA may be used as a Boolean function generator in much the same way as a ROM. As a simple example, Figure 4.32 shows a small PLA layout with six inputs, twelve product terms, and four outputs. The dots in the matrix of the top section can be thought of as AND inputs and those on the bottom part can be interpreted as OR inputs to generate the outputs. The output functions are easily determined as follows:

$$O_1 = \overline{B}D + D + \overline{A}\overline{D} + \overline{B}CF = \overline{A} + D + \overline{B}CF$$
$$O_2 = BD\overline{F} + \overline{D}E + CE + \overline{C}DE\overline{F}$$
$$O_3 = \overline{C}D + AE + \overline{D}E$$
$$O_4 = \overline{C}D + \overline{B}\overline{D}F$$

FIGURE 4.32 **PLA Implementation.**

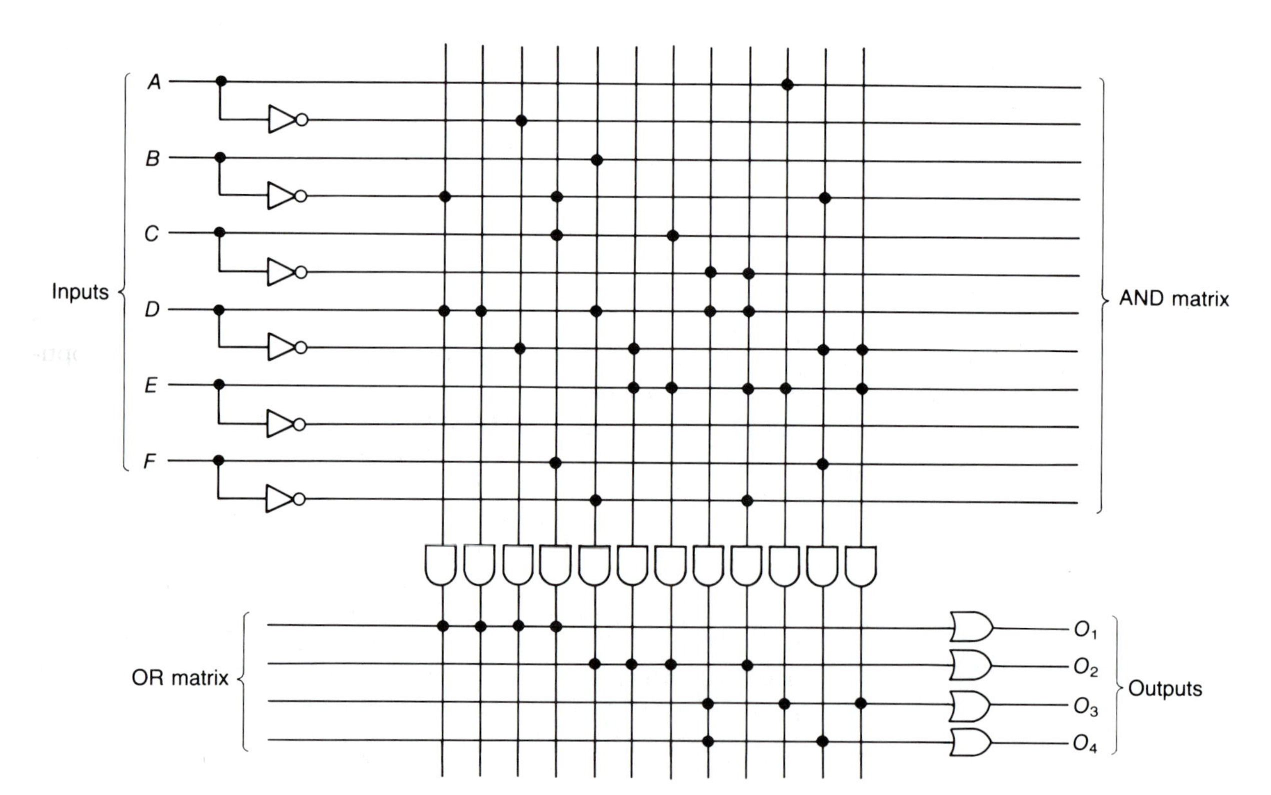

Larger numbers of product terms and/or outputs may be obtained when more than one PLA is cascaded. Some of these expansion schemes are shown in Figure 4.33. The product terms essentially are increased by tying the outputs in parallel. This configuration resembles the wiring of open-collector outputs. Correspondingly, an increase in the word size could be accomplished by unhooking the outputs.

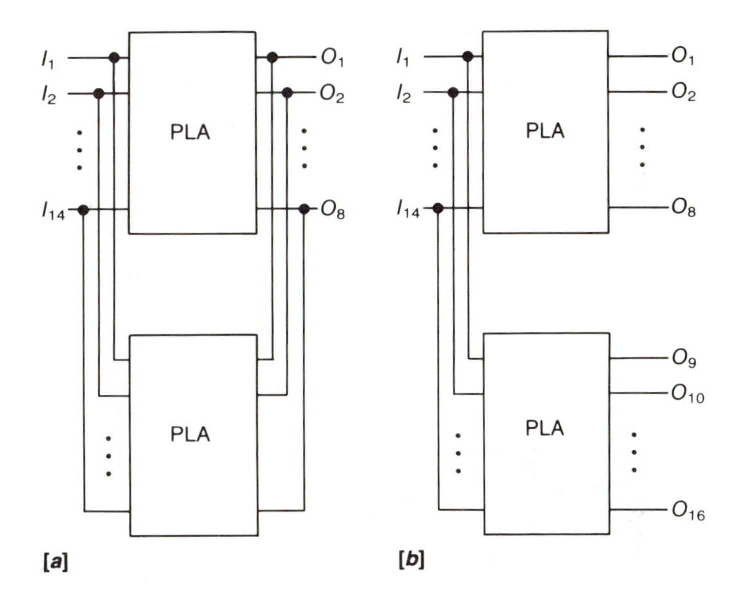

FIGURE 4.33 **PLA Expansion Scheme: [*a*] Product Term Expansion and [*b*] Output Expansion.**

The designer must use care in choosing the minterms to be formed in the AND section of the PLA. In order to use PLAs optimally, it is necessary to have as many output functions as possible that have common minterms. It is not necessary for each function to be minimized; the goal is to minimize the total number of minterms required to implement the set of functions.

It would be appropriate now to discuss an additional programmable device known as *programmable array logic* (*PAL*). It also allows the systems engineer to design his or her "own chip" by fusible links to configure AND and OR gates to perform the desired logic functions. The PAL is basically a programmable AND array driving a fixed OR array. In comparison, both of the arrays of PLA are programmable, while the programmable version of ROM, known commonly as PROM (to be discussed in detail in Chapter 12), has a fixed AND matrix and programmable OR array.

In the PAL circuit, as shown in Figure 4.34, an AND array allows the designer to specify the product terms required and connect them to perform the required SOP logic functions. The PALs are available in a number of different part types, however, that vary the OR gate format. Specifying the OR gate connection, therefore, becomes a task of device selection rather than of programming.

FIGURE 4.34 **Logic Diagram of a Simple PAL.**

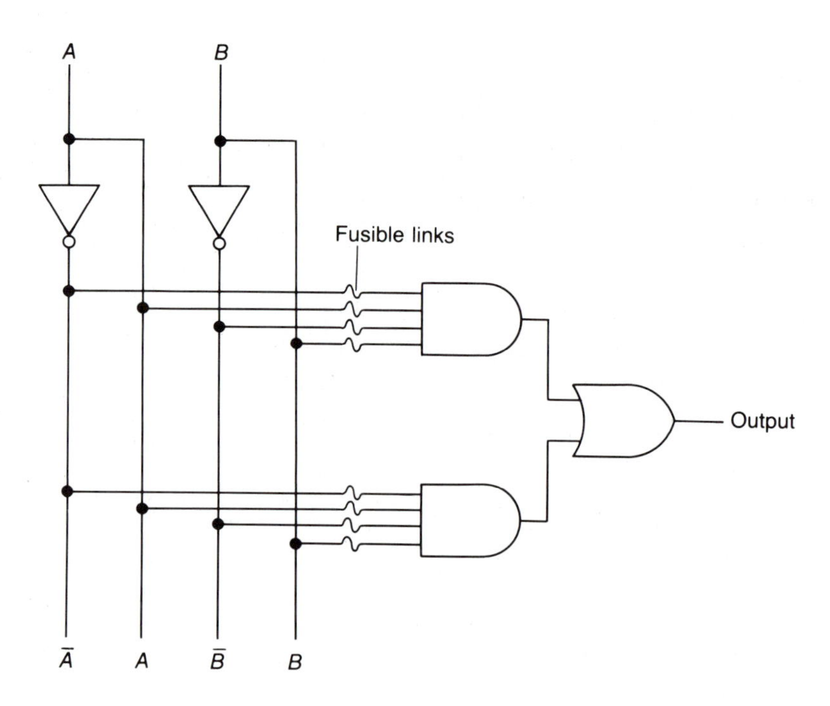

Consequently, PALs totally eliminate the need for a second matrix without any significant loss of flexibility.

In general PALs offer cost-effective capabilities for improving the effectiveness of existing logic designs by expediting and simplifying prototypes and board layouts. Figures 4.35[*a–b*] respectively show thc PLA and PAL configurations of a four-input–four-output AND-OR circuit. The PLA provides the most flexibility for implementing logic functions since the designer is equipped with complete control over all inputs and outputs. However, this flexibility makes PLAs expensive and somewhat formidable to comprehend. In comparison, the PAL combines much of the flexibility of the PLA with the low cost and easy programmability of the PROM.

FIGURE 4.35 **A Four-Input, Four-Output AND-OR Circuit Using [*a*] PLA and [*b*] PAL.**

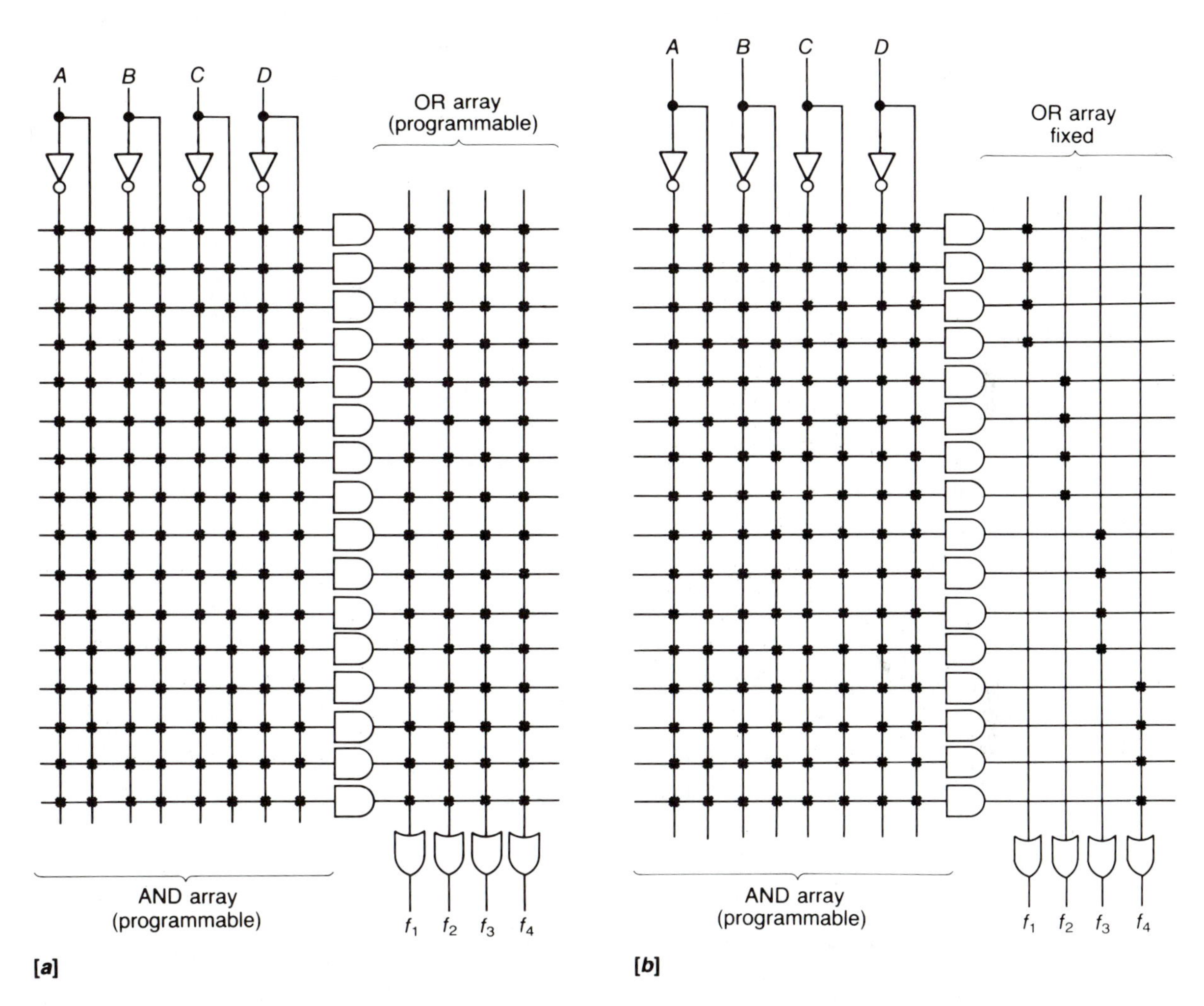

✖ indicates a fusible link. Links are removed where no connection is desired.

4.8 Bridging Technique

The bridging technique is not so much a self-contained design algorithm as it is a way to bend the characteristics of a Boolean function that cannot be reduced further. If after using K-maps or the Q-M technique, the function is still large and unwieldy due to the minterms being logically separated, the function might be *bridged* by using known functions that exhibit similar patterns of logically separated minterms. The X-OR function frequently is used in the bridging process.

Consider the K-maps of several X-OR functions, shown in Figure 4.36. It is obvious that such K-maps are not reducible without the X-OR function. A close examination of these maps reveals that there are equal numbers of 1s and 0s on each half. In addition, the

FIGURE 4.36 **Examples of Several X-OR Functions.**

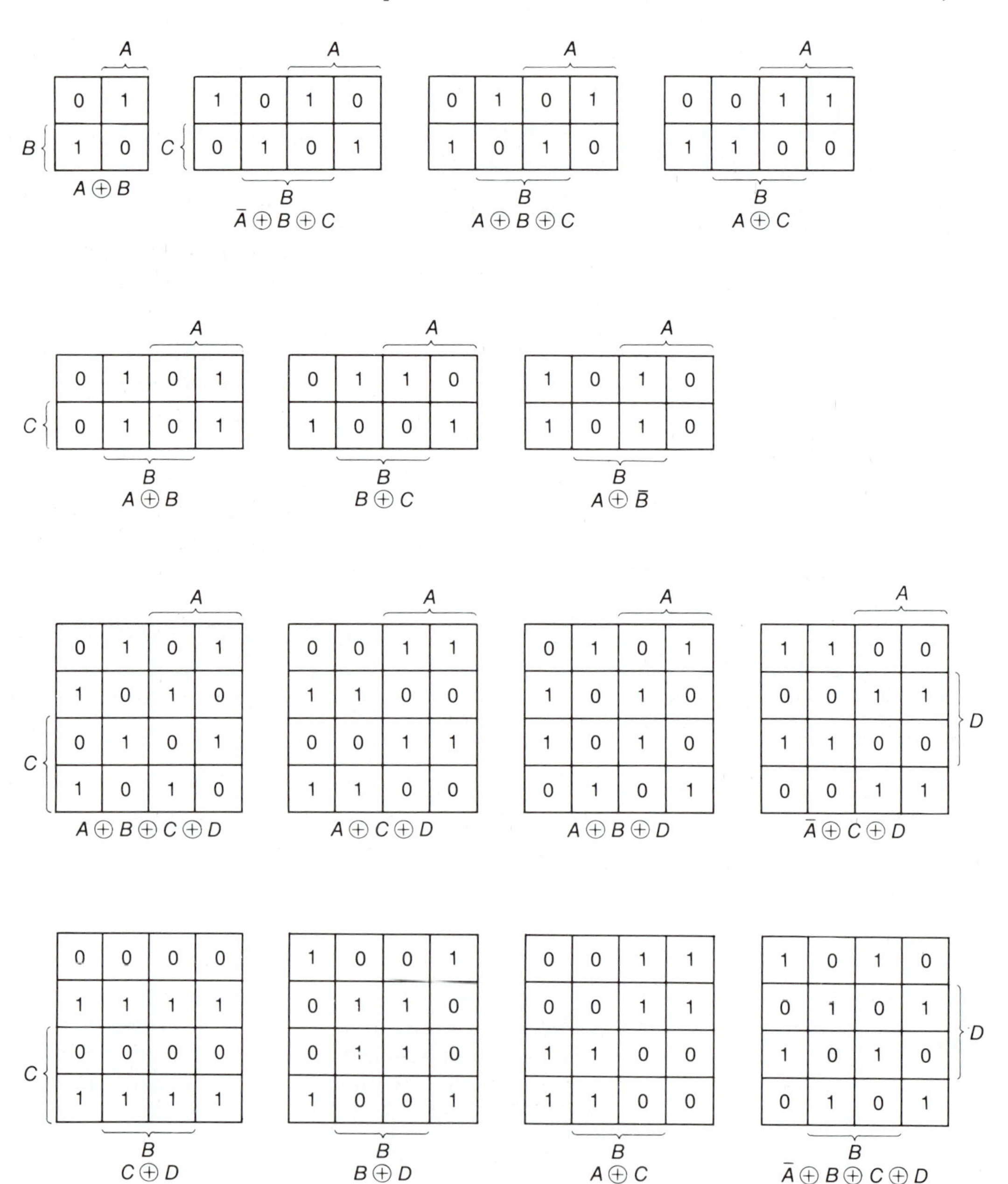

complement of the X-OR function may be obtained either by complementing the whole function or by complementing odd numbers of variables. An X-OR function can be changed to another equivalent X-OR function as long as either (a) an even number of variables have been complemented, or (b) the entire function and an odd number of variables have been complemented. For example,

$$A \oplus B \oplus C \oplus D = \bar{A} \oplus \bar{B} \oplus \bar{C} \oplus \bar{D} = A \oplus \bar{B} \oplus \bar{C} \oplus D$$
$$= A \oplus \bar{B} \oplus C \oplus \bar{D} = \overline{A \oplus \bar{B} \oplus C \oplus D}$$

The X-OR functions are often very useful in implementing functions that have logically isolated minterms.

As long as the function to be implemented bears the characteristic of an X-OR function, it can be realized using one or more X-OR gates. However, the problem becomes more difficult when the K-map closely resembles that of an X-OR function but is not one. The bridging technique then is used to connect the desired function and a closely resembling X-OR function. The technique consists of the following steps:

1. Match the function K-map, F, as closely as possible to a known X-OR K-map, f.
2. Realize the function F using bridging such that $F = f \cdot X + Y$, where X and Y are two separate functions of the same input variables. X and Y are determined by closely comparing K-maps of F and f.

Example 4.9 illustrates the idea behind the bridging scheme.

EXAMPLE 4.9

Using X-OR and other assorted gates, implement the function

$F(A,B,C,D) = \Sigma m(0,1,3,6,9,10,15)$

SOLUTION

The function K-map is obtained as shown in Figure 4.37. By comparing the K-map of Figure 4.37 with those of Figure 4.36, it would become obvious that the closest match occurs with $f = \bar{A} \oplus B \oplus C \oplus D$. However,

FIGURE 4.37

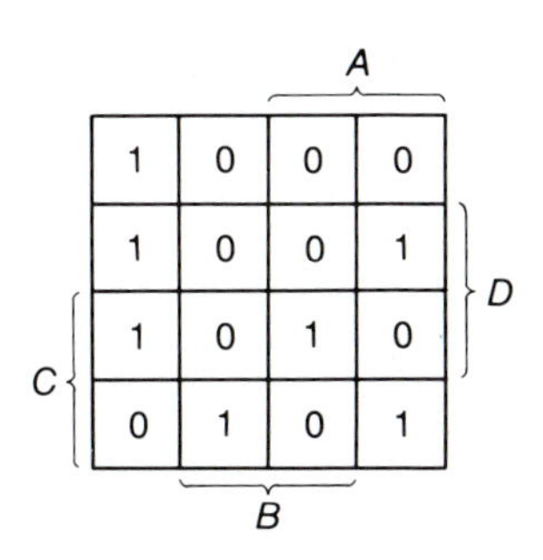

they are not exactly alike. They differ at three minterm locations: 1, 5, and 12. The two functions may be bridged, therefore, as shown in Figure 4.38. The bridging between the two functions F and f required that certain constraints, as listed in Figure 4.39, be met in determining X and Y functions. These constraints follow directly from the equation $F = f \cdot X + Y$.

FIGURE 4.38

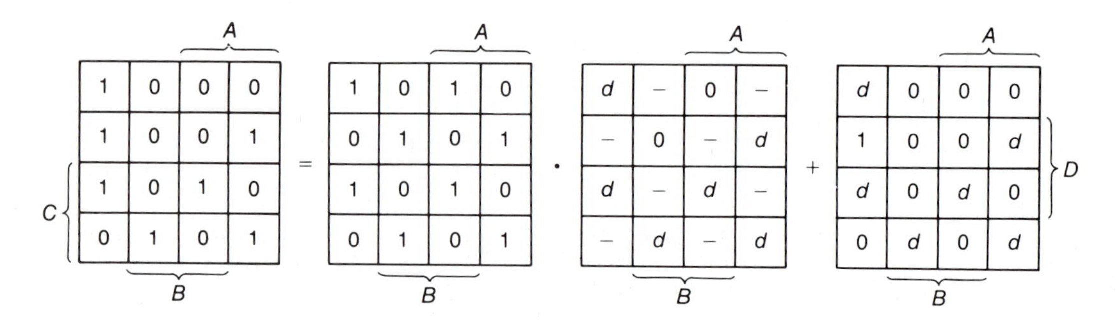

FIGURE 4.39

F	f	X	Y
0	0	–	0
0	1	0	0
1	0	–	1
1	1	d	d

The d's in the table of Figure 4.39 indicate that either X or Y or both must equal 1. In other words, X and Y cannot simultaneously be 0 when $F = f = 1$. This use of d, however, permits many choices for the selection of X and Y. It can be seen that if all d's in X are set equal to 0, then $F = Y$, which is contrary to what is expected in bridging. When $F = Y$ no bridging is needed. On the other hand, if all d's in X are set equal to 1, then

$$Y = \overline{A}\overline{B}\overline{C}$$

$$X = \overline{B} + C$$

Therefore,

$$F = (\overline{A} \oplus B \oplus C \oplus D) \cdot (\overline{B} + C) + \overline{A}\overline{B}\overline{C}$$

The resultant circuit is obtained as shown in Figure 4.40. This bridged circuit is certainly better than the circuit that could be obtained by using only

FIGURE 4.40

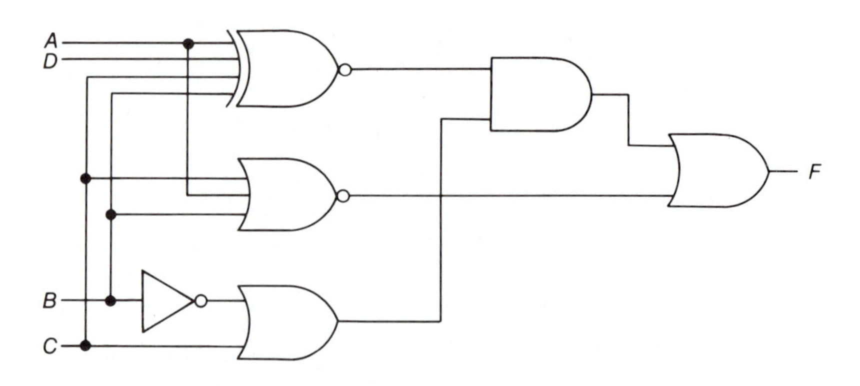

a K-map. If a designer was limited to using only a K-map, the function would have reduced instead to

$$F(A,B,C,D) = \bar{A}\bar{B}\bar{C} + \bar{A}\bar{B}D + \bar{B}\bar{C}D + ABCD + \bar{A}BC\bar{D} + A\bar{B}C\bar{D}$$

The bridge scheme is very general and it does not have to involve only X-OR functions. This technique can be used for generating a complex function when a like function of the same variables already exists (see Problem 16).

4.9 Summary

In this chapter various practical techniques were introduced for realizing combinational circuits. In particular, circuits using only NAND gates, only NOR gates, only MUXs, only ROMs, only PLAs, and only PALs were discussed. In addition, the bridging technique was introduced to handle functions that are otherwise not reducible.

Problems

1. Obtain the circuit for the following functions using only NAND gates:
   a. $f(A,B,C,D) = \Sigma m(1,4,10,11,13,15)$
   b. $f(A,B,C,D) = \Sigma m(1,3,4,9,10,13)$
   c. $f(A,B,C,D) = \Sigma m(1,8\text{–}10,15)$
   d. $f(A,B,C,D,E) = \Sigma m(1,3\text{–}7,11,14\text{–}17,22,24\text{–}27,30)$
   e. $f(A,B,C,D,E) = \Sigma m(1,8\text{–}10,13\text{–}17,21,25\text{–}27,30,31)$
2. Obtain NOR circuits for the functions of Problem 1.
3. a. Use a single level of 1-of-8 MUXs and a few assorted gates (if needed) to obtain a combinational circuit for each of the functions of Problem 1.
   b. For each of your solutions, complement one variable and rearrange the inputs so that the function is still correct.
4. Using 1-of-4 MUXs, obtain a two-level MUX circuit for each of the functions of Problem 1.
5. Obtain the circuit for the function $f(X,Y,Z,U,V) = \Sigma m(0,1,6,7,9,12,13,15,18,20,22,24\text{–}26,28)$ using two levels of 1-of-4 MUXs and a few assorted gates.
6. Use bridging to implement the following functions using X-OR gates:
   a. $f(A,B,C,D) = \Sigma m(1,2,5,7,8,10,13,14)$
   b. $f(A,B,C,D) = \Sigma m(2,3,6,7,9,11,12,13)$
   c. $f(W,X,Y,Z) = \Sigma m(0,2,3,6\text{–}8,10,13)$
   d. $f(W,X,Y,Z) = \Sigma m(0,6,9,10,15)$
7. Implement the functions of Problem 6 using ROMs.
8. Implement the functions of Problem 6 using PLAs.

9. Given the function $f(A,B,C,D) = \Sigma m(0,4,9,10,11,12)$ and $f_1 = B \oplus D$, determine f_2 and f_3 for the circuit of Figure 4.P1.

FIGURE 4.P1

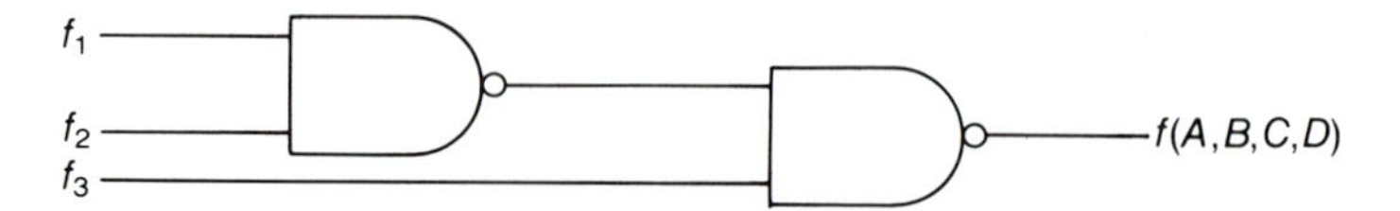

10. Use a ROM to design a binary-to-Gray code converter.
11. Draw the logic diagram of an 8 × 2 ROM that produces the full adder function as described in Chapter 1 (Table 1.4).
12. Use a ROM to achieve four-bit by four-bit binary multiplication.
13. Design a four-input network that squares each of the binary inputs using (a) a ROM, (b) a PLA, and (c) a PAL. Assume that both input and output are unsigned.
14. Design a five-input logic network that finds the 2's complement of a positive number using (a) a ROM, (b) a PLA, and (c) a PAL.
15. Design a network that accepts trigonometric angles in degrees (between 0° and 10° in steps of 1°) and gives out the corresponding tangent value correct up to five significant places. Use (a) a ROM and (b) a PLA.
16. Consider the truth table for the full adder of Table 1.4. Bridge the carry-out function with $A_i \oplus B_i$, where A_i and B_i are the augend and addend, respectively.

Suggested Readings

Bartee, T. C. "Computer design of multiple output logical networks." *IRE Trans. Elect. Comp.* vol. EC-10 (1961): 21.

Cerny, E., and Marin, M. A. "A computer algorithm for the synthesis of memoryless logic circuits." *IEEE Trans. Comp.* vol. C-23 (1974): 455.

Cerny, E., and Marin, M. A. "An approach to unified methodology of combinational switching circuits," *IEEE Trans. Comp.* vol. C-26 (1977): 745.

Culliney, J. N.; Young, M. H.; Nakagawa, T.; and Muroga, S. "Results of the synthesis of optimal networks of AND and OR gates for four-variable switching functions." *IEEE Trans. Comp.* vol. C-27 (1979): 76.

Curtis, H. A. "Short-cut method of deriving nearly optimal arrays of NAND trees." *IEEE Trans. Comp.* vol. C-28 (1979): 521.

Davio, M. "Read only memory implementation of discrete functions." *IEEE Trans. Comp.* vol. C-29 (1980): 931.

Dietmeyer, D. L., and Su, Y. H. "Logic design automation of fan-in limited NAND networks." *IEEE Trans. Comp.* vol. C-18 (1969): 11.

Ektare, A. B., and Mital, D. P. "Multiplexer logic circuit design using cubical complexes." *Elect. Lett.* vol. 16 (1980): 495.

Ektare, A. B., and Mital, D. P., "Probabilistic approach to multiplexer logic circuit design," *Elect. Lett.* vol. 16 (1980): 686.

Ellis, D. T. "A synthesis of combinational logic with NAND or NOR elements." *IEEE Trans. Elect. Comp.* vol. EC-14 (1965): 701.

Fleisher, H., and Maissel, L. "An introduction to array logic." *IBM J. Res. & Dev.* vol. 19 (1975): 98.

Jones, J. W. "Array logic macros." *IBM J. Res. & Dev.* vol. 19 (1975): 120.

Jullien, G. A. "Residue number scaling and other operations using ROM arrays." *IEEE Trans. Comp.* vol. C-27 (1978): 325.

Kambayashi, Y. "Logic design of programmable logic arrays." *IEEE Trans. Comp.* vol. C-28 (1979): 609.

Lai, H. C., and Muroga, S. "Minimum parallel binary adders with NOR(NAND) gates." *IEEE Trans. Comp.* vol. C-28 (1979): 648.

Li, H. F. "Variable selection in logic synthesis using multiplexers." *Int. J. Electron.* vol. 49 (1980): 185.

Liu, T. K.; Hohulin, K. R.; Shiau, L. E.; and Muroga, S. "Optimal one-bit full adders with different types of gates." *IEEE Trans. Comp.* vol. C-23 (1974): 63.

Logue, J. C.; Brickman, N. F.; Howley, F.; Jones, J. W.; and Wu, W. W. "Hardware implementation of a small system in programmable logic arrays." *IBM J. RES. & Dev.* vol. 19 (1975): 110

Lotfi, Z. M., and Tosser, A. J. "Systematic search for minimum synthesis of logical functions with multiplexers." *Int. J. Electron.* vol. 47 (1980): 569.

Nagle, H. T., Jr.; Carroll, B. D.; and Irwin, J. D. *An Introduction to Computer Logic.* Englewood Cliffs, N.J.: Prentice-Hall, 1975.

Nakamura, K. "Synthesis of gate-minimum multi-output two-level negative gate networks." *IEEE Trans. Comp.* vol. C-28 (1979): 768.

Papachristou, C. A. "An algorithm for optimal NAND cascade logic synthesis." *IEEE Trans. Comp.* vol. C-27 (1978): 1099.

Peatman, J. B. *Digital Hardware Design.* New York: McGraw-Hill, 1980.

Preparta, F. P. "On the design of universal Boolean functions." *IEEE Trans. Comp.* vol. C-20 (1971): 418.

Sasao, T. "Input variable assignment and output phase optimization of PLA's." *IEEE Trans. Comp.* vol. C-33 (1984): 879.

Sasao, T. "An algorithm to derive the complement of a binary function with multiple-valued inputs." *IEEE Trans. Comp.* vol. C-34 (1985): 131.

Shannon, C. E. "A symbolic analysis of relay and switching circuits." *Trans. AIEE.* vol. 57 (1938): 713.

Weinberger, A. "Device sharing in array logic." *IBM Tech. Disc. Bull.* vol. 19 (1976): 1357.

Weinberger, A. "High-speed programmable logic array adders." *IBM J. Res. & Dev.* vol. 23 (1979): 163.

Whitehead, D. G. "Algorithm for logic circuit synthesis by using multiplexers." *Elect. Lett.* vol. 13 (1977): 355.

Wood, R. "A high density programmable logic array chip." *IEEE Trans. Comp.* vol. C-28 (1979): 602.

Design of Combinational Circuits

CHAPTER FIVE

5.1 Introduction

A *combinational logic circuit*, as shown by the block diagram of Figure 5.1, is defined as a combination of logic devices whose output is a function of the present values of the input variables and independent of the past values. After propagation time through the circuit, input variable changes cause output changes that are dependent only on the present input values.

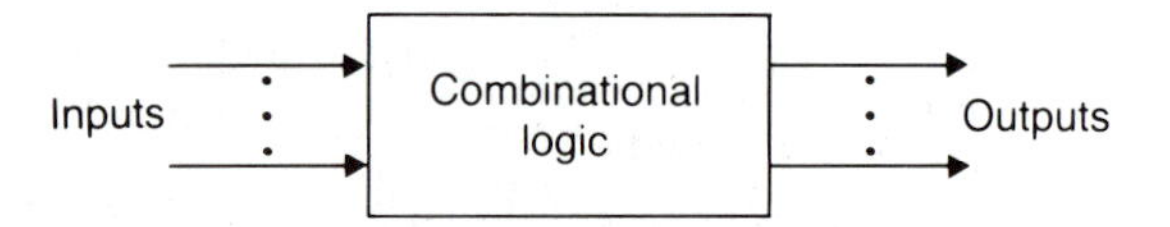

FIGURE 5.1 **Block Diagram of a Combinational Network.**

In Chapters 1, 2, and 4 we introduced the necessary tools to design combinational logic circuits. The design algorithm leading to the realization of a complex combinational circuit consists of the following essential steps:

1. The complex logic problem is intuitively analyzed and decomposed into a set of smaller but nontrivial functional units.
2. The number and characteristics of both input and output variables for each of the functional units are identified.
3. A truth table for each of the functional units is determined.
4. The output functions for each of the functional units are simplified using one of the minimization schemes covered in previous chapters.
5. The circuits corresponding to each of the functional units are assembled and tested individually and then connected to form the desired complex function.

In practice, the designer would have to consider various practical limitations such as the number of logic gates, interconnections, gate inputs, fan-out, and the length of propagation delay. In the not too distant past the only option available to the designer was to assemble the entire logic circuit with a sack full of SSI chips. However, at this time we have better alternatives because many more complex logic circuits are available in IC form. The only limitation to the use of either the MSI or the LSI is that we may not be able to locate a device that exactly meets our requirements. In that event the designer must modify the standard device by externally combining it with other SSI or MSI chips.

In this chapter applications of these combinational logic design tools will be considered. The design and application of MSI devices, including adders, subtracters, decoders, encoders, and error-control logic, is presented. These devices play an important part in the development of more advanced digital systems. The application of the MSI devices considered is not governed by a set of design procedures as well defined as those procedures for individual logic gates. Experience and intuition (horse sense) become important. There is no substitute for understanding exactly what the MSI devices can do. After studying this chapter, you should be able to:

- Break a complex design into manageable subunits;
- Design individual subunits and be able to cascade them together;
- Understand the working principles and design process of various combinational binary adders and/or subtracters;
- Understand the working principles and design process of various code converters;
- Understand the working principles and design of BCD arithmetic circuits;
- Understand the working principles and design of various decoders and encoders;
- Understand the working principles and design of various error-correcting circuits.

5.2 Binary Adders

Most arithmetic operations are reducible to simple addition or subtraction processes that can be performed repetitively for more complex operations such as multiplication and division. If the addition circuit has no carry-in, the addition of the least significant bits involves only two operands, the addend and augend. The addition of the remaining bits, however, requires the carry-in from the addition of the previous column.

A multi-bit adder can be realized in various ways, each having different speed and cost characteristics. A two-level network would obviously prove to be the fastest. However, this network would require a large number of gates and gate inputs. It would be necessary to have 2^{2n} NAND gates of $2n + 1$ inputs and one NAND gate of 2^{2n} inputs to add two n-bit numbers. This number of gates and inputs is quite significant for even small values of n. The alternative to this expensive design is quite straightforward. It follows directly from our observation of the algorithm of an n-bit add operation. Irrespective of the number of bits, the process of adding augend and addend is identical at each of the columns except at the least significant position. The design of a parallel n-bit addition circuit, therefore, is accomplished by designing a total of n single-bit addition circuits. In order to allow n single-bit adders to be connected together to form an n-bit adder, the single-bit adder stages need to be full adders (adders with three inputs). The least significant carry-in is tied to a 0, and each of the remaining carry-in inputs is tied to the carry-out of the previous single-bit addition.

As an introduction to adder design, we shall first consider a half adder (adder with only two inputs, addend and augend). The resultant circuit will have application in the design of a full adder. In fact, an n-bit adder circuit using $n - 1$ full adders and one half adder also can be designed.

5.2.1 Half Adder (HA)

The *half adder* (*HA*) unit is a simple multiple-output combinational circuit used for adding two bits without a carry-in. The truth table for the two inputs, A_i and B_i, and the output sum, S_i, and carry-out, C_i, are shown in Figure 5.2.

FIGURE 5.2 **Half Adder: [*a*] Block Diagram and [*b*] Truth Table.**

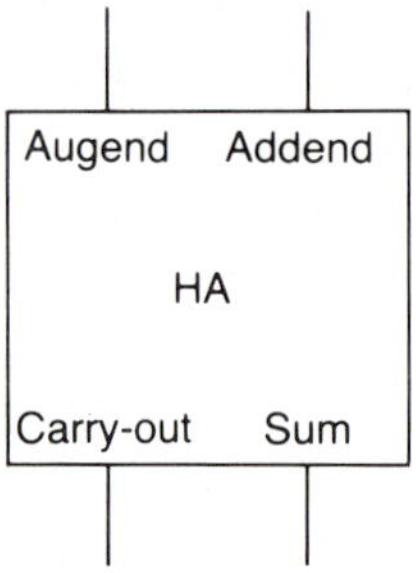

Augend Addend

HA

Carry-out Sum

[*a*]

A_i	B_i	C_i	S_i
0	0	0	0
0	1	0	1
1	0	0	1
1	1	1	0

[*b*]

Using a Karnaugh map, the equations for the sum, S_i, and carry, C_i, are as follows:

$$S_i = A_i\overline{B}_i + \overline{A}_iB_i = A_i \oplus B_i \qquad [5.1]$$

$$C_i = A_iB_i \qquad [5.2]$$

There are several ways to implement these functions. Figure 5.3

illustrates four ways to implement the HA functions. Circuits using [*a*] an AND and an X-OR gate, [*b*] NOR gates, [*c*] NAND gates, and [*d*] multiplexers are shown.

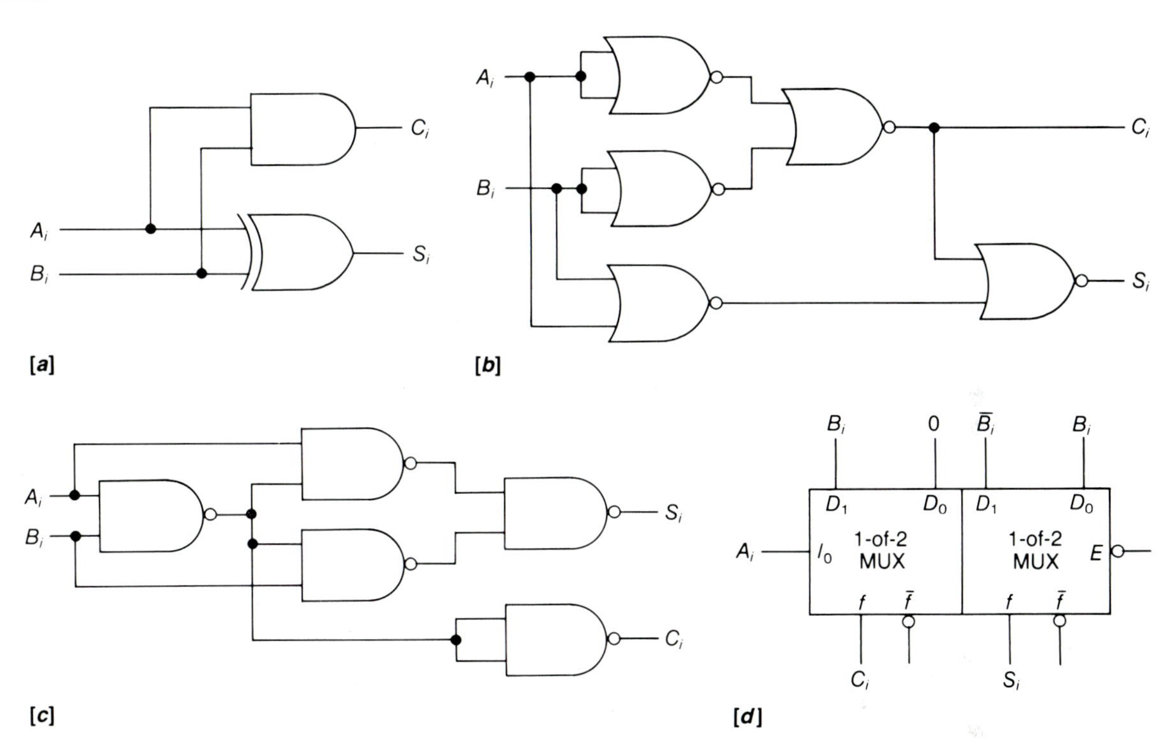

FIGURE 5.3 **HA Circuit: [*a*] Using AND and X-OR Gates, [*b*] Using NOR Gates, [*c*] Using NAND Gates, and [*d*] Using MUXs.**

5.2.2 Full Adder (FA)

A *full adder* (*FA*) is a three-input, two-output logic circuit that adds two binary digits, A_i and B_i, and a carry-in from the $i - 1$ bit position, C_{i-1}. The block diagram and the corresponding truth table are shown in Figures 5.4[*a*–*b*]. The K-maps for the sum bit, S_i, and the carry-out, C_i, are constructed from the truth table and shown in Figure 5.4[*c*]. The equations for the sum and carry-out can then be obtained from the K-maps as follows:

$$S_i(A_i,B_i,C_{i-1}) = \overline{A}_i\overline{B}_iC_{i-1} + \overline{A}_iB_i\overline{C}_{i-1} + A_i\overline{B}_i\overline{C}_{i-1} + A_iB_iC_{i-1} \quad [5.3]$$

$$= \overline{\overline{\overline{A}_i + \overline{B}_i + C_{i-1}} + \overline{\overline{A}_i + B_i + \overline{C}_{i-1}} + \overline{A_i + \overline{B}_i + \overline{C}_{i-1}} + \overline{A_i + B_i + C_{i-1}}} \quad [5.4]$$

and

$$C_i(A_i,B_i,C_{i-1}) = A_iB_i + A_iC_{i-1} + B_iC_{i-1} \quad [5.5]$$

$$= \overline{\overline{A_i + B_i} + \overline{A_i + C_{i-1}} + \overline{B_i + C_{i-1}}} \quad [5.6]$$

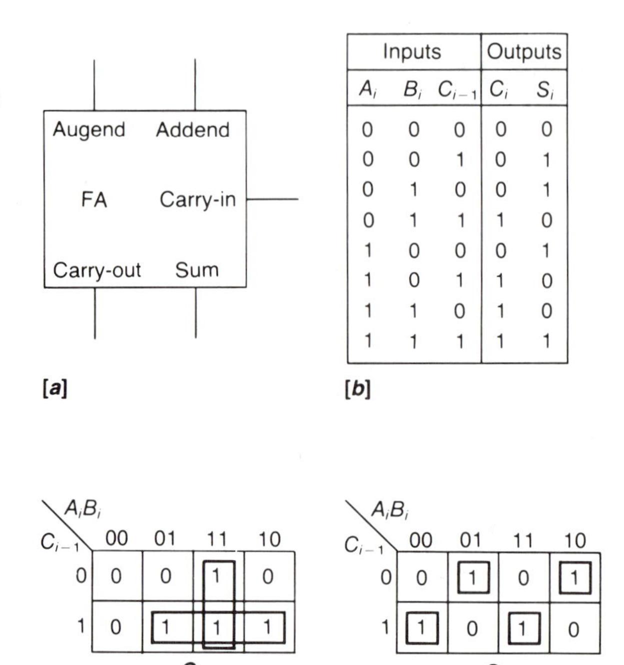

Inputs			Outputs	
A_i	B_i	C_{i-1}	C_i	S_i
0	0	0	0	0
0	0	1	0	1
0	1	0	0	1
0	1	1	1	0
1	0	0	0	1
1	0	1	1	0
1	1	0	1	0
1	1	1	1	1

FIGURE 5.4 **Full Adder: [*a*] Block Diagram, [*b*] Truth Table, and [*c*] K-Maps for the Carry-Out and the Sum.**

There are several ways to implement the FA equations. The direct implementation of Equations [5.3] and [5.5] leads to FA circuits using only NAND gates. If Equations [5.4] and [5.6] are used, the equivalent NOR circuits may be obtained. Figures 5.5[*a–b*] respectively show the typical FA circuits using only NAND and only NOR gates. Each of these circuits requires a total of 12 gates and 31 gate inputs. However, by making use of the bridging technique, the number of gates can be reduced.

Note that the K-map for the sum output is irreducible, producing a cumbersome circuit. Note also that the K-map for S_i is that of a three-input X-OR function. A review of Figure 4.36 indicates that the sum equation may be reduced to

$$S_i = A_i \oplus B_i \oplus C_{i-1} \qquad [5.7]$$

Examining the carry-out K-map reveals that it could be bridged with an X-OR function as follows (see Chapter 4, Problem 16):

$$C_i = (A_i \oplus B_i)C_{i-1} + A_iB_i \qquad [5.8]$$

Now recall that $A_i \oplus B_i$ is the sum output and A_iB_i is the carry-

FIGURE 5.5 **FA Circuit: [*a*] Using Only NAND Gates and [*b*] Using Only NOR Gates.**

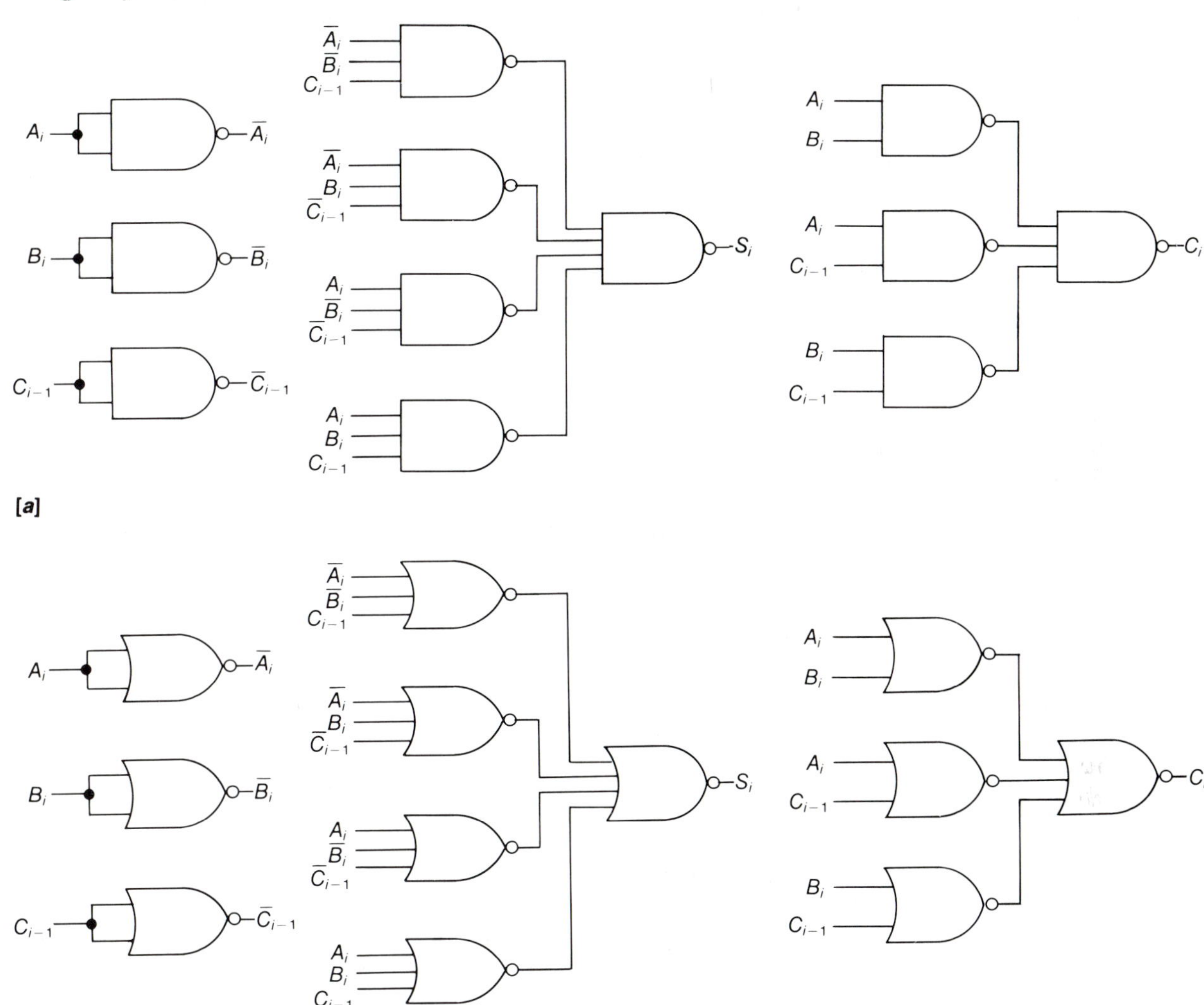

out for an HA. Thus the HAs designed in the last section may be used for realizing an FA. Expressing the FA equations in terms of the HA equations, Equations [5.1] and [5.2], gives

$$S_i = S_{i(\mathrm{HA})} \oplus C_{i-1} \qquad [5.9]$$

$$C_i = S_{i(\mathrm{HA})} \cdot C_{i-1} + C_{i(\mathrm{HA})} \qquad [5.10]$$

where $S_{i(\mathrm{HA})}$ and $C_{i(\mathrm{HA})}$ are the sum and carry-out of the HA. Note also that Equation [5.9] involves an X-OR operation between the carry-in and the HA sum. If the sum output of the HA and the carry-in are fed into a second HA, the final sum output will be the

FA sum, S_i. In addition, if the carry-outs from both of the HAs are ORed together, the FA carry-out, C_i, is obtained. Figure 5.6[*a*] shows the FA circuit using two HAs and an OR gate. The circuit corresponding to Equations [5.7] and [5.8] also may be implemented using only NAND gates or only NOR gates. The resultant NAND equivalent circuit is shown in Figure 5.6[*b*]. This NAND circuit requires only nine NAND gates and a total of 18 gate inputs. We have designed an FA circuit using three fewer gates and 13 fewer gate inputs than would be necessary had we not made use of the X-OR K-map structure and bridging.

FIGURE 5.6 **FA Circuits: [*a*] Using HAs and [*b*] Using Only NAND Gates.**

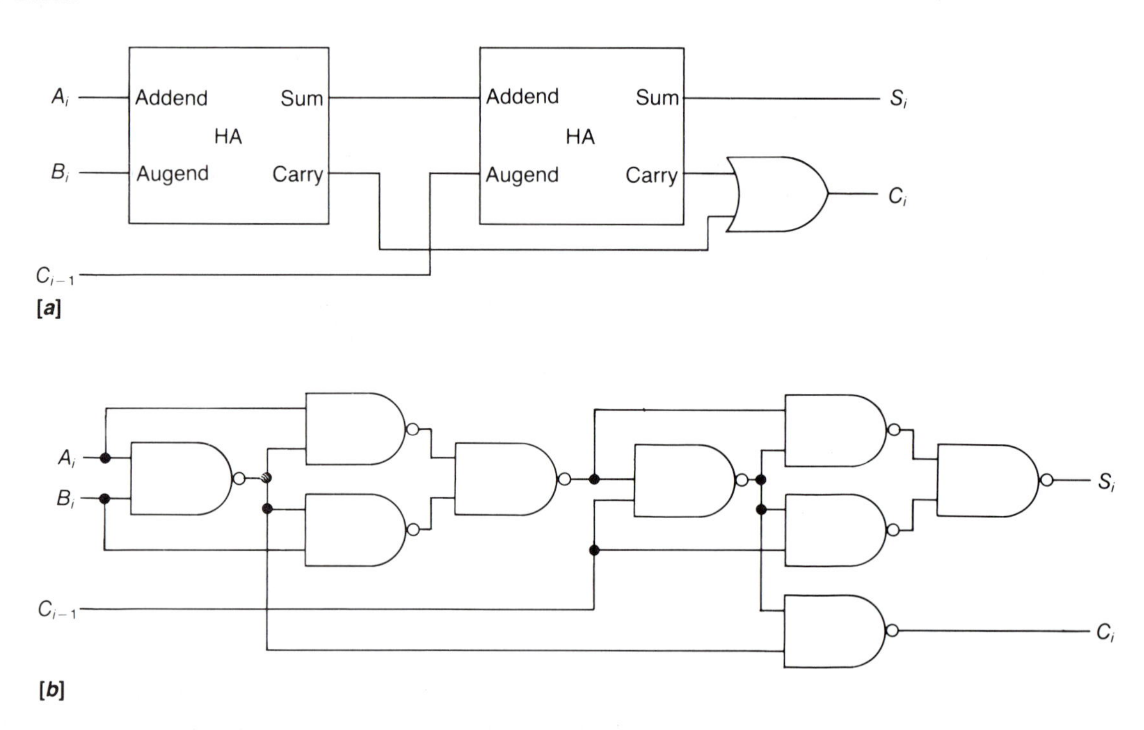

We have completed the design of half and full adders. These devices are also commercially available in the form of MSI devices. Four FAs usually are connected and are commercially available in an MSI four-bit adder IC. Using a commercially available four-bit adder, two quantities, $A_3A_2A_1A_0$ and $B_3B_2B_1B_0$, can be added. The resultant sum is $S_3S_2S_1S_0$, and the carry-out is from the A_3, B_3, and C_2 addition. Figure 5.7 demonstrates the connection of n FAs to

FIGURE 5.7 **n-Bit Parallel Adder Circuit: [*a*] Using n FAs and [*b*] Using $n-1$ FAs and One HA.**

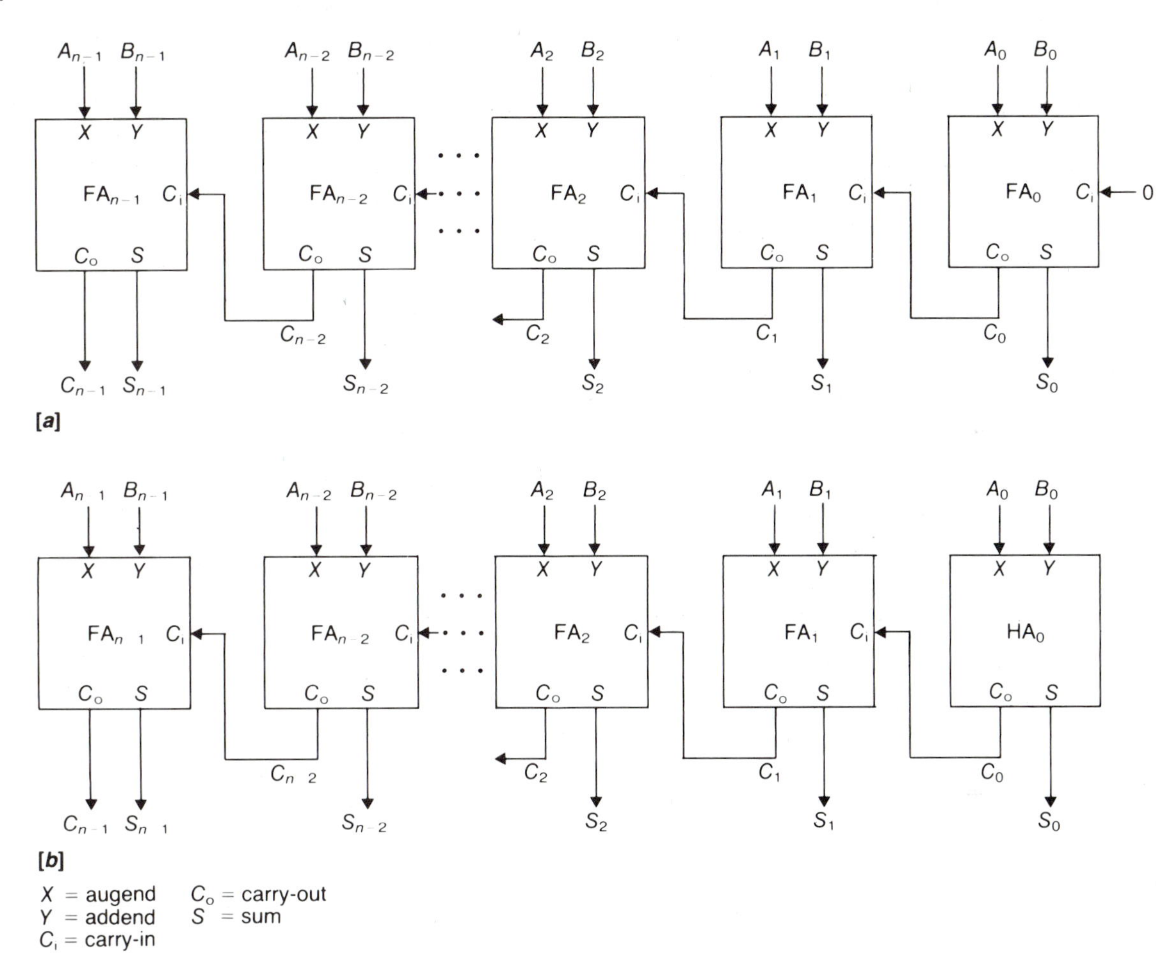

FIGURE 5.8 **Multiplication of Two Three-Bit Numbers.**

			B_2	B_1	B_0
			A_2	A_1	A_0
			A_0B_2	A_0B_1	A_0B_0
		A_1B_2	A_1B_1	A_1B_0	
	A_2B_2	A_2B_1	A_2B_0		
P_5	P_4	P_3	P_2	P_1	P_0

make an n-bit adder. The delay of this n-bit ripple adder is $n\Delta$ where Δ is the propagation delay of the carry-out of a single FA. This delay is accumulated when the carry into the multi-bit adder has to propagate through all of the FAs to get to the final carry-out. Consequently, this delay becomes more and more significant as n becomes larger.

Multiplication of binary numbers makes use of addition just as multiplication of decimal numbers does. The multiplication of two three-bit numbers, $A = A_2A_1A_0$ and $B = B_2B_1B_0$, is symbolically obtained as shown in Figure 5.8, where $P_5P_4P_3P_2P_1P_0$ forms the product. Note that for the multiplication of two n-bit numbers, the product has the possibility of $2n$ bits.

Later in the text sequential design techniques will be presented that will allow designing a multiplier that uses a repetitive algorithm for multiplication. It is possible, however, to design a combinational circuit using FAs that will perform the multiplication of two binary numbers by performing the sum of the three partial products of two numbers. This combinational circuit must be able to add columns of bits. Example 5.1 will demonstrate how FAs can be used to do similar functions.

EXAMPLE 5.1

Design a circuit using FAs that may be used for adding a column of six single-bit numbers.

SOLUTION

Let the column of numbers be a_0, a_1, a_2, a_3, a_4, and a_5. The sum if all were 1 would add up to 110, so there will be three outputs: S_2, S_1, and S_0. However, an FA may add up to only three single bits. Two FAs may be used to obtain two partial sums of the six bits as follows:

$$\begin{array}{rr} & a_0 \\ & a_1 \\ & a_2 \\ \hline x_1 & y_1 \end{array} \qquad \begin{array}{rr} & a_3 \\ & a_4 \\ & a_5 \\ \hline x_2 & y_2 \end{array}$$

where x_1 and x_2 are the carry-outs and y_1 and y_2 are the respective sums. As a next step the least significant bits, y_1 and y_2, may be added as follows:

$$\begin{array}{rr} & y_1 \\ & y_2 \\ & 0 \\ \hline x_3 & S_0 \end{array}$$

where x_3 and S_0 are the carry-out and the sum, respectively. Finally, x_1, x_2, and x_3 could be fed into a fourth FA to yield the carry-out, S_2, and the sum, S_1, as follows:

$$\begin{array}{rr} & x_1 \\ & x_2 \\ & x_3 \\ \hline S_2 & S_1 \end{array}$$

The complete circuit, therefore, would require four FAs. The resulting circuit is obtained as shown in Figure 5.9.

FIGURE 5.9

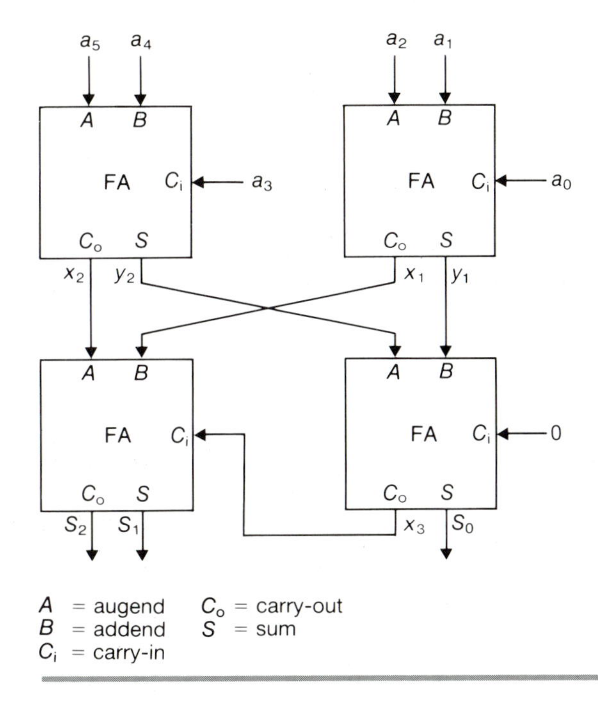

Another example of the application of adders in digital systems is given in Example 5.2. This example also demonstrates how a computer that is designed to handle binary quantities of n bits can perform operations on $2n$-bit quantities. In programming, such an operation is commonly called a *multiple-precision operation.*

EXAMPLE 5.2

Use four-bit multipliers and four-bit binary adders to design a circuit for multiplying two eight-bit numbers. The block diagrams of the multiplier and adder units are provided in Figure 5.10. The four-bit multipliers are assumed to be ROMs. The two four-bit quantities to be multiplied make up an eight-bit address. Each ROM storage location is the eight-bit product of the two four-bit quantities that make up its address.

SOLUTION

FIGURE 5.10

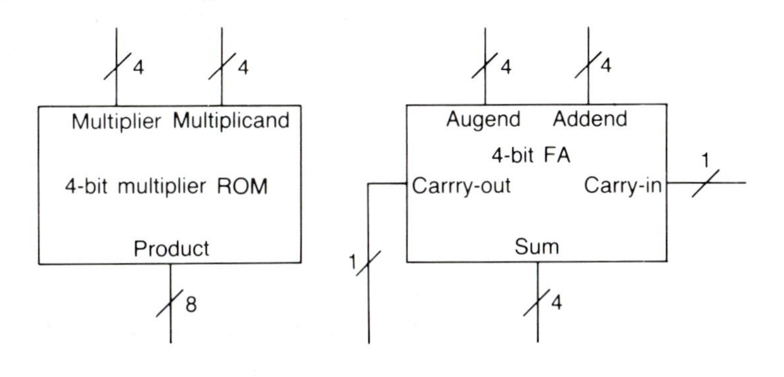

FIGURE 5.11

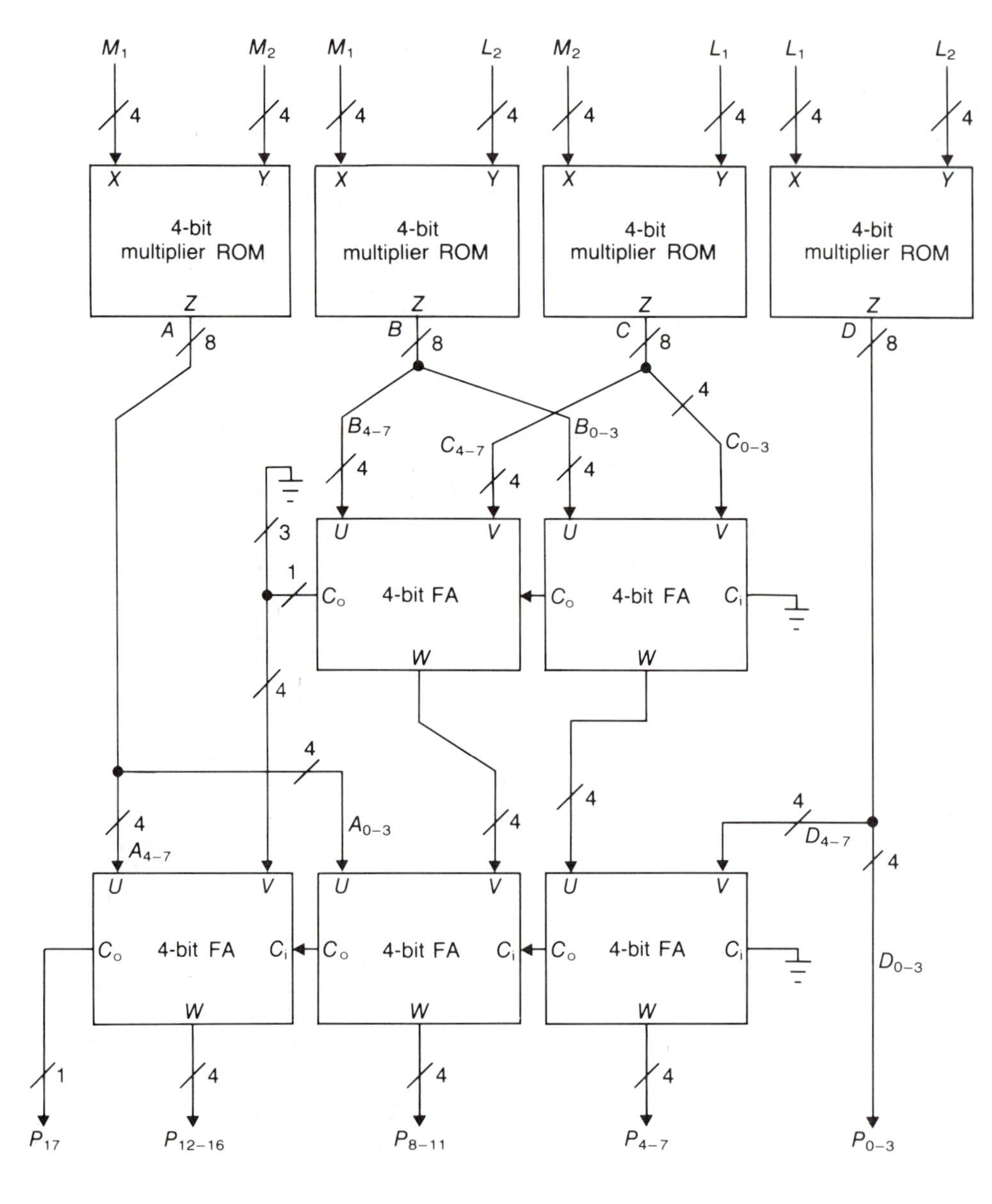

A good approach to any design problem is to break the given problem into several simpler problems. This procedure is necessary in this example in order to make the problem fit the devices that are provided. Consider the multiplication of two eight-bit numbers, X_1 and X_2, each consisting of a least significant four bits, L_i, and a most significant four bits, M_i. The eight-bit number can then be expressed as the sum of the two four-bit parts:

$$X_i = 2^4(M_i) + L_i$$

where M_i is shifted to the left four places (multiplied by 2^4) before being added to L_i. The product, P, may now be expressed as

$$\begin{aligned} P &= [2^4(M_1) + L_1][2^4(M_2) + L_2] \\ &= 2^8(M_1M_2) + 2^4(M_1L_2 + M_2L_1) + (L_1L_2) \end{aligned}$$

Each of these four partial products may be obtained using four four-bit multiplier units. The four multiplier units would respectively have (a) M_1 and M_2, (b) M_1 and L_2, (c) M_2 and L_1, and (d) L_1 and L_2 as inputs. This configuration would result in a total of four eight-bit outputs: A, B, C, and D, respectively. Since two eight-bit quantities are being multiplied, a 16-bit product is expected. Note also that D should be added to the sum of B and C that have been shifted to the left four places, and this in turn should be added to A that have been shifted to the left eight places. A network of six four-bit FAs may be employed, as shown in Figure 5.11, to obtain the 16-bit sum of the shifted partial products. The final product is obtained by adding the partial products. Care must be taken to connect the partial products at the correct bit positions relative to their power of 2. The shifting is implicit in the interconnection pattern of the adder modules.

Note that a single ROM for multiplying two eight-bit numbers would require a total of 1,048,576 bits, that is, 16 bits of address and 16 bits in each location, to store the 16-bit product, or $2^{16} \times 16$ bits. Each ROM that we used in this example had a size of only 2048 bits, giving a total of 8192 bits. This design would probably be less expensive but would be slower due to the time required to perform the additions. Such time and money trade-offs will be a typical design decision that must be made by every engineer.

5.3 Binary Subtracters

Many arithmetic circuits also require a unit for subtraction. A subtracter circuit could be designed from scratch. However, recall from Chapter 1 that subtraction also is possible by adding the complement of the subtrahend to the minuend. Consequently, rather than designing a straightforward subtracter, a multi-bit subtracter can be made using complement arithmetic. This design would involve the use of a multi-bit parallel adder circuit that is fed with the complemented subtrahend and the minuend.

In order to accommodate the multi-bit parallel adder to the requirement of our present design, certain modification is necessary for complementing the subtrahend. If each bit of the subtrahend is individually complemented, the corresponding 1's complement will be obtained. A 2's complement could be formed by adding a one to the LSB of the corresponding 1's complement, that is, by making the carry-in to the LSB position of the adder a 1. To perform the complementing of each bit, we can use an X-OR gate. An X-OR has a useful characteristic that makes it perform as a programmable inverter. Consider a two-input X-OR gate for which one input is a bit of the subtrahend and the other input is tied to a select line, E. When $E = 1$, the output will be the complement of the subtrahend bit. When $E = 0$, the output will be the uncomplemented bit, since from the X-OR truth table,

$$1 \oplus Y = \overline{Y}$$

$$0 \oplus Y = Y$$

The property of an X-OR function provides the possibility of having both an adder and a subtracter out of the same circuit. Such a circuit is shown in Figure 5.12 where the enable, E, allows the circuit to add or to subtract by taking the 1's complement and adding a one to the LSB. When $E = 0$, the circuit adds with a carry-in of 0, and when $E = 1$, the circuit complements B and adds that to A with a carry-in of 1.

FIGURE 5.12 **Four-Bit Adder/ Subtracter Unit.**

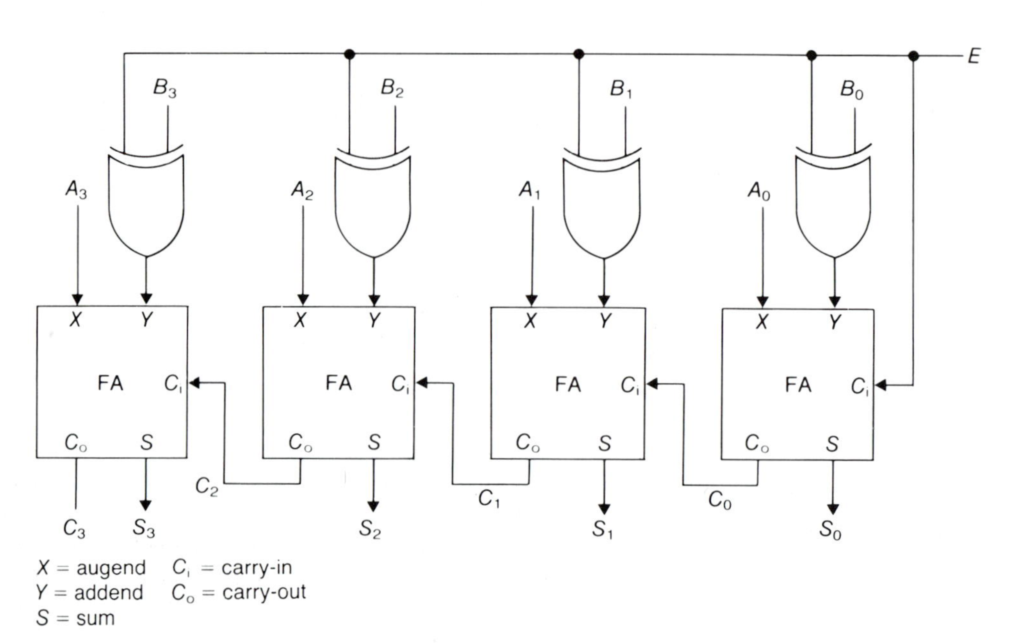

EXAMPLE 5.3

Obtain a combinational circuit for realizing the 2's complement of a five-bit binary number.

SOLUTION

From Section 1.4 we know that the LSB of a number always remains unchanged when obtaining the 2's complement. If the LSB, a_0, is 1, then the next higher bit, a_1, is inverted. An X-OR gate in the form of $a_0 \oplus a_1$

may be used to accomplish this. The next higher bit, a_2, is complemented if either a_1 or a_0 is 1. This line of argument could be carried out for all of the remaining bits. Such logical tests can be implemented by performing $a_0 + a_1 + \cdots + a_{m-1}$, which would indicate whether or not a 1 is present in the least significant m bits.

It is apparent, therefore, that X-OR and OR gates can be used to realize the necessary circuit for performing 2's complement of a number. The OR gates would perform tests and the X-OR gates would complement bits if necessary. The resulting circuit for a five-bit number $a_4a_3a_2a_1a_0$ is obtained as shown in Figure 5.13.

FIGURE 5.13

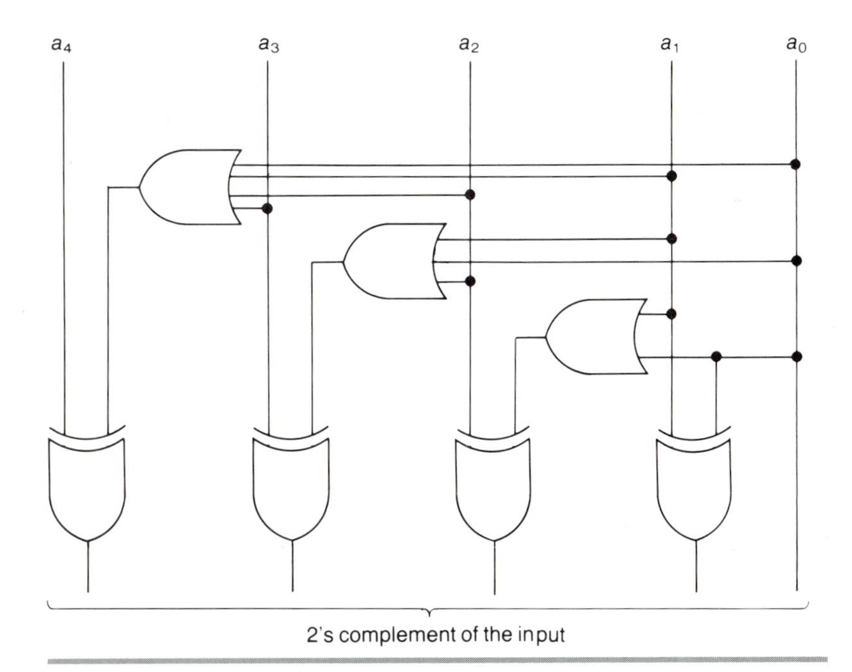

5.4 Carry Look-Ahead (CLA) Adders

The particular multi-bit parallel adder circuit developed in the previous section is sometimes referred to as a *ripple adder* because a carry from one unit of the adder may have to ripple through several units before the sum is obtained. Such ripple adders have also been used to form either 2's complement, or 1's complement sign-and-magnitude binary adder/subtracters. The performance of a ripple adder/subtracter, however, is limited by the time required for the carries to ripple through all of the stages of the circuit. For such devices the maximum delay is directly proportional to the number of FA units.

One particular method of speeding up the combinational addition process is known as *carry look-ahead* (*CLA*). In Figure 5.14 it may be seen that the carry-out is the same as the carry-in as long as one

FIGURE 5.14 **FA Configurations Resulting in a Carry-Out.**

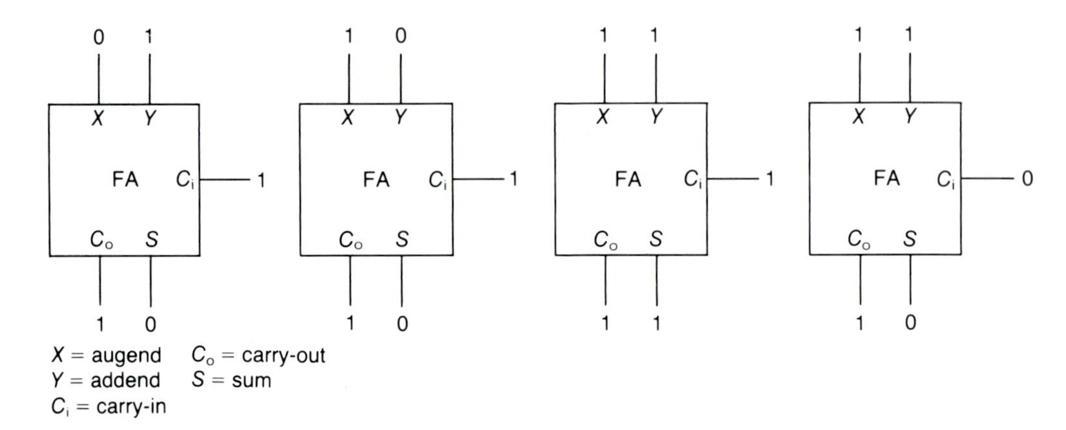

of the other two inputs is a 1. Also, the carry-out is always a 1 independent of the carry-in when both of the other inputs are 1s, and 0 if both are 0. Consequently, two useful functions can be defined: the carry-propagate, P_i, and the carry-generate, G_i

$$P_i = A_i \oplus B_i \quad [5.11]$$

$$G_i = A_i \cdot B_i \quad [5.12]$$

where A_i and B_i are the addend and augend, respectively, of the ith full adder.

The FA equations, Equations [5.7] and [5.8], can then be rewritten as

$$S_i = P_i \oplus C_{i-1} \quad [5.13]$$

$$C_i = G_i + P_i C_{i-1} \quad [5.14]$$

For a four-bit adder the carries for the various stages are as follows:

$$C_0 = G_0 + P_0 C_{-1} \quad [5.15]$$

$$\begin{aligned} C_1 &= G_1 + P_1 C_0 \\ &= G_1 + P_1 G_0 + P_1 P_0 C_{-1} \quad [5.16] \end{aligned}$$

$$\begin{aligned} C_2 &= G_2 + P_2 C_1 \\ &= G_2 + P_2 G_1 + P_2 P_1 G_0 + P_2 P_1 P_0 C_{-1} \quad [5.17] \end{aligned}$$

and

$$\begin{aligned} C_3 &= G_3 + P_3 C_2 \\ &= G_3 + P_3 G_2 + P_3 P_2 G_1 + P_3 P_2 P_1 G_0 + P_3 P_2 P_1 P_0 C_{-1} \quad [5.18] \end{aligned}$$

while the sum bits are

$$S_0 = P_0 \oplus C_{-1} \qquad [5.19]$$

$$S_1 = P_1 \oplus C_0 \qquad [5.20]$$

$$S_2 = P_2 \oplus C_1 \qquad [5.21]$$

and

$$S_3 = P_3 \oplus C_2 \qquad [5.22]$$

In Equations [5.15–18] the carry-out for each of the stages is dependent only on the initial carry, C_{-1}, and the corresponding propagate and generate functions. The equation for C_3 can be implemented with a two-gate level circuit. Since each of the generate and propagate functions can be expressed in terms of the two data bits, C_3 is available after two gate delays, resulting in a fast addition process. The block diagram of a four-bit CLA adder is shown in Figure 5.15. It consists of three sections, each having four subunits. The PG_i section generates the carry-propagate and carry-generate functions. The CL_i section intakes the outputs of the previous section and generates the carries. Finally, the SU_i section generates the sum bits. However, the carry units are different for every bit, and their complexity increases as they move further away from the LSB.

The internal hardware for the four-bit CLA adder is shown in Figure 5.16. Each propagation-generation unit requires five NAND gates, each sum unit requires four NAND gates, and the n-bit carry section requires a total of $(n^2 + 5n)/2$ NAND gates. A four-bit CLA adder, therefore, requires a total of 54 NAND gates and involves a total of eight units of gate delay. In comparison, addition in a four-bit ripple adder requires 12 units of gate delay. The four-bit CLA, therefore, cuts down the time factor by about one-third. Similarly, a 64-bit CLA adder requires almost five times as many NAND gates as a 64-bit ripple adder, but reduces the propagation delay by a factor of 17. It follows, therefore, that the CLA adder will provide a faster addition time, especially when the number of bits is higher.

An inventory of the n-bit CLA adder reveals that the sum and the carry subunits require a total of $9n$ two-input NAND gates. The carry section, however, requires $(2n + 1)$ two-input NAND gates and $(n + 3 - m)$ m-input NAND gates for $3 \leq m \leq n + 1$. Therefore, from a practical standpoint CLA for too large n turns out to be quite problematic. It was seen in Chapter 3 that NAND gates with too many inputs are hard to come by. This limitation is compounded by the fact that the carry-in, C_{-1}, must drive a total of $n + 1$ gates. In addition, the propagate functions will be subjected to a fan-out requirement on the order of $(n + 1)^2/4$. All of these together result in a serious fan-in problem for n too large.

FIGURE 5.15 **Four-Bit CLA Adder: [*a*] Block Diagram and [*b*] Internal Circuitry.**

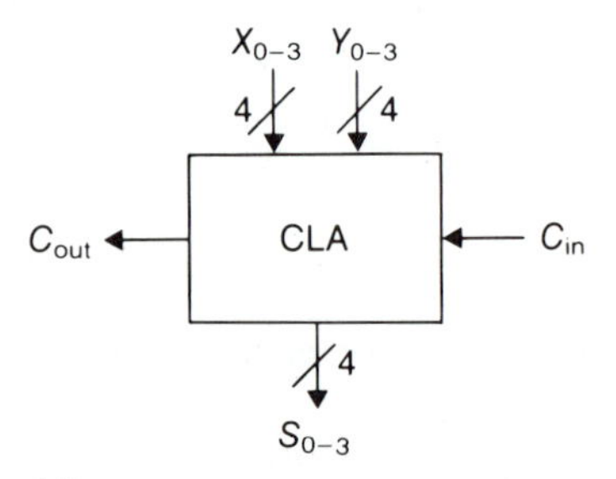

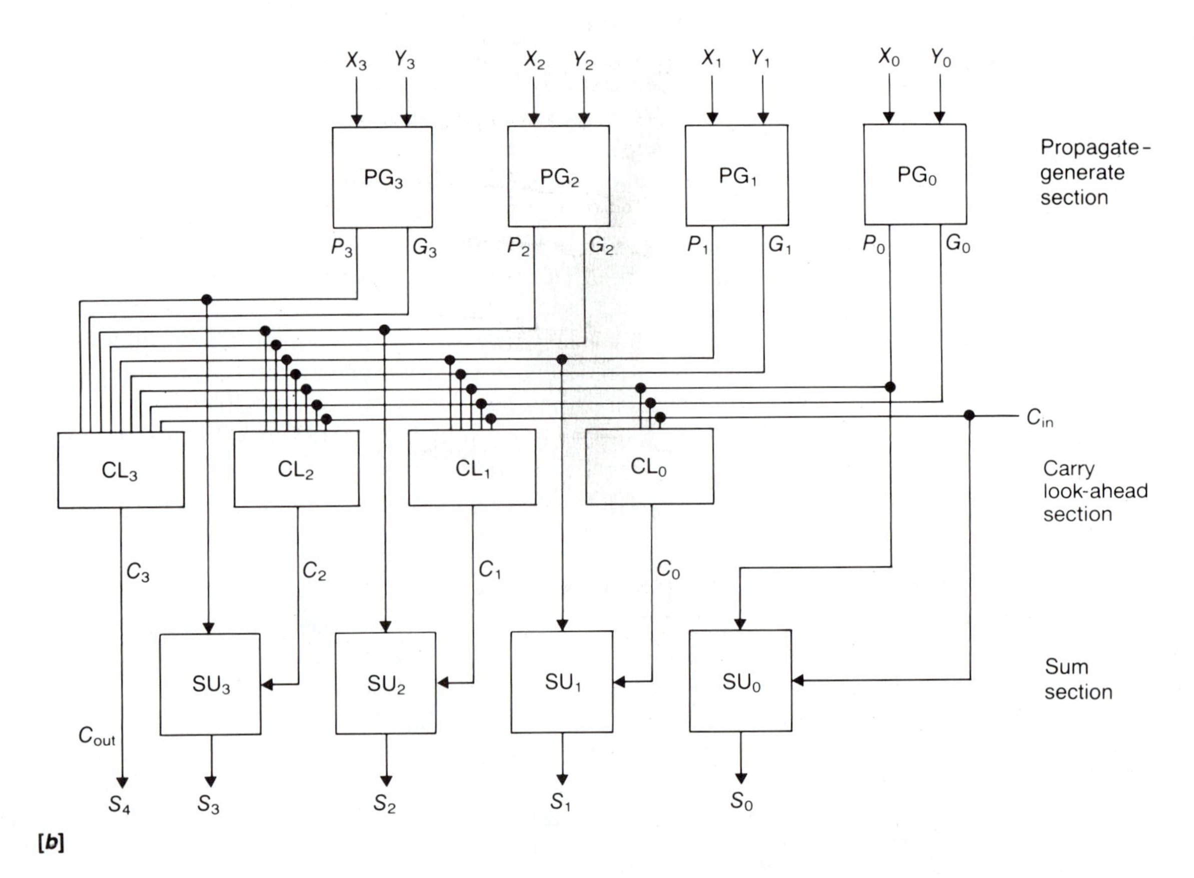

Commercially available four-bit adders perform an internal CLA for a four-bit add. These outputs are available and labeled as *P* for propagate and *G* for generate. These outputs when used with another MSI device called a *look-ahead carry generator* provide significant acceleration in the add operation, particularly for large num-

FIGURE 5.16 **[a] Single Propagation-Generation Unit, [b] Single Sum Unit, and [c] Carry Units of a Four-Bit CLA Adder.**

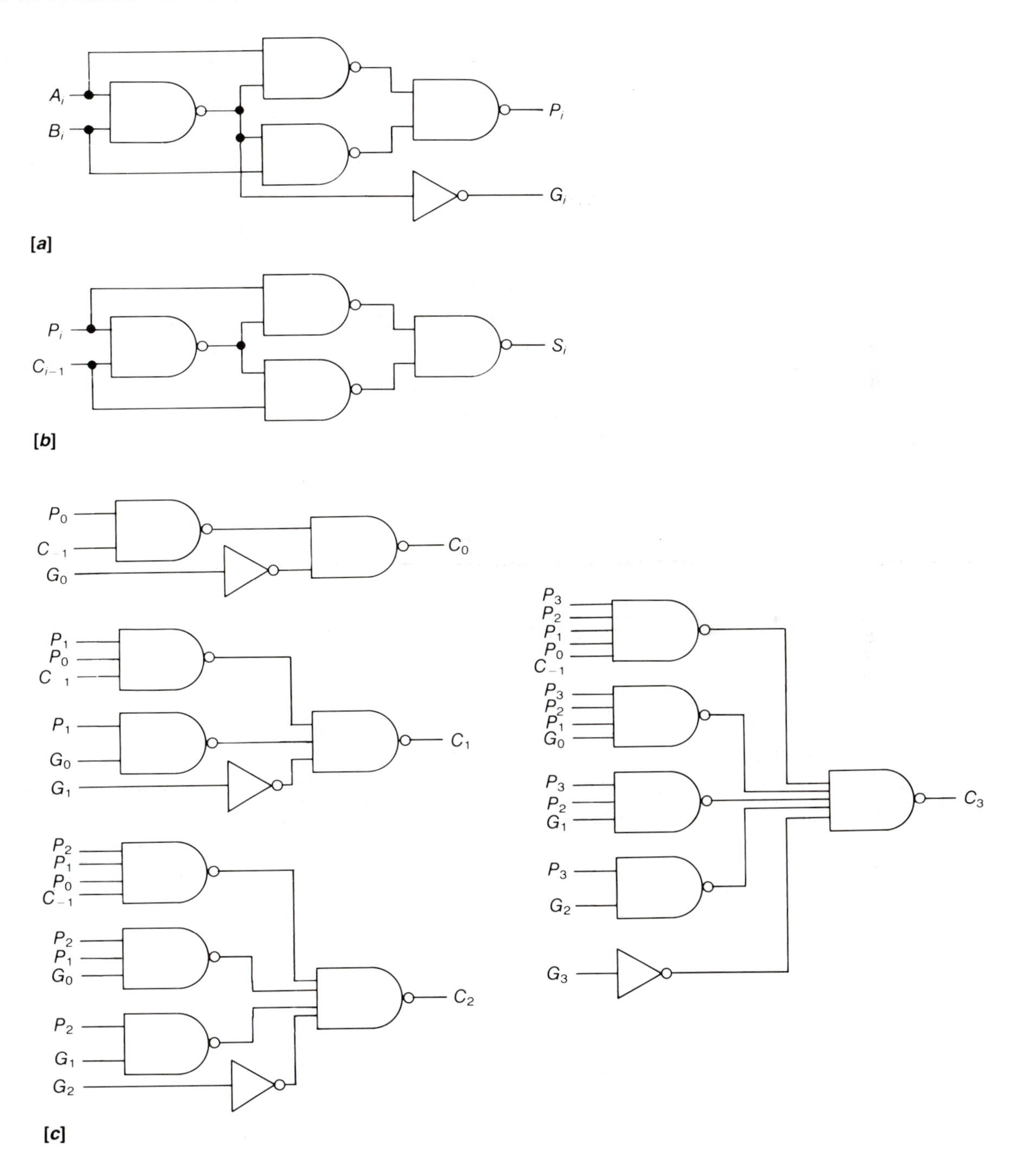

bers of bits. Figure 5.17 shows the connections for performing the addition of two 16-bit numbers with CLA.

Carry-propagate and carry-generate functions for more than four bits can be derived continuing the same process used for deriving C_3 and S_3. Circuit complexity makes such implementation impractical except for special-purpose, high-speed requirements.

FIGURE 5.17 **16-Bit Addition with CLA Modules.**

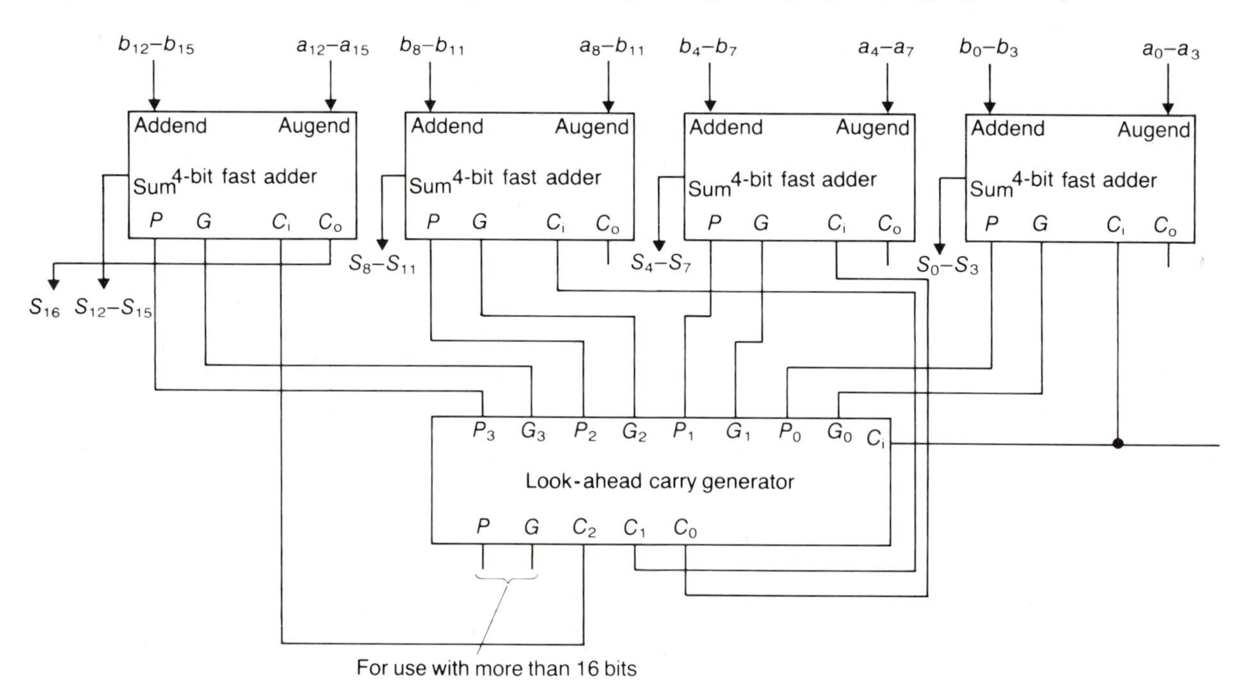

EXAMPLE 5.4

Design an eight-bit fast adder where the fan-in and fan-out problems are avoided by allowing the carries to ripple through after the addition of every four bits.

SOLUTION

The modified carry equations for an eight-bit fast adder are derived as follows:

$$C_0 = G_0 + C_{-1}P_0$$

$$C_1 = G_1 + G_0P_1 + C_{-1}P_0P_1$$

$$C_2 = G_2 + G_1P_2 + G_0P_1P_2 + C_{-1}P_0P_1P_2$$

$$C_3 = G_3 + G_2P_3 + G_1P_2P_3 + G_0P_1P_2P_3 + C_{-1}P_0P_1P_2P_3$$

$$C_4 = G_4 + C_3P_4$$

$$C_5 = G_5 + G_4P_5 + C_3P_4P_5$$

$$C_6 = G_6 + G_5P_6 + G_4P_5P_6 + C_3P_4P_5P_6$$

$$C_7 = G_7 + G_6P_7 + G_5P_6P_7 + G_4P_5P_6P_7 + C_3P_4P_5P_6P_7$$

The first four equations are similar to Equations [5.15–18]. The last four equations have a similar form, but C_3 is treated as the carry-in to the fifth bit. The resultant circuit may now be obtained, as shown in Figure 5.18, by having two separate units for the carry section.

FIGURE 5.18

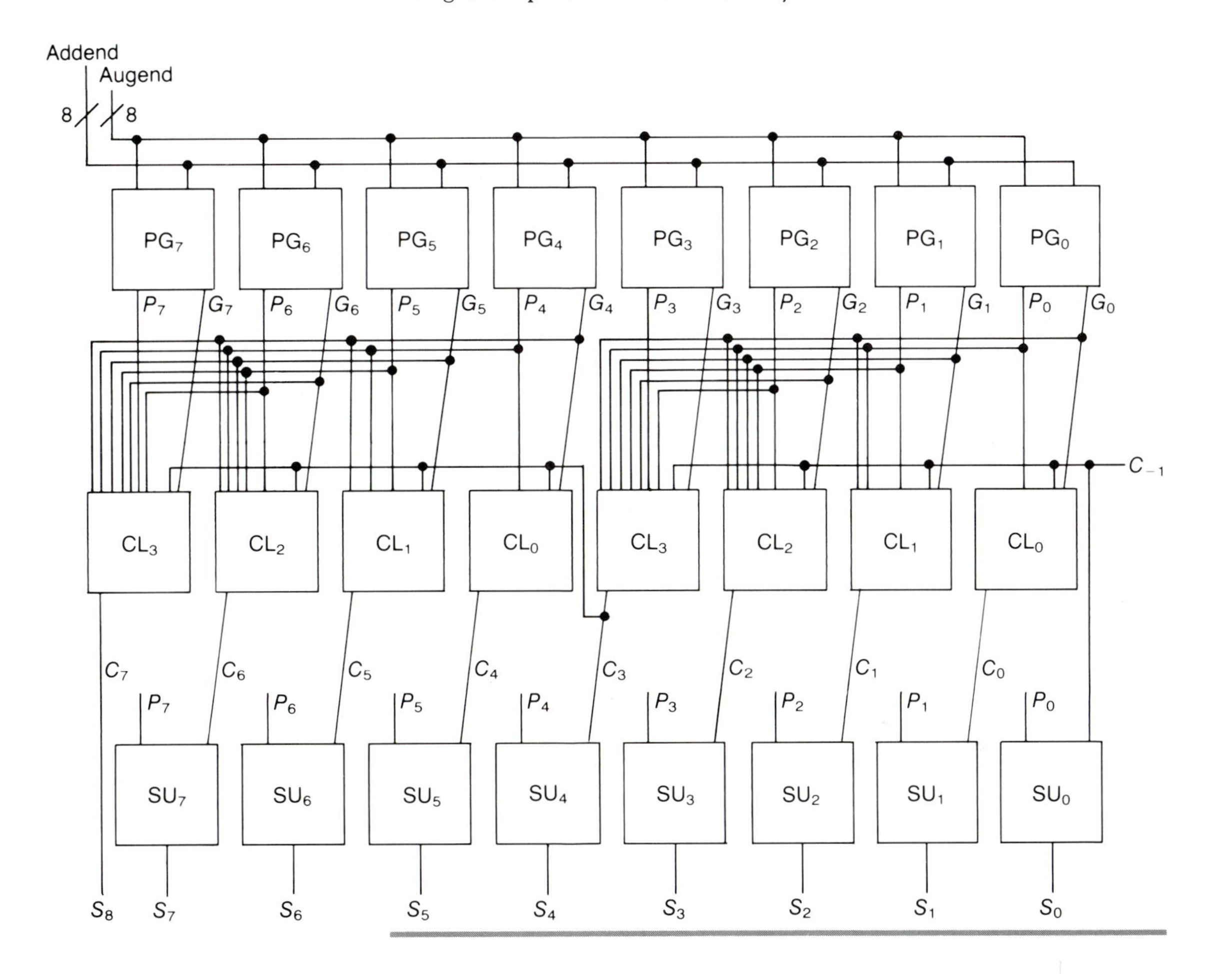

5.5 Code Converters

It was pointed out in Chapter 1 that many different binary codes exist that are used in various digital subsystems. Sometimes it is necessary to transfer data from one subsystem to another. *Code converter circuits* are required to convert one form of binary code to another. Many of these converters use combinational logic, and there are many that use sequential logic as well.

The following examples will illustrate the combinational techniques in the design of various code converters. We shall consider several conversion schemes: Gray-to-binary, binary-to-Gray, binary-to-BCD, and BCD-to-binary.

EXAMPLE 5.5

Design a circuit for converting a five-bit Gray code into its binary equivalent.

SOLUTION

The Gray code is a reflected binary code such that the Gray code for one number differs from the next number in only one bit position. In binary space the codes are one unit distance apart. The truth table of Figure 5.19 shows the corresponding Gray and binary numbers.

FIGURE 5.19

Gray					Binary				
G_4	G_3	G_2	G_1	G_0	B_4	B_3	B_2	B_1	B_0
0	0	0	0	0	0	0	0	0	0
0	0	0	0	1	0	0	0	0	1
0	0	0	1	1	0	0	0	1	0
0	0	0	1	0	0	0	0	1	1
0	0	1	1	0	0	0	1	0	0
0	0	1	1	1	0	0	1	0	1
0	0	1	0	1	0	0	1	1	0
0	0	1	0	0	0	0	1	1	1
0	1	1	0	0	0	1	0	0	0
0	1	1	0	1	0	1	0	0	1
0	1	1	1	1	0	1	0	1	0
0	1	1	1	0	0	1	0	1	1
0	1	0	1	0	0	1	1	0	0
0	1	0	1	1	0	1	1	0	1
0	1	0	0	1	0	1	1	1	0
0	1	0	0	0	0	1	1	1	1
1	1	0	0	0	1	0	0	0	0
1	1	0	0	1	1	0	0	0	1
1	1	0	1	1	1	0	0	1	0
1	1	0	1	0	1	0	0	1	1
1	1	1	1	0	1	0	1	0	0
1	1	1	1	1	1	0	1	0	1
1	1	1	0	1	1	0	1	1	0
1	1	1	0	0	1	0	1	1	1
1	0	1	0	0	1	1	0	0	0
1	0	1	0	1	1	1	0	0	1
1	0	1	1	1	1	1	0	1	0
1	0	1	1	0	1	1	0	1	1
1	0	0	1	0	1	1	1	0	0
1	0	0	1	1	1	1	1	0	1
1	0	0	0	1	1	1	1	1	0
1	0	0	0	0	1	1	1	1	1

The next step in the design process normally would be to produce five five-variable K-maps for determining B_4, B_3, B_2, B_1, and B_0 as functions of G_4, G_3, G_2, G_1, and G_0. Note, however, that G_4 and B_4 are exactly alike. In addition, B_3 is seen to be a 1 when only one of G_4 and G_3 is a 1; B_2 is a 1 when only an odd number of G_4, G_3, and G_2 are 1; B_1 is a 1 when an odd number of G_4, G_3, G_2, and G_1 are 1; and, finally, B_0 is a 1 when an odd number of the five Gray code inputs are 1. These observations eliminate the need for the K-map minimization scheme, although as a general rule, K-maps should be used and will always give a correct, if not minimum, solution. As in this particular case, however, careful observations may at times reduce complex problems to simpler ones. With our prior experience the application of X-ORs to this problem should become obvious. The binary outputs are obtained as follows:

$$B_4 = G_4$$
$$B_3 = G_3 \oplus G_4$$
$$B_2 = G_2 \oplus G_3 \oplus G_4$$
$$B_1 = G_1 \oplus G_2 \oplus G_3 \oplus G_4$$
$$B_0 = G_0 \oplus G_1 \oplus G_2 \oplus G_3 \oplus G_4$$

Four two-input X-OR gates may be cascaded, as shown in Figure 5.20, to realize the desired Gray-to-binary conversion circuit. This design could be extended to an n-bit converter by cascading an additional ($n - 5$) two-input X-OR gates.

FIGURE 5.20

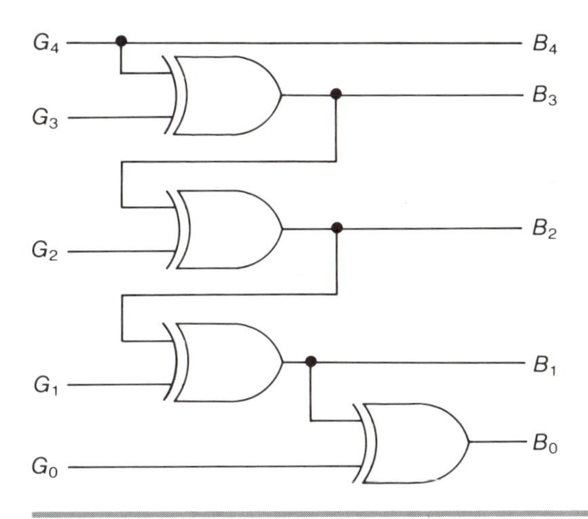

EXAMPLE 5.6

Design an n-bit code converter for converting binary to the equivalent Gray code.

SOLUTION

The truth table of Example 5.5 could be used again to solve for G_0 through G_4 in terms of B_0 through B_4. Checking for patterns in the bit relationships to avoid going through a lengthy minimization process, we note that G_0 is a 1 only when either B_0 or B_1 is a 1. Likewise, G_1 is a 1 when either B_1 or B_2 is a 1; G_2 is a 1 when either B_2 or B_3 is a 1; and, similarly, G_{n-1} is a 1 when B_{n-1} is a 1. Using our knowledge of X-ORs we obtain

$$G_{n-1} = B_{n-1}$$

$$\cdots\cdots\cdots\cdots\cdots$$

$$G_3 = B_3 \oplus B_4$$

$$G_2 = B_2 \oplus B_3$$

$$G_1 = B_1 \oplus B_2$$

$$G_0 = B_0 \oplus B_1$$

The logic circuit of an n-bit binary-to-Gray converter is obtained as shown in Figure 5.21.

FIGURE 5.21

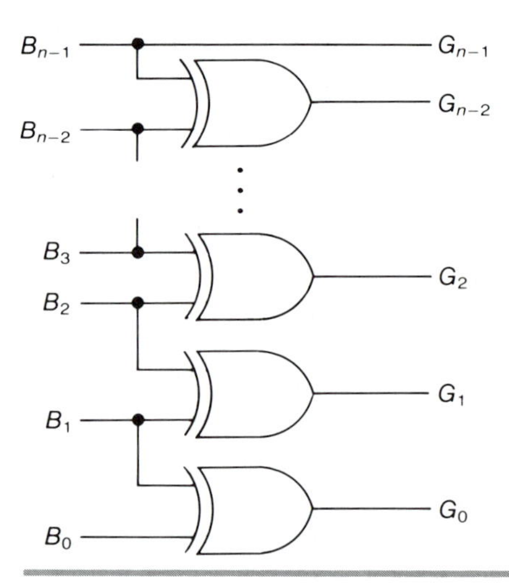

EXAMPLE 5.7

Design a four-bit module that could be used to obtain an n-bit binary-to-BCD conversion circuit.

SOLUTION

The partially complete truth table for the binary-to-BCD conversion is obtained as shown in Figure 5.22. It can be seen that the LSB of both the binary and the BCD numbers are the same. We could, therefore, design a circuit that will convert the remaining binary bits to the corresponding BCD bit: $D_4D_3D_2D_1$. A close examination of the truth table reveals that

$$D_4D_3D_2D_1 = \begin{cases} (B_4B_3B_2B_1) & \text{when } 0 < (B_4B_3B_2B_1) \leq 0100 \\ (B_4B_3B_2B_1 + 0011) & \text{otherwise} \end{cases}$$

The regular design of the conversion module would consist of obtaining the minimized Boolean expressions for D_4, D_3, D_2, and D_1 from the corresponding K-maps as shown in Figure 5.23. In the K-map, however, the output

FIGURE 5.22

Binary					BCD							
B_4	B_3	B_2	B_1	B_0	D_7	D_6	D_5	D_4	D_3	D_2	D_1	D_0
0	0	0	0	0	0	0	0	0	0	0	0	0
0	0	0	0	1	0	0	0	0	0	0	0	1
0	0	0	1	0	0	0	0	0	0	0	1	0
0	0	0	1	1	0	0	0	0	0	0	1	1
0	0	1	0	0	0	0	0	0	0	1	0	0
0	0	1	0	1	0	0	0	0	0	1	0	1
0	0	1	1	0	0	0	0	0	0	1	1	0
0	0	1	1	1	0	0	0	0	0	1	1	1
0	1	0	0	0	0	0	0	0	1	0	0	0
0	1	0	0	1	0	0	0	0	1	0	0	1
0	1	0	1	0	0	0	0	1	0	0	0	0
0	1	0	1	1	0	0	0	1	0	0	0	1
0	1	1	0	0	0	0	0	1	0	0	1	0
0	1	1	0	1	0	0	0	1	0	0	1	1
0	1	1	1	0	0	0	0	1	0	1	0	0
0	1	1	1	1	0	0	0	1	0	1	0	1
1	0	0	0	0	0	0	0	1	0	1	1	0
1	0	0	0	1	0	0	0	1	0	1	1	1
1	0	0	1	0	0	0	0	1	1	0	0	0
1	0	0	1	1	0	0	0	1	1	0	0	1

FIGURE 5.23

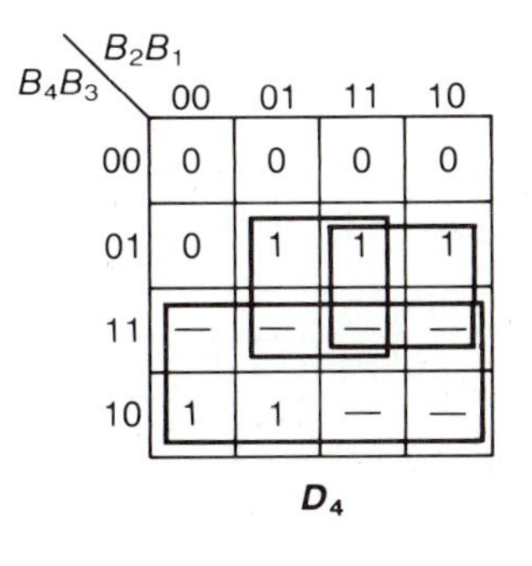

B_4B_3 \ B_2B_1	00	01	11	10
00	0	0	0	0
01	1	0	0	0
11	—	—	—	—
10	0	1	—	—

D_3

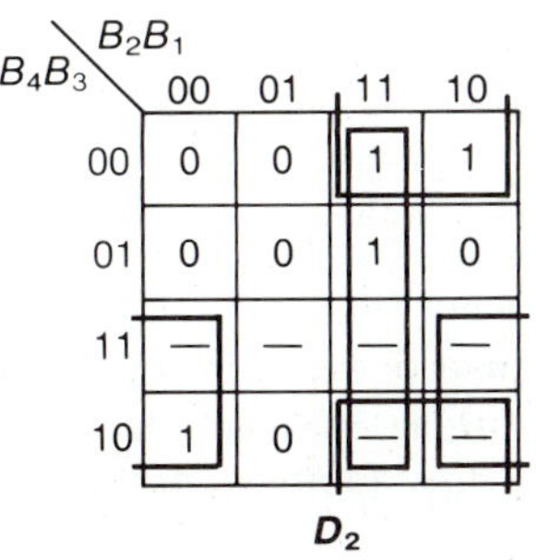

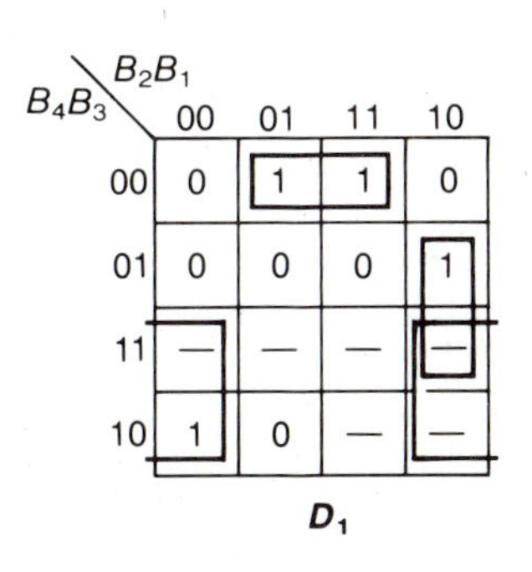

corresponding to the inputs 10–15 may be considered as don't-cares. The equations for the BCD digits are found to be

$$D_4 = B_3B_1 + B_3B_2 + B_4$$
$$D_3 = B_4B_1 + B_3\overline{B}_2\overline{B}_1$$
$$D_2 = B_2B_1 + B_4\overline{B}_1 + \overline{B}_3B_2$$
$$D_1 = \overline{B}_4\overline{B}_3B_1 + B_3B_2\overline{B}_1 + B_4\overline{B}_1$$

The implementation of the conversion module, as shown in Figure 5.24, is now reasonably straightforward using the preceding equations.

FIGURE 5.24

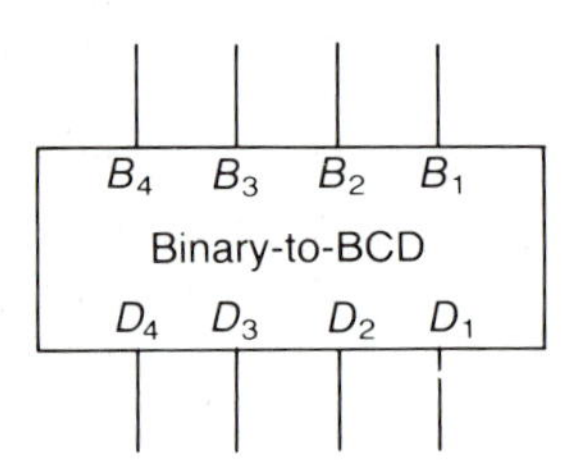

The equations found for D_4, D_3, D_2, and D_1 are the equations that would be implemented in a commercial binary-to-BCD IC. Another solution using an adder would involve designing a circuit that outputs a 1 whenever $B_4B_3B_2B_1 > 0100$. The circuit output is tied to the least significant two addend inputs of a four-bit adder while B_4, B_3, B_2, and B_1 are tied to the corresonding augend inputs. The sum outputs of the four-bit adder unit, as shown in Figure 5.25, would yield the desired BCD output.

FIGURE 5.25

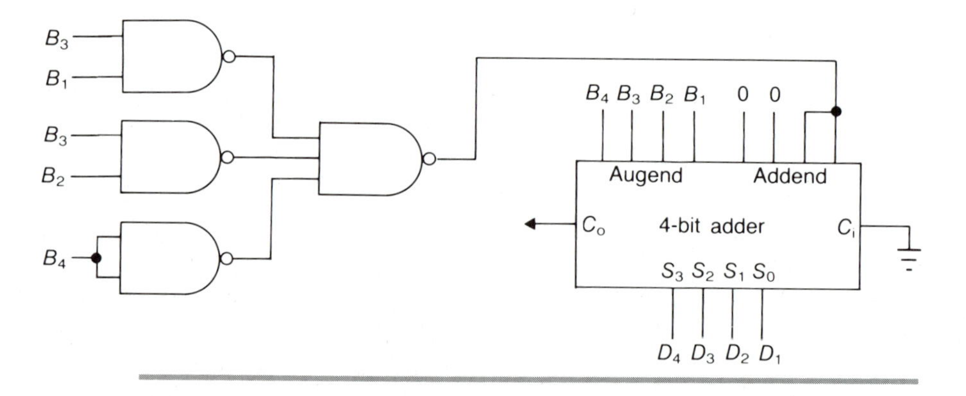

Before the module designed in Example 5.7 is used for conversions of more than five bits, a thorough understanding of the interrelationship between a binary number and its BCD equivalent number is necessary. The decimal value of an n-bit binary number was given by Equation [1.4] as

$$(NI)_{10} = ((\ldots(((a_{n-1})2 + a_{n-2})2 + a_{n-3})2 + \cdots + a_2)2 + a_1)2 + a_0$$

where each of the coefficients, a_j, is either 0 or 1. This nested multiplication by two can be carried out by shifting the binary number $(0.a_{n-1}a_{n-2}a_{n-3} \ldots a_3a_2a_1a_0)$ to the left n times. As these bits are shifted left, each group of four consecutive bits, beginning with the binary point, represents a *decade*. Within each decade each left-shift represents a multiplication by two. However, if a bit passes from the MSB of one decade to the LSB of the next higher decade, its value increases from 8 to 10 only (instead of 16). Figure 5.26 shows the nine possible BCD digits, their values after being shifted left without correction, and their values after being corrected and then shifted.

Note that for values less than or equal to 0100, the shifted value gives the correct BCD digit. If the BCD digit is greater than 0100, the shifted number can be corrected by adding a binary 0110 to the shifted bits. An equivalent correction can also be made by adding 0011 to the BCD digit prior to the shifting. The device already designed in Example 5.7 adds 0011 to the four-bit input if the four bits represent a value greater than binary four.

***FIGURE 5.26* Shifted BCD Digits.**

BCD Digits	BCD Digits Shifted Left One Bit	BCD Digits Corrected and Then Shifted Left One Bit
0000	0 0000	0 0000
0001	0 0010	0 0010
0010	0 0100	0 0100
0011	0 0110	0 0110
0100	0 1000	0 1000
0101	0 1010	1 0000
0110	0 1100	1 0010
0111	0 1110	1 0100
1000	1 0000	1 0110
1001	1 0010	1 1000

In order to provide an understanding of the use of multiple binary-to-BCD converters in converting binary digits of more than five bits, an example involving seven bits will be solved with explanations of each step. The steps will lead us to an algorithm that can be used for the conversion of any n-bit number. The converter module we designed is used to correct a BCD digit prior to the necessary shifting mechanism. The process of shifting bits is achieved by means of hard-wiring between separate stages of converter modules. A separate stage of converter module would be necessary for each of the shift operations.

EXAMPLE 5.8

Convert $127_{10} = 1111111_2$ to BCD.

SOLUTION

Step 1. *Correct-Shift*: Note that for the first two left-shifts of 0.1111111_2 no converter module would be necessary. The third shift, however, results in an integer larger than four. Therefore, the integer needs to be corrected before further shifting of bits:

$0111\,.\,1111 \times 2^4$	uncorrected
$1010\,.\,1111 \times 2^4$	corrected BCD prior to shift
$1\ 0101\,.\,111 \times 2^3$	hybrid number after shift

After completing Step 1 we have a hybrid number; the integer portion is in BCD while the fractional portion is in binary. The value of this hybird number is

$$[15 + (7/8)] \times 2^3 = 127_{10}$$

Step 2. *Correct-Shift*: The four bits on the immediate left of the binary point could be larger than 0100, and in the present case they are. A new correct-shift operation is required, therefore:

$1\ 0101\,.\,111 \times 2^3$	from previous operation
$1\ 1000\,.\,111 \times 2^3$	corrected BCD prior to shift
$11\ 0001\,.\,11 \times 2^2$	hybrid number after shift

Also,

$$[31 + (3/4)] \times 2^2 = 127_{10}$$

Keep in mind that the shift operation just performed brought in a bit from the right of the binary point to be included in the BCD portion. This move is valid since our four-bit input binary-to-BCD converter actually converts five bits, the four connected and the one immediately to the right of those connected.

Prior to shifting, we again add 0011 to those BCD digits that could be greater than 0100. The left-most BCD digit can at most be 0011, so no correction prior to shifting is necessary.

Step 3. *Correct-Shift*:

$11\ 0001\,.\,11 \times 2^2$	from previous step
$11\ 0001\,.\,11 \times 2^2$	corrected BCD prior to shift
$110\ 0011\,.\,1 \times 2^1$	hybrid number after shift

The hybrid number gives $[63 + 1/2] \times 2^1 = 127_{10}$.

Step 4. The next correction prior to shifting requires two devices since there are sufficient bits in both BCD digits to have values greater than 0100. Again, the left-most binary-to-BCD has one input connected to 0 to allow for the possibility of a four-bit output with only three bits in:

0110 0011 . 1 × 2^1	from previous step
1001 0011 . 1 × 2^1	corrected BCD prior to shift
1 0010 0111 . 0	hybrid number after shift

The process is now complete.

You would discover after reviewing the preceding steps that the bit positions can remain constant provided the converter modules are moved right to effect a left-shift. The process gives the circuit of Figure 5.27.

FIGURE 5.27

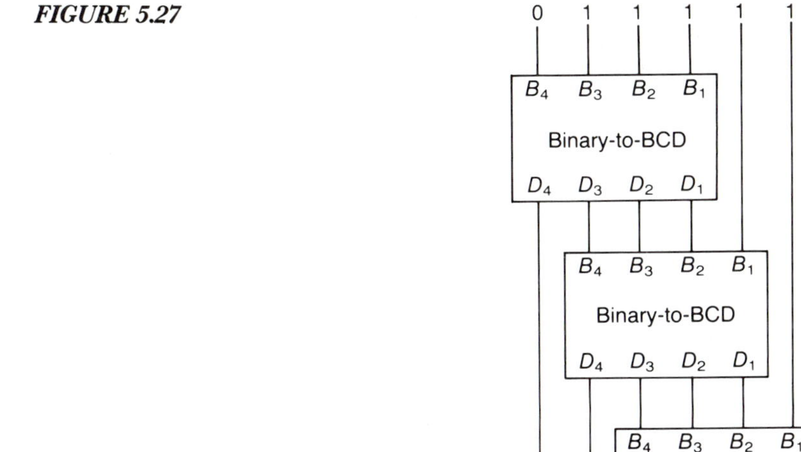

Consider the pattern in the circuit of Figure 5.27. The conversion of an n-bit binary number to BCD can be performed using the converter designed in Example 5.7 as follows:

1. Add a 0 to the left of the MSB position of the binary number. Connect to the inputs of a four-bit converter module a 0 and the three MSBs of the binary number.

2. Take the three least significant processed bits and the most significant unprocessed bit as inputs to a converter module. The *most significant processed bit* (*MSPB*), which is the MSB of the left-most converter module, is considered in the next step.
3. If there are three MSPBs, connect a 0 to the MSB position of a converter module and the three MSPBs to the remaining three inputs; otherwise carry the MSPBs from above operations to the next level.
4. Once a converter module is added to a level, the number of converter modules remains the same until three MSPBs accumulate, at which time another is added.
5. This process is continued until the LSB is the only unprocessed bit.

EXAMPLE 5.9

Use several of the modules designed in Example 5.7 for converting the binary number 11010110100_2 to its BCD equivalent.

SOLUTION

The logic circuit necessary for obtaining the conversion of this 11-bit binary number follows directly from the previous discussion of the *n*-bit binary-to-BCD conversion algorithm. The resulting circuit configuration is shown in Figure 5.28. After every three levels, the number of modules per level increases by one. The output of the network is 1716_{BCD} as expected.

EXAMPLE 5.10

Design a four-bit module suitable for the BCD-to-binary conversion of five bits.

SOLUTION

Since the LSBs of the binary and BCD numbers are equal, the design will consider the four MSBs of the five-bit BCD value. The design involves undoing the conversion of binary-to-BCD numbers considered in the preceding few examples. Intuitively we might assert that this will involve shifting right and then correcting the resultant values by subtracting 0011. Accordingly, the combinational module that could be used for this algorithm has the following characteristics:

$$B_4B_3B_2B_1 = \begin{cases} (D_4D_3D_2D_1) & \text{when } (D_4D_3D_2D_1) < 1000 \\ (D_4D_3D_2D_1 - 0011) & \text{otherwise} \end{cases}$$

FIGURE 5.28

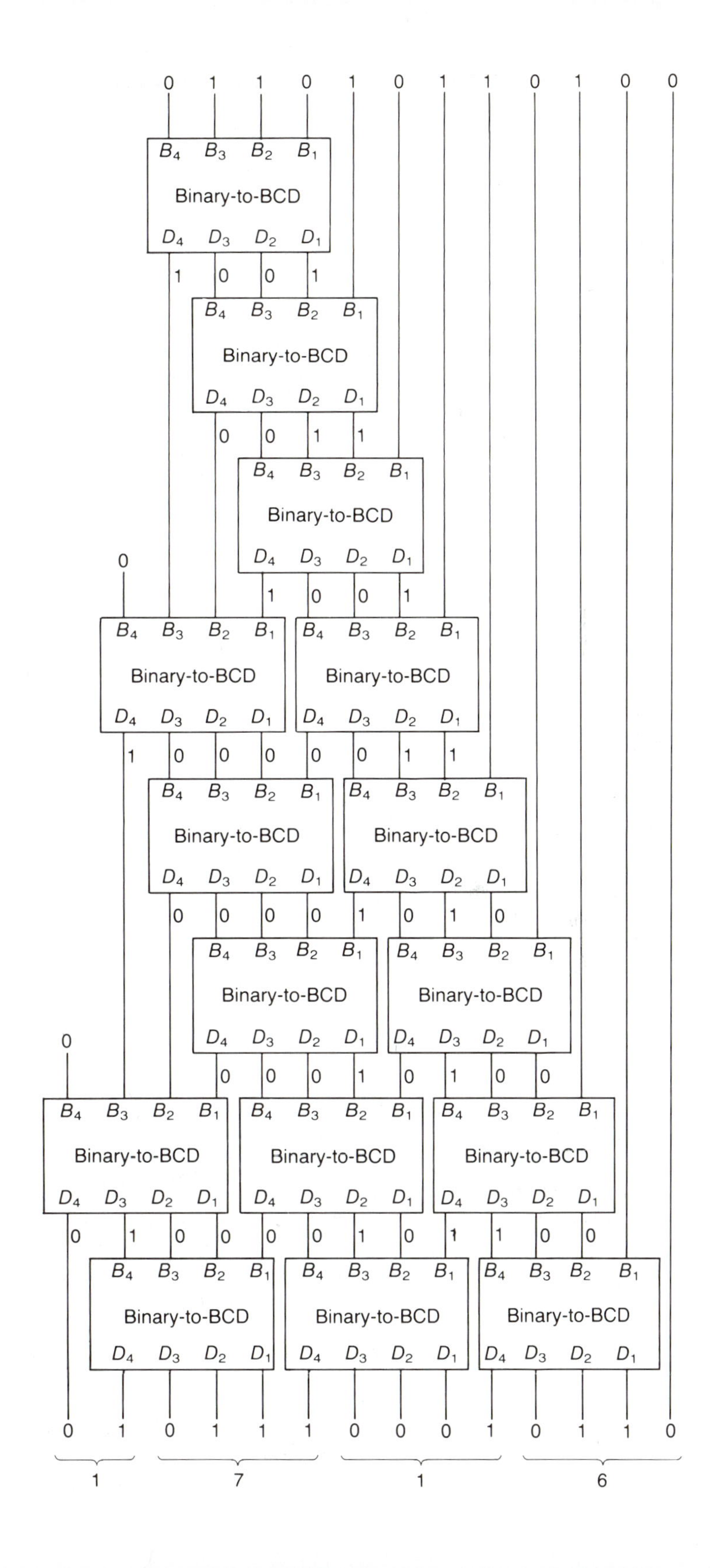
0 1 1 0 1 0 1 1 0 1 0 0
B4 B3 B2 B1
Binary-to-BCD
D4 D3 D2 D1
0 1 0 1 1 1 0 0 0 1 0 1 1 0
1 7 1 6

FIGURE 5.29

D_4	D_3	D_2	D_1	B_4	B_3	B_2	B_1
0	0	0	0	0	0	0	0
0	0	0	1	0	0	0	1
0	0	1	0	0	0	1	0
0	0	1	1	0	0	1	1
0	1	0	0	0	1	0	0
0	1	0	1	—	—	—	—
0	1	1	0	—	—	—	—
0	1	1	1	—	—	—	—
1	0	0	0	0	1	0	1
1	0	0	1	0	1	1	0
1	0	1	0	0	1	1	1
1	0	1	1	1	0	0	0
1	1	0	0	1	0	0	1
1	1	0	1	—	—	—	—
1	1	1	0	—	—	—	—
1	1	1	1	—	—	—	—

The truth table for such a module is obtained as shown in Figure 5.29.

The don't-cares in the binary table occur for six different input values that will never appear as uncorrected BCD digits. For example, a 0101 will never occur since that would imply the presence of an unlikely BCD number, either 1010 or 1011, prior to the shift-right operation. Examination of the truth table indicates that BCD digits equal to or less than 0100 require no modification. However, for BCD digits 8 through 12 a value of 3 must be subtracted for correction. Constructing and minimizing the corresponding K-maps for B_4, B_3, B_2, and B_1 gives the following Boolean equations:

$$B_1 = D_1\bar{D}_4 + \bar{D}_1D_4 = D_1 \oplus D_4$$
$$B_2 = D_2\bar{D}_4 + D_1\bar{D}_2D_4 + \bar{D}_1D_2$$
$$B_3 = D_2\bar{D}_4 + D_1\bar{D}_2D_4 + \bar{D}_1\bar{D}_3D_4$$
$$B_4 = D_3D_4 + D_1D_2D_3$$

The circuit for the BCD-to-binary module may now be obtained as shown in Figure 5.30.

FIGURE 5.30

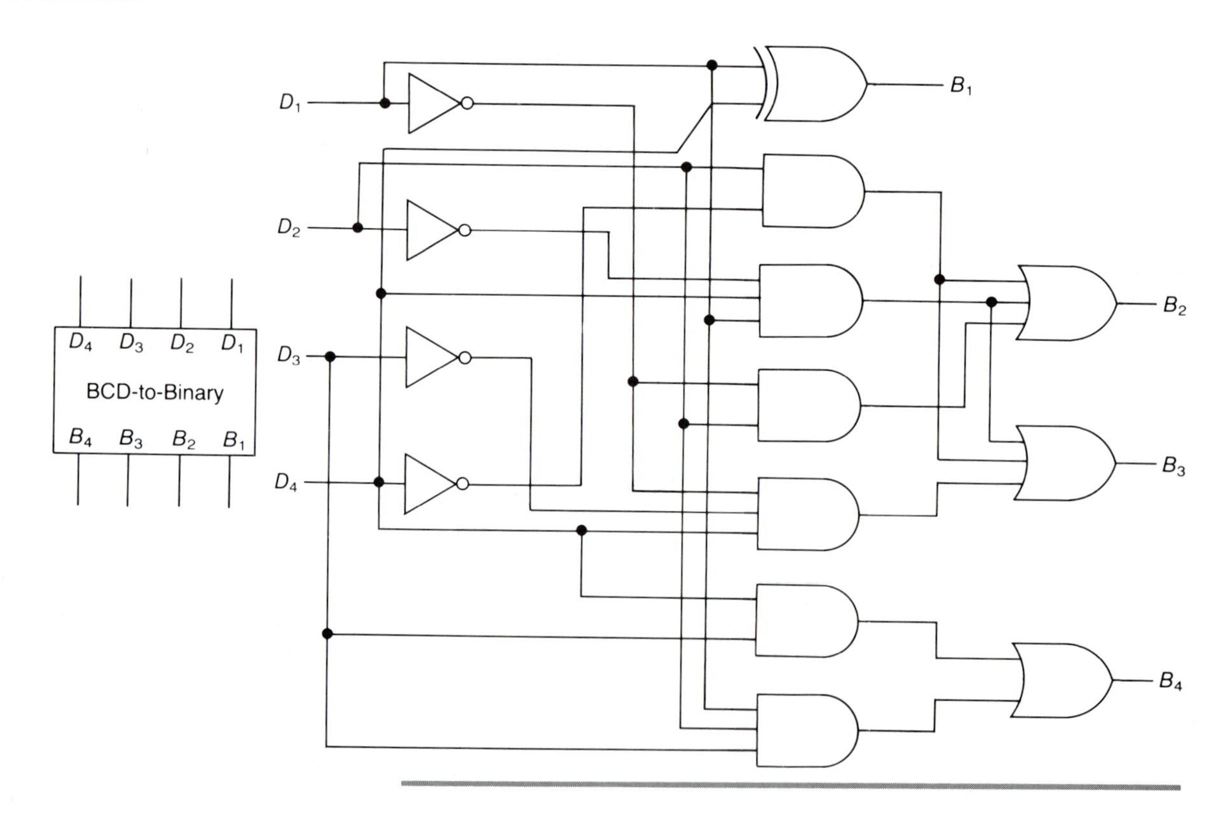

EXAMPLE 5.11

Use the module designed in Example 5.10 to perform the conversion of 127_{10} and consequently develop an algorithm for use in an *n*-bit BCD-to-binary conversion. Note that

$127_{10} = 000100100111_{BCD}$

SOLUTION

As mentioned previously, this process should be the reverse of the binary-to-BCD conversion scheme developed earlier.

Step 1. Shift the BCD quantity to the right by one bit, which will replace one bit to the right of the decimal. We are considering a hybrid number again. The digits to the right of the radix point will be binary and those to the left BCD. The process of shifting results in uncorrected BCD values. The BCD-to-binary converter is used to correct these uncorrected results:

$1001\ 0011\ .\ 1 \times 2^1$ uncorrected BCD

$0110\ 0011\ .\ 1 \times 2^1$ corrected BCD

The hybrid number is $[63 + (1/2)] \times 2 = 127_{10}$.

Step 2. Note that the left-most 0011 needs no correction:

$0011\ 0001\ .\ 11 \times 2^2$ uncorrected BCD

$0011\ 0001\ .\ 11 \times 2^2$ corrected BCD

The hybrid number still gives $[31 + (3/4)] \times 4 = 127_{10}$.

Step 3. *Correct-Shift:*

$0001\ 1000\ .\ 111 \times 2^3$ uncorrected BCD

$0001\ 0101\ .\ 111 \times 2^3$ corrected BCD

And, $[15 + (7/8)] \times 2^3 = 127_{10}$.

Step 4. At this point we have five bits of BCD left to convert to binary:

$1010\ .\ 1111 \times 2^4$ uncorrected BCD

$0111\ .\ 1111 \times 2^4$ corrected BCD

And, $[7 + (15/16)] \times 2^4 = 127_{10}$.

The resultant value 1111111_2 is indeed equal to 000100100111_{BCD}. The resulting circuit using the converter is obtained as shown in Figure 5.31.

An algorithm now may be developed for extending the process just examined to the conversion of *n*-bit BCD numbers:

Step 1. Starting at the right end, skip the LSB and connect four-input, BCD-to-binary converter modules to all remaining bits. If two or less bits would be connected to the left-most converter, leave it off and extend the bits to the next level. If all bits at a level are connected to a converter module, the MSB of the output will be zero and should not be carried to the next level.

Step 2. Skip the least significant processed bit at each level and reassign bits as directed in Step 1. Continue until the MSB (processed or unprocessed) from an upper level is included as the most significant input to a converter. Remember not to bring down bits that would be zeros from upper MSB positions (see Figure 5.31).

FIGURE 5.31

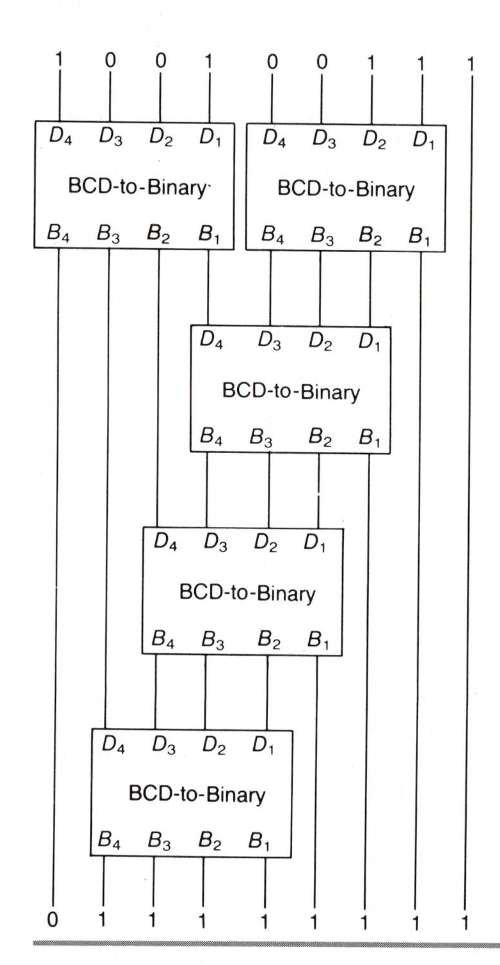

5.6 BCD Arithmetic Circuits

Even though most computers use regular binary numbers for their arithmetic operations, some special-purpose computers and calculators operate in the decimal number system using BCD. BCD-based systems require more memory to store information because of less efficient coding and complex arithmetic circuitry. However, the final results in these systems do not have to be decoded prior to display as decimal digits. BCD arithmetic is usually complicated by the fact that some of the sums or differences are invalid. When two BCD numbers are added on a binary adder, it is possible to obtain 16 different sums, of which six are undefined. In addition, when there is a carry-out, four additional conditions turn out to be undefined as well. Figure 5.32 shows the to-be-corrected and the corresponding corrected BCD sums and carry. The largest uncorrected sum could be 10011, corresponding to the sum of two BCD nines and a carry-in of one. Note that the sums that are greater than 1001 are all undefined and are in need of correction.

FIGURE 5.32 **Table for the Uncorrected and Corrected BCD Sums.**

Uncorrected					Corrected					
C'_3	S'_3	S'_2	S'_1	S'_0	C_3	S_3	S_2	S_1	S_0	
0	0	0	0	0	0	0	0	0	0	No correction needed
• • •			• • •			• • •		• • •		
0	1	0	0	1	0	1	0	0	1	
0	1	0	1	0	1	0	0	0	0	
0	1	0	1	1	1	0	0	0	1	
0	1	1	0	0	1	0	0	1	0	
0	1	1	0	1	1	0	0	1	1	
0	1	1	1	0	1	0	1	0	0	
0	1	1	1	1	1	0	1	0	1	
1	0	0	0	0	1	0	1	1	0	
1	0	0	0	1	1	0	1	1	1	
1	0	0	1	0	1	1	0	0	0	
1	0	0	1	1	1	1	0	0	1	

A BCD adder could be designed using the design techniques we have already studied. Such an adder circuit would have a total of nine inputs and five outputs. Of the nine inputs, four inputs would be for the augend, another four for the addend, and the ninth input would be the carry-in. Of the five outputs, one would be for the carry-out and the remaining four would be for the sums. Therefore, the truth table would consist of $2^9 = 512$ different input combinations, many of which would lead to don't-care conditions. The two obvious design choices for such a circuit would be

1. Make use of a ROM or PAL or PLA.
2. Minimize the function using the Q-M technique and then generate the circuit using combinational gates.

In either case, the design would be too involved to pursue further.

The easiest route to a BCD adder design is to base its operation on a four-bit FA module. If the sum of the corresponding nine inputs exceeds nine, a carry is generated and the sum is corrected by means of a correction circuit. It should be noted from Figure 5.32 that the corrected sum is obtainable from the uncorrected sum simply by adding a $(0110)_2$ to it. Furthermore, it can be seen that C_3 becomes 1 only when correction becomes necessary. The condition for the generation of a carry-out can be found, therefore, from the minimization of the K-map for C_3 shown in Figure 5.33. The Boolean expression for the corrected carry-out becomes

$$C_3 = C_3' + S_1'S_3' + S_2'S_3' \qquad [5.23]$$

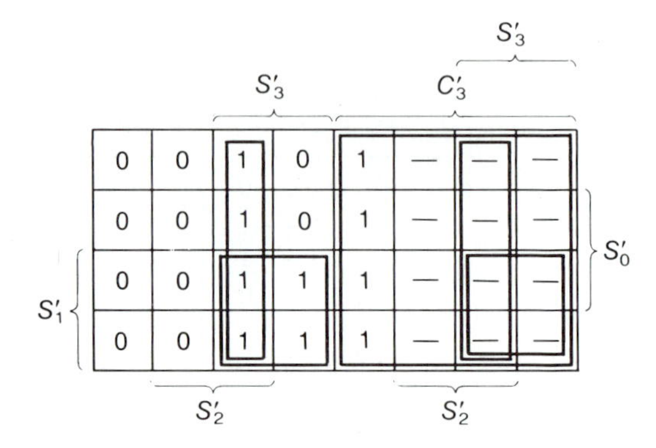

FIGURE 5.33 **K-Map for the Corrected Carry Bit.**

An analysis of our previous discussion would reveal that the single-decade BCD adder would involve the following subunits:

a four-bit FA for addition,

a carry decoder circuit,

a second four-bit FA circuit for correction.

An adder made in this way is shown in Figure 5.34. The sum bits of the first adder unit are introduced at the augend inputs of the second adder unit. When the corrected carry-out, C_3, is a 0, the uncorrected sum remains unchanged. When it is a 1, the uncorrected sum is added to 0110 to give the corrected sum at the output of the second four-bit FA.

At this time the designer may be concerned with the amount of worst-case propagation delays. The propagation delays basically arise from the two FA subunits used in this circuit. A close examination of the circuit, however, reveals that the correction does not need all four of the single-bit FAs. S_0' never needs any correction. An improved version of the BCD adder may be obtained by having only two HAs and one FA in the correction unit. This improved circuit is shown in Figure 5.35. Another approach for designing BCD adders would be first to convert the BCD numbers into binary, add the resulting binary numbers, and then transform the binary sum into the correct BCD sum. This design requires both postcorrection and preconversion circuitries and, therefore, is more expensive than the one already designed.

FIGURE 5.34 **BCD Adder Circuit.**

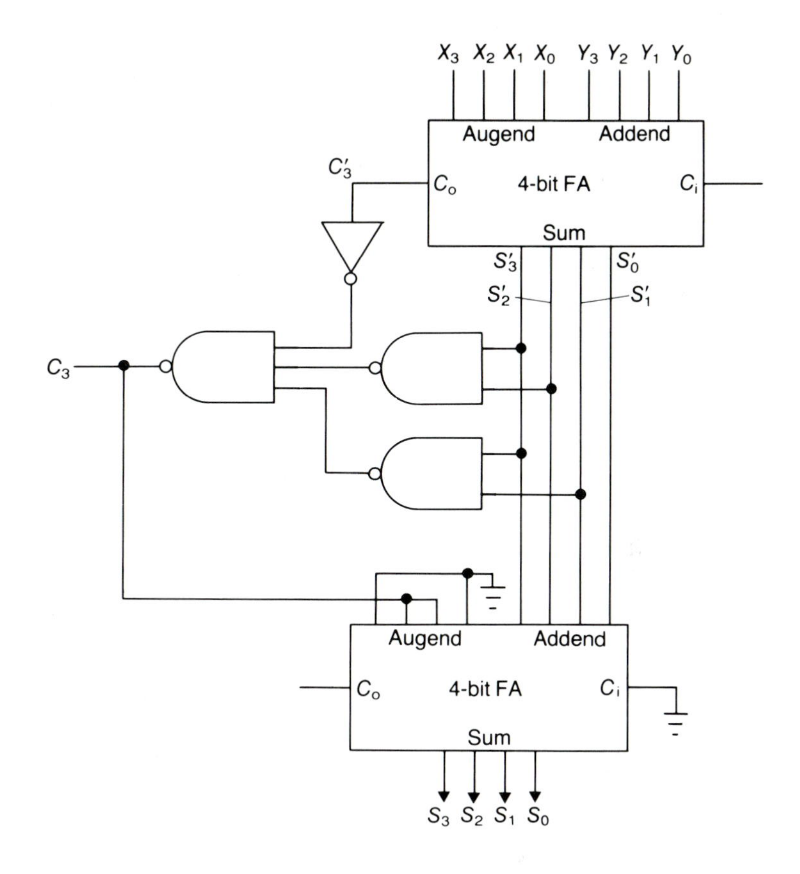

Next consider the design of a BCD subtracter. Subtraction is more complex than BCD addition, since the possibility of having a negative difference exists. Recall from earlier observations made in Sections 1.4, 1.5, and 5.3 that there are at least two ways to handle negative BCD numbers: either by 10's complement or by 9's complement. The technique using 10's complement would result in a circuit similar to that of Figure 5.12. The scheme using 9's complement, however, would require an end-round-carry into the system. The BCD subtracter using 9's complement could be designed using a BCD adder and a 9's complementer circuit. In order to obtain the 9's complementer circuit, consider the truth table in Figure 5.36.

Using the table of Figure 5.36, we would obtain

$$O_3 = \overline{I_1}\,\overline{I_2}\,\overline{I_3} = \overline{I_1 + I_2 + I_3}$$

$$O_2 = I_1 \oplus I_2$$

$$O_1 = I_1$$

$$O_0 = \overline{I_0}$$

FIGURE 5.35 Improved BCD Adder Circuit.

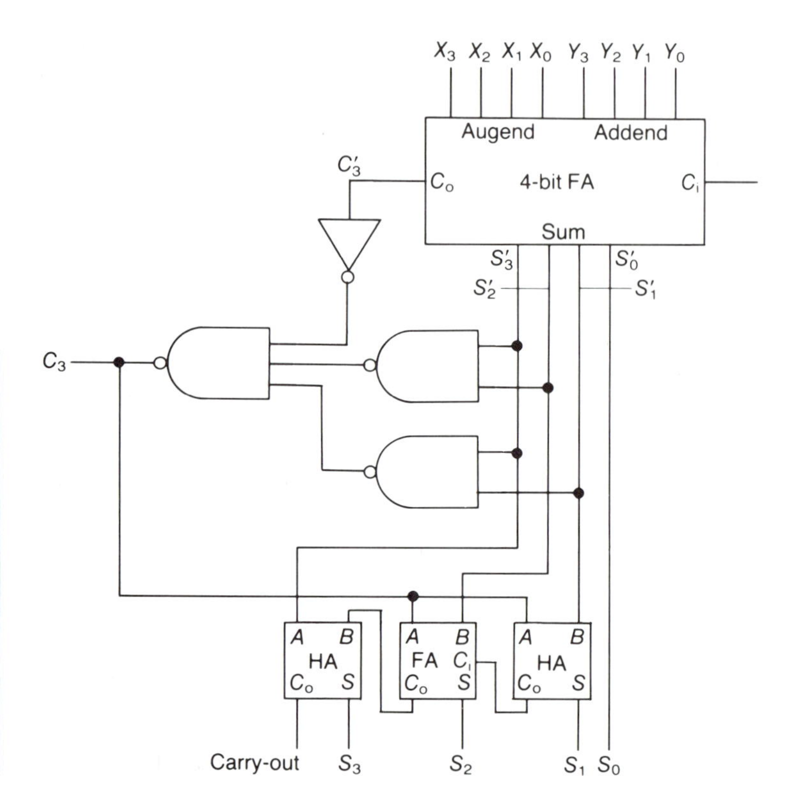

FIGURE 5.36 Truth Table for the BCD 9's Complement.

BCD Input				9's Complement			
I_3	I_2	I_1	I_0	O_3	O_2	O_1	O_0
0	0	0	0	1	0	0	1
0	0	0	1	1	0	0	0
0	0	1	0	0	1	1	1
0	0	1	1	0	1	1	0
0	1	0	0	0	1	0	1
0	1	0	1	0	1	0	0
0	1	1	0	0	0	1	1
0	1	1	1	0	0	1	0
1	0	0	0	0	0	0	1
1	0	0	1	0	0	0	0

A BCD 9's complementer circuit using the preceding equations is shown in Figure 5.37[*a*]. Figure 5.37[*b*] shows how a four-bit adder may be used to achieve the same goal. The latter circuit performs the 1's complement of the BCD digits and then adds to it the 2's complement of six (1010).

FIGURE 5.37 9's Complementer: [*a*] Using Simple Gates and [*b*] Using a Four-Bit FA.

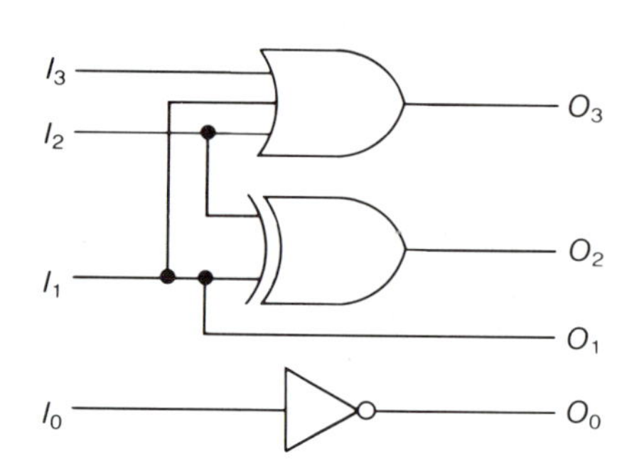

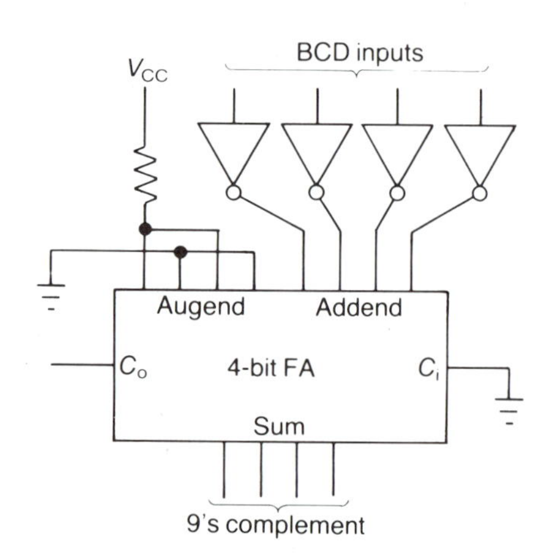

Once the 9's complementer is designed, the rest of the steps are obvious. If X-OR gates are used in place of the inverters of Figure 5.37[*b*], a combined BCD adder/subtracter unit may result. The BCD adder/subtracter unit so designed is shown in Figure 5.38. Each decade unit consists of three four-bit FAs. The first one works

FIGURE 5.38 **A Single-Decade BCD Adder/Subtracter Unit.**

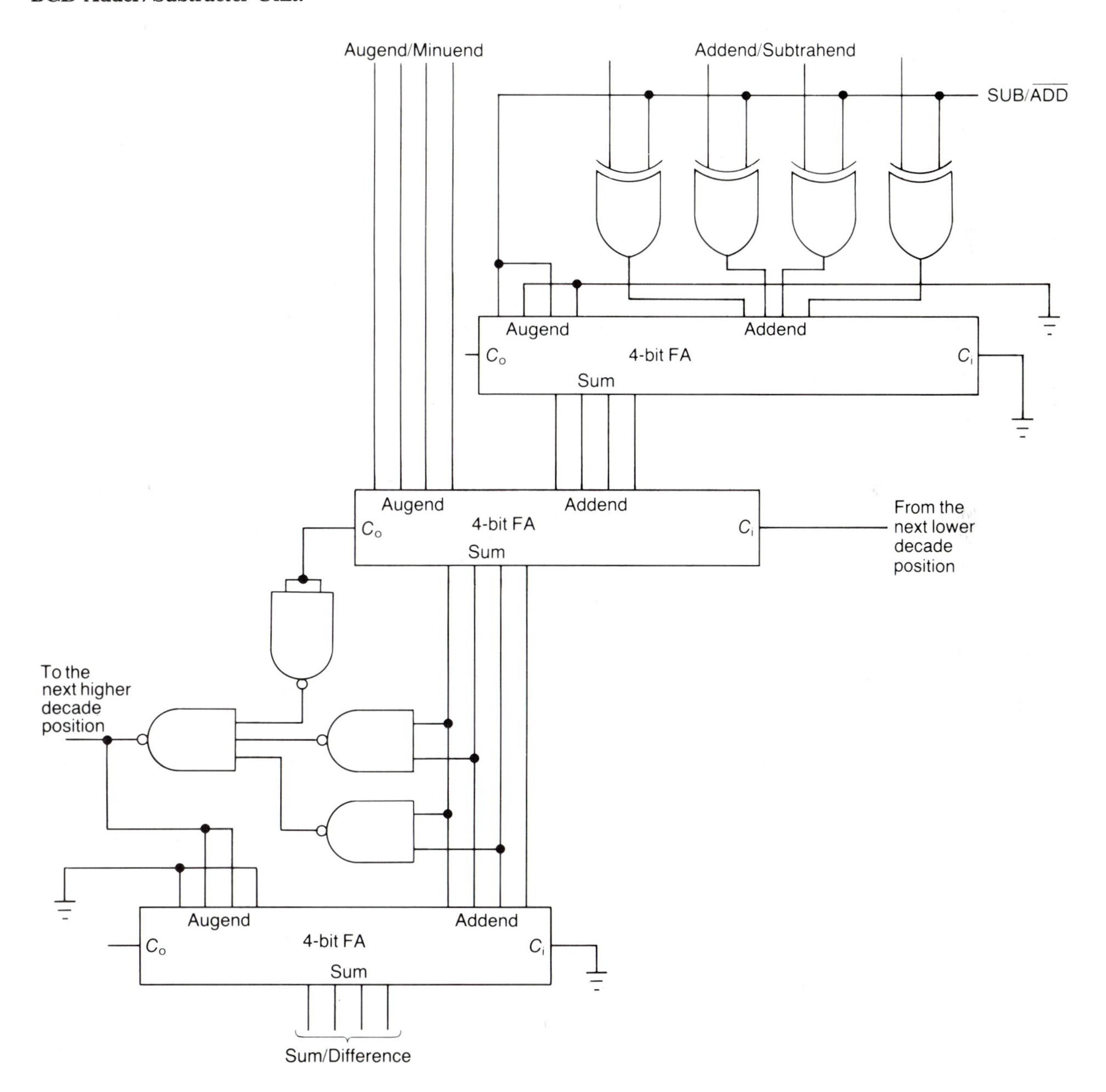

as the 9's complementer unit, the second as the adder unit, and the third works as part of the correction circuit. When the SUB/$\overline{\text{ADD}}$ input is a 0, the unit performs simple addition as the complementer unit leaves the addend unchanged. When SUB/$\overline{\text{ADD}}$ is a 1, the subtrahend is complemented and is then added to the minuend. Consequently, the sum appears at the final output when the SUB/$\overline{\text{ADD}}$ input is a 0 and the difference is obtained otherwise. When several such units are cascaded to make a multi-decade BCD adder/subtracter, the carry-out from the most significant decade unit is fed into the carry-in of the least significant decade unit. This arrangement guarantees the inclusion of an end-around-carry to satisfy the need of the 10's complement system.

EXAMPLE 5.12

Use the XS3 code for designing a BCD adder/subtracter unit.

SOLUTION

As you can see from the truth table in Figure 5.39, the 9's complement of an XS3 number may be obtained simply by taking its 1's complement. Consequently, it is easier to obtain the 9's complement of an XS3 than that of its BCD equivalent.

It would be worthwhile to review Example 1.17 at this time. Notice that correction would be necessary for obtaining the true XS3 sum. Corresponding to the expected decimal numbers 3 and 7, the sum yielded 1001 and 1101, respectively, and no carry-out. However, 1000 and 0000 were respectively obtained along with a carry-out in place of decimal numbers 8 and 0. It can now be concluded that there are two simple rules that must be followed while adding numbers in the XS3 code:

a. If a carry is produced, 0011 is added.

b. If a carry is not produced, 0011 is subtracted. This subtraction is usually done by adding the 2's complement of 0011 (1101) and ignoring any carry if so produced.

These rules should be better understood by examining the following cases.

Case I.

$$\begin{array}{rrl} 47_{10} & 0111\ 1010_2 & \\ +\ 34_{10} & +\ 0110\ 0111_2 & \\ \hline 81_{10} & 1110\ 0001_2 & \\ & +\ 1101\ 0011_2 & \text{correction} \\ \hline & 1\ 1011\ 0100 & \text{XS3} \\ & (8\quad 1)_{10} & \end{array}$$

Case II.

$$\begin{array}{rrl} 47_{10} & 0111\ 1010_2 & \\ -\ 34_{10} & +\ 1001\ 1000_2 & \text{9's complement} \\ \hline 13_{10} & 1\ 0001\ 0010_2 & \\ & \hookrightarrow 1_2 & \\ & 0011\ 0011_2 & \text{correction} \\ \hline & 0100\ 0110 & \\ & (1\quad 3)_{10} & \end{array}$$

FIGURE 5.39

Decimal	XS3	9's Complement of XS3
0	0011	1100
1	0100	1011
2	0101	1010
3	0110	1001
4	0111	1000
5	1000	0111
6	1001	0110
7	1010	0101
8	1011	0100
9	1100	0011

Case III.

34_{10}	$0110\ 0111_2$	
$-\ 47_{10}$	$+\ 1000\ 0101_2$	9's complement
$-\ 13_{10}$	$1110\ 1100_2$	
	$1101\ 1101_2$	correction
	$1011\ 1001_2$	
	$\rightarrow 0100\ 0110_2 = -(13)_{10}$	

A positive result is indicated when adding the 9's complement results in a carry-out of the MSB. This carry-out is used as the end-around-carry. Figure 5.40 shows a BCD adder/subtracter circuit that has incorporated the rules of XS3 addition and subtraction.

FIGURE 5.40

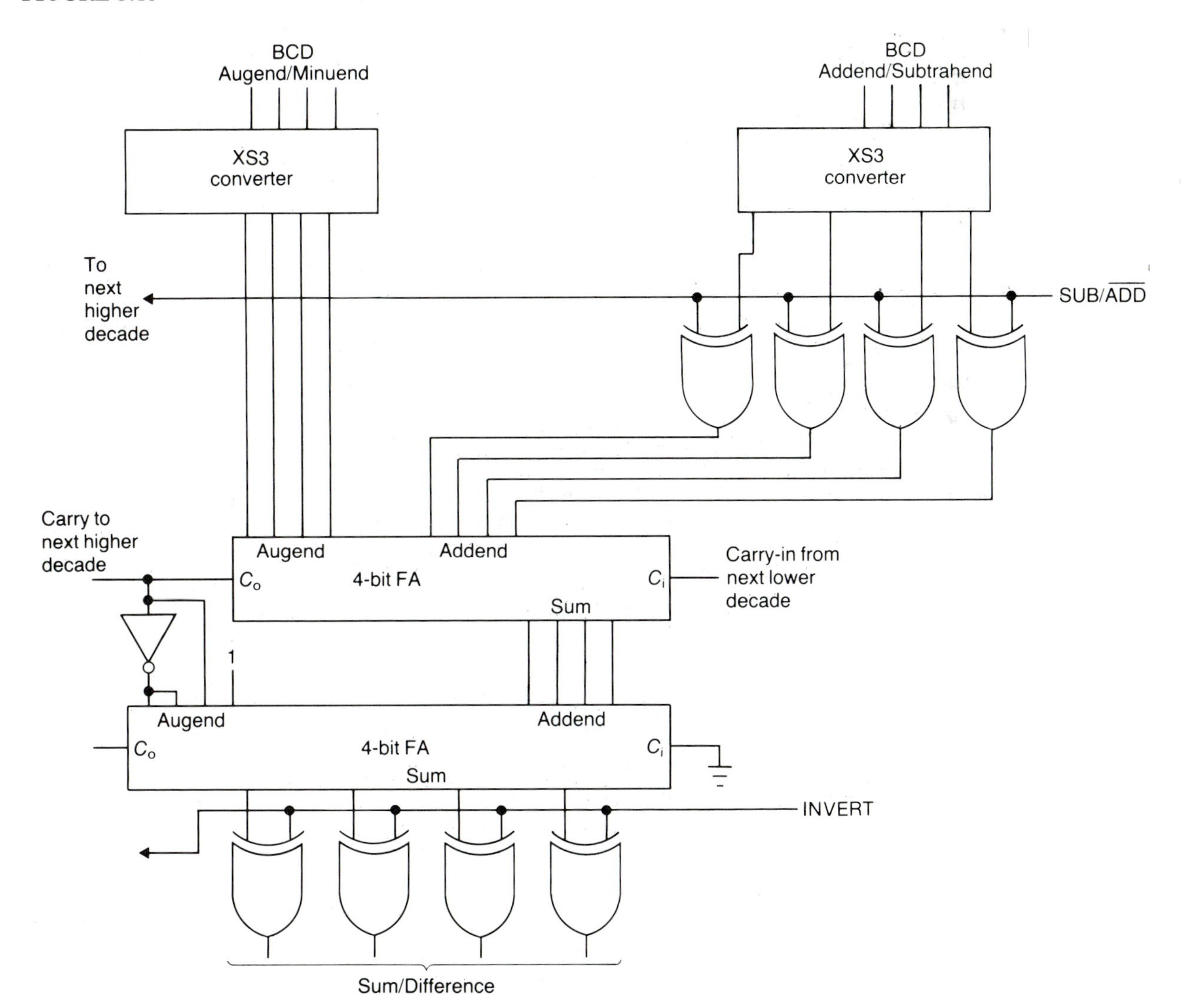

When the SUB/$\overline{\text{ADD}}$ input is a 0, the addend arrives at the first adder along with the augend. Otherwise, the complement of the subtrahend is added to the minuend. The output of the first adder is added to either 0011 or 1101, depending on whether or not a carry is generated. The INVERT input is activated when there is no end-around-carry, and in that event the output is complemented to yield the final result.

5.7 Arithmetic Logic Unit (ALU)

The examples presented in previous sections have shown some of the multi-bit arithmetic functions that are used in digital systems. Many times it may as well be necessary to perform bit-by-bit logic operations between two multi-bit operands. A multi-function circuit that can operate on groups of bits, therefore, proves to be extremely advantageous in many complex digital systems. Combinational design techniques generally are used to design such *multi-function circuits*, otherwise known as *arithmetic logic units* (*ALUs*). The various operations are usually selected by means of several control or select lines.

The various logic operations are realized by routing the inputs to circuits that perform various logic functions and using a MUX to select any one of the possible logic operations. There are commercially available ALUs that operate on two four-bit values with 32 arithmetic and 16 logic operations selectable by the combination of four control inputs, a mode selector, and a carry-in. In this section the design of a relatively less complex ALU will be considered to demonstrate the process and some typical functions.

Consider an ALU with two four-bit inputs, $A_3A_2A_1A_0$ and $B_3B_2B_1B_0$. Let us first consider bit-by-bit logic operations of several types. Between any two inputs, X and Y, there could be a total of 16 different types of logic outputs: 0, 1, X, Y, $\overline{X}$, $\overline{Y}$, $X + Y$, $\overline{X} + Y$, $X + \overline{Y}$, XY, $\overline{X}Y$, $X\overline{Y}$, $\overline{X + Y}$, $\overline{XY}$, $X \oplus Y$, and $\overline{X \oplus Y}$. All of these outputs are achievable by means of logic gates.

Figure 5.41[*a*] shows a logic circuit consisting of a 1-of-8 MUX and only eight logic gates. This circuit is able to perform up to eight different logic operations between its inputs, A_i and B_i. A specific operation is selected by means of the three control inputs: S_2, S_1, and S_0. The enable input, S_3, is set to a 0 to enable the MUX in all of these eight cases. The truth table in Figure 5.41[*b*] lists the possible logic operations and the corresponding control conditions. For example, when control input is 0100, the D_4 input of the 1-of-8 MUX is activated, and consequently $\overline{A_iB_i}$ becomes available at the MUX output. This 1-of-8 MUX could have been replaced with a 1-of-16 MUX for realizing 16 different logic operations. In that case, however, an additional control input would be necessary.

The arithmetic section may be designed around a four-bit ripple or CLA adder circuit. If the carry-in is utilized, it is possible to obtain a larger number of arithmetic operations for the same

FIGURE 5.41 **Logic Section of an ALU: [*a*] Circuit and [*b*] Function Table.**

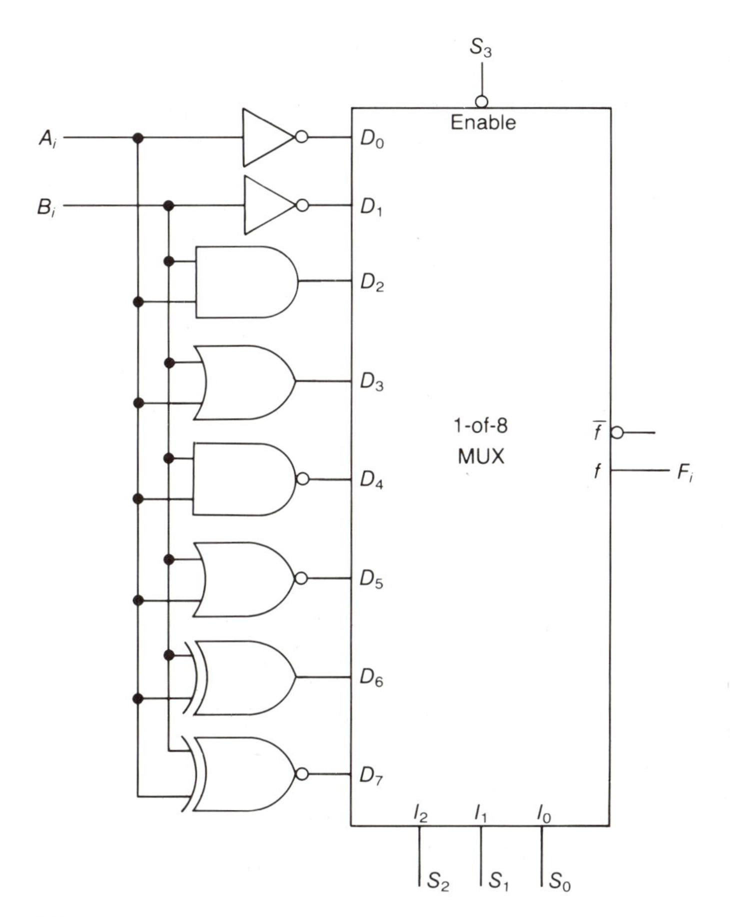

[*a*]

Control Inputs	Output
S_3 S_2 S_1 S_0	F_i
0 0 0 0	$\overline{A_i}$
0 0 0 1	$\overline{B_i}$
0 0 1 0	A_iB_i
0 0 1 1	$A_i + B_i$
0 1 0 0	$\overline{A_iB_i}$
0 1 0 1	$\overline{A_i + B_i}$
0 1 1 0	$A_i \oplus B_i$
0 1 1 1	$\overline{A_i \oplus B_i}$

[*b*]

number of control inputs. For every control condition the arithmetic output when $C_i = 1$ is always one greater than the corresponding output when $C_i = 0$. Figure 5.42[*a*] shows an arithmetic circuit corresponding to a single-bit input. It may be used to obtain a total of 16 different arithmetic operations. This circuit consists of a 1-of-8 MUX, a full adder, and five logic gates. In order to differentiate between the arithmetic and the logic operations, $\overline{S_3}$ is used to select arithmetic operations. The output of the MUX unit is fed into one of the adder inputs while another of the FA inputs is tied to A_i directly.

Figure 5.42[*b*] lists all of the arithmetic operations and the corresponding control inputs needed for operating this arithmetic unit. The MUX output carries into the adder either a function of both A_i

FIGURE 5.42 **Arithmetic Unit of an ALU: [*a*] Circuit and [*b*] Function Table.**

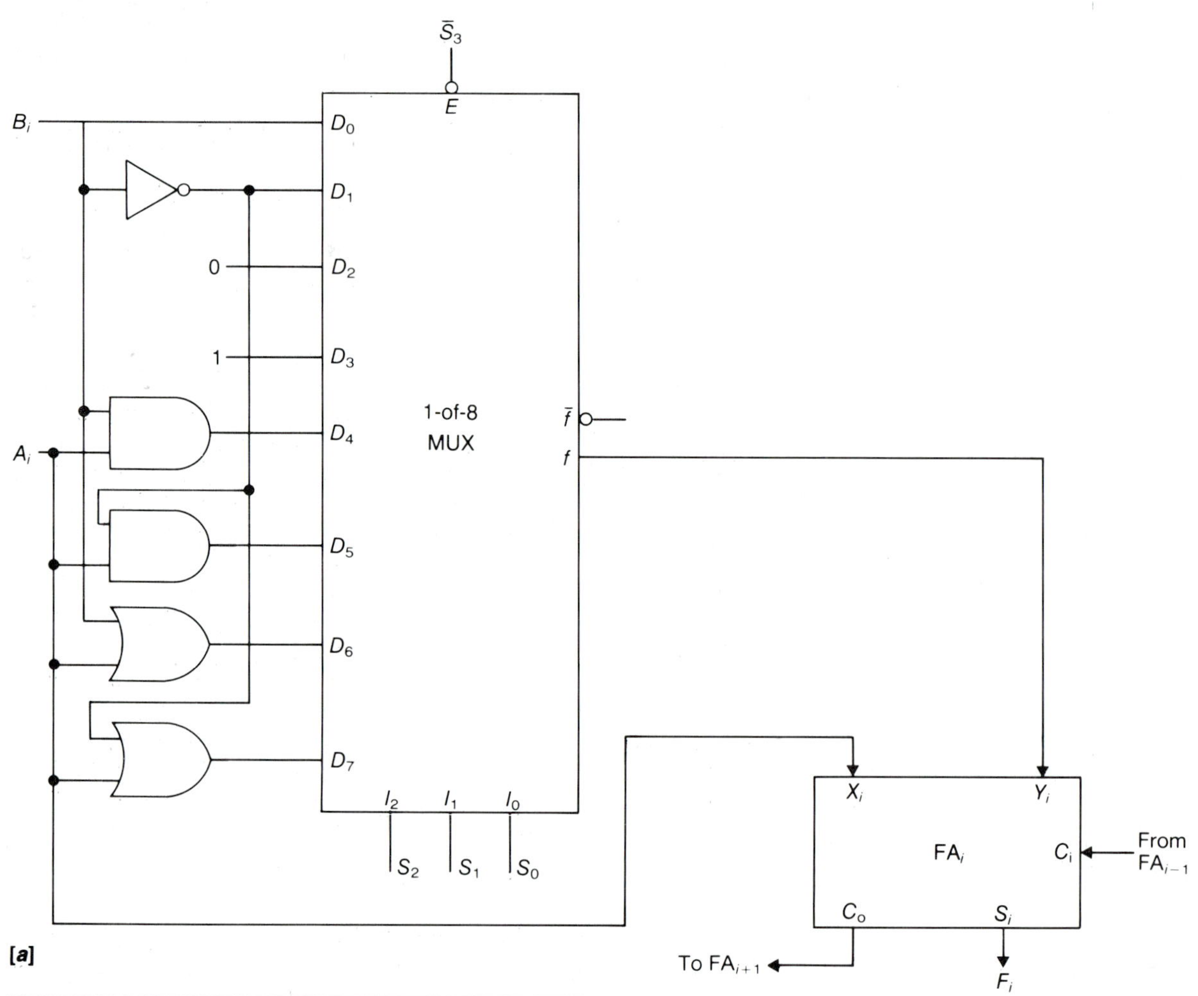

[*a*]

Control				F_i	
S_3	S_2	S_1	S_0	$C_i = 0$	$C_i = 1$
1	0	0	0	$A_i + B_i$	$A_i + B_i + 1$
1	0	0	1	$A_i + \bar{B}_i$	$A_i + \bar{B}_i + 1$
1	0	1	0	A_i	$A_i + 1$
1	0	1	1	$A_i - 1$	A_i
1	1	0	0	$A_i + A_iB_i$	$A_i + A_iB_i + 1$
1	1	0	1	$A_i + A_i\bar{B}_i$	$A_i + A_i\bar{B}_i + 1$
1	1	1	0	$A_i + (A_i + B_i)$	$A_i + (A_i + B_i) + 1$
1	1	1	1	$A_i + (A_i + \bar{B}_i)$	$A_i + (A_i + \bar{B}_i) + 1$

[*b*]

and B_i, or a 0, or a 1. For example, when the control input is 1100, the D_4 input of the MUX is activated, which introduces A_i, A_iB_i, and the carry-in as inputs to the FA. Thus the sum becomes A_i plus A_iB_i in the absence of a carry-in, and A_i plus A_iB_i plus 1 otherwise. Consequently, four 1-of-8 MUXs, twenty (5 × 4) discrete gates, and a four-bit FA may be connected to form an arithmetic unit for processing two four-bit binary inputs. Note when the control input is 1011 and $C_i = 0$, the four bits of A are added to a string of four 1s. This operation is equivalent to adding A to the 2's complement of 0001.

Both of these units, arithmetic and logic, may now be combined together to make an ALU, as shown by the block diagram in Figure 5.43. The two basic units could be internally ORed together to produce the ALU output. The ALU processes four bits of A and four bits of B simultaneously. If $S_3 = 0$, one of the eight logic operations would be realized, and when $S_3 = 1$, one of the 16 arithmetic operations would be available. Consequently the ALU consists of four single-bit arithmetic units and four single-bit logic units, and it could perform a total of 24 different operations. If the word size exceeds four bits, several four-bit ALUs may be cascaded by tying the carry-out of one ALU to the carry-in of the next ALU.

FIGURE 5.43 **Block Diagram of a Four-Bit ALU.**

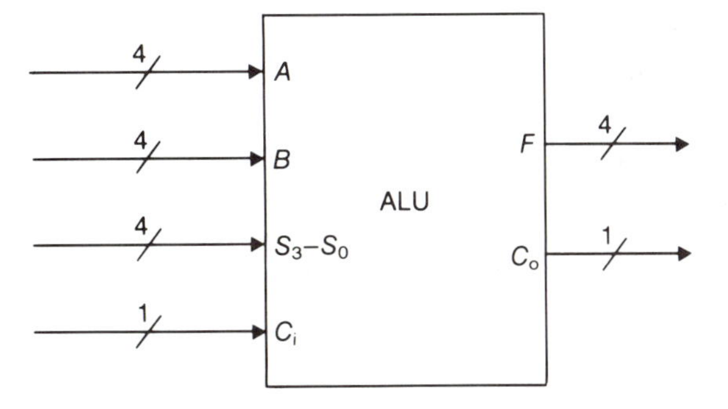

5.8 Decoders and Encoders

Very often in digital systems it is necessary to convert one code to another. The process that determines what character, digit, or number a code represents is called *decoding*. A *decoder* is an integral part of this process. It is a specially organized combinational circuit that translates a code to a more useful or meaningful form. In this section we shall look at just a few of the many decoding functions.

One of the frequently used decoders is a BCD–to–seven-segment decoder. This particular type of decoder accepts as inputs BCD and provides outputs to drive a seven-segment LED display device, as shown in Figure 5.44, in order to decode bits into readable digits. The decimal inputs to a circuit are first changed to equivalent binary form by means of BCD-to-binary converter modules for the desired binary operation. The resultant binary output is finally reconverted back to equivalent BCD output and is usually dis-

FIGURE 5.44 **Seven-Segment Display Device.**

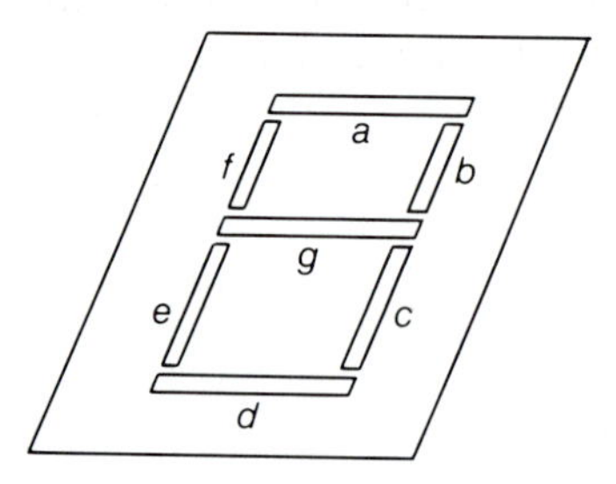

played by means of seven-segment display devices. These devices are used often in calculators.

The display device consists of seven light-emitting segments that represent each of the ten decimal numbers when activated in suitable combination. As an example, segments *a*, *f*, *g*, *c*, and *d* have to be illuminated to represent a 5. There are two choices for representing a 1: either *f* and *e* or *b* and *c*. The normal decimal code for these indicators is shown in Figure 5.45. A Boolean expression corresponding to each segment of the display may now be found. The Boolean equations for the illumination of display segments may be obtained as follows:

$$a(D,C,B,A) = \overline{\overline{(D + C + B + \bar{A})} + \overline{(D + \bar{C} + B + A)} + \overline{(D + \bar{C} + \bar{B} + A)}}$$

$$b(D,C,B,A) = \overline{\overline{(D + C + B + \bar{A})} + \overline{(D + \bar{C} + B + \bar{A})} + \overline{(D + \bar{C} + \bar{B} + A)}}$$

$$c(D,C,B,A) = \overline{\overline{(D + C + B + \bar{A})} + \overline{(D + C + \bar{B} + A)}}$$

$$d(D,C,B,A) = \overline{\overline{(D + C + B + \bar{A})} + \overline{(D + \bar{C} + B + A)} + \overline{(D + \bar{C} + \bar{B} + \bar{A})} + \overline{(\bar{D} + C + B + \bar{A})}}$$

$$e(D,B,C,A) = \overline{\overline{(D + C + \bar{B} + \bar{A})} + \overline{(D + \bar{C} + B + A)} + \overline{(D + \bar{C} + B + \bar{A})} + \overline{(D + \bar{C} + \bar{B} + \bar{A})} + \overline{(\bar{D} + C + B + \bar{A})}}$$

$$f(D,C,B,A) = \overline{\overline{(D + C + \bar{B} + A)} + \overline{(D + C + \bar{B} + \bar{A})} + \overline{(D + \bar{C} + \bar{B} + \bar{A})}}$$

$$g(D,C,B,A) = \overline{\overline{(D + C + B + A)} + \overline{(D + C + B + \bar{A})} + \overline{(D + \bar{C} + \bar{B} + \bar{A})}}$$

FIGURE 5.45 **Truth Table for Seven-Segment Decoding Function.**

Number	Segments						
	a	*b*	*c*	*d*	*e*	*f*	*g*
0 $\bar{D}\bar{C}\bar{B}\bar{A}$	1	1	1	1	1	1	0
1 $\bar{D}\bar{C}\bar{B}A$	0	0	0	0	1	1	0
2 $DCBA$	1	1	0	1	1	0	1
3 $\bar{D}\bar{C}BA$	1	1	1	1	0	0	1
4 $\bar{D}C\bar{B}\bar{A}$	0	1	1	0	0	1	1
5 $\bar{D}C\bar{B}A$	1	0	1	1	0	1	1
6 $\bar{D}CB\bar{A}$	0	0	1	1	1	1	1
7 $\bar{D}CBA$	1	1	1	0	0	0	0
8 $D\bar{C}\bar{B}\bar{A}$	1	1	1	1	1	1	1
9 $D\bar{C}\bar{B}A$	1	1	1	0	0	1	1

These Boolean expressions for the segments lead to the implementation of the desired decoder circuit as shown in Figure 5.46. The NOR circuit requires 16 gates and a total of 59 gate inputs while the NAND circuit requires 17 gates and a total of 87 gate inputs.

FIGURE 5.46 **BCD–to–Seven-Segment Decoder Circuit: [*a*] Using NOR Gates and [*b*] Using NAND Gates.**

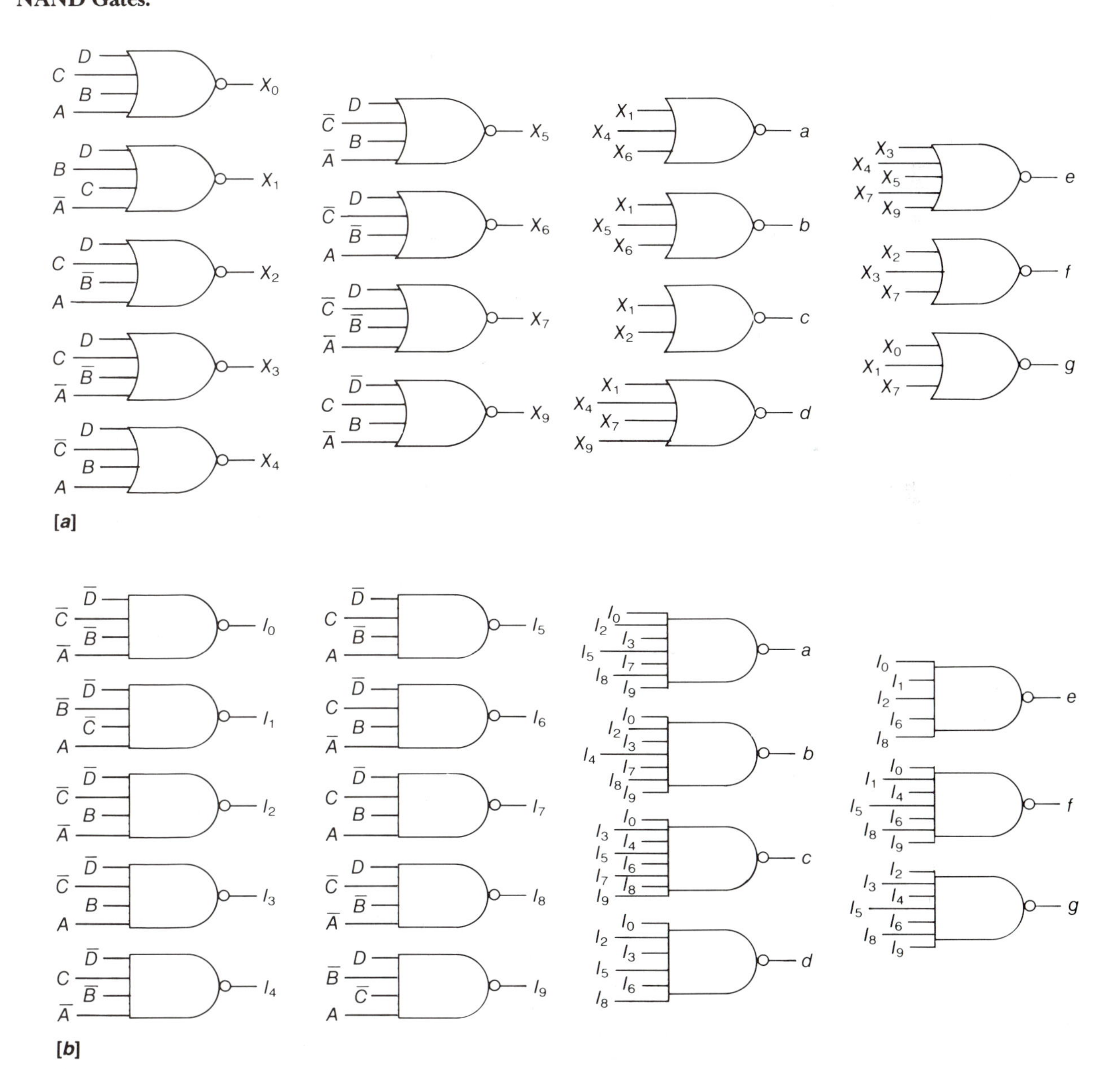

We shall consider next a slightly different decoder scheme that accepts n bits of binary information and converts them to up to 2^n unique outputs. For example, two bits of binary information would result in four unique output values. One such decoder scheme is shown by the circuit of Figure 5.47. This decoder is commonly known as a 2-4 line decoder. If the enable input E is 0, the decoder is enabled, and when E is 1 the decoder is disabled. When E is 0, all outputs are 1 except for the line corresponding to the decimal value of the input bits. For example, when $AB = 00$, only D_0 is 0. Such a scheme would allow us to activate four different circuits by means of the four decoder outputs. In that case the operations of these four peripheral circuits could be controlled by means of two select inputs, A and B.

FIGURE 5.47 **2-4 Line Decoder: [*a*] Block Diagram and [*b*] Circuit.**

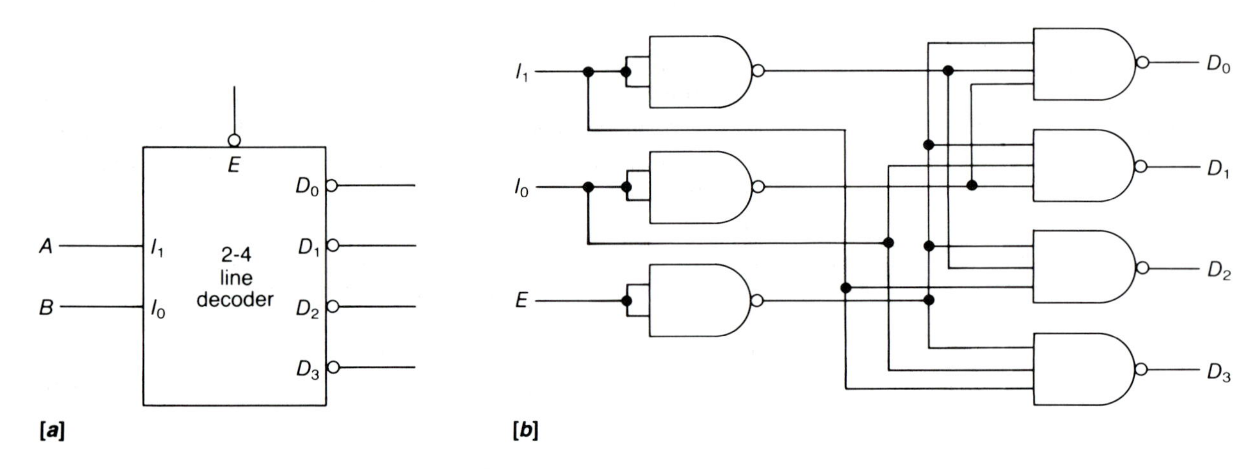

With proper connections a demultiplexer circuit can be made from a decoder. A *demultiplexer* (*DMUX*) receives data on a single entry line and outputs this data on one of its many output lines. Figure 5.48 shows a 1-4 line DMUX circuit where the decoder inputs are treated as the DMUX selects. The enable input E is connected to the input data, which appear at the output specified by the values of A and B.

There are times when several decoder/DMUX circuits may be cascaded together to form a larger decoder/DMUX. Figure 5.49 shows how two 2-4 line decoders are combined by means of their

FIGURE 5.48 **1-4 Line Demultiplexer: [*a*] Block Diagram and [*b*] Circuit.**

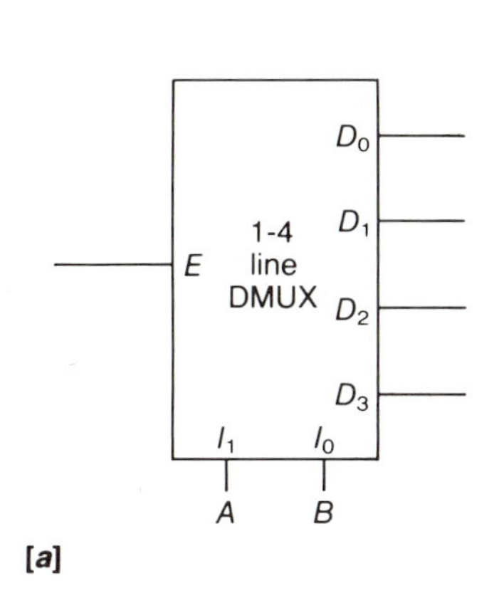

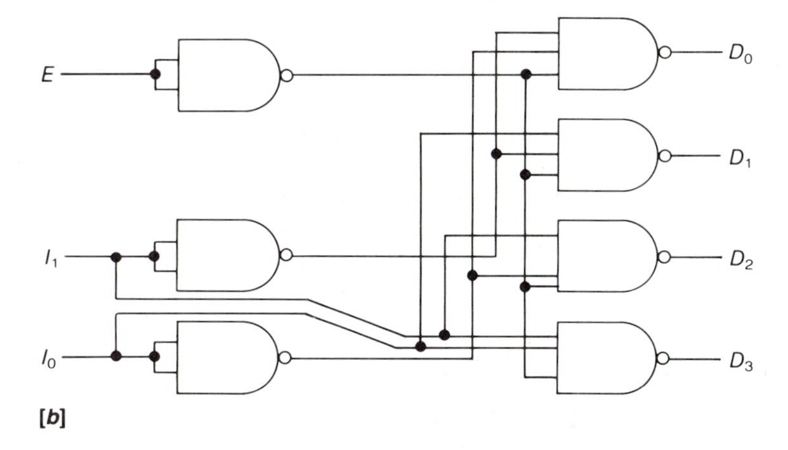

enable inputs. When $C = 1$, only the top decoder is enabled and the lower one is disabled. When $C = 0$, the top decoder is disabled and the bottom decoder is enabled. The combined circuit functions as a 3-8 line decoder.

The decoders that have been discussed thus far are often constructed using transistors in AND formation. The number of transistors used in each of the gates is approximately equal to the number of inputs to each gate, and the number of gates present is on the order of the number of decoder outputs. Therefore, the number of transistors required in a decoder circuit increases as 2^n with increasing inputs. Consequently designers would like to see a decoder scheme whose sum of gates and gate inputs is reasonably small. Figure 5.50 shows a particular configuration for a 4-16 line decoder that uses a reduced number of transistors. Such a configuration is commonly known as the *tree-type* decoding network. An examination would show that 64 transistors are needed to fabricate such a circuit. Comparatively, a regular 4-16 line decoder designed similarly to that shown in Figure 5.47 would require a total of 72 transistors.

The decoder network also may use an innovative scheme called the *balanced* decoding scheme, illustrated in Figure 5.51. It requires only 56 transistors. The significance of this improvement becomes more important as larger decoders are considered. The regular decoder network of Figure 5.47 is still the fastest because it involves only two stages of NAND gates. The inclusion of an enable input in a 4-16 line decoder, however, would involve an additional 18 transistors.

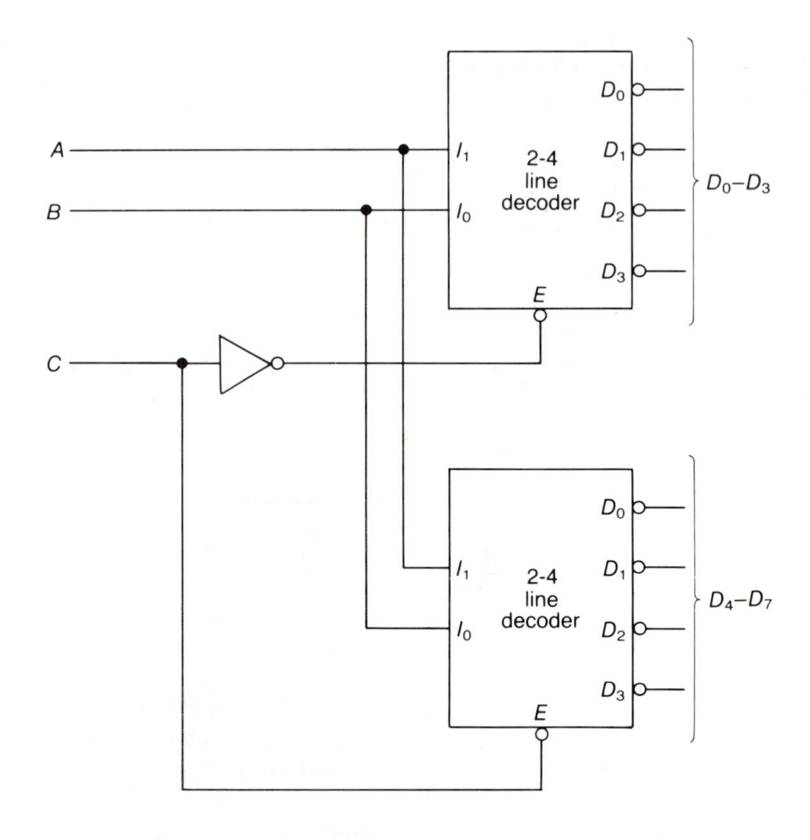

FIGURE 5.49 **3-8 Line Decoder Using Two 2-4 Line Decoder Units.**

An *encoder* is a combinational circuit that accepts a digit on its inputs and converts it to a coded output. In fact, an encoder reverses the function of a decoder. As an example, let us consider the design of a decimal-to-BCD encoder. The device has a total of 10 inputs—one for each decimal digit—and four outputs to represent the corresponding BCD numbers. These decimal inputs could possibly be the keys on a hand-held calculator.

The truth table for the encoder is listed in Figure 5.52[*a*] from which the following expressions may readily be obtained:

$$A = 1 + 3 + 5 + 7 + 9$$

$$B = 2 + 3 + 6 + 7$$

$$C = 4 + 5 + 6 + 7$$

$$D = 8 + 9$$

The resulting circuit is very straightforward, as shown in Figure 5.52[*b*].

FIGURE 5.50 **4-16 Line Tree-Type Decoder: [*a*] Block Diagram and [*b*] Circuit.**

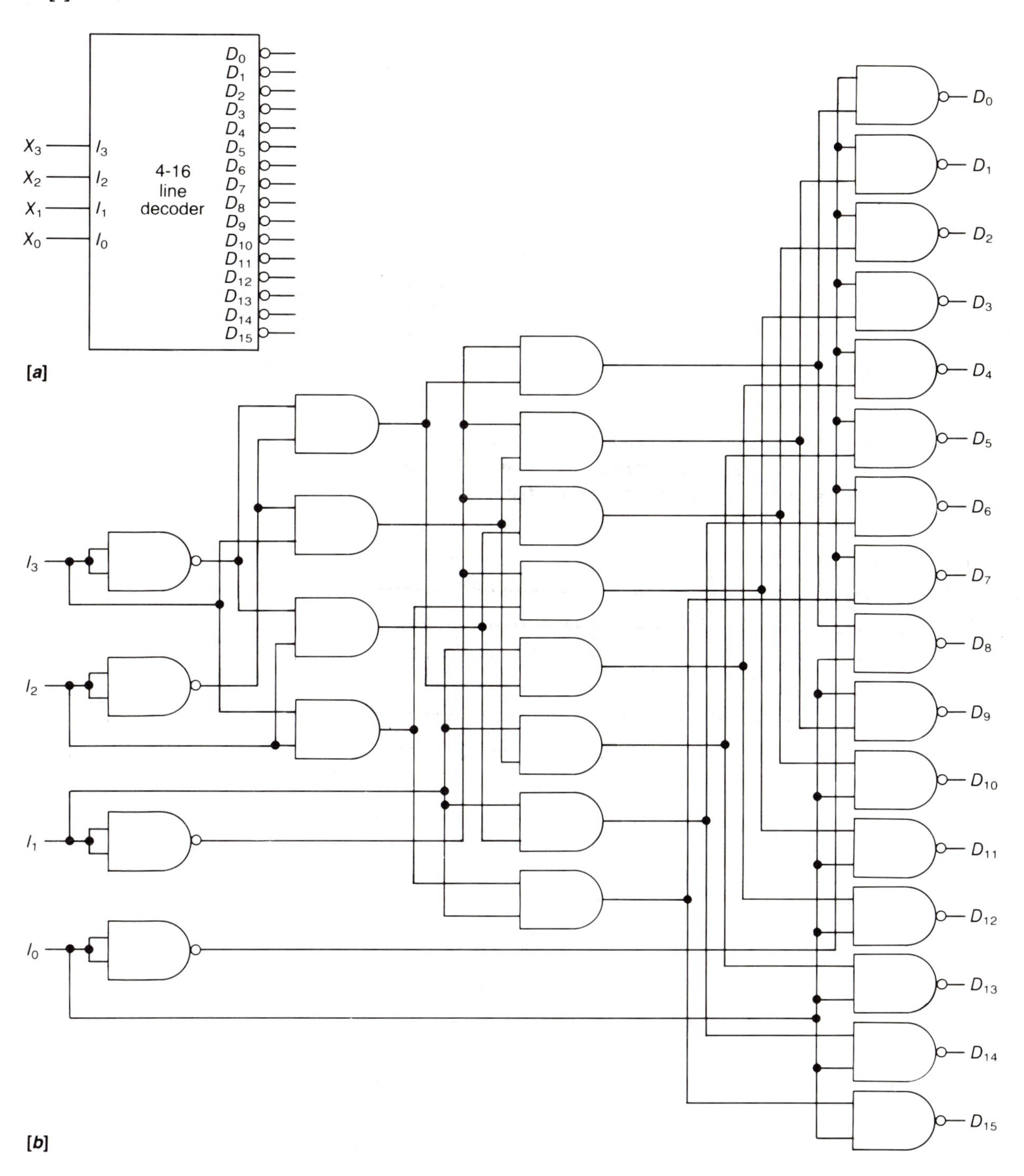

FIGURE 5.51 4-16 Line Balanced Decoder Circuit.

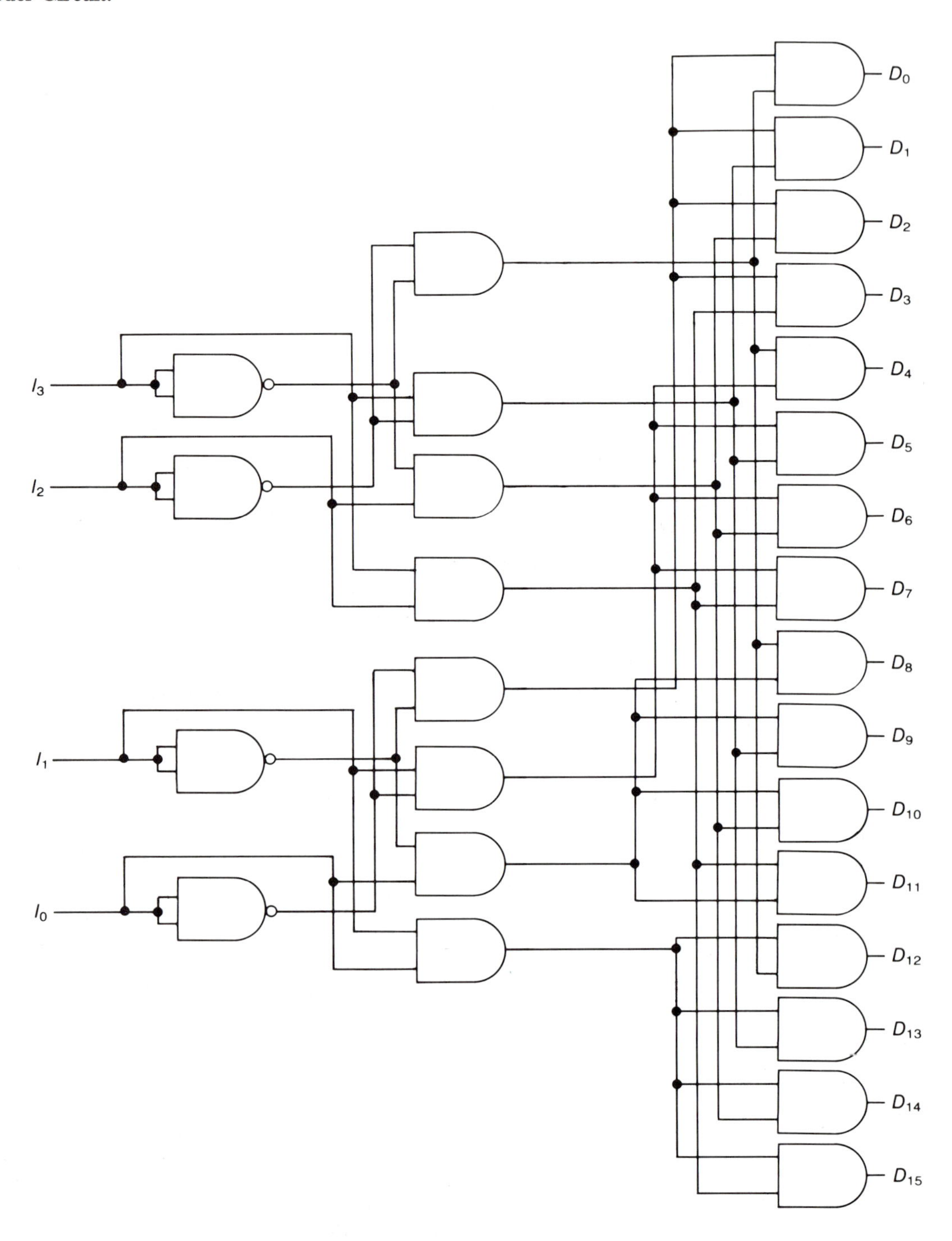

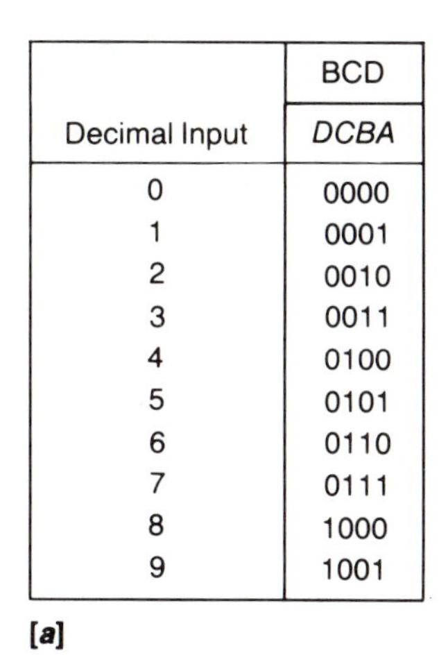

Decimal Input	BCD DCBA
0	0000
1	0001
2	0010
3	0011
4	0100
5	0101
6	0110
7	0111
8	1000
9	1001

[a]

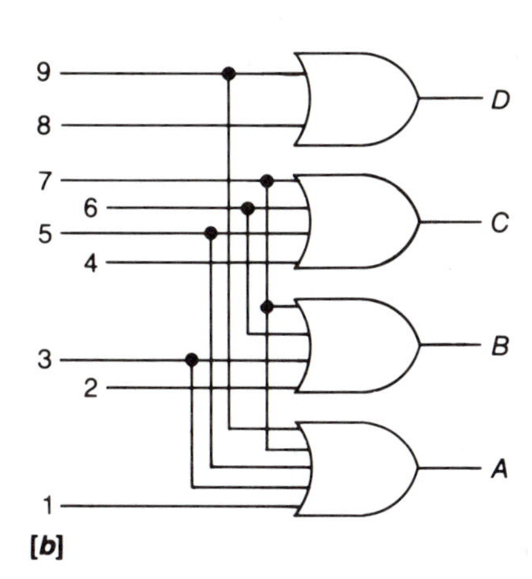

[b]

FIGURE 5.52 **Decimal-to-BCD Encoder: [*a*] Truth Table and [*b*] Circuit.**

EXAMPLE 5.13

Consider the seven-segment display device of Figure 5.44. Design a scheme to store up to five BCD integers such that the left-most BCD zeros are not displayed. For example, 00932 should be displayed as only 932.

SOLUTION

This design may be accomplished by modifying the already-designed BCD-to-LED display of Figure 5.46. Each unit accepts four BCD inputs and outputs seven LED outputs, as shown by the block diagram of Figure 5.53. The circuit of Figure 5.53 requires an added feature that suppresses the display of leading zeros. First it must be determined if the MSB is a 0. If it is, the next significant bit is tested, and so on. The first nonzero bit stops further testing. Accordingly, one may design a circuit for the *i*th bit if it is a zero or not, as shown by the block diagram of Figure 5.54.

FIGURE 5.53

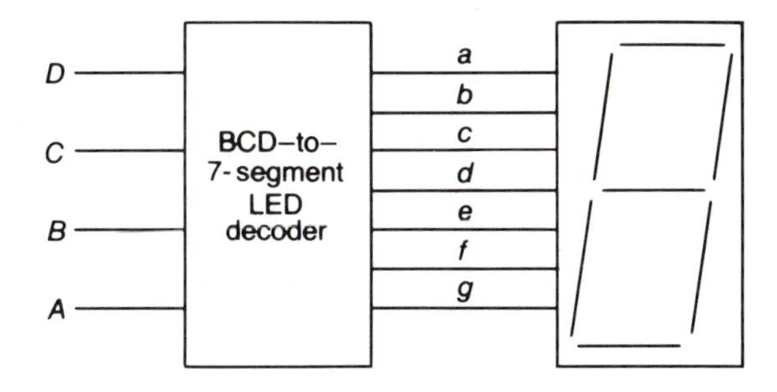

A high *TDS input* to the decoder implies that *This Decoder needs to be Searched.* A high *NDS output* similarly implies that the *Next Decoder needs to be Searched* as well. The NDS output of one input may be introduced to the next unit on the right as the TDS input, and so on. This feature may be accomplished by adding a circuit such as shown in Figure 5.55 to the already available circuit.

FIGURE 5.54

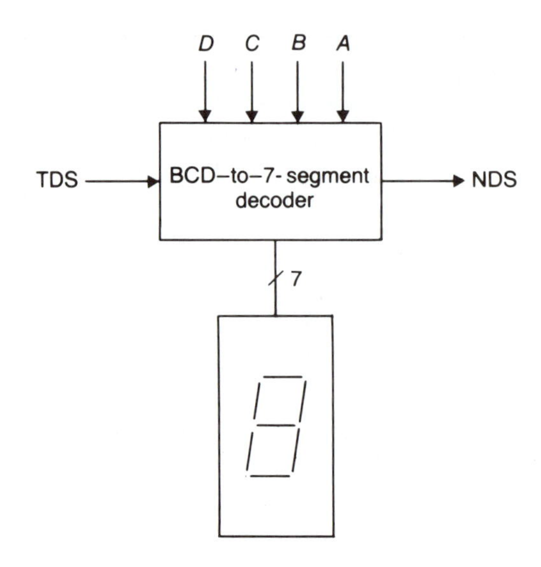

FIGURE 5.55

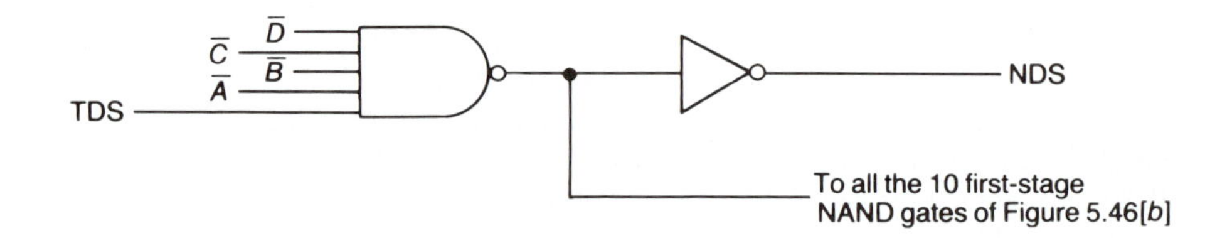

When the BCD digit corresponds to a 0 and the TDS input is high, the above circuit disables the decode circuit and suppresses the display. At the same time the resulting NDS becomes a 1, resulting in further search for zeros if there are any. If the BCD digit is not a 0, the decoder is not disabled and further search is abandoned. The overall five-bit circuit is obtained as shown in Figure 5.56.

FIGURE 5.56

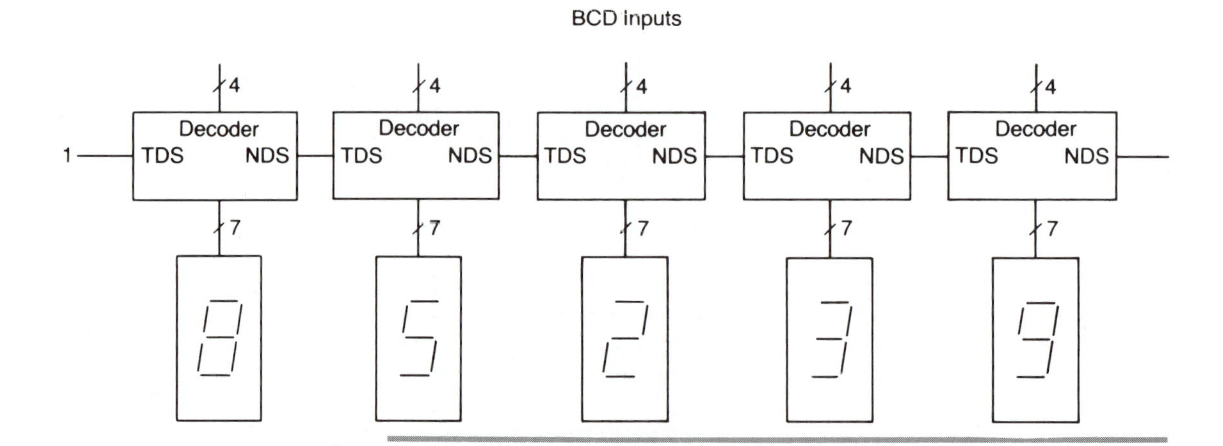

EXAMPLE 5.14

Design a four-input priority encoder such that when two inputs, D_i and D_j, are high simultaneously, D_i has priority over D_j when $i > j$. The encoder produces a binary output code corresponding to the input that has the highest priority.

SOLUTION

The block diagram for such a device may be as shown in Figure 5.57. As a beginning step, the corresponding truth table needs to be known. The truth table for such a device is easily obtained as shown in Figure 5.58. The don't-cares are introduced under input columns whenever appropriate. The Boolean equations for the outputs are obtained directly as follows:

$$f_0 = D_3 + D_1\bar{D}_2$$

$$f_1 = D_2 + D_3$$

The four-input priority encoder circuit is obtained accordingly, as shown in Figure 5.59. The request indicator, M, shows whether or not any of the four inputs are active.

FIGURE 5.57

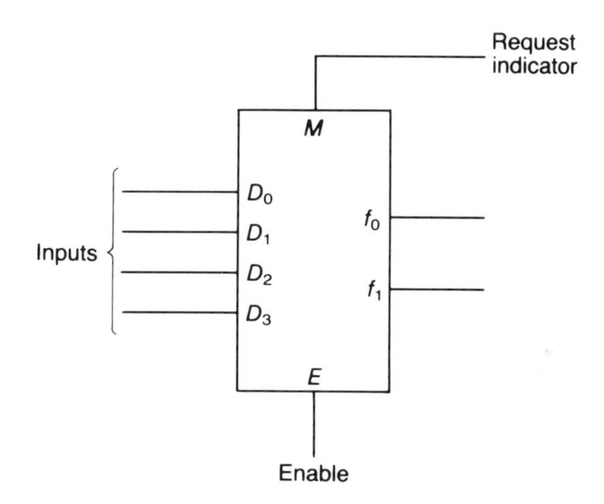

FIGURE 5.58

D_0	D_1	D_2	D_3	f_1	f_0
1	0	0	0	0	0
—	1	0	0	0	1
—	—	1	0	1	0
—	—	—	1	1	1

FIGURE 5.59

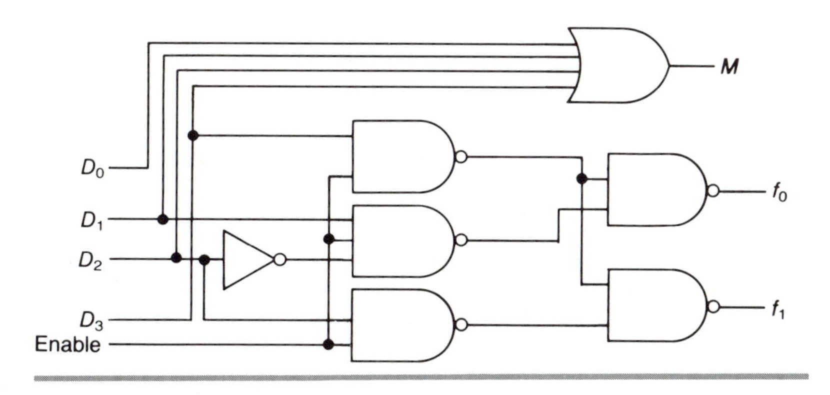

5.9 Error-Control Circuits

Errors may occur as digital codes are transmitted from one system or subsystem to another. There is always a possibility, albeit small, that a random-noise pulse will change a zero to a one or a one to a zero. It is possible, however, to code the data so that the occurrence of an error can be detected after the data have been received. The simplest approach is to add an extra bit, called a *parity bit*, to each of the number codes. If the coded data including the parity bit have an even number of ones, the code is said to have *even parity*. If the coded word including the parity bit has an odd number of ones, the code is said to have *odd parity*. Prior to sending a code, the number of ones are counted and the parity bit is set to make the number of ones odd or even as determined by the parity scheme chosen. At the receiving end, a check is made to see how many ones are present in the coded word. If odd parity is used and an even number of ones are received, it implies that an error has occurred. However, the proposed parity bit scheme cannot detect the occurrence of an even number of errors. For situations where the probability of multiple errors is high, a more sophisticated coding scheme must be used. In computers or communications equipment the possibility of random noise causing changes in more than one bit is low.

In order to check for or generate the proper parity bit in a given code, it is necessary to determine whether an odd or even number of ones are present. An X-OR gate functions in such a way that the output of an even number of ones is always a 0, and the output of an odd number of ones is always a 1. As an example, the circuit of Figure 5.60 makes use of these gates to generate an even parity bit for normal BCD input and to check for a possible error that may have been caused during transmission. The parity generator circuit at the source end examines the contents of the four data lines and accordingly generates a parity bit so that the encoded message (five bits in all) has even parity. At the receiver end the parity-checking circuit determines if an error has occurred or not. A high output at the parity-checking circuit indicates the occurrence of an error during transmission. This circuit could be made suitable for odd parity by simply replacing the final X-OR gates of both generator and checker circuits with X-NOR gates.

Several other schemes are also available, especially for coding decimal digits. The most common of these are the 2-out-of-5 code and 2-out-of-7 code. They are listed along with parity-coded BCD in the table of Figure 5.61. The 2-out-of-7 code is also known as the *biquinary code*. The zeroth through the sixth bit have positional weights of 0, 1, 2, 3, 4, 0, and 5 respectively. In both of these m-out-of-n codes, there are m ones and $(n - m)$ zeros. The advantages of these schemes are understood by comparing the different permissible codes. Two codes are said to be at a distance p if the codes differ from each other in p locations. Clearly, each code in an m-out-of-n

FIGURE 5.60 **Even Parity Generator-Checker Circuit.**

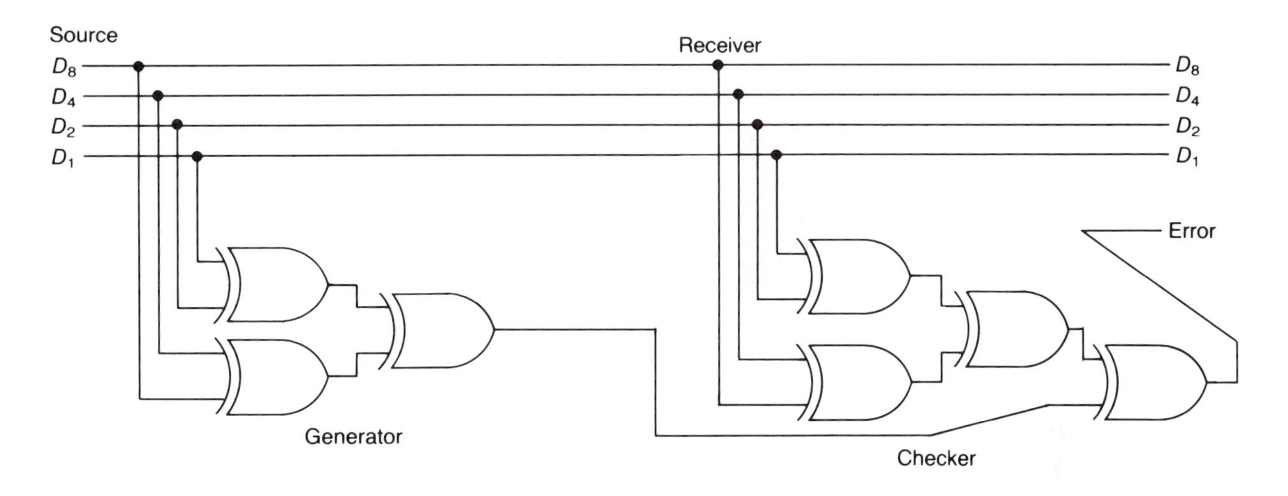

FIGURE 5.61 **Some Codes with Error Control.**

Decimal	BCD with Even Parity	BCD with Odd Parity	2-out-of-5	2-out-of-7
0	00000	00001	00011	0100001
1	00011	00010	00101	0100010
2	00101	00100	00110	0100100
3	00110	00111	01001	0101000
4	01001	01000	01010	0110000
5	01010	01011	01100	1000001
6	01100	01101	10001	1000010
7	01111	01110	10010	1000100
8	10001	10000	10100	1001000
9	10010	10011	11000	1010000

scheme is at least distance two away from the next code. Consequently, these codes can be used to detect single errors.

To correct k errors the minimum distance between two code words must not be smaller than $2k + 1$. A total of k errors would produce an error word k distance away from the correct code word. To be able to correct this error, no other k errors should be able to produce this same error word. The error word, therefore, should be at a distance at least $k + 1$ from any other code word. Accordingly, the minimum distance between two code words should be $2k + 1$. A minimum distance of two provides single-error detectability; any single error moves the code closer to where it was than to any other possible code. A circuit could be designed to move it back to its correct position. Of course, this minimum distance code could also be used instead for double-error detection. A minimum distance of four will provide both single-error correction plus double-error detection. A minimum distance of five would allow double-error correction. Next we will examine a code for which both error detection and error correction are straightforward.

One of the most useful codes is the *Hamming code*. This scheme not only provides for the detection of an error, but also locates the bit position in error so that it may be readily corrected. A block diagram of the implementation of this scheme is shown in Figure 5.62. The m bits of data are encoded with p parity bits before transmission. The received word is tested by a checking circuit to see if any error has occurred. The decoder then locates the exact position of error and, accordingly, the data are corrected by a corrector circuit.

FIGURE 5.62 **Block Diagram of the Hamming Code.**

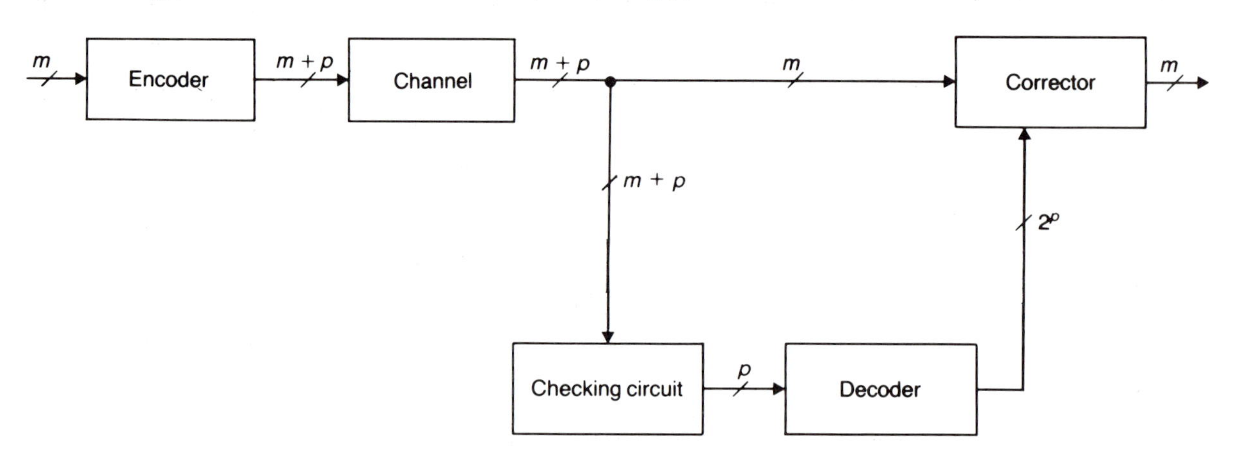

The Hamming code uses multiple parity bits placed at specific locations of the coded word. The general rules for the generation of parity bits are summarized as follows:

1. If m is the number of information bits, then the number of parity bits, p, is equal to the smallest integer value of p that satisfies $2^p \geq m + p + 1$.
2. Parity bits are placed at the locations 1, 2, 4, 8, 16, and so on, of the coded word. Information bits are placed in order at locations 3, 5, 6, 7, 9, 10, and so on.
3. Each parity bit individually takes care of only a few bits of the coded word. To determine the bits of the coded word that are checked by a parity bit, every bit position is expressed in binary. A parity bit would check those bit positions, including itself, that have a 1 in the same loca-

tion of their binary representation as in the binary representation of the parity bit.

Example 5.15 illustrates the coding mechanisms involved.

EXAMPLE 5.15

Determine the Hamming-coded word for the message 10101 using even parity.

SOLUTION

The number of bits is $m = 5$; therefore, $p = 4$. This coding would result in a nine-bit coded word. The corresponding message bits (10101) are positioned respectively in locations 3, 5, 6, 7, and 9. These locations are specified in the table of Figure 5.63 as M_1, M_2, M_3, M_4, and M_5 respectively. To determine the exact value of each parity bit, each of the position designations is expressed at first in binary. The parity bits are generated from the following observations:

a. P_1 checks bit positions 1, 3, 5, 7, and 9 since all of these locations have a 1 in the LSB of their binary representations. The message bits present at four of these locations are, respectively, 1, 0, 0, and 1. The parity bit at location 0001, therefore, should be a 0 to maintain an even parity.

b. P_2 checks bit positions 2, 3, 6, and 7 because they all have a 1 at the same location in their binary representations. Bits 3, 6, and 7 house, respectively, a 1, 1, and 0. Therefore, P_2 must be a 0.

c. P_3 checks bit positions 4, 5, 6, and 7. Bits 5, 6, and 7 house, respectively, 0, 1, and 0, which requires that P_3 be a 1.

d. P_4 checks bit positions 8 and 9 and should be a 1 to maintain an even parity.

Therefore, the coded word is 001101011.

FIGURE 5.63

Bit Designation	P_1	P_2	M_1	P_3	M_2	M_3	M_4	P_4	M_5
Bit Position	1 0001	2 0010	3 0011	4 0100	5 0101	6 0110	7 0111	8 1000	9 1001
Message Bits			1		0	1	0		1
Parity Bits	0	0		1				1	

As mentioned earlier, the Hamming code also provides a means to detect and correct a single error. The *general detection algorithm* consists of the following steps:

Step 1. Check parity on each parity bit P_n and the bits for which it provides parity.

Step 2. If the test indicates the preservation of assumed parity, a 0 is assigned to the test result. A failed test is indicated by a 1.

Step 3. The binary number formed by the score of parity tests indicates the location of the bit in error.

Example 5.16 illustrates the detection mechanisms.

EXAMPLE 5.16

Determine if any bit is in error in the coded word 001101111. The message was coded with even assumption.

SOLUTION

The bit position table, as shown in Figure 5.64, is first prepared and then the coded bits are placed in their proper places. The following observations can be made regarding the coded word:

P_1 checks bits 1, 3, 5, 7, and 9. Consequently the first test fails since there are three 1s →	1	(LSB)
P_2 checks bits 2, 3, 6, and 7. This test also fails since there are three 1s →	1	
P_3 checks bits 4, 5, 6, and 7. This test also fails since there are three 1s →	1	
P_4 checks bits 8 and 9. This is a good check since there are two 1s →	0	(MSB)

The test score is 0111. The bit in error is the seventh bit (0111); therefore, the seventh bit is changed to 0. Therefore, the correct coded word should be 001101011. This result agrees with the earlier findings of Example 5.15.

FIGURE 5.64

Bit Designation	P_1	P_2	M_1	P_3	M_2	M_3	M_4	P_4	M_5
Bit Position	1 0001	2 0010	3 0011	4 0100	5 0101	6 0110	7 0111	8 1000	9 1001
Received Message	0	0	1	1	0	1	1	1	1

The hardware implementations of the Hamming code scheme are relatively easy to achieve. Consider as an example a logic circuit for processing five bits of data, just as in the last two examples. The parity bit generator circuit is to generate four additional bits: P_1, P_2,

P_3, and P_4. The rules for the generation of parity bits, as stated earlier, may be used to obtain

$$P_1 = M_1 \oplus M_2 \oplus M_4 \oplus M_5$$
$$P_2 = M_1 \oplus M_3 \oplus M_4$$
$$P_3 = M_2 \oplus M_3 \oplus M_4$$
$$P_4 = M_5$$

The generator circuit may be implemented with X-OR gates. The resultant circuit is shown in Figure 5.65. Correspondingly, for an odd parity assumption the final X-OR gates of all stages need to be replaced with X-NOR gates and M_5 inverted to generate P_4. It is important to note that the minimum distance between two code words is three because one change in the data bits produces at least two changes in the parity bits.

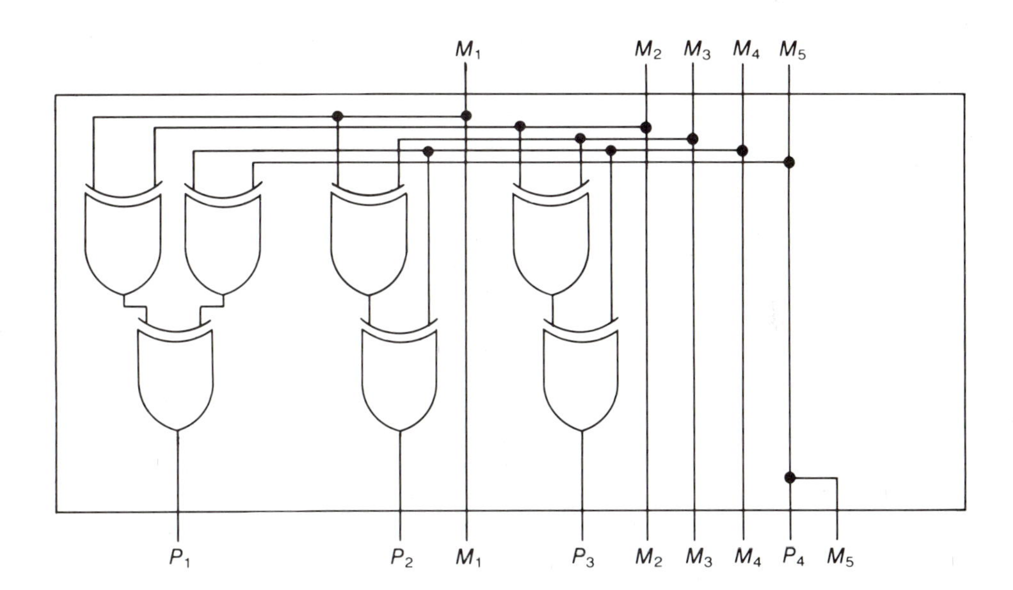

FIGURE 5.65 **Hamming-Coded Message Generator.**

Using similar reasoning the parity-checking circuit may also be designed. The parity of each parity bit and its corresponding data bits are tested. The Boolean equations for the nine-input parity-testing circuit are readily obtained as follows:

$$C_1 = Y_1 \oplus Y_3 \oplus Y_5 \oplus Y_7 \oplus Y_9$$
$$C_2 = Y_2 \oplus Y_3 \oplus Y_6 \oplus Y_7$$
$$C_3 = Y_4 \oplus Y_5 \oplus Y_6 \oplus Y_7$$
$$C_4 = Y_8 \oplus Y_9$$

where $Y_1, Y_2, \ldots, Y_8$, and Y_9 correspond, respectively, to P_1, P_2, M_1, P_3, M_2, M_3, M_4, P_4, and M_5 of the generator circuit. The parity-checking circuit is again realized using X-OR gates and is shown in Figure 5.66[*a*]. The value $C_4C_3C_2C_1$ points to the bit in error and

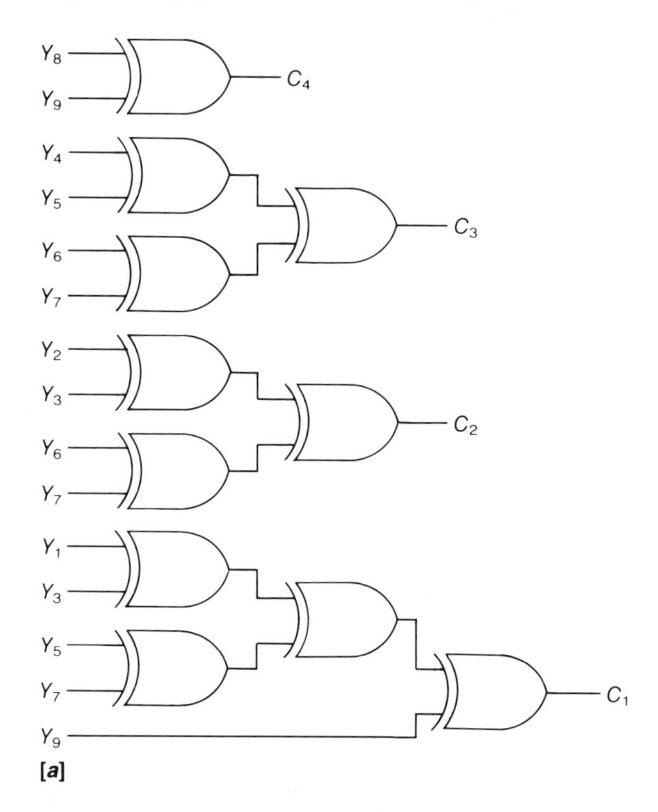

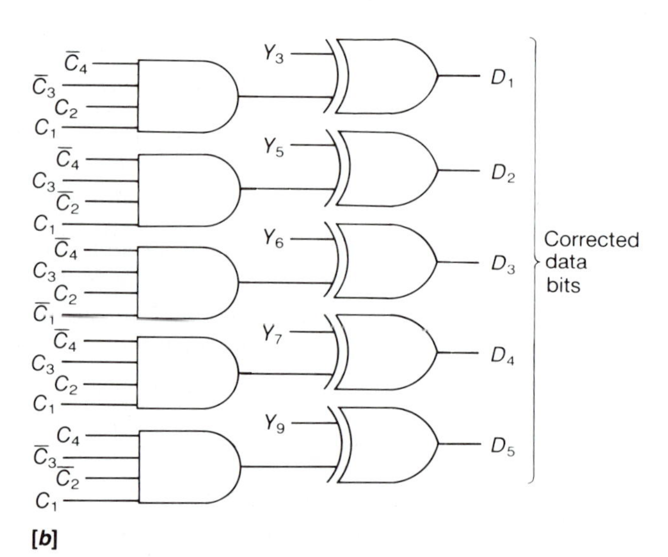

FIGURE 5.66 **[*a*] Parity-Checking Circuit and [*b*] Correction Circuit.**

may be used to correct the bit. Our interest is in recovering the corrected data bits. Making use of the X-OR programmable inverter function, the equations for the corrected bits are obtained as follows:

$$D_1 = (\bar{C}_4 \cdot \bar{C}_3 \cdot C_2 \cdot C_1) \oplus Y_3$$
$$D_2 = (\bar{C}_4 \cdot C_3 \cdot \bar{C}_2 \cdot C_1) \oplus Y_5$$
$$D_3 = (\bar{C}_4 \cdot C_3 \cdot C_2 \cdot \bar{C}_1) \oplus Y_6$$
$$D_4 = (\bar{C}_4 \cdot C_3 \cdot C_2 \cdot C_1) \oplus Y_7$$
$$D_5 = (C_4 \cdot \bar{C}_3 \cdot \bar{C}_2 \cdot C_1) \oplus Y_9$$

For example, if the parity test results are 0111, then the fourth data bit (i.e., the seventh bit of the coded word) is in error. The bit in error is corrected by complementing Y_7. All of the other bits remain unchanged. The corresponding error-correcting circuit is shown in Figure 5.66[*b*].

5.10 Summary

In this chapter different aspects of combinational circuit design were presented. Primary emphasis was placed on designing smaller and manageable modules, several of which were then cascaded together to realize a robust system. In particular, the design and working principles of binary and nonbinary adders/subtracters, code converters, decoders, encoders, and various error-correcting circuits were explored. Many of these devices will be used time and again throughout the rest of this text for developing more advanced concepts.

Problems

1. Design an FA circuit using logic gates suitable for adding two bits of addend, two bits of augend, and carry-in input.
2. Obtain a single-bit FA using only MUXs.
3. Design a single-bit FA using only NOR gates.
4. Use the FAs designed in Problem 1 to perform addition of six-bit numbers. Show the configuration of the setup for adding $(110110)_2$ and $(000010)_2$.
5. Design a four-bit FA using combinational logic.
6. Design a four-bit FA using ROM technology.
7. Use bridging to implement a standard full subtracter circuit (three inputs and two outputs) using X-OR gates.
8. Verify Equations [5.4] and [5.6].
9. Design a circuit for dividing a four-bit number by a four-bit number.
10. The following message needs to be transmitted using the Hamming code under even parity assumption. Determine the

parity bits and the order in which the coded message will be sent. The to-be-coded message is 1010111001011. Show the corresponding circuit.

11. Design a half subtracter circuit using (a) only NOR gates and (b) only MUXs.
12. The Hamming-coded message received under odd parity assumption is 1010111001011. Determine if the message has any error and write out the correct message bits only. Obtain the corresponding correction circuit.
13. Design a full subtracter using half subtracter modules.
14. Using only a four-bit binary adder, design decimal code converters for the following conversions:
    a. 8-4-2-1 to XS3
    b. XS3 to BCD
    c. XS6 to XS3
    d. BCD to XS3
15. Remove the combinational FAs from the circuit of Example 5.2 and replace these with equivalent ROMs. Show the ROM logic for one of these units and determine the total ROM size needed for the complete circuit.
16. Design a combinational circuit capable of comparing two eight-bit binary integers (without sign bits) X and Y. The output Z should be a 1 whenever $X \geq Y$.
17. Design a controllable, dual-purpose, four-bit converter that converts binary to Gray and also Gray to binary.
18. Use the module of Example 5.7 for obtaining the following conversions:
    a. 15-bit
    b. 20-bit
    c. 25-bit

    Justify your designs using exemplary nontrivial binary inputs.
19. Use the module of Example 5.10 for obtaining the following conversions:
    a. 15-bit
    b. 20-bit
    c. 25-bit

    Justify your designs using exemplary nontrivial BCD inputs.
20. Design an adder/subtracter using cascaded ALUs. Show how it works when adding and when subtracting if $A = 84$ and $B = 32$. Repeat the problem using base-16 equivalents of the numbers.
21. Design a 12-bit FA in which carries are allowed to ripple after the first six bits of addition.

22. Show how the ALU can be used to (a) subtract one from and (b) add one to a number. Show the setup if the number is 76_{10}.
23. Show how a 3-8 line decoder could be used to generate $f(A,B,C) = \Sigma m(0,1,3,5)$.
24. Design a logic circuit that multiplies an input decimal digit (in BCD) by five. The output is also in BCD form. Show that the outputs can be obtained from the input lines without using any logic gates.
25. Implement the FA circuit of Problem 1 using MUXs.
26. Obtain the most minimal circuit that squares a three-bit binary number.
27. Design a special-purpose unit using FAs (a) for adding 12 single-bit binary numbers and (b) for adding 17 single-bit binary numbers.
28. Design a four-bit CLA circuit where the propagate function is defined as $P_i = A_i + B_i$ instead of $P_i = A_i \oplus B_i$. How does the current design differ from the one discussed in Section 5.4?
29. Use the techniques considered in Section 5.4 to obtain a four-bit fast subtracter.
30. Obtain the CLA carry equations when $n > 13$ and show that the maximum fan-out is dependent on variable $P_{(n-2)/2}$ and is equal to $\{[(n + 1)^2/4] + 2\}$ for odd n. Also show that for even n, the maximum fan-out is dependent on both $P_{(n/2)-1}$ and $P_{(n/2)}$ and is equal to $\{[n(n + 2)]/4 + 2\}$.
31. Design an n-bit binary comparator circuit to test if an n-bit number A is equal to, larger than, or smaller than a second n-bit number B. The problem could be broken into one unit of a half comparator module and $n - 1$ units of full comparator modules, as shown in Figure 5.P1. Each of the modules gives out two outputs: G_n and L_n, such that
    a. $G_n = 1$ and $L_n = 0$ if $A_n > B_n$
    b. $G_n = 0$ and $L_n = 1$ if $A_n < B_n$
    c. When $A_n = B_n$, then $G_n = G_{n-1}$ and $L_n = L_{n-1}$ for a full comparator and $G_n = L_n = 0$ for a half comparator.

FIGURE 5.P1

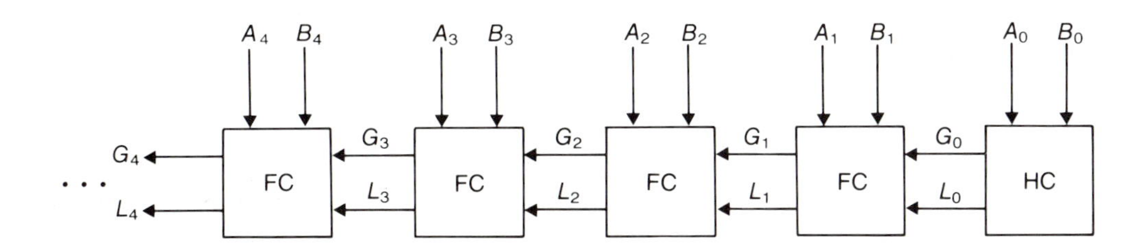

Describe the working principles of the design and show how you could improve upon this design.

32. Design an n-bit comparator module different from that of Problem 31 such that the comparison process begins from the MSB and moves toward the LSB until the final decision is made.
33. Obtain the circuit for a 16-input, 4-output priority encoder.
34. Design a serial-to-parallel converter circuit that routes a long sequence of binary digits into four different output lines as specified by external control signals.
35. Design a BCD adder circuit that first converts the BCD numbers into binary, then adds the resulting binary numbers, and finally converts the binary sum to the correct BCD sum.

Suggested Readings

Agrawal, D. P. "Negabinary carry look ahead adder and fast multiplier." *Elect. Lett.* vol. 10 (1974): 312.

Arazi, B. "An electrooptical adder." *Proc. IEEE.* vol. 73 (1985): 162.

Bartee, T. C. *Digital Computer Fundamentals.* New York: McGraw-Hill, 1985.

Benedek, M. "Developing large binary to BCD conversion structures." *IEEE Trans. Comp.* vol. C-26 (1977): 688.

Bin Nun, M. A., and Woodward, M. E. "Halfadders modulo-2 using read-only memories." *Elect. Lett.* vol. 10 (1974): 213.

Bywater, R. E. H. *Hardware/Software Design of Digital Systems.* Englewood Cliffs, N.J.: Prentice-Hall International, 1981.

Daws, D. C., and Jones, E. V. "Hardware-efficient bit sequential adders and multipliers using mode-controlled logic." *Elect. Lett.* vol. 16 (1980): 434.

Feldstein, A., and Goodman, R. "Loss of significance in floating point subtraction and addition." *IEEE Trans. Comp.* vol. C-31 (1982): 328.

Floyd, T. L. *Digital Fundamentals.* 2d ed. Columbus, Ohio: Charles E. Merrill, 1982.

Gaitanis, N. "Single error correcting and multiple unidirectional error detecting cyclic AN arithmetic codes." *Elect. Lett.* vol. 20 (1984): 638.

Greer, C. R., and Thompson, R. A. "Combinational logic design with decoders." *IEEE Trans. Comp.* vol. C-27 (1978): 869.

Hamming, R. W. *Coding and Information Theory.* Englewood Cliffs, N.J.: Prentice-Hall, 1980.

Lai, H. C., and Muroga, S. "Minimum parallel binary adders with NOR(NAND) gates." *IEEE Trans. Comp.* vol. C-28 (1979): 648.

Langdon, G. G., Jr., and Tang, C. K. "Concurrent error detection for group look-ahead binary adders." *IBM J. Res. & Dev.* vol. 14 (1970): 563.

Ling, H. "High-speed binary adder." *IBM J. Res. & Dev.* vol. 25 (1981): 156.

Liu, T. K.; Hohulin, K. R.; Shiau, L. E.; and Muroga, S. "Optimal one-bit full adders with different types of gates." *IEEE Trans. Comp.* vol. C-23 (1974): 69.

Majerski, S. "On determination of optimal distributions of carry skips in adders." *IEEE Trans. Elect. Comp.* vol. EC-16 (1967): 45.

Peterson, W. W. "Error correcting codes." *Sci. Am.* vol. 215 (1962): 96.

Schmookler, M. S. "Design of large ALUs using multiple PLA Macros." *IBM J. Res. & Dev.* vol. 24 (1980): 2.

Wakerley, J. *Error Detecting Codes, Self-Checking Circuits and Applications.* Amsterdam: North-Holland, 1978.

CHAPTER SIX

Sequential Devices

6.1 Introduction

A circuit is known as combinational as long as its steady-state outputs depend only on its current inputs. If, on the other hand, the present value of the outputs are dependent on both the present values of the inputs and the past values of the inputs, the circuit is considered to be a *sequential circuit.* One of the important applications of digital techniques is where digital signals are received and interpreted by the system, and control outputs are generated in accordance with the sequence in which the input signals are received. Therefore, such systems require circuits that respond to the past history of the inputs. In general, sequential circuits have the capability of storing information. Consequently, sequential circuits find wide application in digital systems as counters, registers, control logic, memories, and other complex functions.

The most common sequential circuit is the flip-flop. A *flip-flop* (*FF*) is an electronic device that has two stable states. One state is assigned the logic 1 value and the other the logic 0 value. The output of the FF can assume either of the stable states based on input events, and the output can be checked to determine what event occurred in the pair. There are a number of FFs in common usage in digital circuits, and they differ from one another in the number of inputs they have and in the manner in which the binary state is affected by the inputs. The possible changes in the FF outputs generally have a direct correspondence to the frequency with which the input is changing value. However, there is a type of sequential circuit memory device, known as a *monostable multivibrator*, that produces circuit output independent of the input frequency. This chapter introduces the logical behavior and control of various types of FFs. After studying this chapter, you should be able to:

- Understand the design and working principles of latches;
- Understand the design and working principles of FFs;

- Understand the design and working principles of the monostable multivibrator;
- Understand the importance and significance of sequential circuits in general.

6.2 Latches

A *latch* is a bistable circuit that is the fundamental building block of a flip-flop. The latch is basically a combinational circuit that has one of its outputs fed back as an input. It can be realized from an OR gate with its complemented output fed back as one of its inputs, as shown in Figure 6.1. We have considered Δt to be the *lumped* gate delay (total of all propagation delays) of the gates that are used. If the input I_1 is held at logic 1, the OR output results in a 1 and, therefore, the complemented output, O_2, is a 0. This O_2 output is fed back to the OR gate after a time equal to the lumped delay. As a result the output of the OR gate is held at logic 1. Once the OR output is set to this condition, the gate output will remain in this same state. This phenomenon is commonly known as the *latching effect*. Consequently, this circuit could be used for the storage of logic 1. This latching condition will prevail until the feedback path is broken.

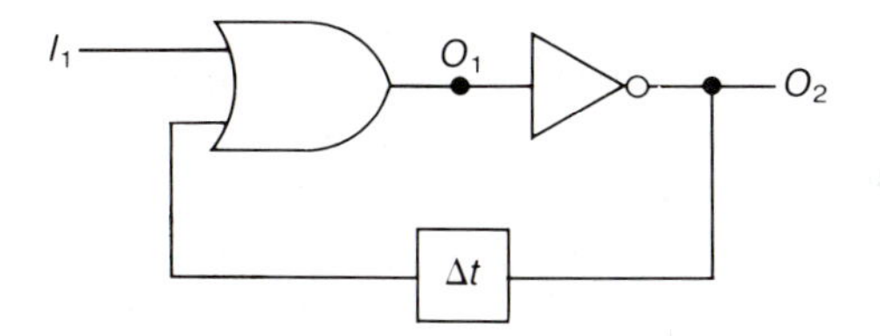

FIGURE 6.1 **Latch for Storing Logic 1.**

The NOT gate of Figure 6.1 may now be replaced by an AND gate to provide for the storage of logic 0. Such provision is known as the *unlatching* of the gate, which is illustrated in Figure 6.2. This circuit, however, is able to store both logic 1 and 0. Each of the two latch outputs, O_1 and O_2, is a logical complement of the other. If one holds I_2 input to logic 0, the feedback route between the input and the output of the OR gate will be logically broken and the OR gate

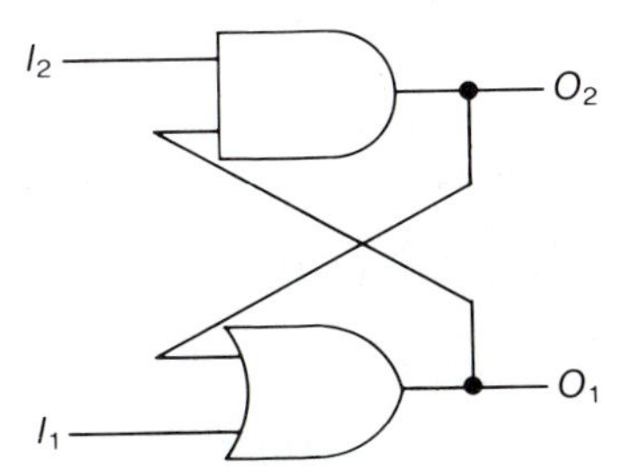

FIGURE 6.2 **AND-OR Latch.**

will return to its initial condition. When the input I_2 is a 0, the output O_2 is maintained at logic 0, and when I_1 is held at logic 1, the output O_1 remains at logic 1.

The latching concept developed in this section will be used next to come up with a standardized latch suitable for subsequent development of FFs. An FF has the capability of storing a single binary bit of information. When the values stored in the FFs change, we say that the sequential circuit *changes state*. Generally, however, an FF should have two outputs, called Q and $\overline{Q}$, that are complements of each other. The characteristic table of Figure 6.3 details the pertinent working principles of one such basic latch unit, where t is used to denote the time variable and Δt is the short time duration between a change in the input and a possible change in the output. The interval Δt is equivalent to the lumped delay of the circuit. The two inputs S (set) and R (reset) are used to control the output based on the current state of the output. If $R = 0$ and $S = 1$, the output is turned on if not already on. If $R = 1$ and $S = 0$, the output is turned off if not already off. When $S = R = 0$, no output change occurs. The to-be-designed latch circuit, however, manifests an undesirable condition when both inputs go to 1 simultaneously. When $S = R = 1$, the two outputs, Q and $\overline{Q}$, would no longer be complements of each other. In addition, the behavior of the latch would become unpredictable once the inputs returned to 0. Consequently, the simultaneous existence of $S = R = 1$ is forbidden.

FIGURE 6.3 **Characteristic Table for a Basic Latch.**

Inputs			Output
$R(t)$	$S(t)$	$Q(t)$	$Q(t + \Delta t)$
0	0	0	0
		1	1
0	1	0	1
		1	1
1	0	0	0
		1	0
1	1	0	—
		1	—

The circuit may be determined accordingly, using the K-map of Figure 6.4[a]. The equation for $Q(t + \Delta t)$ is obtained as follows:

$$\begin{aligned} Q(t + \Delta t) &= S(t) + \overline{R(t)}Q(t) \\ &= \overline{\overline{S(t)} \cdot \overline{\overline{R(t)}Q(t)}} \end{aligned} \qquad [6.1]$$

FIGURE 6.4 **RS Latch: [a] K-map and [b] Circuit.**

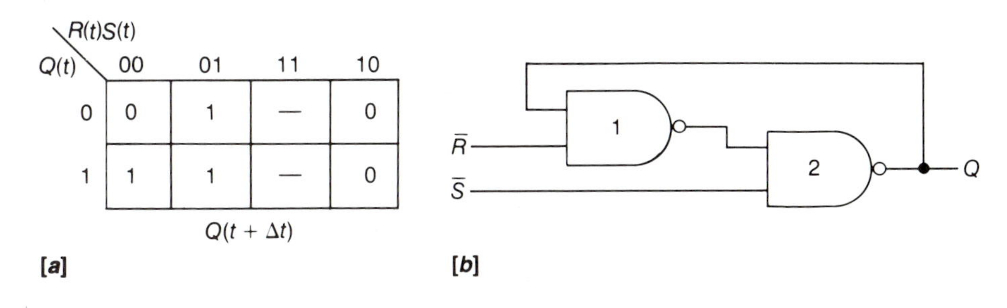

This equation, known also as the *next-state equation*, states that after a short time, Δt, the new value of Q is determined by the values of Q, R, and S at time t. The corresponding circuit is shown in Figure 6.4[*b*]. If R and S values change at time t, a new value of Q will result Δt time later. Δt time in this circuit is the total gate delay of the two NAND gates. The output of NAND gate 1 is traditionally known as the $\bar{Q}$ output since the outputs of two NAND gates are complements of each other.

FIGURE 6.5 **Revised RS Latch Characteristic Table.**

$Q(t)$	$R(t)$	$Q(t + \Delta t)$
0	0	$S(t)$
0	1	0
1	0	1
1	1	0

The revised characteristic table of Figure 6.5 shows the corresponding characteristics of the *RS latch* (also called reset-set latch) where the don't-care of the forbidden state is assumed to be equal to a 0. The equation for $Q(t + \Delta t)$ may also be obtained as follows:

$$
\begin{aligned}
Q(t + \Delta t) &= \bar{Q}(t)\bar{R}(t)S(t) + Q(t)\bar{R}(t) \\
&= \bar{R}(t)[Q(t) + \bar{Q}(t)S(t)] \\
&= \bar{R}(t)[Q(t) + S(t)] \\
&= \overline{R(t) + [\overline{Q(t) + S(t)}]}
\end{aligned} \qquad [6.2]
$$

This NOR form of latch could also be derived by grouping the zeros of the K-map of Figure 6.4[*a*]. The two circuits, NAND and NOR latches, are respectively known as *RS* and *SR* flip-flops or more commonly as latches. The corresponding latch circuits are obtained as shown in Figure 6.6.

FIGURE 6.6 **SR Latch: [a] Block Diagram, [b] NAND Circuit, and [c] NOR Circuit.**

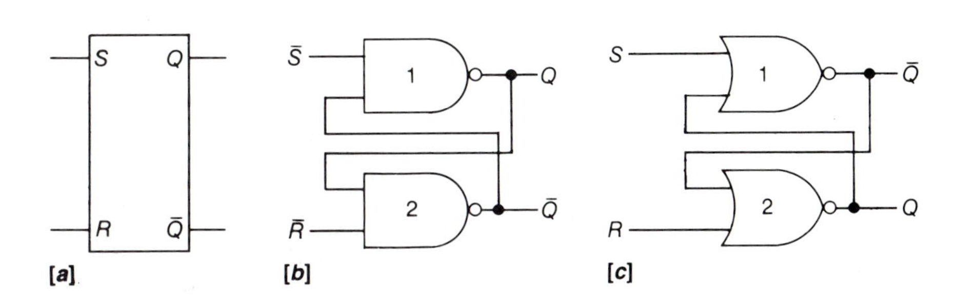

Note that the latch outputs are always complements of each other: When Q is a 1, $\overline{Q}$ is 0, and when Q is a 0, $\overline{Q}$ is 1. The forbidden state occurs when $S = R = 1$ at the same time. As long as S and R are both set at 1, Q and $\overline{Q}$ are forced to be at the same logic value simultaneously, thus violating the basic complementary nature of the outputs.

EXAMPLE 6.1

Obtain the response of a periodic square wave of period 8 units, when it is fed into the sequential circuit of Figure 6.7. Assume that the NAND gate has a total delay of 1 unit.

SOLUTION

FIGURE 6.7

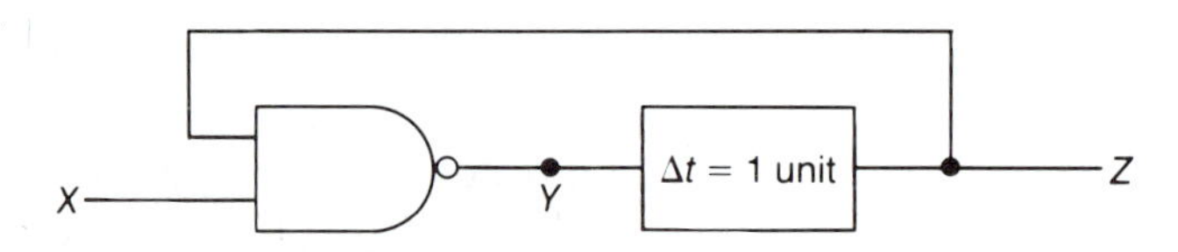

The output, Z, is given by

$$Z(t + \Delta t) = \overline{X(t)Z(t)} = \overline{X}(t) + \overline{Z}(t)$$

Consequently, the timing diagram is obtained as shown in Figure 6.8. Whenever X changes from a 0 to a 1, the output Z starts oscillating with a period of $2\Delta t$ (2 units in this case). However, when X changes from a 1 to a 0, the output Z becomes 1 after a time delay of Δt.

FIGURE 6.8

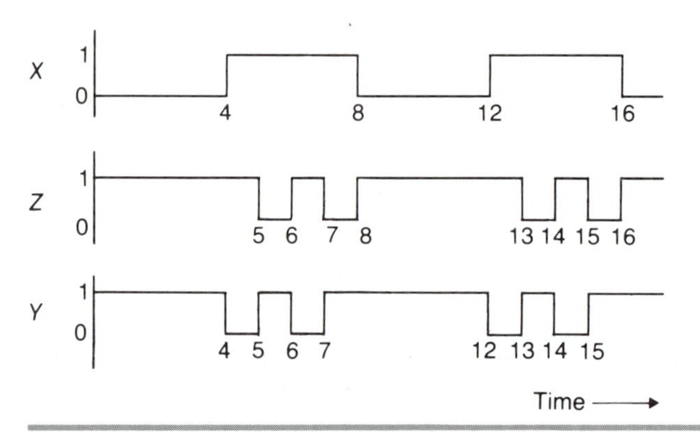

The *SR* latch described earlier has its time delay lumped together. However, there is another form of the latch model known commonly as the *distributed gate delay model.* Consider, for example, the NAND latch of Figure 6.6[*b*] where the gates numbered 1 and 2 are assumed to have gate delays t_1 and t_2, respectively. Accordingly,

$$Q(t + t_1) = \overline{\overline{S}(t) \cdot \overline{Q}(t)} = S(t) + Q(t) \qquad [6.3]$$

$$\overline{Q}(t + t_2) = \overline{\overline{R}(t) \cdot Q(t)} = R(t) + \overline{Q}(t) \qquad [6.4]$$

These two equations now may be combined to yield

$$\overline{Q}(t + t_1 + t_2) = \overline{\overline{R}(t + t_1) \cdot Q(t + t_1)}$$
$$= \overline{\overline{R}(t + t_1)\,[S(t) + Q(t)]} \quad [6.5]$$

Therefore,

$$Q(t + t_1 + t_2) = \overline{R}(t + t_1)[S(t) + Q(t)] \quad [6.6]$$

Assuming equal gate delays for a total of Δt delay, the resulting equation becomes

$$Q(t + \Delta t) = \overline{R}\left(t + \frac{\Delta t}{2}\right)[S(t) + Q(t)] \quad [6.7]$$

The timing diagrams of Figure 6.9 show the behavior pattern of a NAND latch corresponding to the distributed gate delay model. For simplicity it has been assumed that both of the gates have 1 unit length of gate delay and also that Q and $\overline{Q}$ at time $t = 0$ are, respectively, 0 and 1. This timing diagram shows that the latch works as intended. However, under various input conditions the latch may have problems. Example 6.2 will illustrate one of these input conditions and its consequences.

FIGURE 6.9 **Timing Diagram of a NAND Latch.**

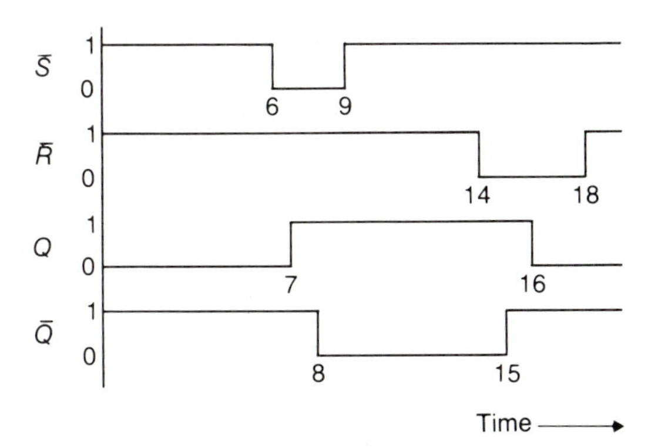

EXAMPLE 6.2

Obtain the timing diagram for the latch of Figure 6.10 when the input I_1 changes from 1 to 0 for a duration much shorter than the total gate delay.

SOLUTION

FIGURE 6.10

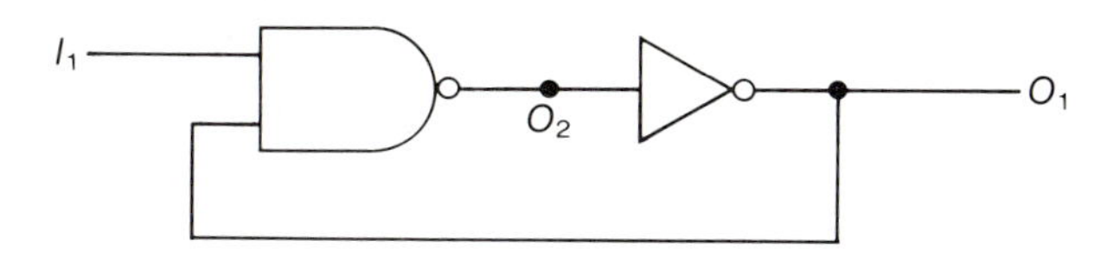

For this example the gate delays for the NAND and NOT gates are chosen to be 3 units and 2 units, respectively. Accordingly, the timing diagrams of Figure 6.11 are obtained. The output is oscillatory in nature and is not latched to any fixed value.

FIGURE 6.11

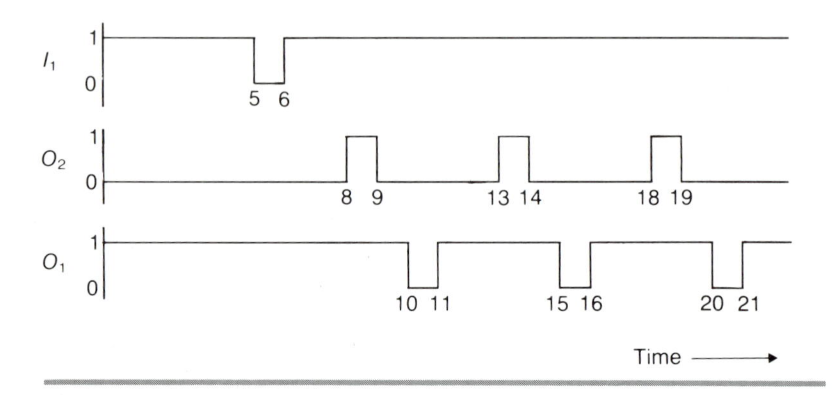

6.3 Clocked *SR* Flip-Flop

In the last section we introduced circuits for the *SR* FF. For dependable operation of such devices one must attempt to prevent transient pulses from appearing on either of the inputs. It is advantageous to control the times when the *SR* FF output is allowed to change by means of an additional input. This additional signal is commonly called a *clock*. The clock pulses (CK) can be periodic or a set of random pulses. Almost always, however, they are periodic.

The purpose of the clock input is to force the FF to remain in its rest (or hold) state while changes occur on the set and reset inputs. CK is set to logic 1 once the inputs have settled. The NAND and NOR latches with clock input are shown in Figure 6.12. In order to operate these devices effectively, the following conditions must be met:

1. The FF inputs should be allowed to change only when $CK = 0$.
2. The clock input should be long enough so that the outputs will be able to reach steady states.
3. The condition $S = R = 1$ must not be allowed to occur when CK is equal to logic 1. For proper operation, therefore, $S(t)R(t)$ should always equal zero.

It can be seen that the circuit action can occur only when the CK signal is high. When $CK = 0$, the FF outputs do not change. The S and R inputs may, however, be simultaneously high when the clock is absent since the FF will be inhibited. The overall functioning of the gated *SR* FF is illustrated by the characteristic table of Figure 6.13. Note in the timing diagram shown in Figure 6.14 that it is

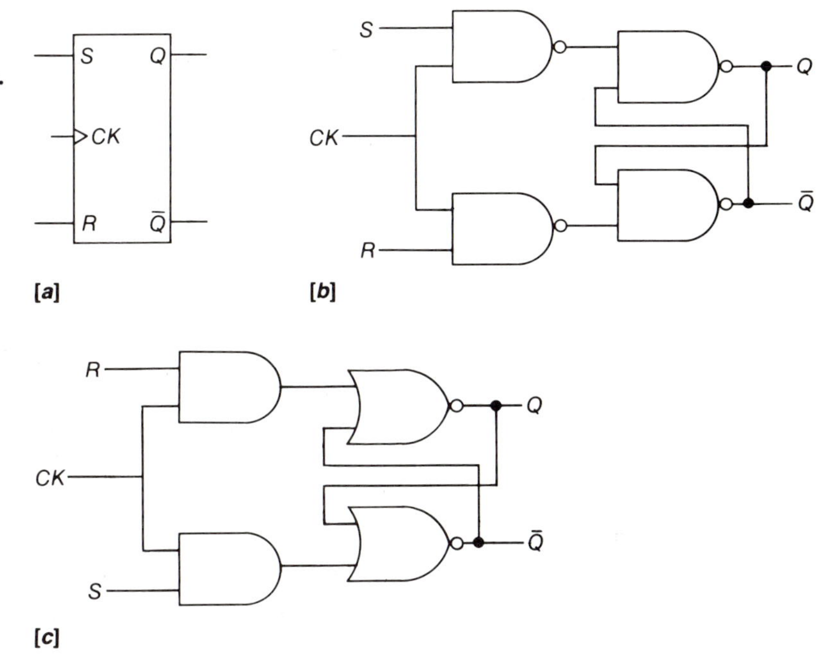

FIGURE 6.12 **Clocked *SR* FF: [*a*] Logic Symbol, [*b*] NAND Gate Circuit, and [*c*] NOR Gate Circuit.**

FIGURE 6.13 **Characteristic Table for the Clocked *SR* FF.**

Inputs			Mode	Outputs	
$CK(t)$	$S(t)$	$R(t)$		$Q(t+\Delta t)$	$\bar{Q}(t+\Delta t)$
0	—	—	No action	$Q(t)$	$\bar{Q}(t)$
⎍	0	0	Hold	$Q(t)$	$\bar{Q}(t)$
⎍	0	1	Reset	0	1
⎍	1	0	Set	1	0
⎍	1	1	Invalid	—	—

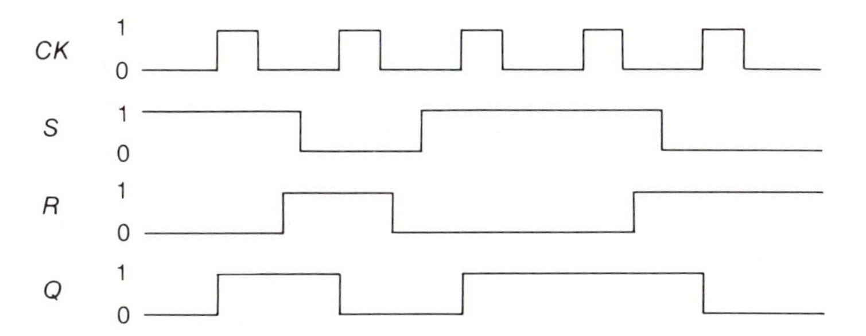

FIGURE 6.14 **Timing Diagram of a Clocked *SR* FF.**

necessary to consider the circuit only at the time *CK* changes from low to high to see if the output changes.

It is now appropriate to introduce several operational characteristics that are commonly associated with the FF usages. Figure 6.15 shows some of these specifications, of which setup and hold time are the most important ones. The *setup time*, t_s, is the time necessary for

FIGURE 6.15 **Timing Characteristics under Worst-Case Condition for Maximum Clock Frequency Determination.**

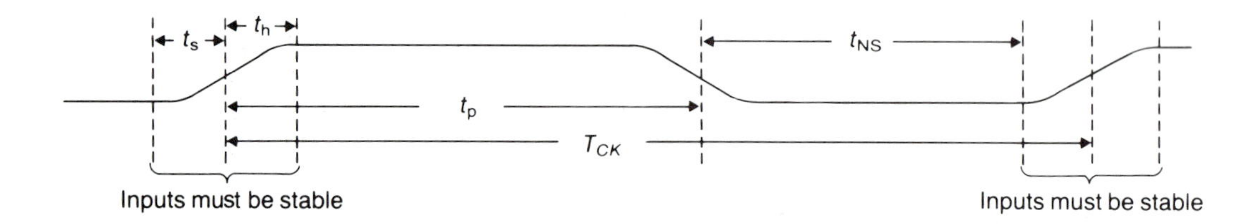

the input data to stabilize before the triggering edge of the clock. Its value is extremely critical since it manifests itself either by ignoring actions or by resulting in partial transient outputs, commonly referred to as *partial set* and *partial reset* outputs. Consequently, it is possible to begin a set or reset mode, causing the output to start to change, but to withdraw back to its initial state. In some cases the output might even end up in a metastable state in which the FF is neither set nor reset. Again, the *hold time*, t_h, is the time necessary for the data to remain stabilized beyond the triggering edge of the clock. This is also a critical parameter in determining the correct behavior of a FF.

The maximum allowable clock frequency for an FF is usually determined from a knowledge of setup time; hold time; FF propagation delay, t_p; and propagation delay of the next-state decoder, t_{NS}. The maximum clock frequency, f_{CK}, under worst-case condition is obtained from

$$f_{CK} = \frac{1}{T_{CK}} \leq \frac{1}{t_s + t_p + t_{NS}} \qquad [6.8]$$

The constraint of Equation [6.8] must be met when using any FF, integrated or not.

EXAMPLE 6.3

Obtain $Q(t + \Delta t)$ as a function of the inputs and $Q(t)$ in the circuit of Figure 6.16.

SOLUTION

FIGURE 6.16

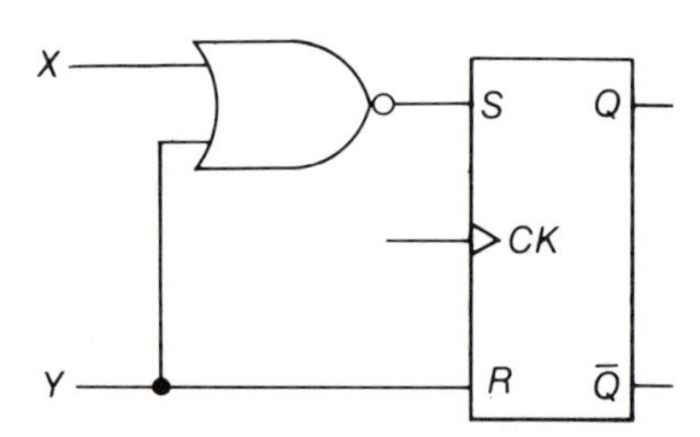

FIGURE 6.17

$X(t)$	$Y(t)$	$Q(t)$	$Q(t + \Delta t)$
0	0	0	1
0	0	1	1
0	1	0	0
0	1	1	0
1	0	0	0
1	0	1	1
1	1	0	0
1	1	1	0

From Equation [6.1] the next-state equation follows as

$$Q(t + \Delta t) = S(t) + \overline{R}(t)Q(t)$$

In this circuit $S(t) = \overline{X(t) + Y(t)} = \overline{X}(t) \cdot \overline{Y}(t)$ and $R(t) = Y(t)$. Note also that $S(t)R(t) = \overline{X}(t) \cdot \overline{Y}(t) \cdot Y(t) = 0$. Hence, $S(t)$ and $R(t)$ are not simultaneously equal to 1 and, therefore, the circuit meets all conditions necessary for a perfect operation. Consequently, we may obtain

$$Q(t + \Delta t) = \overline{X}(t)\overline{Y}(t) + \overline{Y}(t)Q(t)$$

The table, as shown in Figure 6.17, lists all possible combinations of the inputs and the corresponding outputs.

6.4 *JK* Flip-Flop

We saw in the last section that the clocked *SR* FF has an indeterminate state. When using clocked *SR* FFs the designer is required to be cautious about the FF inputs. This troublesome restriction can be removed by modifying the *SR* FF; the refined FF is known as the *JK* FF. This modification involves feeding the outputs of the FF back into the inputs of the circuit shown in Figure 6.12[*a*]. The resulting circuit, its block diagram, and its functional behavior are shown in Figures 6.18[*a*–*c*].

FIGURE 6.18 *JK* **FF: [*a*] Block Diagram, [*b*] Logic Circuit, and [*c*] Characteristic Table.**

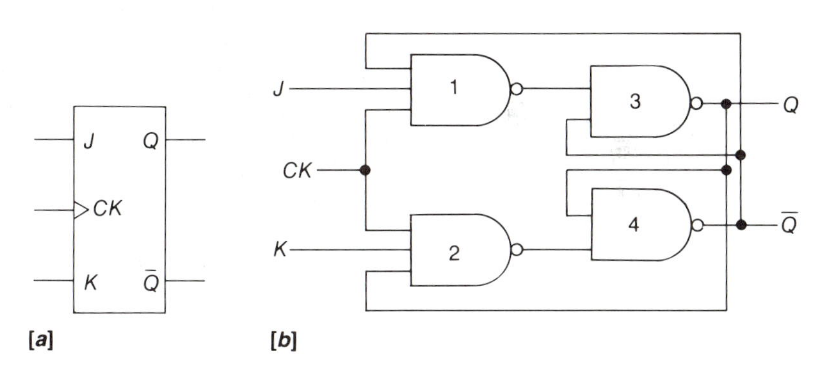

Mode	$J(t)$	$K(t)$	$Q(t)$	$Q(t + \Delta t)$
Hold	0	0	0	0
			1	1
Reset	0	1	0	0
			1	0
Set	1	0	0	1
			1	1
Toggle	1	1	0	1
			1	0

[*c*]

Even when the J and K inputs are both 1, the outputs of NAND gates 1 and 2 cannot simultaneously be 0. With $Q = 0$, NAND gate 2 outputs a 1, and when $Q = 1$, NAND gate 1 outputs a 1. Consequently, the input restriction of the SR FF is automatically eliminated. The additional feedback provides for an additional switching mode, called *toggle*, to the FF. The characteristic table of Figure 6.18[*c*] describes in detail the actions of the FF. The next-state equation may accordingly be obtained as follows:

$$Q(t + \Delta t) = J(t)\overline{Q}(t) + \overline{K}(t)Q(t) \qquad [6.9]$$

If $J = 1$ and $K = 0$, the FF is set to an output of 1 if not already set. Similarly when $J = 0$ and $K = 1$, the FF resets to 0 if not already reset. If $J = K = 1$, the FF output is complemented (toggled), and when $J = K = 0$, no change takes place.

In spite of many advantages the JK FF still has a serious limitation, which is illustrated in Figure 6.19 and should be understood by the designer. When the clock input goes high, the FF responds according to the J and K inputs. The flowchart of Figure 6.19[*a*] shows the desired circuit operation and the consequence of having a clock pulse that is too long. If the Q output changes before the termination of the clock input, then the input conditions to NAND gates 1 and 2 change again, and this leads to subsequent change in the Q output. As a consequence, Q may be indeterminate at the termination of the clock input. Such a possibility exists as long as Δt is less than T_{CK}. It is desired that only one FF change occur during each clock input. This may be easily accomplished by maintaining the clock width much smaller than the total delay time.

FIGURE 6.19 **[*a*] *JK* FF Flowchart and [*b*] Pulse Width Problem in *JK* FF.**

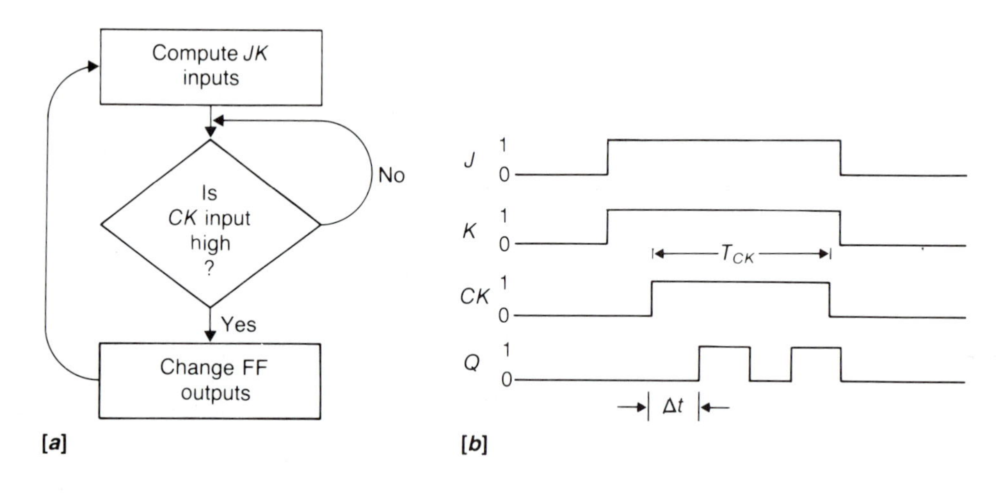

Another way to eliminate the problem caused by a clock pulse width that is too long is to design the FFs to respond to only transitions of the clock, either $1 \rightarrow 0$ or $0 \rightarrow 1$. Edge-triggered FFs are provided with a pulse-narrowing circuit, as shown in Figure 6.20. As you can see, there is a small delay on one of the NAND gate inputs so that the inverted clock pulse arrives at the gate input a couple of nanoseconds later than the true clock pulse. This results in an output spike of an extremely small time duration at the very beginning of the clock pulse. This narrow pulse is then used for the clock input of the *JK* FF, eliminating the necessity for narrow clock pulses. It is appropriate to consider the effect this narrow pulse might have on the FF actions. Depending on the parameters of the gates that are being used in the circuit of Figure 6.20, the resulting pulse could be too narrow to trigger an FF. In such an event more than one, but only an odd number of, NOT gates could be used in place of the first NOT gate. Again the number of NOT gates should not be too large, because a clock input that is too wide might result.

FIGURE 6.20 **Pulse-Narrowing Circuit.**

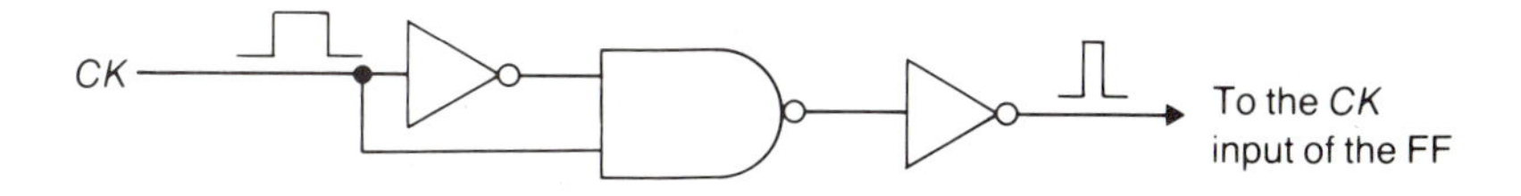

Edge-triggered devices are of two types. Positive edge-triggered devices respond when the clock input makes the transition $0 \rightarrow 1$, and the negative edge-triggered devices respond when the clock input makes the transition $1 \rightarrow 0$. Figure 6.21 shows the logic symbols for both positive edge-triggered and negative edge-triggered *SR* and *JK* FFs where the arrowhead input corresponds to the *CK* input. Another way to achieve edge triggering is to use a special FF known as the *master-slave FF* that appears to trigger only on the clock edge.

FIGURE 6.21 **Logic Symbols for Edge-Triggered FFs.**

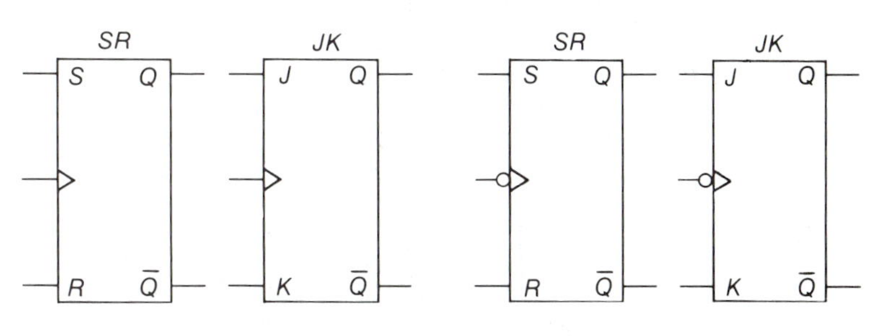

EXAMPLE 6.4

Obtain the timing diagram for the sequential circuit shown in Figure 6.22 for at least six clock cycles. Assume that $Q_1(0)Q_2(0) = 00$.

SOLUTION

FIGURE 6.22

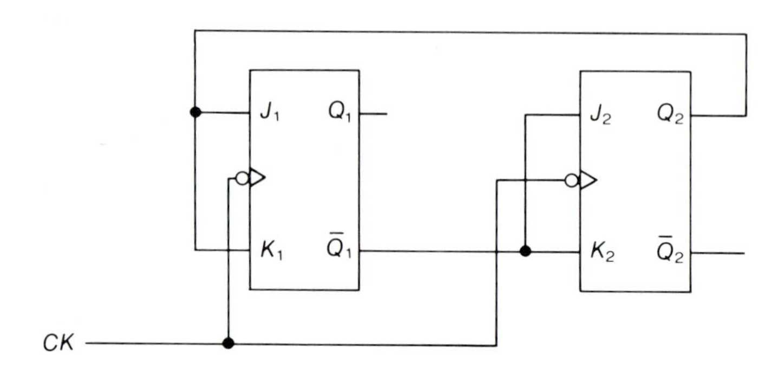

Figure 6.23 shows the timing diagram that is obtained readily by making use of the function table of a *JK* FF. It may be seen that within two cycles Q_1 is set and Q_2 is reset. The waveform will not change until Q_1 is reset externally.

FIGURE 6.23

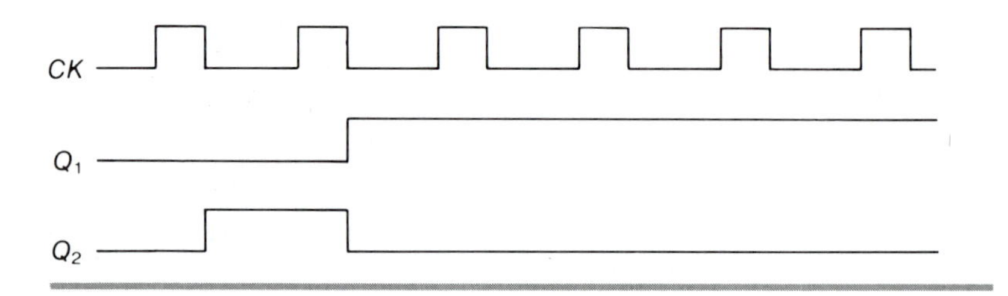

EXAMPLE 6.5

Obtain the response of the circuit of Figure 6.24, where each of the gates is assumed to have 1 unit of gate delay. The input x remains high for a duration longer than 4 units.

SOLUTION

FIGURE 6.24

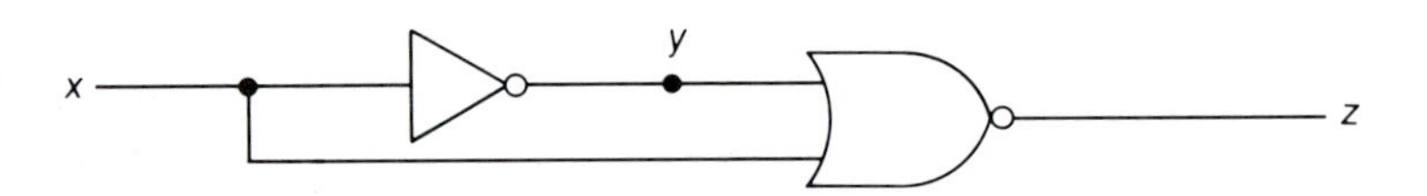

The input, x, is assumed to be 5 units wide. The timing diagram is then obtained as shown in Figure 6.25. This circuit locates the trailing edge of the input pulse. Note that the circuit of Figure 6.20 functions likewise but locates the leading edge of an input.

FIGURE 6.25

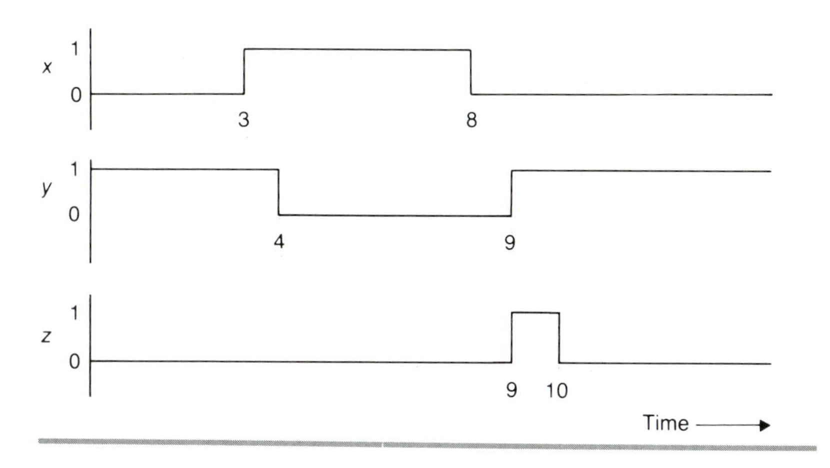

6.5 Master-Slave Flip-Flop

The master-slave concept is introduced into the FF circuitry to eliminate the requirement for limiting the clock width of a value determined by the circuit gate delays. This type of FF is composed of two sections: the master section and the slave section. This device is dependent not on the synchronous clocking of both units, but rather on their alternate turn-on and turn-off characteristics. The logic circuit of an *SR* master-slave FF is shown in Figure 6.26. It consists of a master FF, a slave FF, and an inverter for achieving out-of-phase clocking of the two units.

FIGURE 6.26 **Master-Slave *SR* FF.**

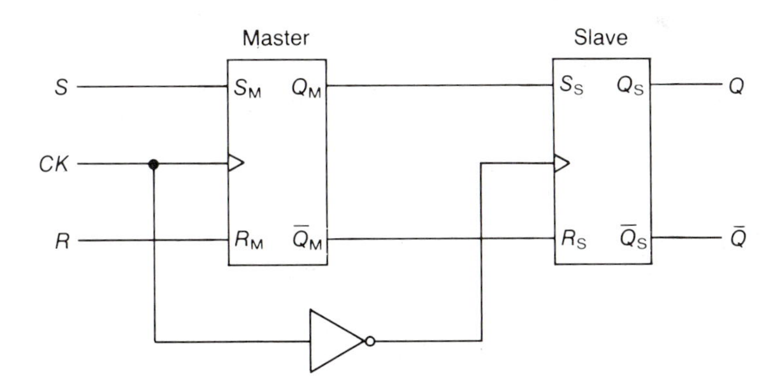

As a result of the presence of the inverter, the master unit is turned on and the slave unit is turned off when $CK = 1$. When $CK = 0$, the master unit is turned off and the slave unit is turned on. The circuit works as follows: for all inputs of S and R, except when $S = R$, $Q_M = S$ and $\overline{Q}_M = R$ when $CK = 1$. At this time the slave unit remains turned off. When the clock input goes to 0, $Q_S = Q_M$, $\overline{Q}_S = \overline{Q}_M$, and the master unit is turned off.

The timing diagram shown in Figure 6.27 illustrates the sequence of operations that takes place in a master-slave *SR* FF. The overall master-slave outputs appear to change at the negative edge of the clock input. However, there are many IC FFs that are the positive edge-triggered type. The master-slave cascading may be accomplished for any FF by similar introduction of an inverter between the two sections. As another example, Figure 6.28 shows the logic diagram of a master-slave *JK* FF. This is slightly different from that of a master-slave *SR* FF in that the outputs of NAND gates 7 and 8 are introduced as inputs to NAND gates 2 and 1, respectively. We have learned from Figure 6.19 that the Q and $\overline{Q}$ outputs might change several times during a wide clock pulse, leading to an unpredictable FF condition. Similar clock inputs would still cause the master outputs, Q_M and $\overline{Q}_M$, to change; but the slave outputs, Q_S and $\overline{Q}_S$, would not change because the inverted clock pulse disables the slave section. The values for J and K are still determined by the preclock values of Q_S and $\overline{Q}_S$. When the clock input to the master section goes low, the clock input to the slave section goes high, transferring Q_M and $\overline{Q}_M$ to the slave section. We have thus eliminated the problem of clock pulses that are too wide by the master-slave principle. A representative timing diagram for the master-slave *JK* FF is shown in Figure 6.29.

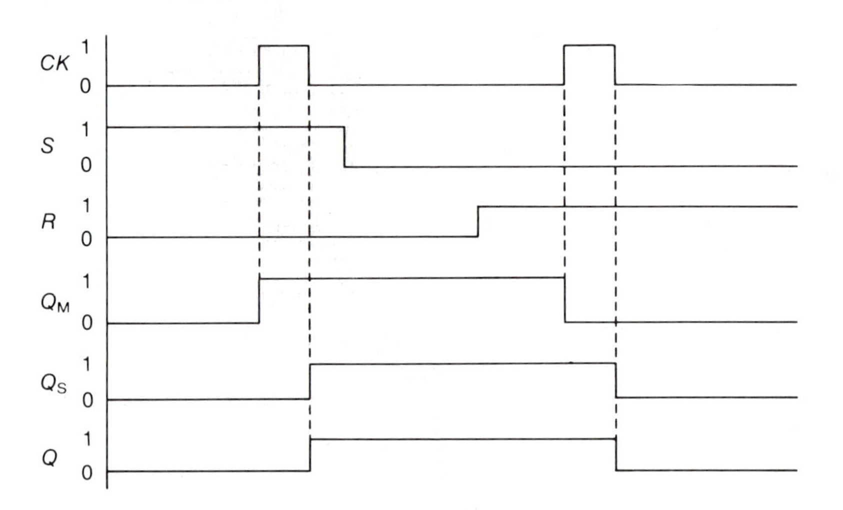

FIGURE 6.27 **Timing Diagram of a Master-Slave *SR* FF.**

A master-slave *JK* FF is also not without problems. The master section of the FF is vulnerable during the period when the clock is high and, therefore, may be set or reset by appropriate changes of the input. This results in "1s and 0s catching" problems. When $Q_S = 0$, $\overline{Q}_S = 1$, and the clock is high and while still high, the J input becomes high, Q_M is set, and consequently Q_S "catches" a 1 on the trailing edge of the clock input. Again when $Q_S = 1$, $\overline{Q}_S = 0$, and

FIGURE 6.28 **Master-Slave *JK* FF.**

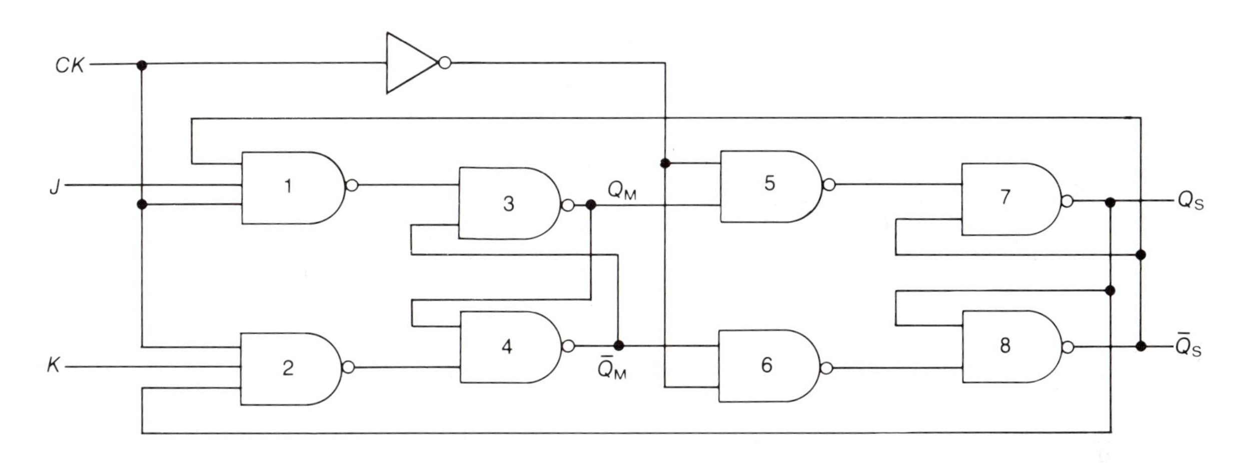

FIGURE 6.29 **Master-Slave *JK* FF Timing Diagram.**

K becomes a 1 after the clock has already become high, Q_M is reset, and Q_S "catches" a 0 on the trailing edge of the clock input. It is important, therefore, to make sure that no such input changes can gain entry into the FF.

Often an edge-triggered *JK* FF is also provided with two additional control inputs: preset and clear. The *preset* (*PR*) and *clear* (*CLR*) inputs allow initializing the FF to either a set ($Q = 1$) or a reset ($Q = 0$) condition. Addition of these two control inputs requires alteration of only the slave section of the FF. Figure 6.30 shows the logic diagram and the corresponding slave section of the FF that allows preset and clear inputs. Throughout this text both preset and clear inputs are considered to be active when low. Often these two FF control inputs are not labeled in the FF logic diagram. In such cases preset and clear inputs are always indicated by verti-

FIGURE 6.30 **Complete Master-Slave *JK* FF: [*a*] Logic Diagram and [*b*] Slave Circuit.**

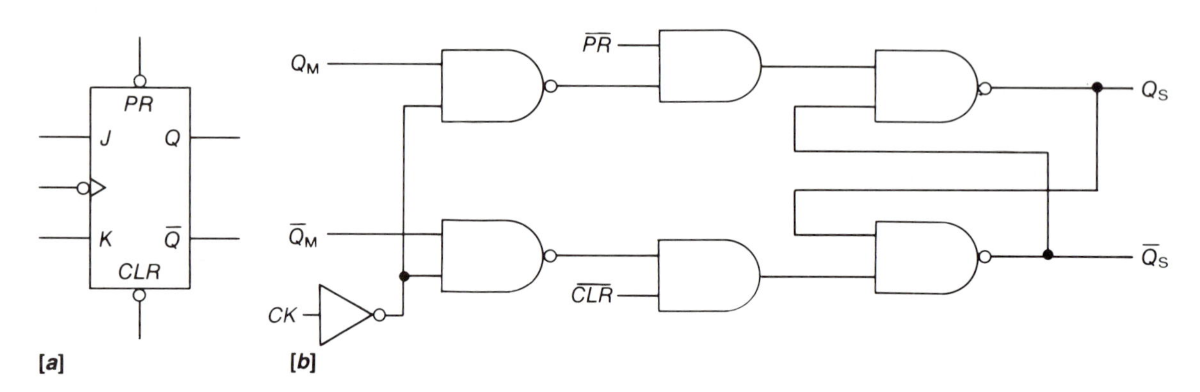

cal inputs (with a bubble) respectively at the top and bottom of the corresponding FF logic diagram.

6.6 Delay and Trigger Flip-Flops

There are two other types of flip-flops that are commonly used: the *delay* (D) and the *trigger* (T) FFs. Unlike those in the previous sections, these two FFs have only one control input line besides the clock (excluding set and preset). Both of these FFs can be realized by externally manipulating the inputs of a *JK* FF.

Often it is necessary to have a sequential device that simply retains the input data value between clock pulses. The D FF performs this function. The FF output follows the FF input whenever a clock pulse is 1 and holds the value the input had when the clock changed to 0. The logic diagram and the characteristic table for a D FF are shown in Figures 6.31[*a–b*]. A comparison of this characteristic table with that for the *JK* FF (Figure 6.18[*c*]) reveals that a D FF is realizable from a *JK* FF by making $K = \bar{J}$ and using J as the D input, as illustrated in Figure 6.31[*c*]. The next-state equation of the D FF is given by

$$Q(t + \Delta t) = D(t) \qquad [6.10]$$

FIGURE 6.31 ***D* FF: [*a*] Logic Diagram, [*b*] Characteristic Truth Table, and [*c*] Circuit Implementation.**

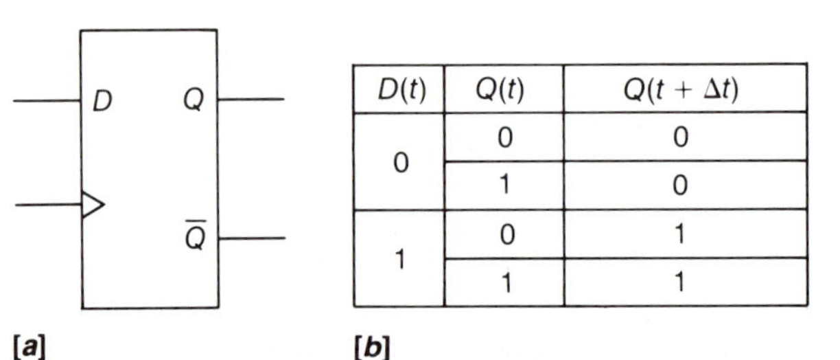

$D(t)$	$Q(t)$	$Q(t + \Delta t)$
0	0	0
	1	0
1	0	1
	1	1

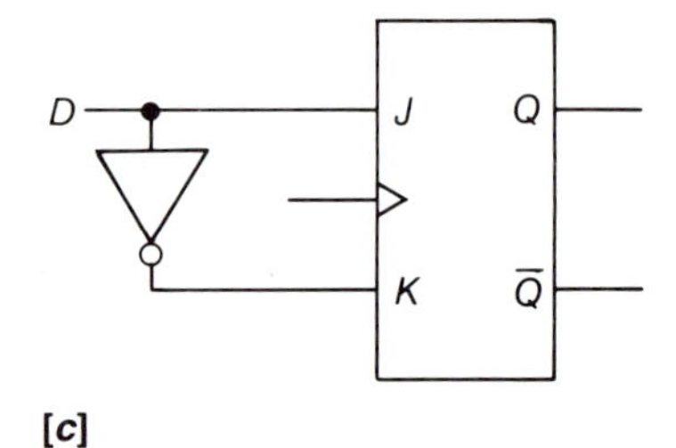

If severe restrictions placed on the clock input of an *SR* FF pose no problem, the *SR* FF can also be used to produce a *D* FF. The resulting circuit is shown in Figure 6.32[*a*]. Figure 6.32[*b*] shows a slight variation of the circuit of Figure 6.32[*a*] where advantage is taken of the special properties of NAND gates to eliminate one gate and still retain the characteristics of a *D* FF.

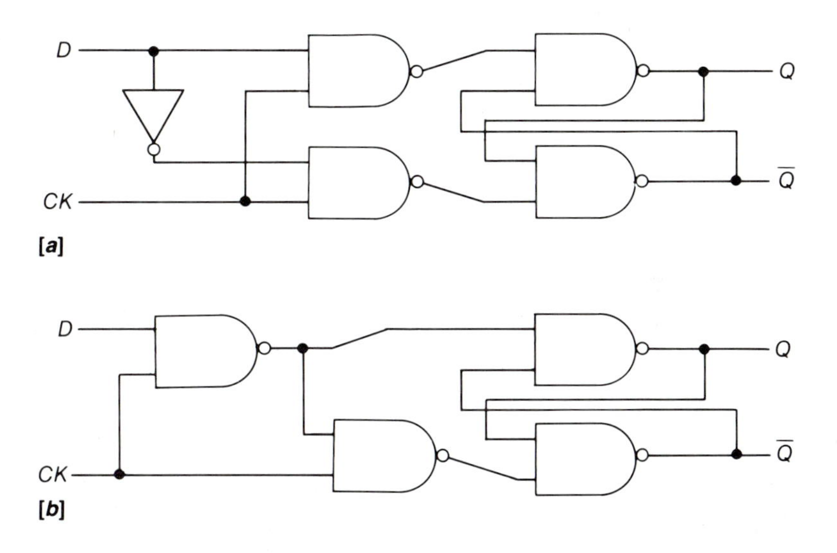

FIGURE 6.32 ***D*** **FF: [*a*] Using *SR* FF and an Inverter and [*b*] Using Modified NAND Latch.**

The *T* (trigger) FF, often called a toggle FF, has a single input that causes the output to change each time a pulse occurs at the input. The output remains unchanged as long as $T = 0$. The logic diagram and the characteristic table for a *T* FF are shown in Figures 6.33[*a–b*]. It should be noted that the *JK* FF has this mode available. The *JK* FF can be reorganized for realizing a *T* FF, as shown in Figure 6.33[*c*]. As long as both the *T* input and the *CK* input are high, the FF output will change. Its next-state equation, therefore, is obtained as follows:

$$Q(t + \Delta t) = Q(t) \oplus T(t) \qquad [6.11]$$

FIGURE 6.33 ***T*** **FF: [*a*] Logic Diagram, [*b*] Characteristic Table, and [*c*] Circuit Implementation.**

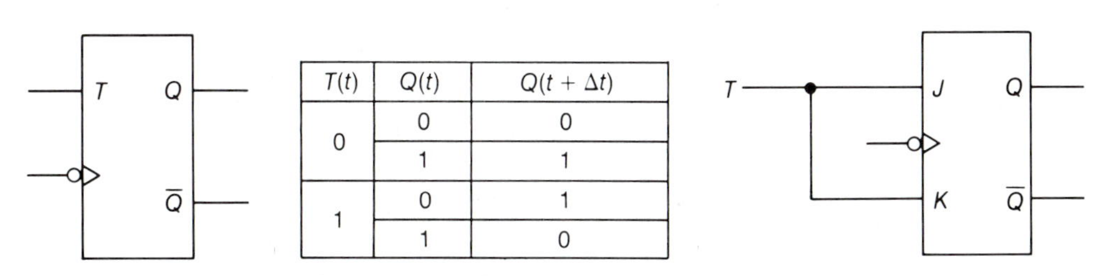

$T(t)$	$Q(t)$	$Q(t + \Delta t)$
0	0	0
	1	1
1	0	1
	1	0

A different version of the *T* FF involves a one-input device. Both *J* and *K* inputs of the *JK* FF are tied to a 1 to realize this unclocked

T FF. The input data are then introduced at the original clock input. The corresponding circuit for the unclocked *T* FF is shown in Figure 6.34.

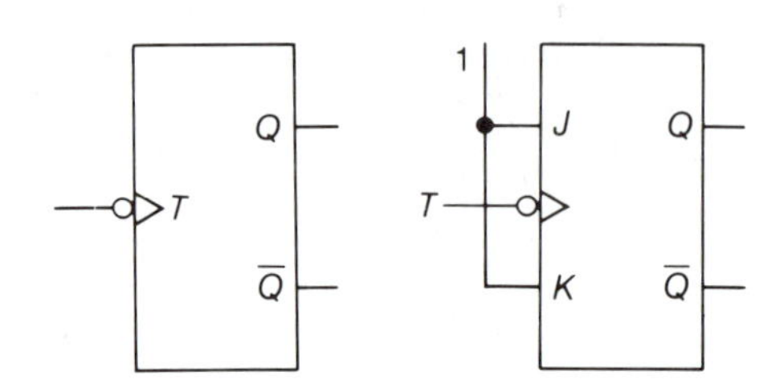

FIGURE 6.34 **Unclocked *T* FF: [*a*] Block Diagram and [*b*] Logic Circuit.**

The unclocked *T* FFs are very important but are not made commercially. The logic usually is obtained using a *JK* FF as shown in Figure 6.34. One may even obtain this function from a *D* FF. In fact it is also easy to transform an unclocked *T* FF back to a *JK* FF. Such a conversion circuit is shown in Figure 6.35.

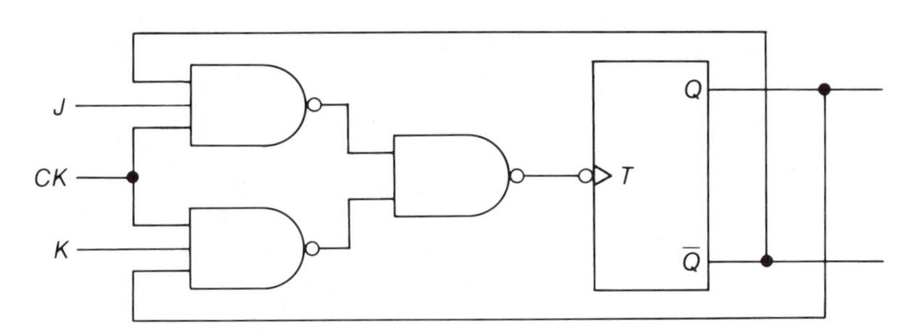

FIGURE 6.35 **Conversion of an Unclocked *T* FF to a Regular *JK* FF.**

EXAMPLE 6.6

Analyze the circuit of Figure 6.36.

SOLUTION

FIGURE 6.36

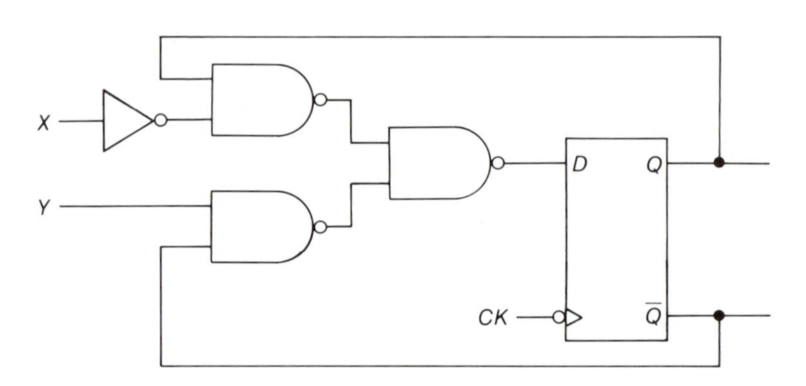

The next-state equation of this circuit is as follows:

$$Q(t + \Delta t) = D(t)$$
$$= Y(t)\overline{Q}(t) + \overline{X}(t)Q(t)$$

A close examination of this equation reveals that it is very similar to Equation [6.9] in that Y acts like the J input and X acts like the K input. In other words, this circuit functions exactly like a JK FF.

6.7 Monostable Flip-Flop

The *monostable* FF, also known as a *one-shot*, is an edge-triggered device used for producing output pulses of a duration independent of the input frequency. It produces an output pulse of specified width that is initiated by an input trigger signal. After a specified period of time the output returns to its quiescent state. The pulse duration is determined by the parameters of the resistor-capacitor network located external to the one-shot.

One-shots are of two types: nonretriggerable and retriggerable. In the nonretriggerable one-shot, if the device receives two successive trigger pulses of separation ζ less than the width δt of the output pulse generated by a single trigger, the second trigger input is ignored by the device. The retriggerable one-shot would be activated by the second trigger pulse, resulting in an output pulse of width approximately $\zeta + \delta t$. By applying a succession of trigger pulses separated by $\zeta < \delta t$, the output of a retriggerable one-shot could be maintained high (logic 1) as long as desired.

A one-shot may be designed using basic logic gates and a resistor-capacitor network. However, it is more convenient to use an IC one-shot because they are widely available and relatively inexpensive. Figure 6.37 shows the logic diagram and the function table of a standard retriggerable one-shot. The four inputs, A_1, A_2, B_1, and B_2, are available to provide flexibility of operation. The capacitor, C,

[a]

A_1	A_2	B_1	B_2	T
$1 \to 0$	1	1	1	$0 \to 1$
1	$1 \to 0$	1	1	$0 \to 1$
—	0	$0 \to 1$	1	$0 \to 1$
0	—	$0 \to 1$	1	$0 \to 1$
—	0	1	$0 \to 1$	$0 \to 1$
0	—	1	$0 \to 1$	$0 \to 1$

[b]

FIGURE 6.37 **Retriggerable One-Shot FF: [*a*] Logic Diagram and [*b*] Trigger Conditions.**

and the resistor, R, are external to the IC one-shot and are used to control the duration of the output pulse. Note that the trigger pulse, T, is given by $\overline{A_1 \cdot A_2} \cdot B_1 \cdot B_2$. The triggering conditions, as shown in Figure 6.37[*b*], cause T to change from a 0 to a 1.

The duration of the output pulse, δt, is determined by the resistor-capacitor network. Adjustable resistors and/or capacitors may be used to trim the output pulse to the desired width. In general, δt is given by

$$\delta t = f(R,C) \qquad [6.12]$$

where $f(R,C)$ is a function of the resistor and capacitor. The manufacturer provides the exact numerical relationship or curves, giving the output pulse width as a function of the timing resistors and capacitors. The minimum output pulse usually is realized using no external capacitor. Note, however, that there will be some stray capacitance existing between the terminals even in the absence of the external capacitor.

The retriggerable one-shot may be transformed into a nonretriggerable one-shot by feeding the Q output as one of the NAND gate inputs, say, A_2, while the other input A_1 is treated as the only triggering input. The remaining two inputs, B_1 and B_2, should be tied to a 1.

It is advisable to use monostable FFs only when no other solution can be found. Circuits with a number of monostables are very difficult to troubleshoot. Monostables can be falsely triggered by noise in the power supply voltage, causing serious circuit malfunctions.

6.8 Sequential Circuits

The general form of a sequential circuit is shown in Figure 6.38. The circuit in consideration has p inputs, q outputs, and r FFs used as memory. The combinational part of the circuit monitors the

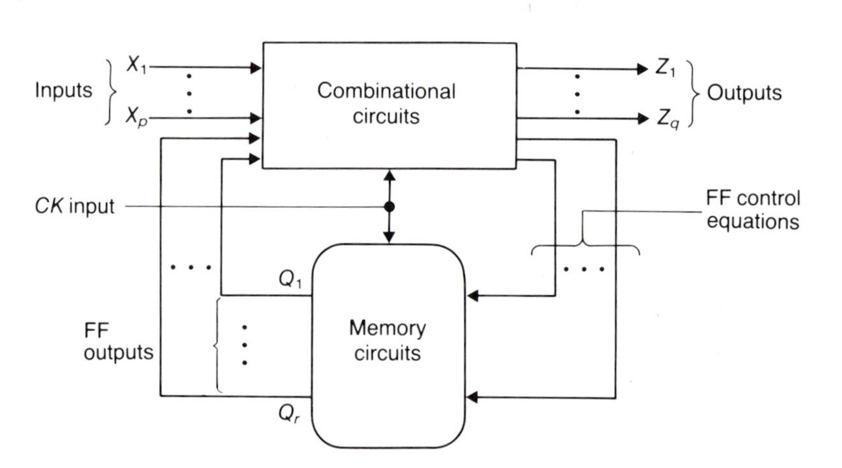

FIGURE 6.38 **Block Diagram of a General Sequential Circuit.**

input values, X_j, checks the FF states, Q_k, and computes the FF control variables to assure that the next initiating action causes the correct changes to be made in the FF values. In addition the combinational part also computes the correct outputs, Z_l, for the circuit. Thus the current inputs and the previous-state information stored in the circuit's memory (FFs) are used to generate the current outputs and to determine the next state in the sequential circuits. The clock input is used only in clocked sequential circuits (the predominant type of sequential circuit).

The memory part of the circuit may be provided by using bistable devices such as FFs, relays, magnetic devices, switches, and so on. The most commonly used bistable device, however, is the FF. The control characteristics of various FFs are summarized in Figure 6.39, which provides the FF excitation inputs necessary to cause change in the FF output, Q. For example, the output of a JK FF can be changed from 1 to 0 by setting $K = 1$ while the J input could be tied either to a 1 or to a 0. The corresponding state transitions between $Q = 0$ and $Q = 1$ for each of the four FFs are shown in Figure 6.40 where the conditions for transitions are indicated next

FIGURE 6.39 **FF Control Characteristics.**

$Q(t)$	$Q(t + \Delta t)$	S	R	J	K	D	T
0	0	0	—	0	—	0	0
0	1	1	0	1	—	1	1
1	0	0	1	—	1	0	1
1	1	—	0	—	0	1	0

FIGURE 6.40 **Transition Diagrams: [*a*] *SR* FF, [*b*] *JK* FF, [*c*] *D* FF, and [*d*] *T* FF.**

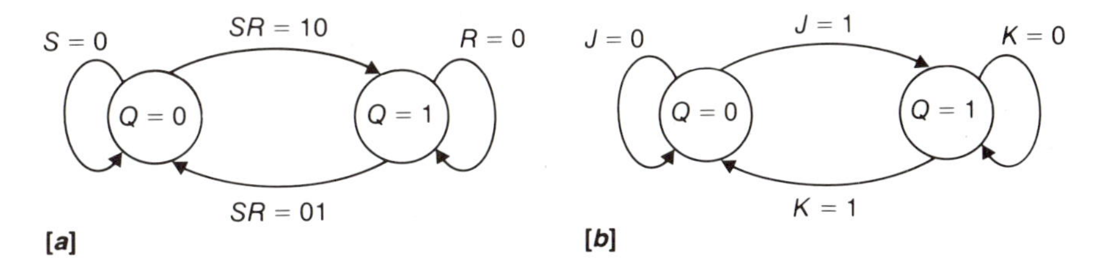

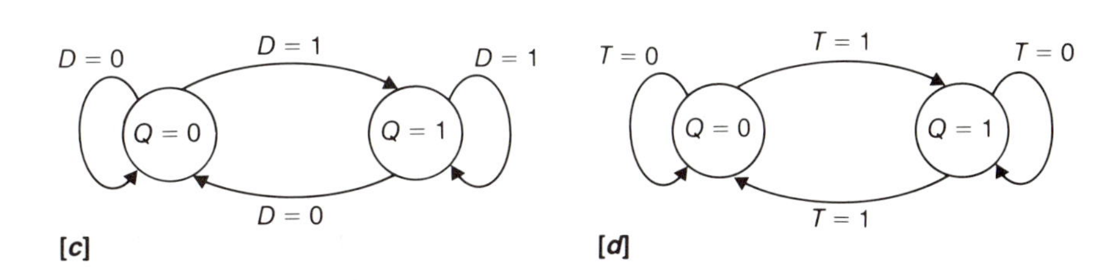

to the transition lines. These characteristics will play an extremely important role in the design of complex sequential circuits.

One of the advantages of sequential systems over purely combinational systems is that circuit savings may be possible through the repetitive use of the same logic circuit. Some examples are multi-bit adder/subtracter circuits, multi-bit comparator circuits, and multi-bit code-converter circuits. It was shown in Chapter 5 that n different combinational circuit units are needed to accomplish an n-bit parallel operation. However, the price we will pay for using the same logic circuit repetitively is in the circuit operation speed.

Consider the four-bit ripple adder circuit shown in Figure 6.41. This addition operation is called parallel since two four-bit numbers are fed as inputs simultaneously to the adder circuit, and after the gate delays, the resultant bits become available simultaneously. However, addition can also be implemented serially. A *sequential circuit* that can perform such an operation is illustrated in Figure 6.42. The process can be started by introducing A_0 and B_0 to the FA. The resultant sum bit is stored in a multi-bit storage device, made up of several FFs and called a *register*, and the carry-out is stored in an FF and fed back into the FA. The process is repeated until all of the bits have been considered. The final sum would consist of the last carry-out and the sum bits stored in the register (see Problem 1 at the end of the chapter). The characteristics of the registers and how the input bits are fed sequentially to the adder will be considered in a later chapter. For the time being it will suffice to say that the register is a storage area where the bits can be moved slowly to the right as the storage process continues.

However, in spite of all their merits sequential systems are not devoid of deficiencies. Some of these disadvantages are very critical to the functioning of the system; therefore, they must be considered very carefully. Consider a very simple circuit such as the one shown in Figure 6.43. When the input I is a 0, the output O becomes a 1. This output is then fed back again to the NAND gate. As a consequence, the values of the output disagree with that of the input,

FIGURE 6.41 **Four-Bit Parallel Adder Circuit.**

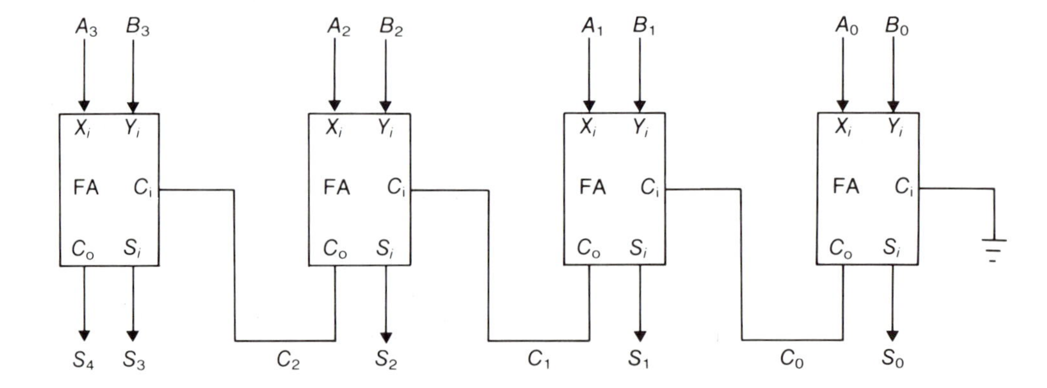

FIGURE 6.42 **Primitive Block Diagram for the Parallel Adder.**

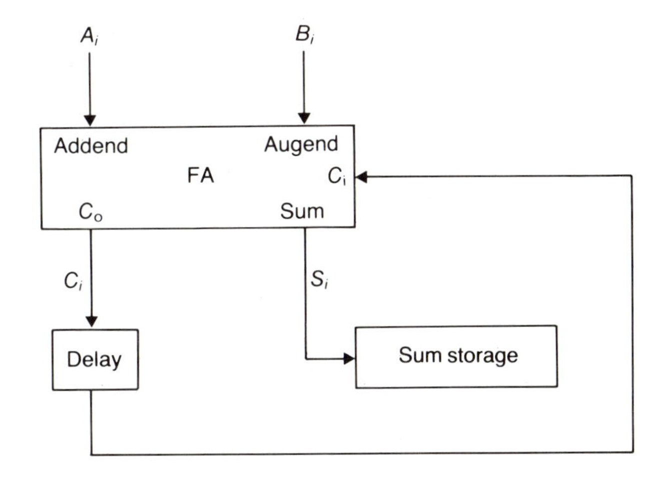

FIGURE 6.43 **Indeterminate Feedback Circuit.**

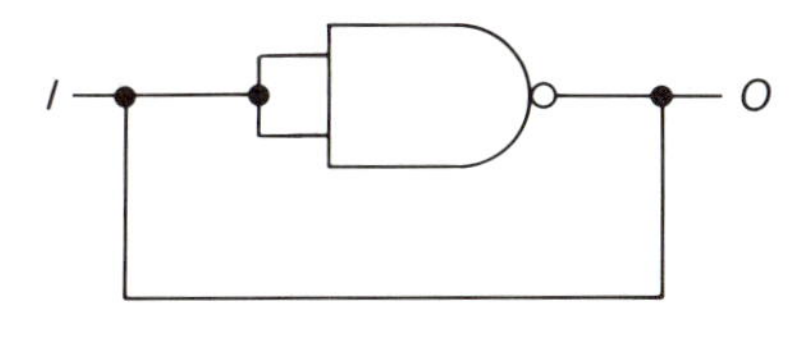

resulting in an indeterminate feedback system. Such problems, however, may be avoided by eliminating the conditions of continuing oscillations.

Next consider the combinational circuit of Figure 6.44[*a*] where the gates numbered 1, 2, and 3, respectively, have gate delays of Δt_1, Δt_2, and Δt_3. Assume that at time t_0 the inputs are $A = B = C = 1$, and at time $t > t_0$ the input A changes from 1 to 0. Assume further that $\Delta t_1 < \Delta t_2$. The resulting timing diagram is illustrated in Figure 6.44[*b*]. It is apparent that the combinational circuit output results in a transient error pulse for a duration of $\Delta t_2 - \Delta t_1$. This error pulse is small but not negligible. If the output of this circuit is introduced as the clock, preset, or clear inputs to an FF, we can expect to see additional problems in the sequential circuit. However, errors will not occur until the pulse width exceeds the time required to trigger the FF. These and other problems are tackled by taking proper precautions either during the design or during the operation of a sequential system.

FIGURE 6.44 **[*a*] Combinational Circuit and [*b*] Its Corresponding Timing Diagram When $B = C = 1$.**

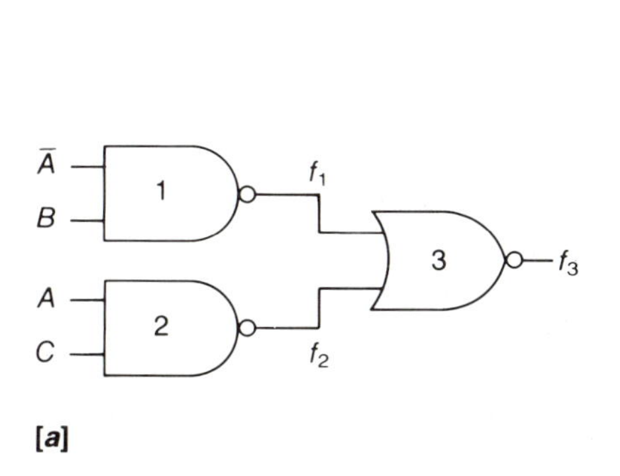

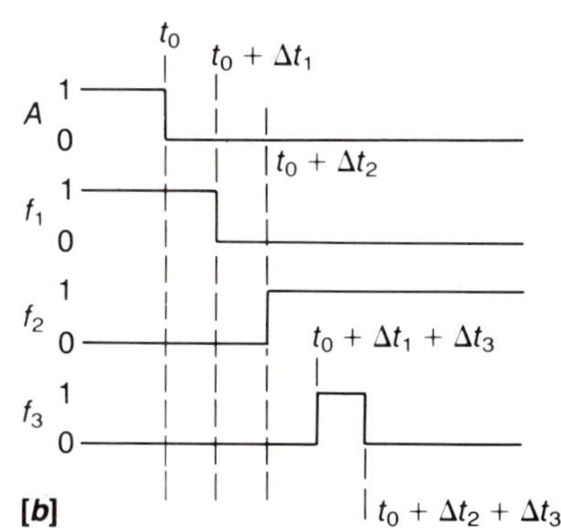

There are three different types of sequential circuits. They are classified according to the characteristics of the inputs and memory types.

Synchronous sequential circuits: Synchronous sequential circuits involve FF action that occurs in synchronization with the clock input. The values of the external variables that control the FF states may change while the clock is not present. All transients due to the previous clock must have disappeared prior to the next clock for correct circuit action.

Pulse-mode circuits: The input variables of pulse-mode circuits can have only mutually exclusive pulses. In addition, the FFs that are used are not clocked since no clock is present in this mode.

Fundamental-mode circuits: Fundamental-mode circuits involve level inputs and asynchronous memory devices. The FFs change state whenever an input variable logic level changes.

While the first type listed, also known as a *clocked* sequential circuit, is synchronous, the other two are asynchronous in character. However, synchronous sequential circuits account for the overwhelming majority of sequential circuits.

6.9 Summary

In this chapter the concept of a sequential circuit was introduced. The design and working principles of latches, various FFs, and the monostable multivibrator were discussed. Particular emphasis was placed on the various practical limitations that these devices have. Finally, the possibility of having different classes of sequential circuits was explored. The design and the characteristics of these sequential systems will be presented respectively in the next three chapters.

Problems

1. The FA receives two external inputs X and Y; the third input Z comes from the output of a D FF as shown in Figure 6.P1. The carry-out is transferred to the FF at every clock pulse.

FIGURE 6.P1

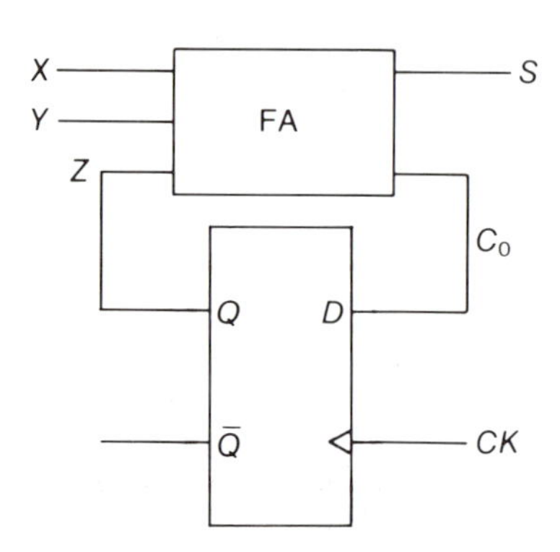

The S output represents the sum. Obtain the state equations for S and C_o.

2. Draw the timing diagram for the given input signal and circuit of Figure 6.P2. Assume the starting value of $Q_2Q_1 = 00$.

FIGURE 6.P2

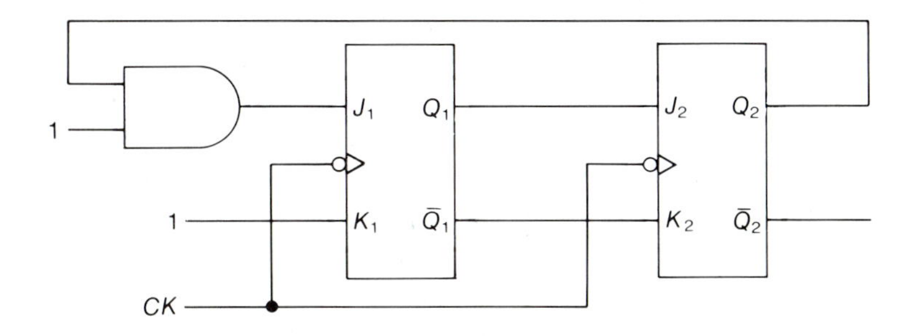

3. What sequence should repeat for the sequential circuit of Figure 6.P3 for the following initial inputs:
   a. $Q_3Q_2Q_1 = 001$ b. $Q_3Q_2Q_1 = 100$

FIGURE 6.P3

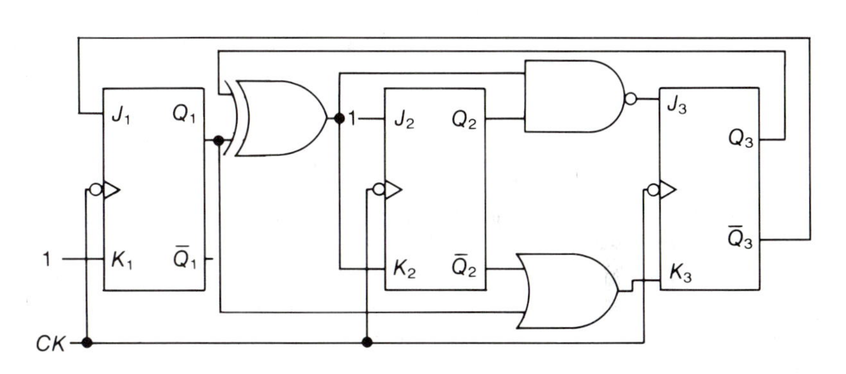

4. Obtain a T FF from a D FF.
5. Explain the behavior of the circuit of Figure 6.P4.

FIGURE 6.P4

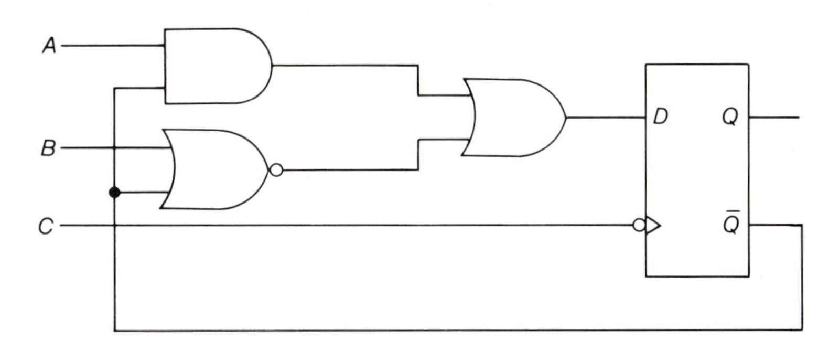

6. A sequential circuit has two inputs, X and Y, and one output, Z, such that

$$J_1 = XQ_2 + \overline{Y}\overline{Q}_2$$
$$J_2 = X\overline{Q}_1$$
$$K_1 = X\overline{Y}\overline{Q}_2$$
$$K_2 = X\overline{Y} + Q_1$$
$$Z = XYQ_1 + \overline{X}\overline{Y}Q_2$$

Obtain the logic diagram and state equations.

7. Find the output and state sequences for the circuit of Figure 6.P5 if the initial state is $Q = 0$ and the input sequence is
   a. $x = 101101100$ b. $x = 111011101$

FIGURE 6.P5

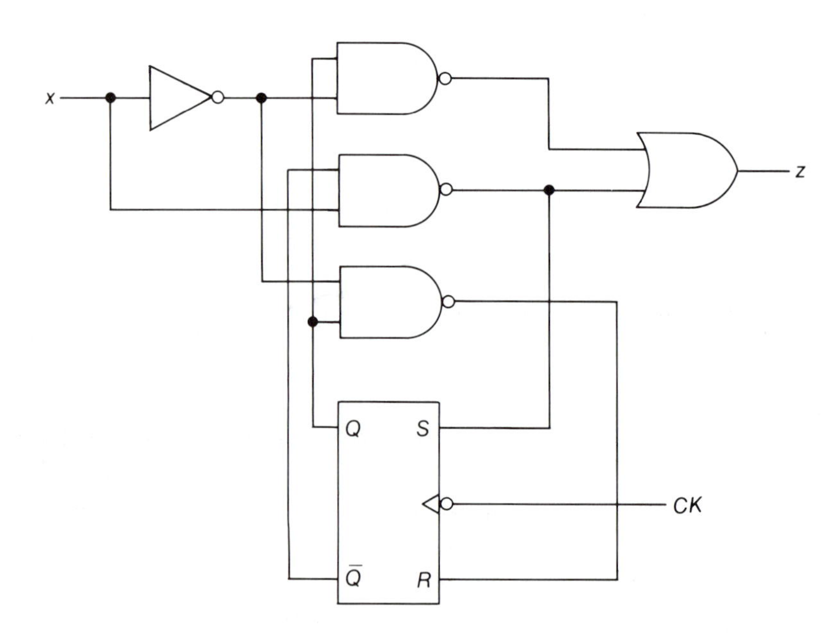

8. Show the detailed working of the circuit of Figure 6.35.
9. Repeat Example 6.2 when the NAND gate has a delay of 4 units and the NOT gate has a delay of 3 units.
10. Repeat Example 6.2 when the gate delays are lumped together.
11. Repeat Example 6.2 when the NAND gate has a delay of 4 units and the NOT gate has a delay of 3 units, but assume a lumped model for the circuit.
12. Comment on the behaviors of the two circuits of Figure 6.P6.
13. Analyze the circuit of Figure 6.P7.
14. Obtain a sequential system for performing a multi-bit, BCD-to-binary conversion. Describe the general working principles of your circuit.

FIGURE 6.P6

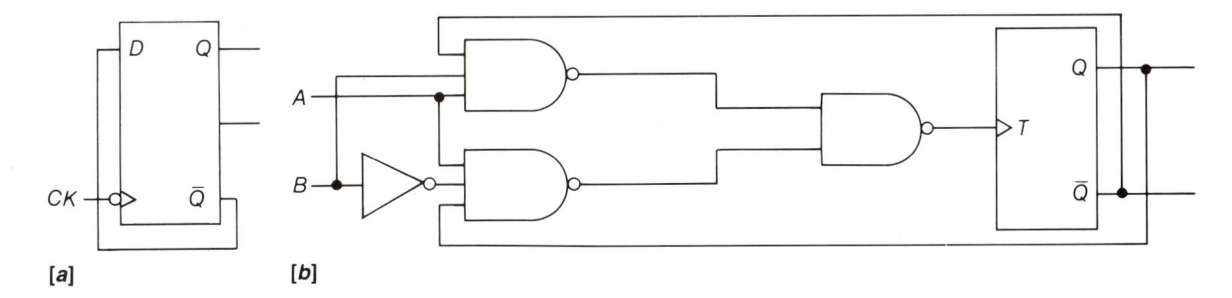

FIGURE 6.P7

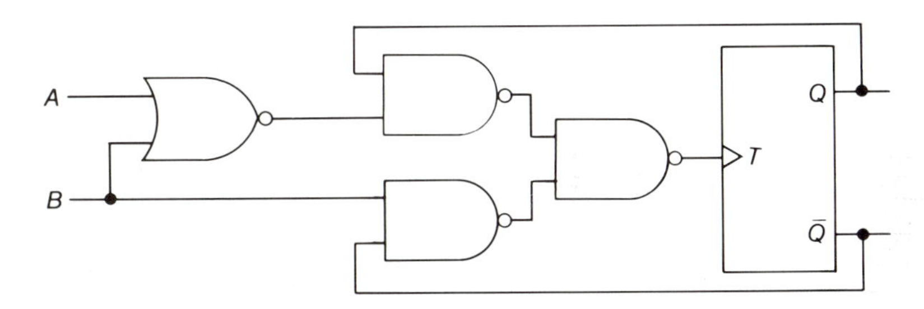

15. Obtain a sequential system for performing a multi-bit, binary-to-BCD conversion. Describe the general working principles of your circuit.
16. Obtain a sequential system for performing a multi-bit, Gray-to-binary conversion. Describe the general working principles of your circuit.
17. Obtain a sequential system for performing a multi-bit, binary-to-Gray conversion. Describe the general working principles of your circuit.
18. Obtain a sequential system for performing a multi-bit comparison between two numbers. Describe the general working principles of your circuit.

Suggested Readings

Almaini, A. E. A. "Sequential machine implementations using universal logic modules." *IEEE Trans. Comp.* vol. C-27 (1978): 951.

Beraru, J. "A one-step process for obtaining flip-flop input logic equations." *Comp. Des.* vol. 6 (1967): 58.

Chen X., and Hurst, S. L. "A comparison of universal-logic-module realizations and their application in the synthesis of combinatorial and sequential logic networks." *IEEE Trans. Comp.* vol. C-31 (1982): 140.

Fleischhammer, W., and Dortok, O. "The anomalous behavior of flip-flops in synchronizer circuits." *IEEE Trans. Comp.* vol. C-28 (1979): 273.

Fletcher, W. I. *An Engineering Approach to Digital Design.* Englewood Cliffs, N.J.: Prentice-Hall, 1980.

Floyd, T. L. *Digital Fundamentals.* 2d ed. Columbus, Ohio: Charles E. Merrill, 1982.

Gaitanis, N., and Halatsis, C. "Negative failsafe sequential circuits." *Elect. Lett.* vol. 16 (1980): 615.

Huffman, D. A. "The synthesis of sequential switching circuits." *J. Frank. Inst.* vol. 257 (1954): 275.

Joseph, J. "On easily diagnosable sequential machines." *IEEE Trans. Comp.* vol. C-27 (1978): 159.

Kline, R. *Structured Digital Design Including MSI/LSI Components and Microprocessors.* Englewood Cliffs, N.J.: Prentice-Hall, 1983.

Manning, F. B., and Fenichel, R. R. "Synchronous counters constructed entirely of *J-K* flip-flops." *IEEE Trans. Comp.* vol. C-25 (1976): 300.

Mealy, G. H. "A method for synthesizing sequential circuits." *Bell Syst. Tech. J.* vol. 34 (1955): 1045.

Nanda, N. K., and Bennetts, R. G. "Reconvergence phenomenon in synchronous sequential circuits." *Elect. Lett.* vol. 16 (1980): 303.

Noe, P. S., and Rhyne, V. T. "Optimum state assignment for *D* flip-flop." *IEEE Trans. Comp.* vol. C-25 (1976): 306.

Taub, D. M. "Hardware method of synchronizing processes without using a clock." *Elect. Lett.* vol. 19 (1983): 772.

Unger, S. H. "Self-synchronizing circuits and nonfundamental mode operation." *IEEE Trans. Comp.* vol. C-26 (1977): 278.

Voith, R. P. "Minimum universal logic module sequential circiuts with decoders." *IEEE Trans. Comp.* vol. C-26 (1977): 1032.

Wilkens, E. J. "Realization of sequential machines using random access memory." *IEEE Trans. Comp.* vol. C-27 (1978): 429.

Witte, H.-H., and Moustakas, S. "Simple clock extraction circuit using a self-sustaining monostable multivibrator output signal." *Elect. Lett.* vol. 19 (1983): 897.

Design of Synchronous Sequential Circuits

CHAPTER SEVEN

7.1 Introduction

In this chapter we will examine clocked sequential circuits. These circuits will employ combinational circuits and flip-flops. All circuit action will take place under the control of a periodic sequence of pulses called a clock. Each clock pulse will permit the circuit to either remain in its present state (present set of FF values) or move to another state (a new set of FF values). The advantage of clocked sequential circuits is that glitches that occur due to the imperfect nature of the logic devices will have no effect. This is possible only if we choose the clock period such that all glitches due to multiple delay paths end before the FFs encounter future changes.

The synthesis of sequential circuits consists of obtaining a table or diagram for the time sequence of inputs, outputs, and internal states. Boolean expressions are then derived by incorporating the behavior patterns of FF memory elements. In the following sections we will introduce these design sequences along with several synchronous sequential circuit examples. After studying this chapter, you should be able to:

- Obtain a state diagram for a synchronous sequential machine;
- Eliminate redundant states;
- Realize a sequential circuit from the state table;
- Differentiate between Mealy and Moore circuits.

7.2 State Diagrams and State Tables

The functional interrelationship that exists among the input, the output, the present state, and the next state is best illustrated by the state diagram or the state table. The *state diagram* is a graphical representation of a sequential circuit in which the states are represented by circles and transitions between states shown by arrows. We have already encountered some examples of two-state state diagrams in Figure 6.40.

FIGURE 7.1 **State Diagram.**

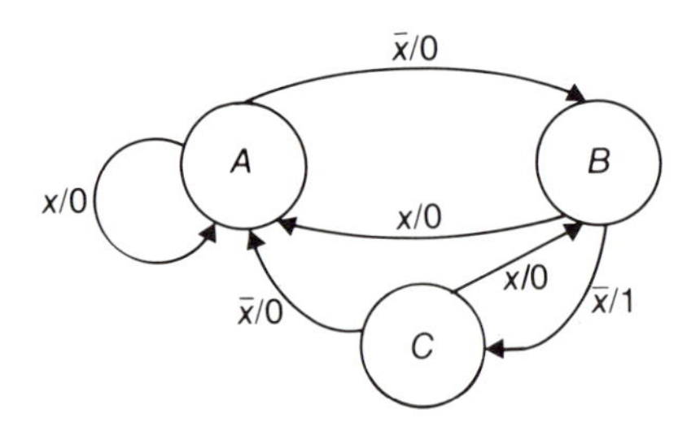

The state diagram in Figure 7.1 represents a synchronous circuit with three states, A, B, and C, and an input variable, x. In each state it is necessary for the circuit to be able to determine which state it is in and what the current value of x is, and then to set up the FF inputs such that the correct state is entered when the clock input occurs. The arrows connecting the states represent the occurrence of a clock input, and the variables alongside the arrows show the input condition that causes that path to be followed.

If the circuit is currently in state A and $x = 1$, the circuit will remain in state A when the clock occurs ($x/0$). If $x = 0$ the circuit enters into state B when the clock occurs ($\bar{x}/0$). In both of these cases the circuit yields an output of 0. The output value and the input condition are both indicated next to the corresponding transition paths. If the circuit is in state B and $x = 1$ when the clock occurs, the circuit returns to state A ($x/0$). However, if $x = 0$ when the clock occurs, the state C is entered, resulting in an output of 1 ($\bar{x}/1$). And finally the circuit moves coincident with the clock from state C to states A and B, respectively, when $x = 0$ and $x = 1$. In either case the output remains 0. The state diagram must show each of the states of the circuit and all conditions necessary for entering or exiting the states. In this case two transition lines leave each of the states.

The implementation of a sequential circuit with n states will require m FFs where $2^m \geq n$. The outputs of these FFs are called the *state variables* and are used to identify which state the circuit is in. An additional design tool that contains the same information as the state diagram in tabular form is the *state table*. The state table of the system shown in the state diagram of Figure 7.1 is given in Figure 7.2. It can be seen that the outputs are associated with the transition paths only and are not functions of any transition states. Circuits such as this are generally known as Mealy-type machines. The *Mealy outputs*, in most cases, are pulses coincident with the input pulse causing the state transition. An alternate output type, called the *Moore output*, is associated with the present state only. The general forms of the Mealy and Moore circuits are shown in Figure 7.3. These circuits take their names, respectively, from G. H. Mealy and E. F. Moore, two of the most famous pioneers in sequential design. The outputs from Moore-type circuits are independent of the inputs. The Moore outputs change their values only when the states change because of a change of the inputs.

FIGURE 7.2 **State Table Corresponding to the Diagram of Figure 7.1.**

Present State (PS)	Next State (NS) Output (Z)	
	$x = 0$	$x = 1$
A	B,0	A,0
B	C,1	A,0
C	A,0	B,0

There are many systems that possess both Mealy and Moore outputs; in other words, some outputs are conditional on both the inputs and the state of the circuit, while others are dependent only on the state of the circuit. Note, however, that the Mealy output is

FIGURE 7.3 **General Model of Sequential Machines: [*a*] Mealy and [*b*] Moore.**

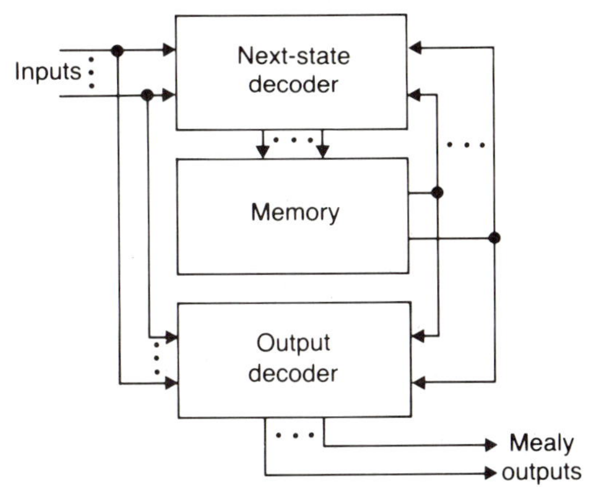

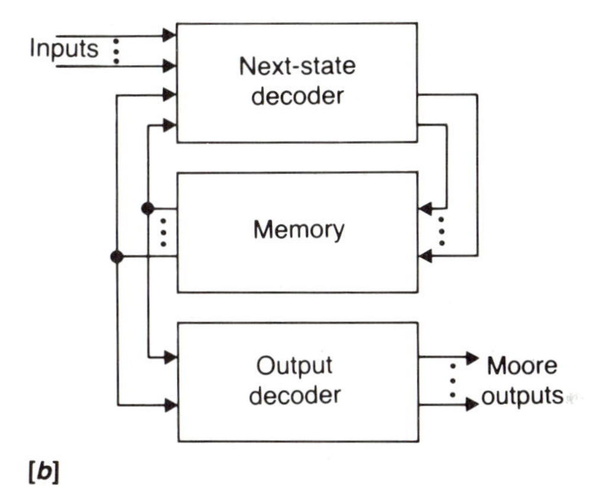

easily convertible to equivalent Moore outputs and vice versa. (More about this conversion will be said in one of the worked-out examples.) Figure 7.4[*a*] shows the format for a state diagram where the Moore-type outputs are circled along with the corresponding present states. Figure 7.4[*b*] shows the state table corresponding to

FIGURE 7.4 **Moore Model for a Sequential Circuit.**

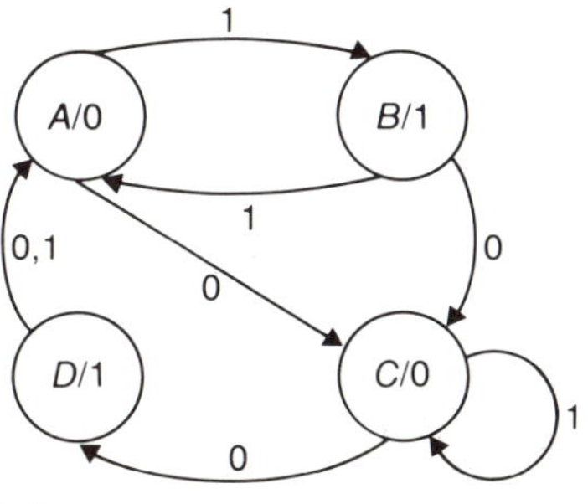

PS	NS $x = 0$	NS $x = 1$	Z
A	C	B	0
B	C	A	1
C	D	C	0
D	A	A	1

[*b*]

the state diagram of part [*a*]. It is important to point out that both Mealy- and Moore-type circuits are equally applicable to both synchronous and asynchronous circuits; and the minimum number of external inputs to any one of these circuits is one. For a synchronous circuit, that one input must be the system clock.

EXAMPLE 7.1

Obtain the state diagram of a controller for a serial machine that performs the 2's complement operation (see Example 5.3 for the equivalent parallel scheme).

SOLUTION

The realization of the state diagram for the controller is very straightforward, as shown in Figure 7.5. This follows from Rule 2(b) of Section 1.4. The 2's complement of a number is obtained by complementing all bits to the left of the least significant 1 in that number. State *A* takes care of the situation when none of the serial inputs are changed, whereas state *B* corresponds to the changing (1's complement) of inputs. The controller remains in state *A* as long as the low-order 0s of the input are encountered. The first input of 1 moves the machine to state *B* so that all subsequent inputs are complemented. To begin a new conversion, the machine needs to be reset back to state *A* (indicated by the broken line).

FIGURE 7.5

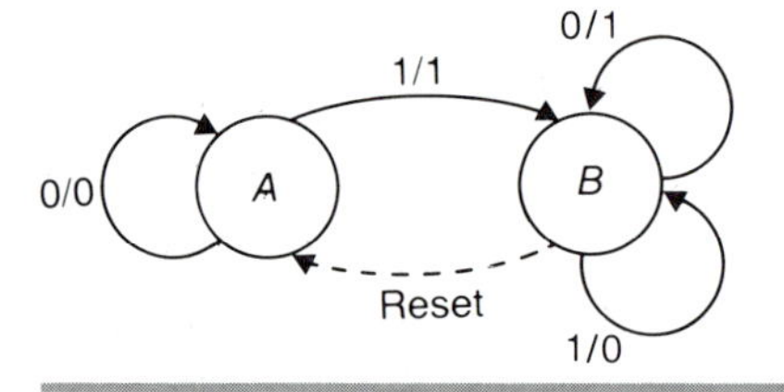

EXAMPLE 7.2

Obtain the state table for a synchronous sequential machine that detects a 01 sequence. The detection of sequence sets the output, $Z = 1$, which is reset only by a 00 input sequence.

SOLUTION

The state diagram for this machine is obtained as shown in Figure 7.6. The machine resides in state *A* as long as the sequence does not begin. This situation would include two distinct cases: either (a) the machine is yet to see a single bit, or (b) the machine has so far examined a 1 or a string of 1s. However, once the first bit, 0, of either sequence, 01 or 00, has been detected, the machine moves to state *B*. Finally, state *C* is reached if either (a) the complete sequence, 01, has been located, or (b) the resetting

FIGURE 7.6

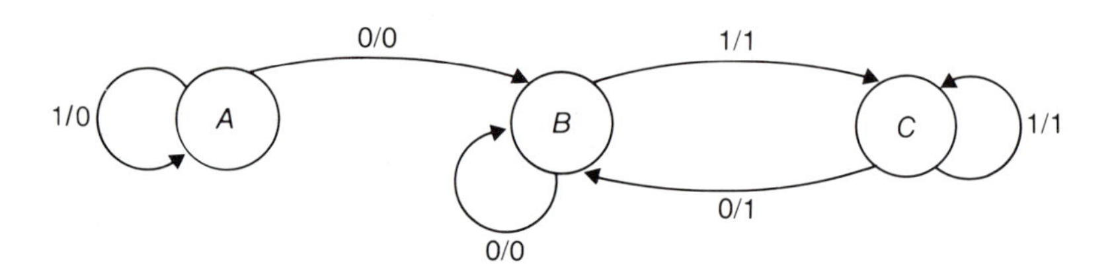

sequence, 00, is yet to begin. The state table of the machine readily follows from the state diagram. It is shown in Figure 7.7.

FIGURE 7.7

PS	NS, Z	
	x = 0	x = 1
A	B,0	A,0
B	B,0	C,1
C	B,1	C,1

EXAMPLE 7.3

Obtain the Moore equivalent state table for the Mealy machine of Figure 7.8.

SOLUTION

FIGURE 7.8

PS	NS, Z	
	x = 0	x = 1
A	C,0	A,0
B	B,0	A,0
C	D,1	C,1
D	D,0	B,0
E	C,1	A,0

The state table of Figure 7.8 is a Mealy type since the outputs are not associated only with the states. It is the desired goal of this problem to associate each of these outputs with a particular state. It may be seen that the two next states, C and D, are associated with two different outputs, 0 and 1. The next states, A and B, are associated always with an output of 0 only. Accordingly, four new states, C', C'', D', and D'', may be introduced to replace the states C and D. The states C' and D' correspond, respectively, to states C and D when the output is a 0. Similarly, the states C'' and D'' correspond, respectively, to states C and D when the output is a 1. The state table obtained after the introduction of only C' and C'' is shown in Figure 7.9. Next, D' and D'' may be included to obtain the state table as shown in Figure 7.10.

FIGURE 7.9

PS	NS, Z	
	x = 0	x = 1
A	C′,0	A,0
B	B,0	A,0
C′	D,1	C″,1
C″	D,1	C″,1
D	D,0	B,0
E	C″,1	A,0

FIGURE 7.10

PS	NS, Z	
	x = 0	x = 1
A	C′,0	A,0
B	B,0	A,0
C′	D″,1	C″,1
C″	D″,1	C″,1
D′	D′,0	B,0
D″	D′,0	B,0
E	C″,1	A,0

At this time each of the states has only one output associated with itself. Accordingly one could obtain the equivalent Moore machine as shown in Figure 7.11. This table has been constructed in a way so that it resembles the format of Figure 7.4[*b*].

FIGURE 7.11

PS	NS		Z
	$x = 0$	$x = 1$	
A	C'	A	0
B	B	A	0
C'	D''	C''	0
C''	D''	C''	1
D'	D'	B	0
D''	D'	B	1
E	C''	A	—

Note in Example 7.3 that the Moore machine of Figure 7.11 is equivalent to the original Mealy machine of Figure 7.8 only in the sense that its output appears as pulses. The outputs that occur in C'' and D'' are pulses that are high (1) for a full clock period. In the Mealy circuit the pulses are high (1) while the clock is high. Moreover, two additional states were necessary to make this conversion complete. In fact, the Moore equivalent machine generally consists of more states than the corresponding Mealy machine.

7.3 Equivalent States

When constructing a state diagram from the word statement of a design problem, a state that is identical to another state may inadvertently be included. The redundant state or states will increase the number of total states and may require the addition of another FF, making the circuit more expensive. Redundant states also decrease the number of don't-cares or unused states and thus increase the overall complexity of the circuit equations. In addition, fault diagnostic techniques used for failure analysis are based on the assumption that no redundant states exist. One of the design goals, therefore, is to eliminate all redundant states from the state diagram and/or table.

Sometimes redundant states are obvious. In Figure 7.12 a state diagram and a state table are given that include several redundant states. Two states are defined as *equivalent* if they have identical outputs and make transitions to the same states for a given control variable value. In Figure 7.12 no state can be equivalent to state D because it is the only state with an output of 1. We can make the statement that state E is equivalent to state F only if state A is equivalent to B and state F is equivalent to state G. Again, states F and G are equivalent only if state G is equivalent to state E and also

FIGURE 7.12 **Sequential Circuit with Redundant States: [*a*] State Diagram and [*b*] State Table.**

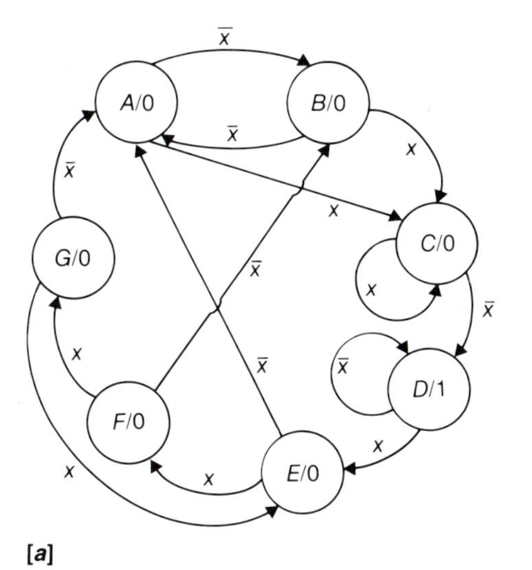

PS	NS		Z
	$x = 0$	$x = 1$	
A	B	C	0
B	A	C	0
C	D	C	0
D	D	E	1
E	A	F	0
F	B	G	0
G	A	E	0

[*b*]

if state A is equivalent to B. This line of reasoning becomes extremely confusing for large systems.

A systematic way of looking for redundant states usually is accomplished by means of a new tool called an *implication table*. The numbers of rows and columns in this table are both equal to $n - 1$ where n is the number of states in the to-be-reduced state table. An implication table that is used to locate the possible redundant states of Figure 7.12[*b*] is illustrated in Figure 7.13[*a*]. This table provides a bookkeeping technique that allows a systematic way to find states that are equivalent. The algorithm employed for the equivalency search in the implication table is as follows:

1. All states except the very first one are used as row labels of the table, and all states except the very last one are used as column labels.
2. The entries at each of the cells are made by comparing the next-state columns of the two states that are used to identify the cell in question. No entries are made in a cell if it corresponds to states that have the same output and the same next states for all control variables. If the equivalency of the two states is dependent on whether or not the states P and Q are equivalent, then the pair (PQ) is entered in the corresponding cell. A cross "X" is placed over those cells for which outputs of the two states are different for any one input condition.

FIGURE 7.13 State Reduction by Implication: [*a–b*] Implication Tables and [*c*] Equivalence Partition Table.

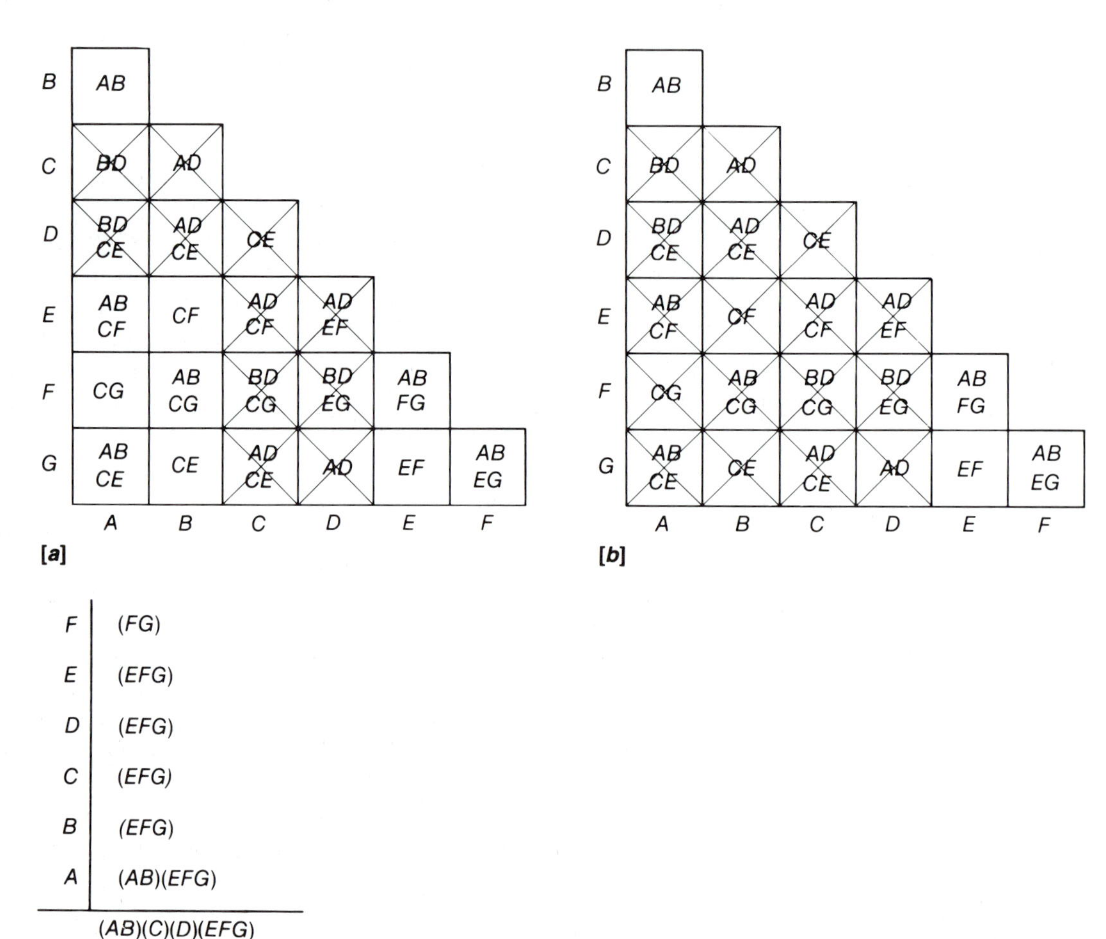

3. The next step involves elimination of as many cells as possible based on the respective cell entries. If the entries in any cell include at least one pair of states that are nonequivalent (i.e., the cell corresponding to that pair has already been crossed out), then that cell is eliminated by marking a X on it.
4. Successive passes are then made through the entire table to determine if any more states should be crossed out to indicate nonequivalency. The cell having an entry *PQ* should be crossed out only if the particular cells corresponding to the labels *P* and *Q* have already been eliminated. This pro-

cess of elimination is continued until no other cells can be eliminated.

5. Redundancy is then determined by examining the surviving cells of the implication table. Each surviving cell corresponds to an equivalency condition between the two states that are used to label that cell.

The role of the preceding algorithm will become meaningful when we use it to investigate the state table of Figure 7.12[*b*].

The entries in the table of Figure 7.13[*a*] are made by comparing each of the states with the others. For example, *CG* is entered in the cell corresponding to *A* and *F* since the equivalency of these two states is dependent on the equivalency of states *C* and *G*. Likewise, both *CG* and *AB* are entered in the cell corresponding to *B* and *F* since these two states would be equivalent only if both (a) *C* and *G* and (b) *A* and *B* are equivalent. Similar reasonings are made in completing the remaining cells.

Referring to Figure 7.13[*a*] and the state table of Figure 7.12[*b*], *D* is the only state that has an output of 1, so it cannot be equivalent to any other state. Therefore, all cells that have either a row or a column designated by *D* are crossed out. Next all the cells that have an entry composed of *D* are crossed out. As a consequence we find that all cells corresponding to the row and column designated by *C* also have been eliminated. This indicates that state *C*, like *D*, is different from all other states as well. We next cross out all cells that have at least one pair of entries involving *C*. Continuing this process, we come up with the table of Figure 7.13[*b*]. Only four surviving cells are left in the implication table.

An *equivalence partition table*, as shown in Figure 7.13[*c*], is next obtained by listing all horizontal labels (in the reverse order) as its row labels. A check is now made of each column of the final implication table (Figure 7.13[*b*]), from right to left, making note of the cells that have not been eliminated. The row and column labels of the surviving cell form an equivalent pair that is noted in the equivalent column identifier in the partition table. Equivalent pairs such as (*PQ*) and (*QR*) imply the presence of a larger group of equivalent states (*PQR*). For each column of the implication table an entry is made in the equivalence partition table, provided there is at least one surviving cell in that column. The entries from the previous rows of the partition table are entered as long as there is no repetition of the entry. In the example in question, the cell corresponding to *G* and *F* is not crossed out, and therefore the equivalent pair (*FG*) is listed next to *F* in the equivalence partition table. There are two cells corresponding to rows *F* and *G* and column *E* that result in (*FE*) and (*EG*) or the equivalent group (*EFG*). Therefore, we write (*EFG*) next to *E*, and since (*EFG*) already contains (*FG*), nothing from the previous row is repeated here. Since columns B, C, and D have no surviving cells, the entry (EFG) is repeated three times. No new equivalencies result until we come to consider A, for which the entry (AB) is added to the list. The last line of the partition table reveals that there are two sets of equivalent states: (AB) and (EFG). This implies that state A is equivalent to state B and states E, F, and G are equivalent to each other. Consequently, the redundant states are now eliminated by considering only one state from each equivalent group. The resultant state diagram and the state table are shown in Figure 7.14 where B, F, and G have been removed. The state B has been replaced by A and states F and G have been replaced with E.

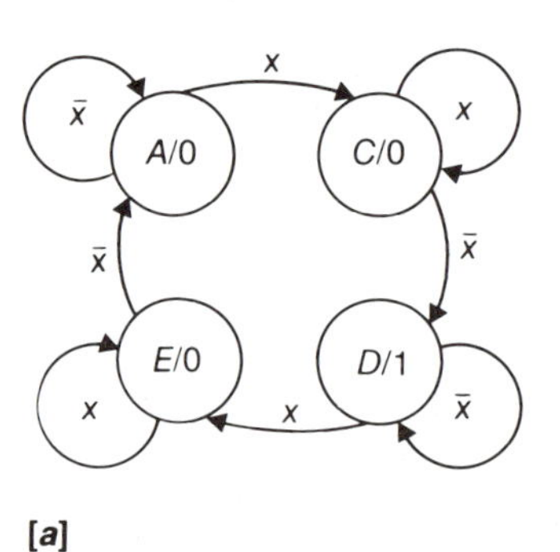

[a]

PS	NS		Z
	$x = 0$	$x = 1$	
A	A	C	0
C	D	C	0
D	D	E	1
E	A	E	0

[b]

FIGURE 7.14 **[*a*] State Diagram with No Redundancy and [*b*] State Table with No Redundancy.**

7.4 State Assignments

State assignment is the process of adopting a binary coding scheme for the symbolic states of the state table so that it is possible for the circuit to remember which state it is in. Each bit in the code represents the output of an FF and is called a *state variable*. For n number of states, a total of m FFs will be necessary such that m is the smallest integer satisfying the relationship $2^m \geq n$. Any unique assignment is valid; however, it is always better if an attempt is made to assign codes in such a way that the number of cases where more than one bit in a code changes when states change is kept to a minimum. When the symbolic states are replaced with the binary coding scheme, a binary state table, commonly known as the *transition table*, results.

The state diagram of Figure 7.1 has three unique states, and, therefore, at least two FFs are necessary for designing the corresponding logic circuit. This implies that of the four different codes—00, 01, 10 and 11—only three can be used. For this example, if one chooses to assign A = 00, B = 01, and C = 11, the resulting transition table of Figure 7.15 is obtained.

FIGURE 7.15 **Transition Table for the Circuit of Figure 7.1.**

PS	NS	
Q_1Q_2	$x = 0$	$x = 1$
00	01,0	00,0
01	11,1	00,0
11	00,0	01,0
10	– –,–	– –,–

We may see that the present state Q_1Q_2 = 01, upon receiving the input x = 0, moves to the next state 11 with the resultant output of 1. During this transition the Q_1 value changes from 0 to 1, as shown highlighted in the table. In the same present state when x =

1, Q_2 changes from 1 to a 0. In either case the binary state changes only one of its bits. However, the present state $Q_1Q_2 = 11$ is different from the others, because when $x = 0$ both of the FF bits need to change from 1 to 0. There are situations where such a condition could cause problems, as we shall discover later.

EXAMPLE 7.4

Obtain a transition table for the sequential machine of Figure 7.14[*a*].

SOLUTION

There are four states, and, therefore, only two FFs need to be considered. The transition table results when $A = 00$, $C = 01$, $D = 11$, and $E = 10$, as shown in Figure 7.16. For the state assignments made, the transition table shows that there are no transitions for which both state variables change. Consequently, this assignment of states is considered to be very good.

FIGURE 7.16

PS	NS		
Q_1Q_2	$x = 0$	$x = 1$	Z
00	00	01	0
01	11	01	0
11	11	10	1
10	00	10	0

7.5 Excitation Maps

Up to this point when considering FFs we have been concerned with how they respond to various inputs. We will now encounter the design problem of determining their inputs such that the proper values are present to cause the next state to result when the clock input occurs. This input control is accomplished by deriving the respective excitation equations. The output equations and the state variable excitation equations are derived separately, as shown in Figures 7.17[*a–b*]. The FF input maps are usually called *excitation maps*.

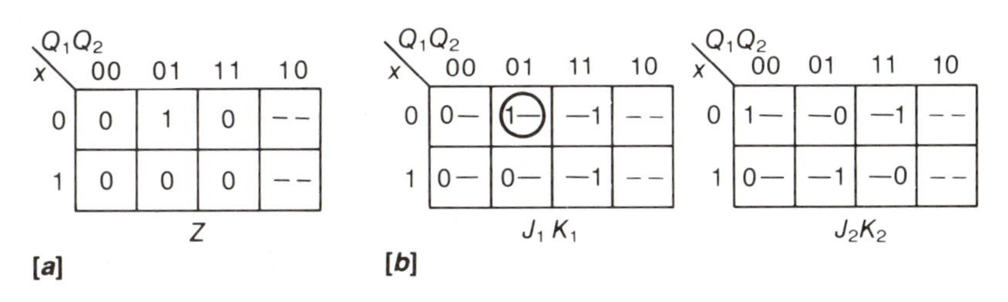

FIGURE 7.17 **[*a*] Output Table and [*b*] Excitation Maps.**

The combinational circuit corresponding to the transition table shown in Figure 7.15 has three variables—x, Q_1, and Q_2—that generate the FF inputs. Q_1 and Q_2 represent the current state of the circuit, and x is the external control input that in combination with the present state determines which action is to occur. The excitation table entries are the values of FF inputs that would cause the transition to next-state variables when the clock input occurs. We have chosen JK FFs, as an example only, for generating the excitations. One particular transition is emphasized for illustration both in the transition table and in the corresponding excitation map of Figure 7.17. The transition of Q_1, as shown in Figure 7.15, from 0 to 1 requires the FF input condition $J_1 = 1$ and $K_1 = -$. This information is entered in the corresponding cell, $x = 0$ and $Q_1Q_2 = 01$, of the excitation map. Continuing this procedure, the excitation and output equations may be obtained from the K-maps of Figure 7.17:

$$J_1 = \bar{x}Q_2$$
$$K_1 = 1$$
$$J_2 = \bar{x}$$
$$K_2 = \bar{Q}_1x + Q_1\bar{x} = Q_1 \oplus x$$
$$Z = \bar{Q}_1Q_2\bar{x} \cdot CK$$

Note that the clock input, CK, is ANDed with $\bar{Q}_1Q_2\bar{x}$ to produce the desired output of a synchronous machine. The resultant sequential circuit is obtained as shown in Figure 7.18.

FIGURE 7.18 **Circuit Implementation of the Example.**

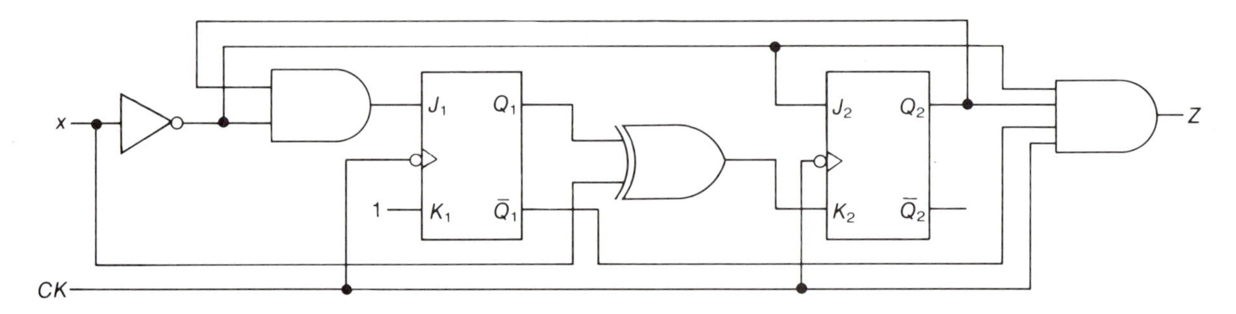

Consider a circuit similar to that of Figure 7.1 but having a Mealy output, Z_1, and a Moore output, Z_2, as shown in Figure 7.19[*a*]. There is to be an output coincident with the clock input when the circuit moves from state B to state A, and another output whenever the circuit is in state C. For the Z_1 output, assuming negative edge-triggered FFs, the circuit must be in state 01 and both x and clock must occur. The Mealy output is given by $Z_1 = \bar{Q}_1Q_2x \cdot CK$. If the FFs being used for the state variables were positive edge-triggered, the Z_1 output would occur in the state following 01 since the positive edge of the clock would move the circuit into the next state. The next state's state variables, 00, and the inputs would be ANDed to form Z_1 in this case. However, the Moore output is a 1

FIGURE 7.19 **Circuit with Mealy and Moore Outputs: [*a*] State Diagram, [*b*] Transition Table, [*c*] Output Tables, and [*d*] Circuit.**

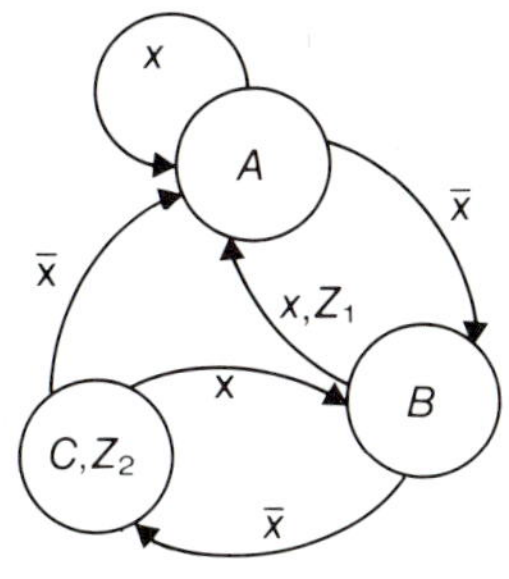

[*a*]

PS	NS, Z_1		
Q_1Q_2	$x = 0$	$x = 1$	Z_2
A 00	01,0	00,0	0
B 01	11,0	00,1	0
C 11	00,0	01,0	1
— 10	--,-	--,-	—

[*b*]

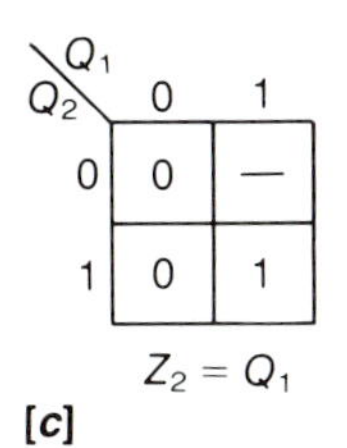

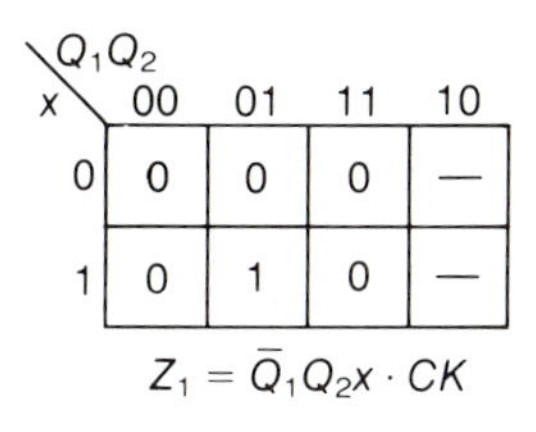

[*c*]

$$Z_2 = Q_1 \qquad Z_1 = \bar{Q}_1 Q_2 x \cdot CK$$

[*d*]

only when the circuit is in state *C*. Figures 7.19[*b–c*] show the steps involved in obtaining the final circuit of Figure 7.19[*d*].

7.6 Design Algorithm

We have examined the steps of the design of a synchronous sequential machine in the last few sections. Figure 7.20 gives a comprehensive flowchart of an algorithm for the design of sequential machines.

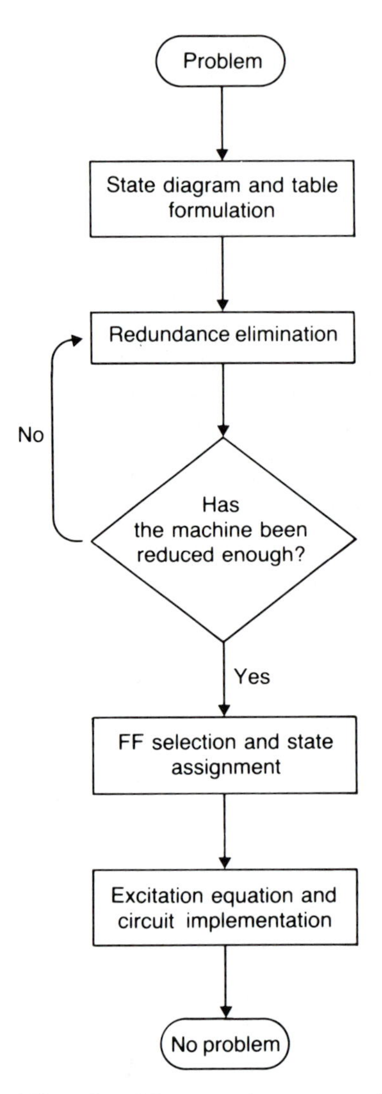

FIGURE 7.20 **Sequential Circuit Design Flowchart.**

The algorithm can be summarized by the following steps:

Step 1. Obtain the state diagram from the word statement of the problem.

Step 2. Obtain the state table from the state diagram.

Step 3. Eliminate the redundant states.

Step 4. Make state assignments.

Step 5. Determine the type of FFs to use and obtain the corresponding excitation maps.

Step 6. Determine the output and FF equations.

Step 7. Construct the logic circuit.

These design steps often lead to a rather lengthy process that varies from problem to problem. The following examples illustrate the implementation of the sequential design algorithm.

EXAMPLE 7.5

Design a two-bit clocked sequential counter circuit that counts clock pulses.

SOLUTION

Steps 1–2. We can visualize such a device as the one that receives clock pulses as input. Each time a clock pulse is received, the counter should count up. No control variable is needed; only the occurrence of the clock pulse is necessary for a state change. We shall assume further that the states change at the trailing edge of the clock input. The counter could be designed such that the outputs go through the sequence 00 → 01 → 10 → 11 and repeat. Since the counter is only a two-bit device, it would have to reset at the fourth clock. The state diagram and the corresponding state table for the counter are obtained as shown in Figure 7.21. We might choose to have the outputs directly from the FF outputs, in which case such outputs would be classed as Moore type.

FIGURE 7.21

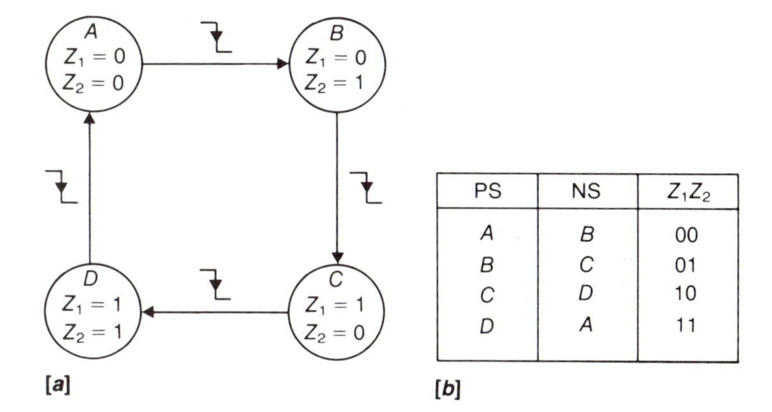

PS	NS	Z_1Z_2
A	B	00
B	C	01
C	D	10
D	A	11

Step 3. The four states—*A, B, C,* and *D*—are all different since they stand for completely different events. Consequently, we may conclude without any doubt that none of these are redundant states.

Step 4. Making state assignments is an important step for at least one reason. The assignments of the states are crucial in determining the simplicity, or for that matter complexity, of the resultant circuit. The output circuits can be eliminated totally if the state assignments for the states *A, B, C,* and *D* are made the same as the corresponding Moore outputs of each state. Accordingly, the chosen assignments are $A = 00$, $B = 01$, $C = 10$, and $D = 11$.

Step 5. The number of FFs are indeed two. This fact was also given in the initial word statement of the problem. We might choose *JK* FFs, for example. Consequently both the transition table and excitation map are obtained as shown in Figure 7.22.

FIGURE 7.22

PS Q_1Q_2	NS	Z_1Z_2
00	01	00
01	10	01
10	11	10
11	00	11

[*a*]

J_1K_1

Q_2 \ Q_1	0	1
0	0—	—0
1	1—	—1

J_2K_2

Q_2 \ Q_1	0	1
0	1—	1—
1	—1	—1

[*b*]

Step 6. The excitation maps are minimized to give

$$J_1 = Q_2$$
$$K_1 = Q_2$$
$$J_2 = 1$$
$$K_2 = 1$$

The outputs are easily realizable directly from Figure 7.22[*a*]. They are as follows:

$$Z_1 = Q_1\bar{Q}_2 + Q_1Q_2 = Q_1(\bar{Q}_2 + Q_2) = Q_1$$
$$Z_2 = \bar{Q}_1Q_2 + Q_1Q_2 = Q_2(\bar{Q}_1 + Q_1) = Q_2$$

Step 7. The resulting circuit obtained from these excitation and output equations is shown in Figure 7.23.

FIGURE 7.23

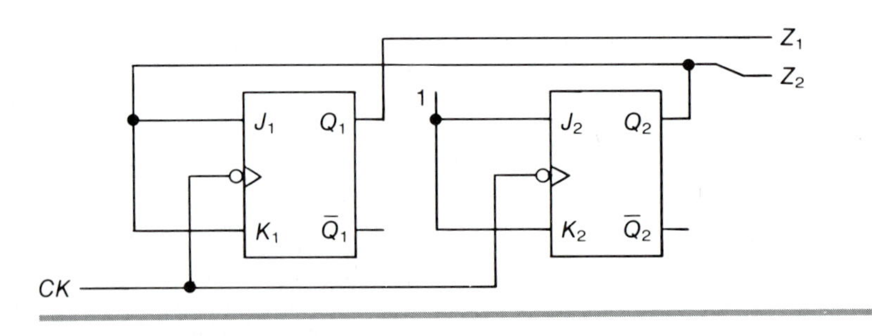

EXAMPLE 7.6

Repeat the design of Example 7.5 by assigning A = 00, B = 01, C = 11, and D = 10. Construct the corresponding timing diagram as well.

SOLUTION

Since the FF state assignments are different from the corresponding Moore outputs, Z_1 and Z_2, the circuit will require additional gates. The corresponding transition table, the *JK* excitation maps, and the output map are shown in Figure 7.24. The corresponding Boolean equations are obtained as follows:

$$J_1 = Q_2$$
$$K_1 = \overline{Q}_2$$
$$J_2 = \overline{Q}_1$$
$$K_2 = Q_1$$
$$Z_1 = Q_1$$
$$Z_2 = Q_1\overline{Q}_2 + \overline{Q}_1Q_2 = Q_1 \oplus Q_2$$

FIGURE 7.24

PS Q_1Q_2	NS	Z_1Z_2
00	01	00
01	11	01
11	10	10
10	00	11

[a]

J_1K_1

$Q_2 \backslash Q_1$	0	1
0	0—	—1
1	1—	—0

J_2K_2

$Q_2 \backslash Q_1$	0	1
0	1—	0—
1	—0	—1

Z_1Z_2

$Q_2 \backslash Q_1$	0	1
0	00	11
1	01	10

[b]

The resulting logic circuit diagram is obtained as shown in Figure 7.25. It is obvious that the circuit in Example 7.5 is simpler. The timing diagram of Figure 7.26 illustrates the operation of the circuit of Figure 7.25.

FIGURE 7.25

FIGURE 7.26

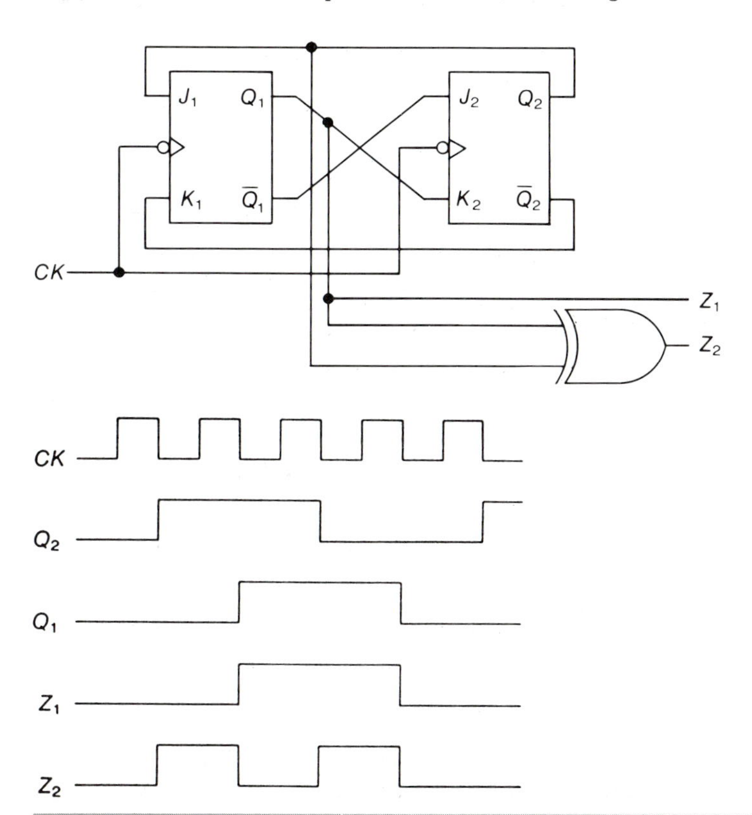

EXAMPLE 7.7

Complete the design of a clocked sequential circuit that recognizes the input sequence 1010, including overlapping such that for input x = 00101001010101110 the corresponding output Z is 00000100001010000.

SOLUTION

The state diagram consists of five states, A–E. States B, C, D, and E represent, respectively, the occurrence of the first, second, third, and fourth bits of the sequence 1010. State E has a Moore output $Z = 1$ indicating the completion of a sequence. A subsequent input of 0 would move the circuit to state A, which indicates the input is out of sequence. An input of 1 while in E moves the circuit from state E to state D since sequence overlapping is allowed. The corresponding state diagram showing the transitions for each value of x is provided in Figure 7.27.

FIGURE 7.27

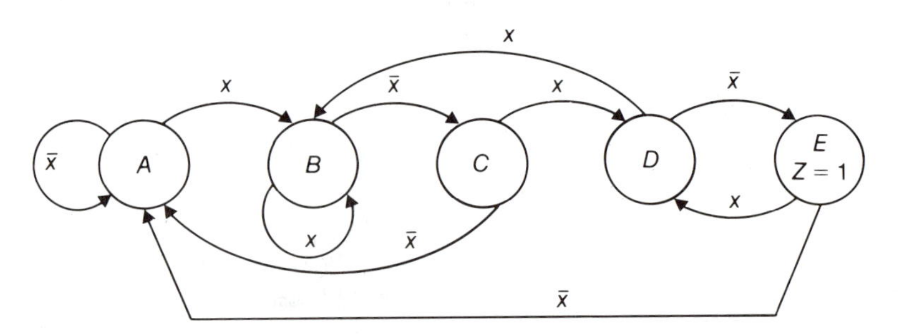

Upon power-up, the circuit begins from state A. As long as the string of input is devoid of 1 (i.e., the first bit of a 1010 sequence), the circuit remains at this beginning state. Once a 1 has been located the circuit moves to state B, indicating that the first bit has already been detected. Subsequent detection of 0, 1, and 0, in that order, would amount to moving the circuit to states C, D, and E, respectively. Once the state E is reached, the circuit gives an output indicating the completion of a 1010 sequence. However, while at state B if the circuit detects a 1, the circuit re-enters state B. This is due to the possibility that the most recently observed 1 might be the beginning of a 1010 sequence. For similar reasons, the detection of a 1 at state D causes the circuit to move to state B also. Again at state C, a detection of 0 eliminates the possibility of having the desired sequence, 1010. So, the circuit resets back to state A. Likewise, the circuit resets from state E to state A if it locates a 0. However, an interesting case happens when the circuit is at state E and it has just detected a 1. This time the circuit moves back to state D. This is due to the fact that detection of a 1 at state E is equivalent to detecting the third bit of a newer 1010 sequence.

The problem involves five states requiring three FFs. Three of the eight possible states will remain unused. The state table and transition table corresponding to the arbitrary assignments of A = 000, B = 001, C = 011, D = 111, and E = 101 are shown in Figure 7.28. The output and excitation maps corresponding to the use of JK FFs may now be obtained as shown in Figure 7.29. Proper grouping of the K-map cells results in the following equations:

FIGURE 7.28

	NS		
PS	$x = 0$	$x = 1$	Z
A	A	B	0
B	C	B	0
C	A	D	0
D	E	B	0
E	A	D	1

PS	NS		
$Q_1Q_2Q_3$	$x = 0$	$x = 1$	Z
000	000	001	0
001	011	001	0
011	000	111	0
111	101	001	0
101	000	111	1

$$Z = Q_1\bar{Q}_2 \cdot CK$$

$$J_1 = Q_2x$$

FIGURE 7.29

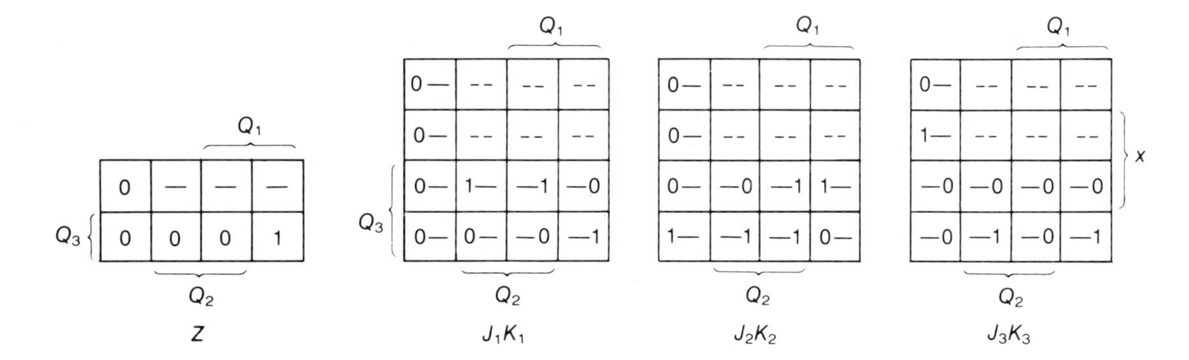

$$K_1 = Q_2x + \bar{Q}_2\bar{x} = \overline{Q_2 \oplus x}$$
$$J_2 = Q_1x + \bar{Q}_1Q_3\bar{x}$$
$$K_2 = \bar{x} + Q_1$$
$$J_3 = x$$
$$K_3 = \bar{Q}_1Q_2\bar{x} + Q_1\bar{Q}_2\bar{x} = (Q_1 \oplus Q_2)\bar{x}$$

The sequential circuit of the 1010 sequence detector is obtained using the above equations and is shown in Figure 7.30[*a*]. The timing diagram of Figure 7.30[*b*] shows the relationship between the output, the clock, and the input.

FIGURE 7.30a

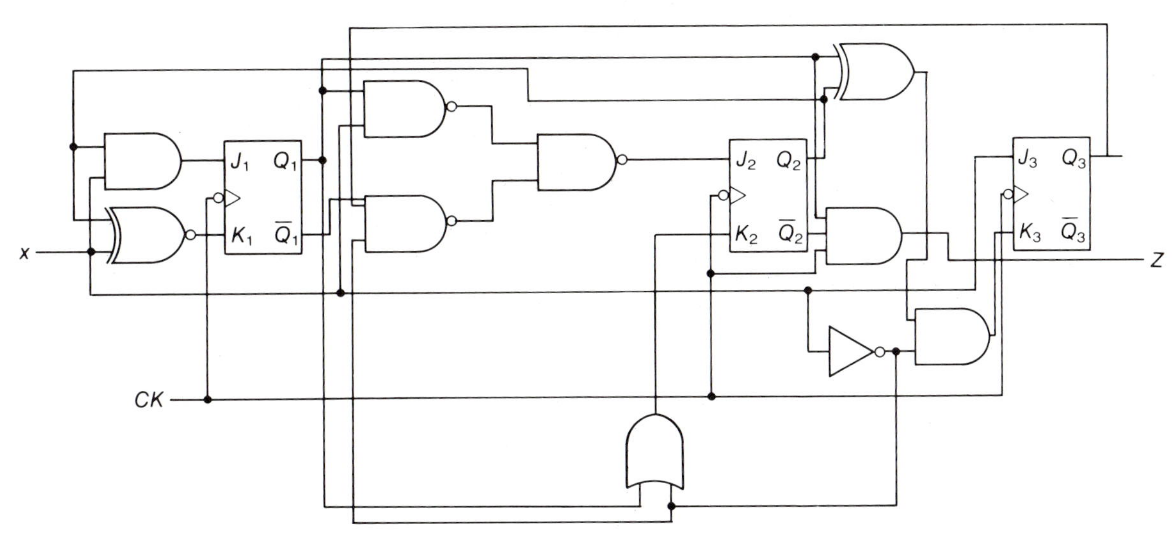

[a]

FIGURE 7.30b

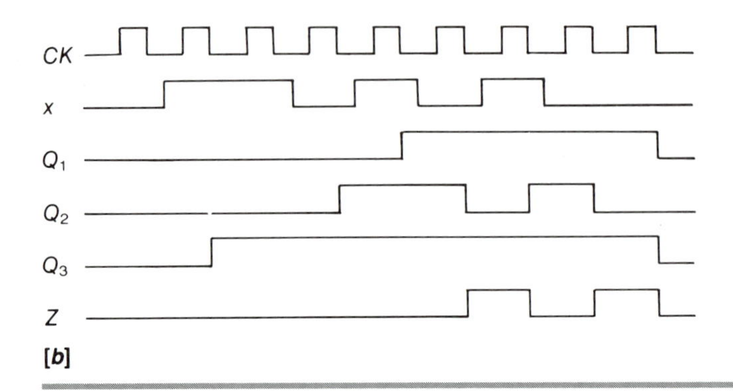

In the examples considered thus far we have used *JK* FFs and gates. *D* FFs and other logic devices may also be used. The excitation table for all of the FFs is given in Figure 6.39. In Examples 7.8 and 7.9, *D* FFs are used.

EXAMPLE 7.8

Obtain a scale-of-seven up-counter, as shown in the state diagram of Figure 7.31, using *D* FFs and PLA. Assume that the counter is tied to a seven-segment display device.

SOLUTION

FIGURE 7.31

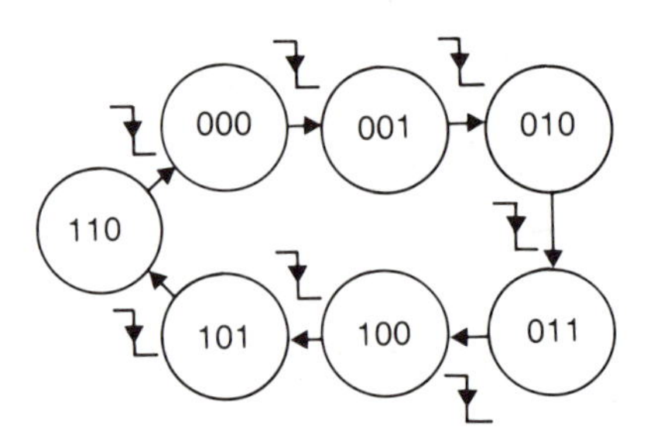

The state table of the counter may be obtained as shown in Figure 7.32.

FIGURE 7.32

PS $Q_3Q_2Q_1$	NS
000	001
001	010
010	011
011	100
100	101
101	110
110	000

The excitation K-maps corresponding to D FFs are next obtained as shown in Figure 7.33. The Boolean equations, therefore, are as follows:

$$D_3 = \overline{Q}_2Q_3 + Q_1Q_2$$
$$D_2 = Q_1\overline{Q}_2 + \overline{Q}_1Q_2\overline{Q}_3$$
$$D_1 = \overline{Q}_1\overline{Q}_2 + \overline{Q}_1\overline{Q}_3$$

FIGURE 7.33

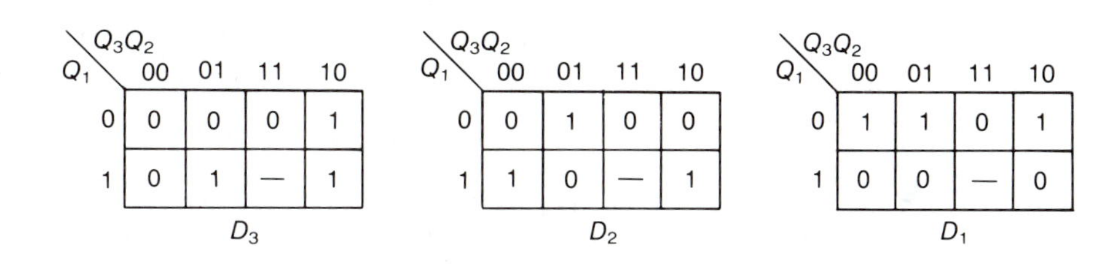

The PLA circuit configuration follows as shown in Figure 7.34. The dots in the intersection matrix correspond to either an OR or an AND operation. The segment allocation for the LED display has already been defined in Example 4.8.

FIGURE 7.34

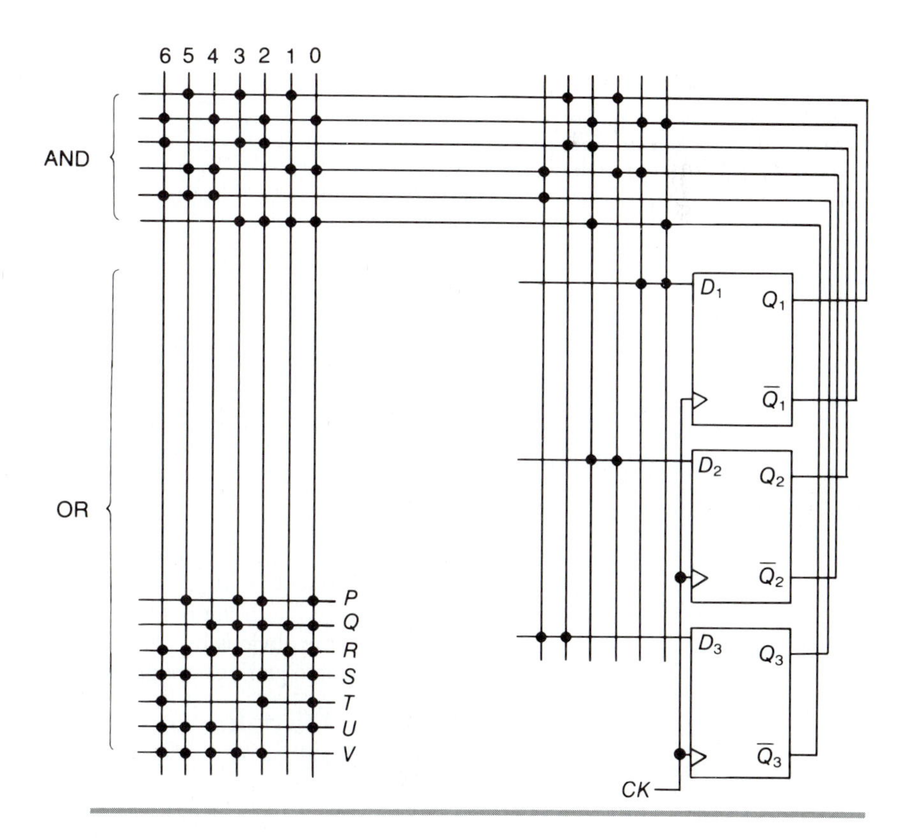

EXAMPLE 7.9

Using D FFs and assorted gates, design a sequential comparator that is to determine which of the two multi-bit numbers, X and Y, of equal length is larger.

SOLUTION

The MSB of both numbers may be fed as inputs to the comparator. Two outputs, Z_1 and Z_2, may be assumed to accompany the circuit. If $X > Y$, then $Z_1 = 1$; if $X < Y$, then $Z_2 = 1$; and if $X = Y$, then $Z_1 = Z_2 = 0$.

The state diagram of the comparator is obtained as shown in Figure 7.35, where X_i and Y_i are the ith bits of X and Y, respectively. The transition table may then be obtained as shown in Figure 7.36. For simplicity,

FIGURE 7.35

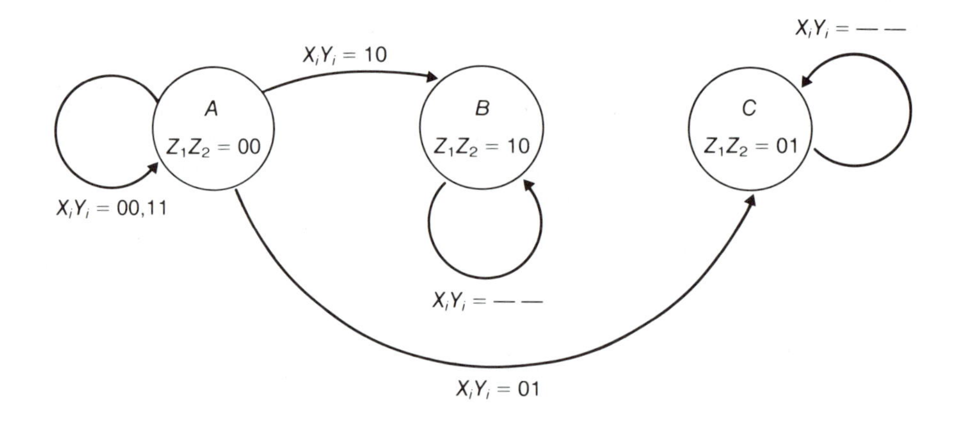

FIGURE 7.36

PS	NS $X_iY_i = 00$	01	11	10	Z_1Z_2
A	A	C	A	B	00
B	B	B	B	B	10
C	C	C	C	C	01

we may assign $A = 00$, $B = 10$, and $C = 01$. These assignments would allow us to derive circuit outputs directly from the FFs. That this is possible would become obvious by comparing Examples 7.5 and 7.6. Accordingly, the excitation equations are given by

$$D_1 = X_i\overline{Y}_i\overline{Q}_2 + Q_1$$
$$D_2 = \overline{X}_iY_i\overline{Q}_1 + Q_2$$

The resultant circuit for the sequential comparator may now be readily obtained. The circuit is illustrated in Figure 7.37.

FIGURE 7.37

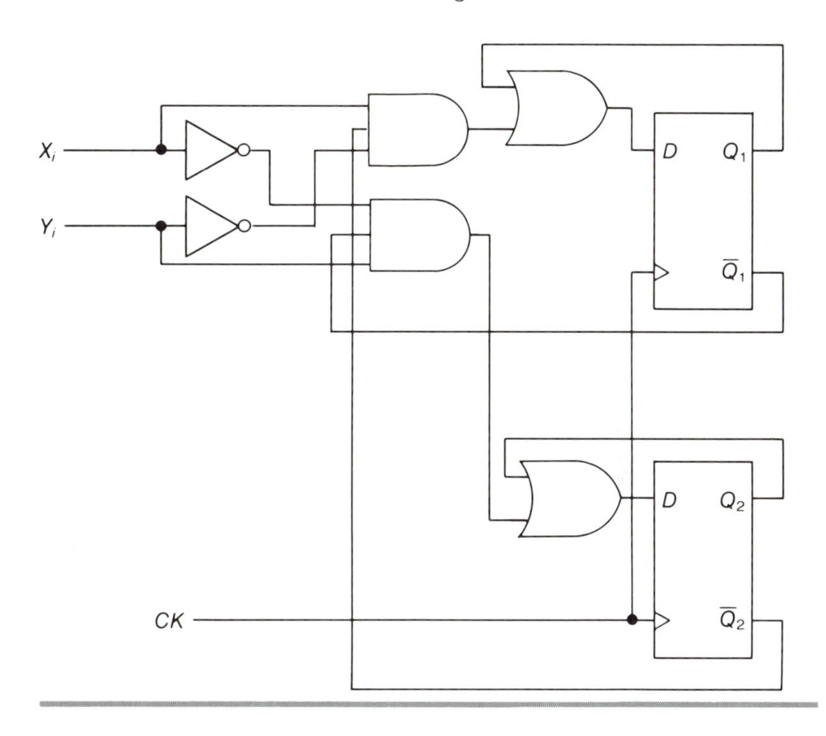

7.7 Incompletely Specified Diagrams

All problems considered thus far in this chapter have been *completely specified*; that is, all next-state and output values were completely defined in their state diagrams and state tables. In this section we shall consider state diagrams and/or state tables that are termed *incompletely specified.* Such sequential circuits include don't-care outputs in their respective state tables and state diagrams. These circuits have an added advantage over the completely specified circuits since the presence of don't-cares may contribute to simpler Boolean expressions.

The minimization process of state tables that contain don't-cares is tedious and requires special consideration. Implication tables are used for removing state redundancies, but the steps involved are different from those for the completely specified state tables (see Section 7.3 for details). The steps for incompletely specified state tables involve the following variations:

1. The entries in the implication table are made exactly as before, but a don't-care in the output is considered to be a 1 or a 0 depending on whether this particular choice aids the formation of an equivalent group.

2. Once the successive passes and crossing out of the cells have been completed, the designer should make entries in the equivalence partition table as before. However, it must be understood that two equivalent pairs like (AB) and (AC) do not automatically imply the existence of a larger equivalent group (ABC) unless there exists an equivalent group (BC) or (BCX). This extra condition is necessary because the don't-cares may have been treated as both 0 and 1 under different conditions.
3. The maximum number of states in the reduced circuit is equal either to the number of sets of maximal compatibles or to the number of states in the original circuit, whichever is less.
4. A *closure table* is obtained by considering the maximal compatibles as states and grouping their next states under respective input columns. The reduced state table is constructed by renaming the sets of maximal compatibles. Note, however, that the resulting reduced state table might still be incompletely specified.

The application of these rules is illustrated in Example 7.10.

EXAMPLE 7.10

Obtain the reduced state table for the sequential machine shown in Figure 7.38.

SOLUTION

FIGURE 7.38

PS	NS, Z	
	$x = 0$	$x = 1$
A	A,—	B,1
B	G,—	D,0
C	B,1	B,—
D	A,1	B,—
E	C,—	A,—
F	F,—	C,—
G	G,—	G,—

The implication table is as obtained as shown in Figure 7.39. The intersection of A and B is crossed out, which indicates that A and B are unequal. This is because when $x = 1$, the next states result in different outputs. More boxes are crossed out as repeated passes are made through the table. To start, the cells with the entry AB are crossed out. At the end of such search the equivalence partition table is obtained as shown in Figure 7.40. For better understanding of the operations, consider row B. In row B there are seven groups—(BG), (BE), (BD), (BC), (CG), (EG), and (DFG)—that reduce to four possible groups of three states: (BEG), (DFG), (BCG), and

FIGURE 7.39

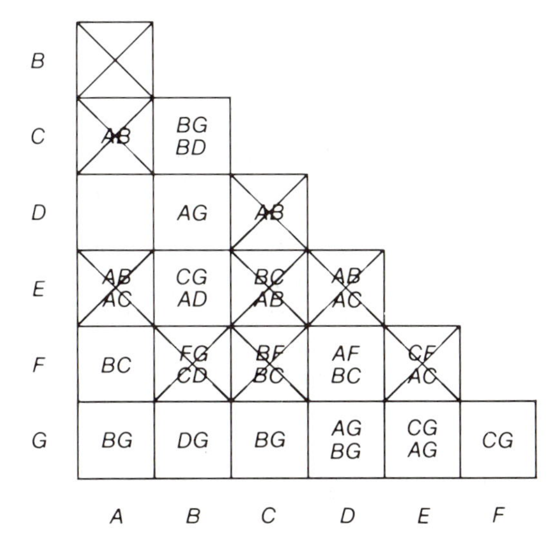

FIGURE 7.40

F	(FG)	
E	(EG)(FG)	
D	(DG)(DF)(EG)(FG)	
D	(EG)(DFG)	Using step 2
C	(CG)(EG)(DFG)	
B	(BG)(BE)(BD)(BC)(CG)(EG)(DFG)	
B	(BEG)(DFG)(BCG)(BDG)	Using step 2
A	(AD)(AF)(AG)(BEG)(DFG)(BCG)(BDG)	
A	(ADFG)(BEG)(BCG)(BDG)	Using step 2

FIGURE 7.41

PS	NS, Z	
	$x = 0$	$x = 1$
ADFG	AFG,1	BCG,1
BEG	CG,—	ADG,0
BCG	BG,1	BDG,0
BDG	AG,1	BDG,0

FIGURE 7.42

PS	NS, Z	
	$x = 0$	$x = 1$
P	P,1	R,1
Q	R,—	P,0
R	QRS,1	S,0
S	P,1	S,0

(*BDG*). (*BG*), (*BE*), and (*EG*) yield (*BEG*); (*BG*), (*BC*), and (*CG*) yield (*BCG*); and (*BG*), (*BD*), and (*DFG*) yield (*BDG*). Note that (*BG*) has been used in all three determinations, and such manipulations are valid. In the final form we have five possible sets renamed as follows:

$$P = (ADFG)$$
$$Q = (BEG)$$
$$R = (BCG)$$
$$S = (BDG)$$

The closure table can now be constructed as shown in Figure 7.41. In the first row of the closure table, *AFG* are the next states for states *ADFG* when $x = 0$. For $x = 1$, *BCG* are the next states for states *ADFG*. Now we may construct the reduced state table by using the variables *P*, *Q*, *R*, and *S*. The table can be organized as shown in Figure 7.42. Corresponding to $x = 0$, state *R* could move to any one of the three states, *Q*, *R*, and *S*. This is because (*BG*) is present in all of those three sets of compatibles. Note that the reduced table is still incompletely specified.

7.8 Ideal State Assignments

In the previous sections state assignments were made arbitrarily with no consideration of the consequences. It will be seen that the combinational circuit complexity is different for different sets of state assignments. The number of possible state assignments for any given problem is impressive. For n present states and p flip-flops, there are $2^p!/[n!(2^p - n)!]$ ways of selecting n out of the 2^n possible combinations. For each of these ways there are $n!$ permutations of assigning the n combinations to the n states. Again, for each of these assignments there are 2^p ways of interchanging logic 0 and logic 1 and there are $p!$ ways of interchanging the FFs. Consequently, there may be a total of $[(2^p - 1)!]/[(2^p - n)!p!]$ unique assignments. For example, the number of unique assignments for a nine-state system can be calculated to be 10,810,800.

The *optimum state assignment* is one that reduces the amount of combinational logic of a sequential system when compared to other assignments. Many different approaches to this state assignment problem have been developed. The complexity and cost of the circuit will differ for different combinations of state assignments. The identification of the best state assignments has been the subject of a considerable amount of research. We can attempt to locate the best set by generating those output and excitation tables that allow the formation of large clusterings of ones. Use of the following guidelines will probably result in simpler circuits:

1. Adjacent assignments should be given to those states that have the same next state for any given input.
2. Two or more states that are the next states of the same state, under adjacent inputs, should be given adjacent assignments.
3. States that have the same output for a given input should be given adjacent assignments.

The term *adjacent assignments* means that the states appear next to each other on the mapped representation of the state table. The assignment guidelines work best with D and JK FFs. These rules usually lead to a good, but not necessarily to the optimum, solution. It may not always be possible to satisfy all of the guidelines at the same time. In case of conflicts, rule 1 is preferable. An attempt should be made to satisfy the maximum number of suggested adjacencies. However, remember that an ideal state assignment may not always reduce the cost, and it is true also that the cost of the devices is often an insignificant part of the overall cost of a digital system.

7.9 Summary

In this chapter all aspects of the design of a synchronous sequential circuit were considered. It is possible to design numerous types of digital systems using synchronous sequential design. However, there are many digital systems that are of the asynchronous type as well. We will investigate the nature of both pulse-mode and fundamental-mode circuits prior to elaborating an additional application of sequential circuits. Chapter 10 presents a variety of such applications that include sequential circuits of all three types and some involving combinations of all three. Next, in Chapter 8, we shall consider pulse-mode sequential circuits.

Problems

1. Design a three-bit counter that counts up when a control variable $E = 0$, and counts down when $E = 1$.
2. Design a four-bit binary up-counter using JK FFs.
3. Design a synchronous sequential circuit using SR FFs that results in an output of 1 whenever each the following sequences occurs:
   a. 0001 e. 10010
   b. 0101 f. 11011
   c. 1101 g. 10011
   d. 1011 h. 110110
4. Repeat Problem 3 using JK FFs.
5. Repeat Problem 3 using T FFs.
6. Assume a two-bit binary counter that counts up when $A = 1$ and $B = 0$; counts down when $A = 0$ and $B = 1$; halts when $A = 0$ and $B = 0$; and is forbidden to operate when $A = B = 1$. Obtain the state diagram and the JK equations.
7. Obtain the equivalent Mealy state table from the machine of Figure 7.14[*b*].
8. Obtain the equivalent Mealy circuit for the circuit of Figure 7.14[*b*].
9. Given the state tables of Figure 7.P1, find the logic equations and logic diagrams for each table using JK FFs.

FIGURE 7.P1

PS	NS, Z	
	x = 0	x = 1
A	A,0	C,0
B	D,1	A,0
C	F,0	F,0
D	E,1	B,0
E	G,1	G,0
F	C,0	C,0
G	B,1	H,0
H	H,0	C,0

[*a*]

PS	NS, Z	
	x = 0	x = 1
A	A,0	B,0
B	C,0	B,0
C	D,0	B,0
D	A,1	B,0

[*b*]

10. Repeat Problem 9 using *SR* FFs.
11. Repeat Problem 9 using *D* FFs.
12. Design a synchronous sequence detector that produces an output of 1 whenever any one of the sequences 1100, 1010, and 1001 occurs. The circuit resets to its initial state after a 1 has been generated.
13. Find the Moore equivalent circuits of the two machines given in Problem 9. Use *JK* FFs.
14. Analyze the sequential circuit shown in Figure 7.P2. Obtain the state equations and the state diagram.

FIGURE 7.P2

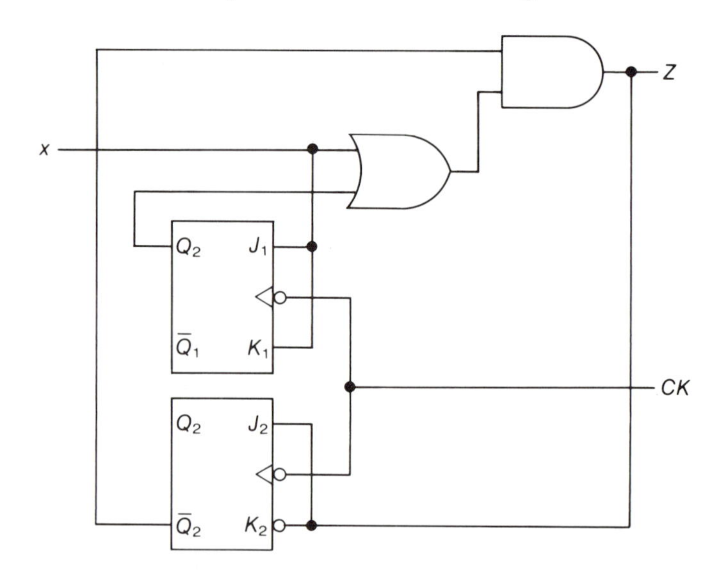

15. Construct the state diagram for the following synchronous equations:

$$D_1 = \bar{x}Q_2 + \bar{Q}_1Q_2 + xQ_1\bar{Q}_2$$
$$D_2 = \bar{x}Q_2 + x\bar{Q}_1Q_2$$
$$z = Q_2 \cdot CK$$

16. Design a BCD counter with (a) *JK* FFs and (b) *D* FFs.
17. Design a four-bit Gray code up-counter using (a) *JK* FFs and (b) *D* FFs.
18. Design counters that follow each of the following binary sequences. For example, "0, 1, 3, 2, 5, 7, 4, and repeat" implies that the counter repeats the sequence 0, 1, 3, 2, 5, 7, 4, 0, 1, 3, 2, 5, 7, 4, 0, 1, 3, 2, 5, 7, 4, 0, 1 and so on.
    a. 0, 1, 3, 2, 5, 7, 4, and repeat. Use *SR* FFs.
    b. 0, 1, 3, 2, 6, 4, 5, and repeat. Use *T* FFs.
    c. 0, 2, 4, 6, 1, and repeat. Use *JK* FFs.

19. A synchronous sequential circuit is shown with its state diagram in Figure 7.P3. Determine the state diagram for the circuit if the primary and secondary variables are interchanged as shown in Figure 7.P4.

FIGURE 7.P3

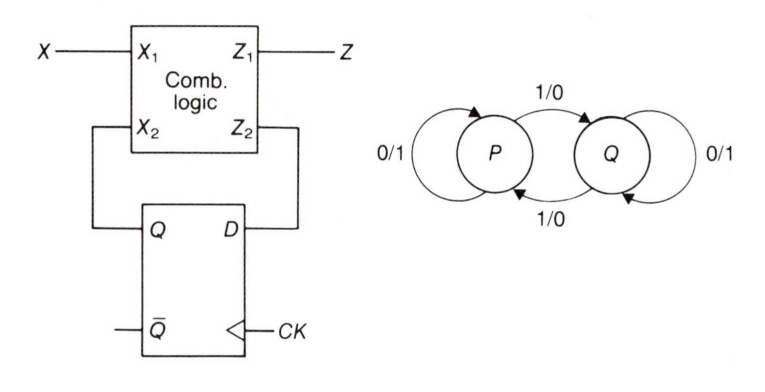

FIGURE 7.P4

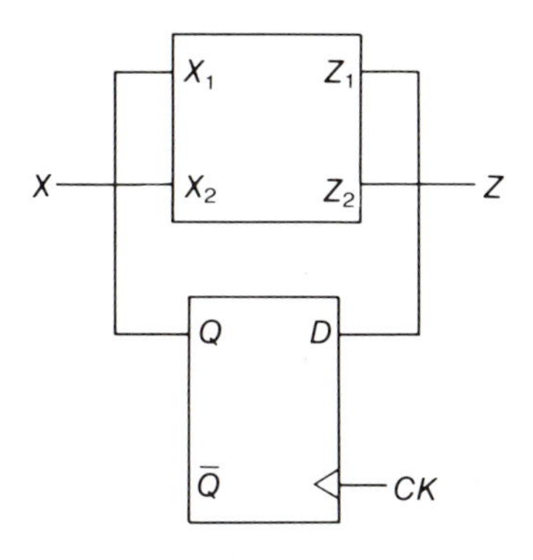

20. Obtain the reduced state machine from the state tables shown in Figure 7.P5. Obtain the corresponding sequential circuits using *D* FFs.

FIGURE 7.P5

PS	NS, Z	
	$x = 0$	$x = 1$
A	F,0	D,1
B	C,1	F,1
C	F,1	B,1
D	E,1	G,1
E	A,1	D,1
F	G,0	B,1
G	A,0	D,1

[*a*]

PS	NS, Z	
	$x = 0$	$x = 1$
A	C,0	B,1
B	C,1	A,1
C	E,0	B,1
D	F,0	A,0
E	A,0	G,1
F	D,1	C,1
G	E,1	C,1

[*b*]

PS	NS, Z	
	$x = 0$	$x = 1$
A	E,0	B,1
B	F,0	D,1
C	E,0	B,1
D	F,0	B,0
E	C,0	F,1
F	B,0	C,0

[*c*]

21. Repeat Problem 20 using *JK* FFs.
22. Repeat Problem 20 using *SR* FFs.

Suggested Readings

Almaini, A. E. A. "Sequential machine implementations using universal logic modules." *IEEE Trans. Comp.* vol. C-27 (1978): 951.

Armstrong, D. B. "A programmed algorithm for assigning codes to sequential machines." *IRE Trans. Elect. Comp.* vol. EC-11 (1962): 466.

Dunworth, A., and Hartog, H. V. "An efficient state minimization algorithm for some special classes of incompletely specified sequential machines." *IEEE Trans. Comp.* vol. C-28 (1979): 531.

Fojo, J. M., and Barreiro, F. D. "A method for reducing the number of input variables to synchronous sequential circuits." *IEEE Trans. Comp.* vol. C-24 (1975): 1029.

Hartmanis, J. "On the state assignment problem for sequential machines I." *IRE Trans. Elect. Comp.* vol. EC-10 (1961): 157.

Joseph, J. "On easily diagnosable sequential machines." *IEEE Trans. Comp.* vol. C-27 (1978): 159.

McCluskey, E. J., and Unger, S. H. "A note on the number of internal variable assignments for sequential switching circuits." *IRE Trans. Elect. Comp.* vol. EC-8 (1959): 439.

Nagle, H. T., Jr.; Carroll, B. D.; and Irwin, J. D. *An Introduction to Computer Logic.* Englewood Cliffs, N.J.: Prentice-Hall, 1975.

Nanda, N. K., and Bennetts, R. G. "Reconvergence phenomenon in synchronous sequential circuits." *Elect. Lett.* vol. 16 (1980): 303.

Noe, P. S., and Rhyne, V. T. "Optimum state assignment for *D* flip-flop." *IEEE Trans. Comp.* vol. C-25 (1976): 306.

Pattavina, A., and Trigila, S. "Combined use of finite-state machines and Petri nets for modelling communicating processes." *Elect. Lett.* vol. 20 (1984): 915.

Paull, M. C., and Unger, S. H. "Minimizing the number of states in incompletely specified sequential switching functions." *IEEE Trans. Elect. Comp.* vol. EC-8 (1959): 356.

Rao, C. V. S., and Biswas, N. N. "Minimization of incompletely specified sequential machines." *IEEE Trans. Comp.* vol. C-24 (1975): 1089.

Stearns, R. E., and Hartmanis, J. "On the state assignment problems for sequential machines II." *IRE Trans. Elect. Comp.* vol. EC-10 (1961): 593.

Unger, S. H. "Self-synchronizing circuits and nonfundamental mode operation." *IEEE Trans. Comp.* vol. C-26 (1977): 278.

Waxman, J., and Rootenberg, J. "Logic circuit for cyclic detection in a state diagram." *IEEE Trans. Comp.* vol. C-26 (1977): 303.

Yamamoto, M. "A method for minimizing incompletely specified sequential machines." *IEEE Trans. Comp.* vol. C-29 (1980): 732.

Yang, C. C. "Closure partition method for minimizing incomplete sequential machines." *IEEE Trans. Comp.* vol. C-22 (1973): 1109.

Design of Pulse-Mode Circuits

CHAPTER EIGHT

8.1 Introduction

In the preceding chapter circuit states were traversed under the influence of control variables and in synchronization with a clock pulse. These circuits are easily designed and dependable. The control inputs are themselves pulses but are considered as levels since they are either high or low for one or more clock periods. However, asynchronous circuits generally require closer attention since there is no clock signal to regulate the events. The control inputs in pulsed sequential circuits will be considered as simply pulses. The resulting circuits respond immediately to input changes rather than responding to the input only in the presence of clock inputs. This characteristic definitely is troublesome. With this in the background, we will consider only input variables that are mutually exclusive pulses to avoid most timing problems.

Keep in mind that the overwhelming majority of the sequential circuits that are being designed today are not necessarily the pulse-mode type. But for systems such as computer keyboards and self-serve vending machines where inputs are generally nonoverlapping, pulse-mode design is the simplest choice. Even though this type of circuit is treated separately in this chapter, the design mechanism is very similar to that covered in the last chapter with some minor changes. After studying this chapter, you should be able to:

- Understand the pulse-mode model of sequential circuits;
- Design and analyze pulse-mode circuits.

8.2 Pulse-Mode Model

A pulse-mode circuit is a type of sequential circuit that involves input signals that are pulses. The state FFs will be unclocked. Figure 8.1 shows the general pulse-mode circuit model, which is the same as that of Figure 6.38 with the clock input eliminated. Because of the asynchronous nature of the pulse-mode model, the resulting circuit could show erratic behavior if not designed properly or if restrictions are not placed on the pulses.

To avoid the problems of instability and unpredictability, the pulse-mode operation requires that the system be designed in a specific way, and although there is no requirement that the inputs be periodic, there are restrictions on their duration and interval. The inputs x_i in the circuit shown in Figure 8.1 must meet the following restrictions:

1. Simultaneous input pulses on two or more lines are forbidden. The interval between any two pulses must be greater than the time required for the system to return to a stable state.
2. Pulse widths must be sufficient to allow the circuit components to respond to them. For level-sensitive FFs the pulse width must be short enough to have come and gone before the FF outputs have changed to their new values. For edge-sensitive FFs, however, the pulse width is not critical as long as the first restriction is met.

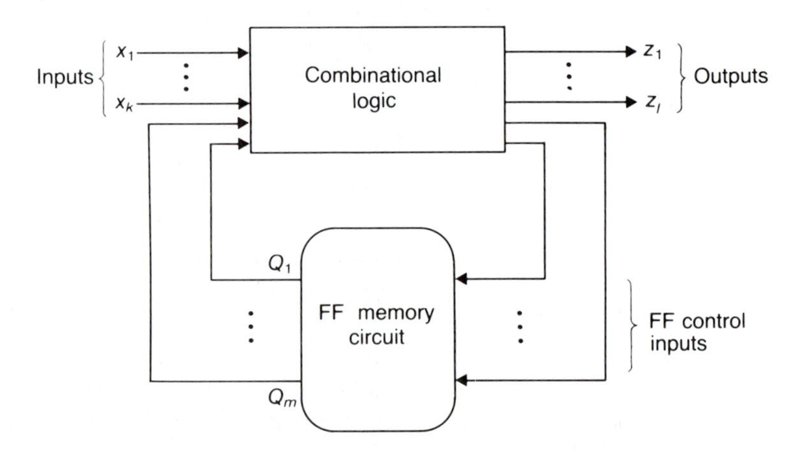

FIGURE 8.1 **General Form of Pulsed Sequential Circuit.**

As long as these stipulations are met we will see that the design of pulse-mode circuits is very similar to that of clocked sequential circuits. Figure 8.2 shows an example of a set of three nonoverlapping pulse trains suitable for the inputs to any pulse-mode circuit. Note that for every n pulse trains going into a pulse-mode machine, the circuit has a maximum of $n + 1$ unique input conditions. The additional input condition corresponds to the simultaneous absence of all pulses.

FIGURE 8.2 **Nonoverlapping Pulse Trains.**

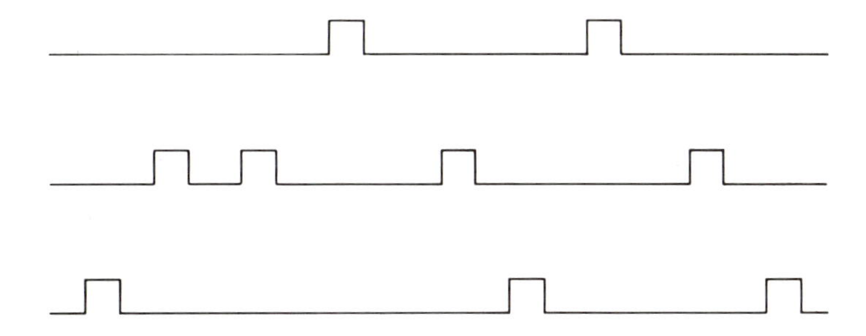

8.3 Design Algorithm

The pulse-mode sequential design algorithm is very similar to that for the synchronous sequential circuit. However, since only pulses provide information to the FFs, the input variables are used only in the uncomplemented form. The design algorithm for pulse-mode circuits consists of the following steps:

1. Obtain the state diagram and/or state table. For n inputs there must be n transition paths leaving each of the states.
2. Remove redundancies and obtain the reduced state table.
3. Generate a transition table using proper state assignments.
4. Obtain Boolean equations from the excitation table.
5. Complete the circuit diagram.

Details of some of the steps are different from the steps listed in Section 7.6. The following examples illustrate the pulse-mode design considerations and applications.

EXAMPLE 8.1

Design a circuit that receives two inputs, x_1 and x_2, and gives an output coincident with the third consecutive x_2 pulse following at least one x_1 pulse.

SOLUTION

A good technique for constructing a state diagram is to consider the sequence of inputs that results in the output, and then determine what happens to each state for all other possible inputs. The state diagram, as shown in Figure 8.3, includes all actions resulting from the occurrence of each pulse. State B corresponds to the occurrence of x_1, while states C and D, respectively, correspond to the occurrence of subsequent x_2 pulses. Once the circuit is at state D an additional x_2 pulse causes the circuit to return to state A and output pulse z. If an x_1 pulse is received in any of the states, the circuit always goes to state B.

The state table can now be constructed from the state diagram, as shown in Figure 8.4. Two FFs are required to define the four states shown in the state table.

Note that for the synchronous mode the column headings in the state table were all the combinations of the control variable values. In the pulse mode, however, $x_1x_2 = 00$ implies that no pulses have occurred. The input

FIGURE 8.3

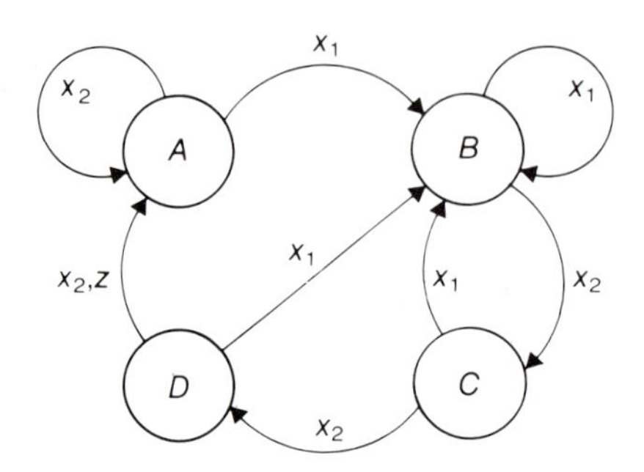

FIGURE 8.4

PS	NS, z	
	x_2	x_1
A	A	B
B	C	B
C	D	B
D	A,z	B

condition $x_1x_2 = 01$ implies that only pulse x_2 has occurred, and $x_1x_2 = 10$ implies that only x_1 has occurred. The existence of $x_1x_2 = 11$ implies that both pulses occurred simultaneously, and such occurrence violates the restrictions imposed upon pulse inputs. Pulsed sequential circuits are most easily realized by using edge-triggered T FFs. The transition/output table, as shown in Figure 8.5, is obtained by arbitrarily assigning $A = 00$, $B = 01$, $C = 11$, and $D = 10$.

The T FF output and excitation maps are shown in Figure 8.6. These excitation maps can be treated as reduced four-variable K-maps. Columns corresponding to $x_1 = x_2 = 0$ and $x_1 = x_2 = 1$ are eliminated because they do not possess any significance. The column corresponding to x_1 is equivalent to $x_1 = 1$ and $x_2 = 0$, and the column corresponding to x_2 is equivalent to $x_1 = 0$ and $x_2 = 1$ in the corresponding four-variable maps.

FIGURE 8.5

PS	NS, z	
Q_1Q_2	x_2 (01)	x_1 (10)
00	00	01
01	11	01
11	10	01
10	00, z	01

FIGURE 8.6

Q_1Q_2	x_1	x_2
00	0	0
01	0	1
11	1	0
10	1	1

T_1

Q_1Q_2	x_1	x_2
00	1	0
01	0	0
11	0	1
10	1	0

T_2

Q_1Q_2	x_1	x_2
00	0	0
01	0	0
11	0	0
10	0	1

z

Note that these two columns of the reduced K-map are in fact not adjacent, so we cannot make any group between the two columns. The pulse-mode minimization leads to the following Boolean equations:

$$T_2 = x_1\overline{Q}_2 + x_2Q_1Q_2$$
$$T_1 = x_1Q_1 + x_2(Q_1 \oplus Q_2)$$
$$z = x_2Q_1\overline{Q}_2$$

The output equation just obtained is of the Mealy type for negative edge-triggered FFs.

The timing diagram for this example problem is shown in Figure 8.7. The circuit resulting from the equations is illustrated in Figure 8.8.

FIGURE 8.7

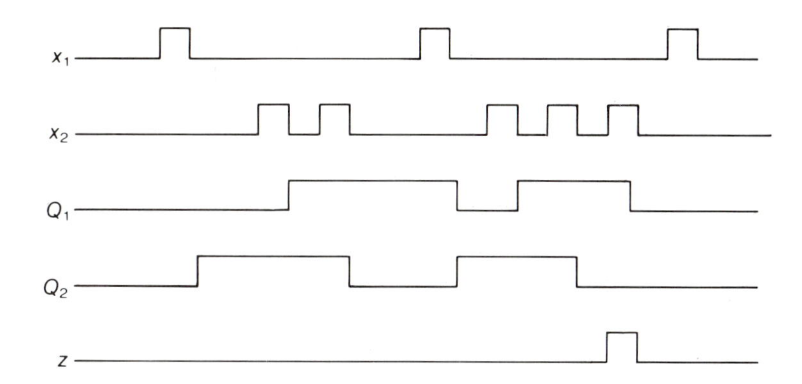

FIGURE 8.8

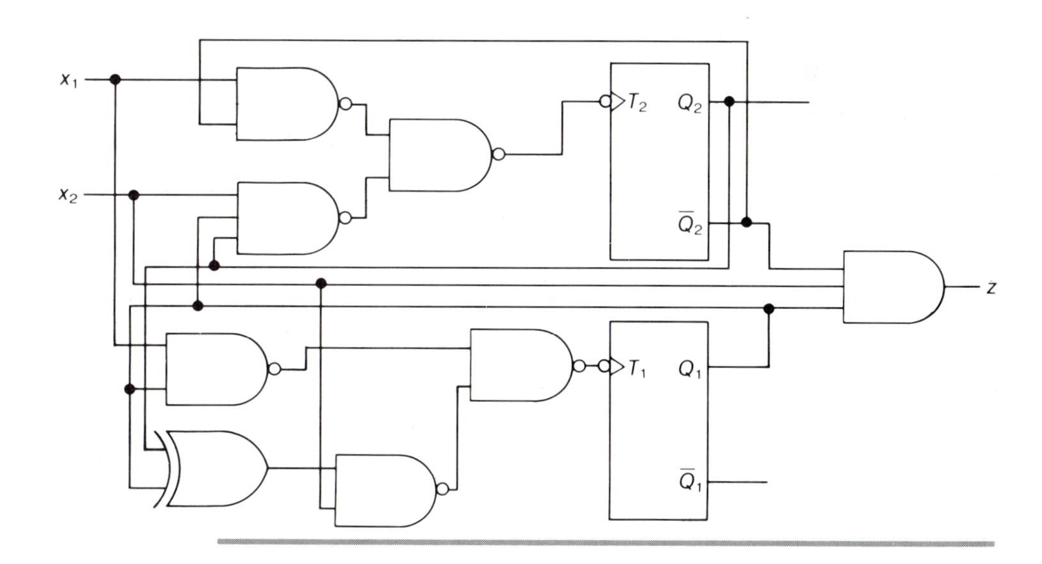

EXAMPLE 8.2

Design a pulse-mode circuit that satisfies the following requirements. The circuit has two inputs, x_1 and x_2, and one output, z. The output pulse is to be produced simultaneously with the last of a sequence of three input pulses if and only if the sequence contained at least two x_1 pulses. After each output a new sequence is looked for.

SOLUTION

The state diagram is obtained as shown in Figure 8.9. The diagram of Figure 8.9 requires a total of at least six states where the states are defined as follows:

A: Reset state.

B: Have seen x_1 out of one bit seen so far.

C: Have seen two x_1 pulses out of two bits seen so far.

D: Have seen x_2 following state B.

E: Have seen x_2 out of one bit seen so far.

F: Have seen x_1 following state E.

FIGURE 8.9

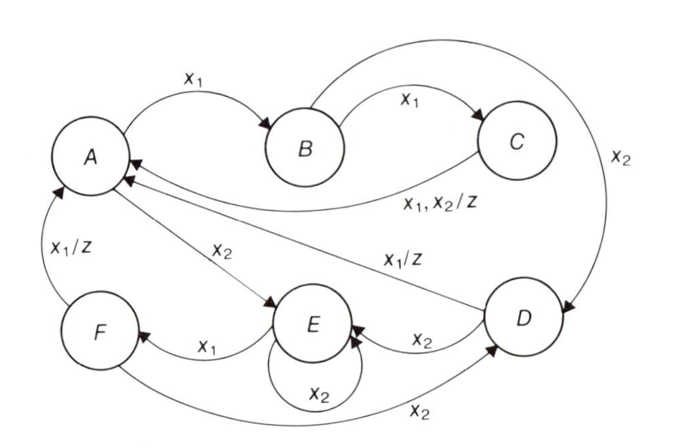

FIGURE 8.10

PS	NS, z	
	x_1	x_2
A	B	E
B	C	D
C	A,z	A,z
D	A,z	E
E	F	E
F	A,z	D

FIGURE 8.11

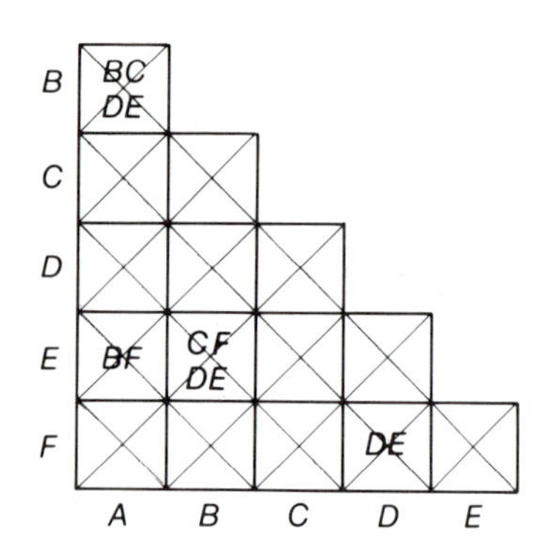

FIGURE 8.12

PS	NS, z	
$Q_1Q_2Q_3$	x_1	x_2
000	001	101
001	010	100
010	000,z	000,z
100	000,z	101
101	110	101
110	000,z	100

We may readily obtain the state table, as shown in Figure 8.10, from the state diagram of Figure 8.9.

A redundancy check is now made using an implication table, as shown in Figure 8.11, to make sure that no unnecessary states have been included. The check reveals that there is no redundancy. Since there are six states, three FFs are necessary to handle the design. We arbitrarily assign A = 000, B = 001, C = 010, D = 100, E = 101, and F = 110. Accordingly, the transition table now takes the form of Figure 8.12.

If we decide to use T FFs, the resultant output and excitation maps are obtained as shown in Figure 8.13. Note that both the excitation and output maps have been split into two parts. This division poses no problem since grouping between x_1 and x_2 variables is not permitted due to the fact that $x_1x_2 \neq 11$. The resulting Boolean equations, therefore, can be obtained as follows:

$$z = x_1(Q_2 + Q_1\bar{Q}_3) + x_2(\bar{Q}_1Q_2)$$
$$T_1 = x_1(Q_1\bar{Q}_3) + x_2(\bar{Q}_1\bar{Q}_2)$$
$$T_2 = x_1(Q_2 + Q_3) + x_2(Q_2)$$
$$T_3 = x_1(Q_3 + \bar{Q}_1\bar{Q}_2) + x_2(\bar{Q}_1\bar{Q}_2 + \bar{Q}_2\bar{Q}_3)$$

The realization of these equations is shown in the circuit of Figure 8.14. Note that a JK FF with $J = K = 1$ will perform as a T FF as long as the circuit outputs corresponding to the respective T equations are introduced at the JK clock inputs.

FIGURE 8.13

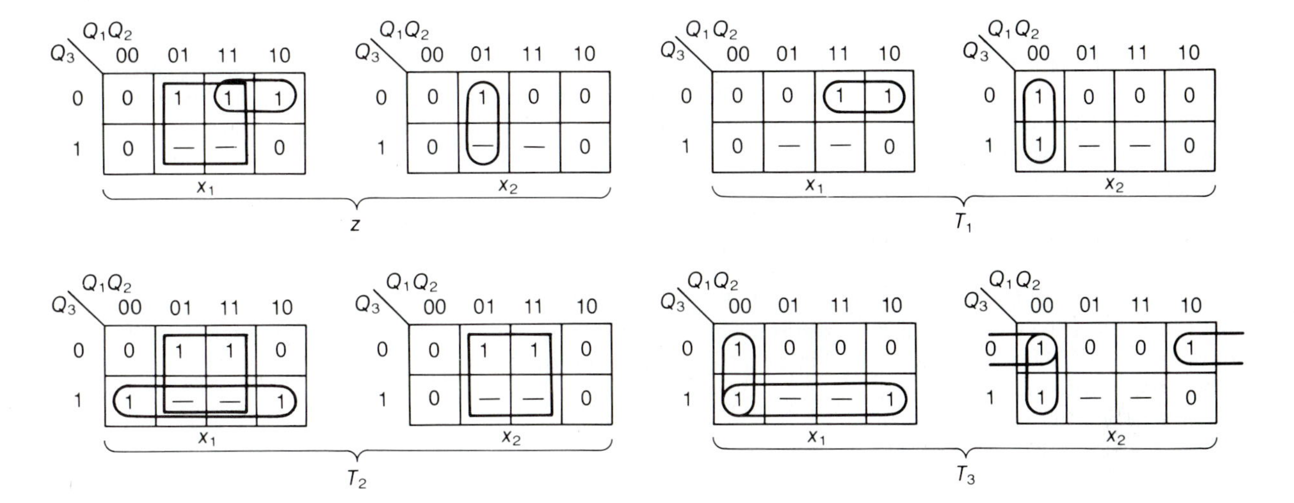

FIGURE 8.14

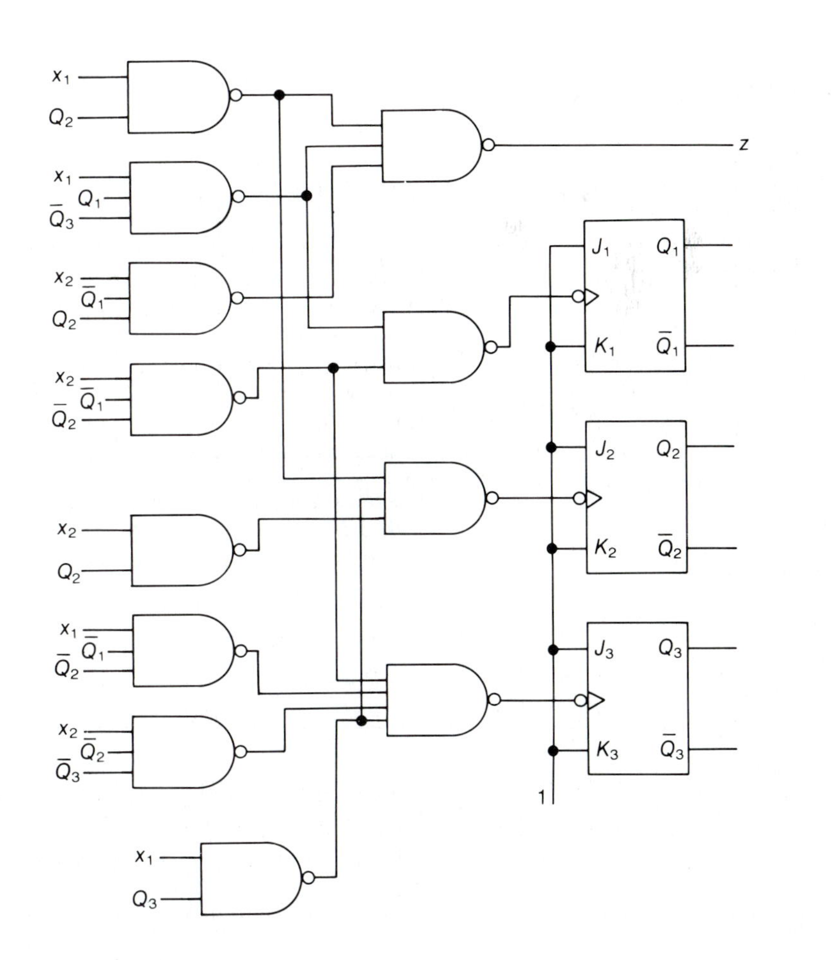

EXAMPLE 8.3

Design an automatic vending machine. The candy bars inside the machine cost 25 cents, and the machine accepts nickels, dimes, and quarters only. An electromechanical system accepts the coins sequentially. The circuit produces a level output that delivers the candy whenever the amount received by the machine is 25 cents or more (this is an old machine). After the candy selection is made, a reset pulse automatically turns off the choice indicator and resets the sequential circuit.

SOLUTION

The circuit accepts any one of the four pulses x_1, x_2, x_3, and x_4 corresponding to nickel, dime, quarter, and reset pulse, respectively. Six states could be used where the states *A, B, C, D, E,* and *F* correspond to the accumulation of 0 cents, 5 cents, 10 cents, 15 cents, 20 cents, and 25 cents or over. The machine does not return overpayment. State *F* is associated with an output of 1 corresponding to the receipt of 25 cents or more.

The reduced state table is obtained as shown in Figure 8.15. To demonstrate their use, we will use three *SR* FFs to identify these six states. The states *A, B, C, D, E,* and *F* are assigned 000, 001, 011, 010, 110, and 111, respectively. The *SR* excitation tables are accordingly obtained as shown in Figure 8.16, which results in the following output and excitation equations:

$$S_1 = x_1(Q_2\overline{Q}_3) + x_2(Q_2) + x_3$$

$$R_1 = x_4$$

$$S_2 = x_1(Q_3) + x_2 + x_3$$

$$R_2 = x_4$$

$$S_3 = x_1(\overline{Q}_2 + Q_1) + x_2(\overline{Q}_3) + x_3$$

$$R_3 = x_1(\overline{Q}_1Q_2) + x_2(\overline{Q}_1Q_3) + x_4$$

$$z = Q_1Q_3$$

FIGURE 8.15

PS	NS				z
	x_1	x_2	x_3	x_4	
A	*B*	*C*	*F*	*A*	0
B	*C*	*D*	*F*	*A*	0
C	*D*	*E*	*F*	*A*	0
D	*E*	*F*	*F*	*A*	0
E	*F*	*F*	*F*	*A*	0
F	*F*	*F*	*F*	*A*	1

If a pulsed output is desired instead, the same design process would be used with a Mealy-type state table. Each pulse causing a transition to state *F* would have an output pulse generated coincident with it.

FIGURE 8.16

PS	NS											
	x_1			x_2			x_3			x_4		
$Q_1Q_2Q_3$	S_1R_1	S_2R_2	S_3R_3	S_1R_1	S_2R_2	S_3S_3	S_1R_1	S_2R_2	S_3R_3	S_1R_1	S_2R_2	S_3S_3
A 000	0—	0—	10	0—	10	10	10	10	10	0—	0—	0—
B 001	0—	10	—0	0—	10	01	10	10	—0	0—	0—	01
C 011	0—	—0	01	10	—0	01	10	—0	—0	0—	01	01
D 010	10	—0	0—	10	—0	10	10	—0	10	0—	01	0—
E 110	—0	—0	10	—0	—0	10	—0	—0	10	01	01	0—
F 111	—0	—0	—0	—0	—0	—0	—0	—0	—0	01	01	01

EXAMPLE 8.4

Analyze the pulse-mode circuit of Figure 8.17.

SOLUTION

FIGURE 8.17

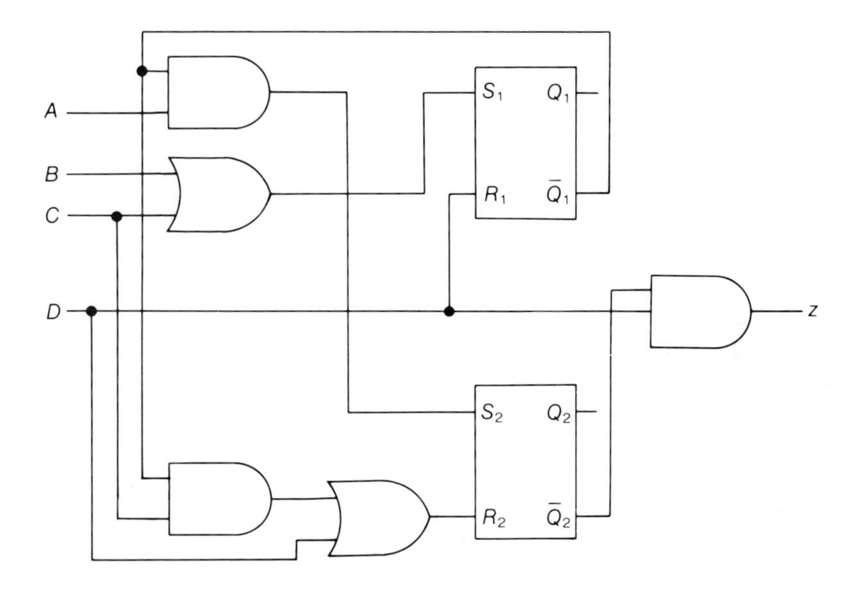

The state and output equations can be readily obtained from the circuit diagram.

$$S_1 = B + C$$
$$R_1 = D$$
$$S_2 = A\overline{Q}_1$$
$$R = C\overline{Q}_1 + D$$
$$Z = D\overline{Q}_2$$

Since it is a pulse-mode circuit the only input conditions that need to be considered are $ABCD$ = 1000, 0100, 0010, and 0001.

The state table may now be obtained using the excitation conditions and the state equations. This is shown in Figure 8.18. The state diagram follows as illustrated in Figure 8.19.

FIGURE 8.18

PS	NS, z			
$Q_1 Q_2$	A	B	C	D
P 00	01,0	10,0	10,0	00,1
Q 01	01,0	11,0	10,0	00,0
R 11	11,0	11,0	11,0	00,0
S 10	10,0	10,0	10,0	00,1

FIGURE 8.19

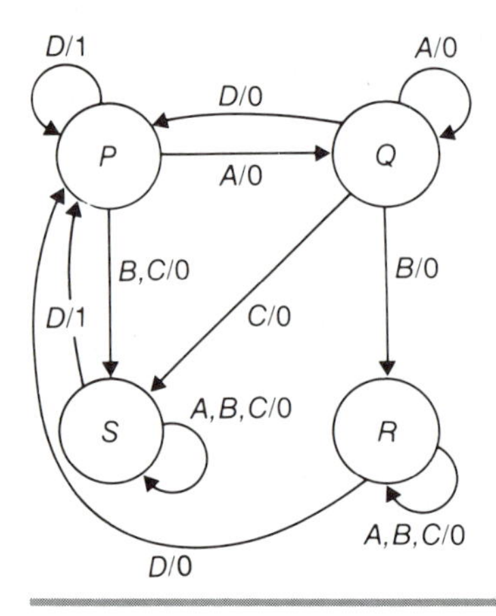

EXAMPLE 8.5

Design a digital lock. Assume that two push-button switches, A and B, are available that output a single pulse for each activation. The two switches are interlocked mechanically so that simultaneous pulses are not possible. The lock should have the following features:

a. The opening sequence is *AABABA*.

b. Three B pulses should give an absolute reset.

c. Any incorrect use of the A switch will cause an output to ring a bell to warn that the lock is being tampered with.

SOLUTION

The state diagram, as shown in Figure 8.20, is obtained to meet the requirements of this lock. State W is reached any time a mistake is made in the prescribed sequence *AABABA*. Once the sequence is correctly completed, the circuit reaches state V with a Moore output of $z_1 = 1$, which actuates the bolt lock. The Mealy output z_2 is generated whenever the A switch is used incorrectly. Three consecutive B pulses would always take the circuit to the reset state P. The lock can be closed by pushing once either A or B, and the circuit is left at state W.

The state assignments are now made with unit distance between successive states in the correct operating sequence. Consequently the transition table is obtained as shown in Figure 8.21. We may decide to use T FFs for the circuit design. Accordingly the excitation and output maps may be obtained as shown in Figure 8.22. The excitation and output equations are as follows:

$$T_1 = \bar{Q}_1Q_2A + (\bar{Q}_1Q_2\bar{Q}_3 + Q_1\bar{Q}_2\bar{Q}_3 + \bar{Q}_1\bar{Q}_2Q_3)B$$

$$T_2 = (\bar{Q}_1Q_3 + Q_1Q_2 + Q_2Q_3)A + (\bar{Q}_1Q_2\bar{Q}_3 + Q_1Q_2Q_3)B$$

$$T_3 = (\bar{Q}_1\bar{Q}_2\bar{Q}_3 + \bar{Q}_1Q_2Q_3 + Q_1\bar{Q}_2Q_3)A + (Q_1Q_2 + Q_3)B$$

$$z_1 = Q_1\bar{Q}_2Q_3$$

$$z_2 = (Q_1\bar{Q}_3 + \bar{Q}_1Q_2Q_3)A$$

FIGURE 8.20

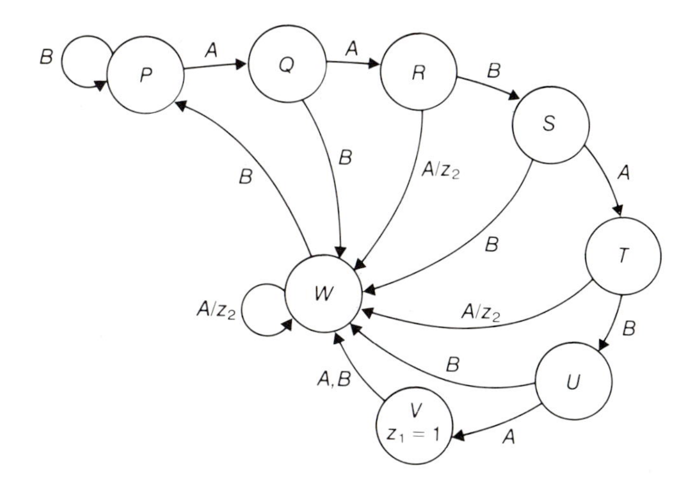

FIGURE 8.21

PS	NS		
$Q_1Q_2Q_3$	A	B	z_1
000	001	000	0
001	011	100	0
011	100,z_2	010	0
010	110	100	0
110	100,z_2	111	0
111	101	100	0
101	100	100	1
100	100,z_2	000	0

FIGURE 8.22

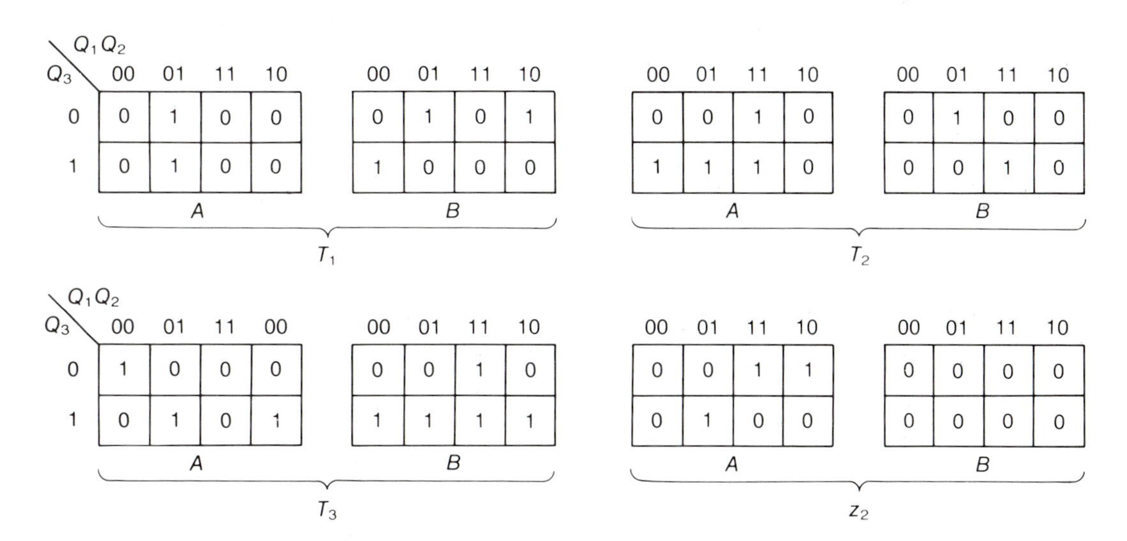

The next step would be to draw the circuit schematic and then construct a prototype circuit.

8.4 Summary

In this chapter the pulse-mode circuit model was introduced and a discussion of the corresponding pulse-mode design algorithm followed. A total of five different examples were considered to demonstrate how such a system could be efficiently designed. By now we are well prepared to study and appreciate the last type of sequential circuit. The design presented in the next chapter removes most of the constraints of pulse-mode circuits and, accordingly, is the most general of all design schemes.

Problems

1. Design a pulse-mode circuit using *SR* FFs that has two inputs, x_1 and x_2, such that an output pulse will be produced simultaneously with the last of a sequence of four input pulses, if and only if the sequence contained at least two x_1 pulses. Sequences may not overlap.
2. Design a pulse-mode circuit using *SR* FFs that has two inputs, x_1 and x_2, such that an output pulse will be produced simultaneously with the last of a sequence of four input pulses, if and only if the sequence contained at least three x_1 pulses. Sequences may not overlap.
3. Design a pulse-mode circuit using *SR* FFs that has two inputs, x_1 and x_2, such that an output pulse will be produced simultaneously with the last of a sequence of four pulses, if and only if the sequence contained at most two x_1 pulses. Sequences may not overlap.
4. Design a pulse-mode circuit using *SR* FFs that has two inputs, x_1 and x_2, such that an output pulse will be produced simultaneously with the last of a sequence of four input pulses, if and only if the sequence contained at most three x_1 pulses. Sequences may not overlap.
5. Design a pulse-mode circuit using only *JK* FFs such that the output pulse, z, is coincident with the last pulse of a sequence of three input pulses. $z = 1$ when the sequence $x_1x_2x_1$ occurs, where x_1 and x_2 are inputs to the circuit.
6. Repeat Problem 5 using only *T* FFs.
7. Repeat Problem 5 using only *D* FFs.
8. Obtain the *JK* equations for the pulse-mode machine of Figure 8.P1.
9. Obtain the *JK* equations for the pulse-mode machine of Figure 8.P2.
10. Design a pulse-mode circuit with inputs A, B, and C and output z. The output changes from 0 to 1 if and only if the sequence ABC occurs while $z = 0$. The output resets only after a B input occurs.

FIGURE 8.P1

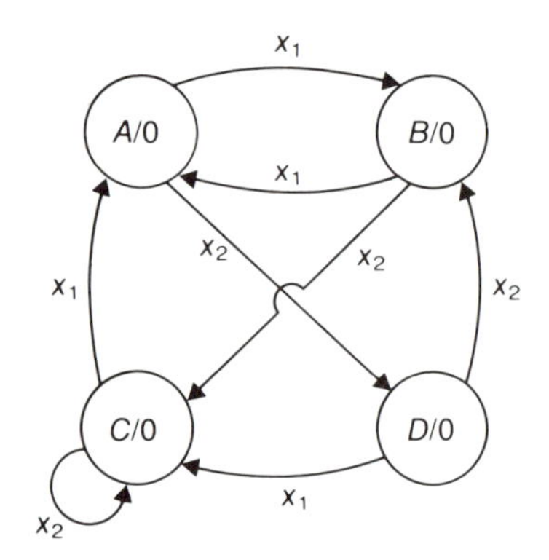

FIGURE 8.P2

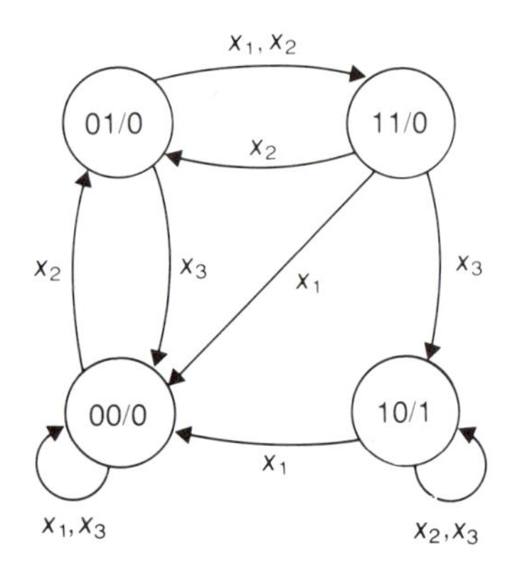

11. Analyze the pulse-mode circuit of Figure 8.P3.
    a. Determine the state table.
    b. Determine the output response to the input sequence *ABAAAABB* if the starting state is 00.
    c. What form (level or pulse) will an output $z = 1$ be? Why?

FIGURE 8.P3

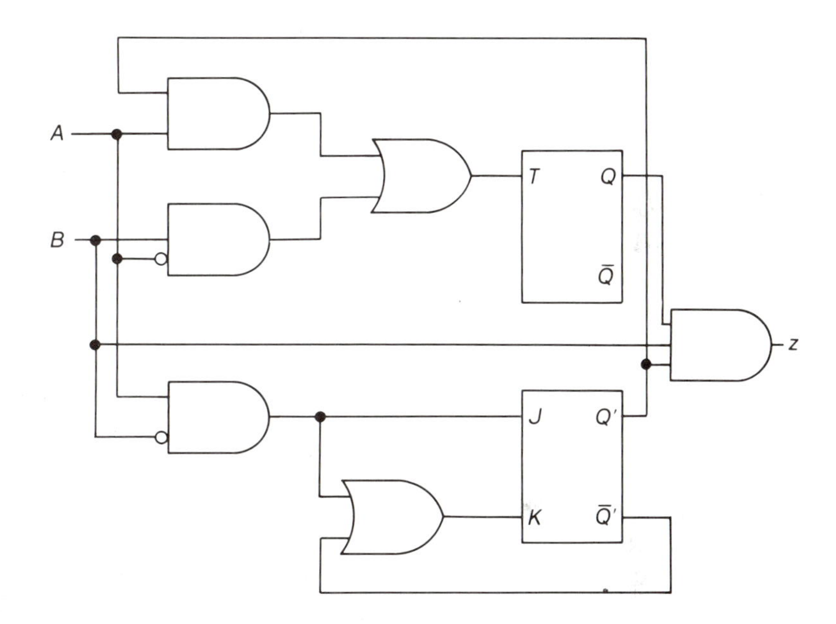

FIGURE 8.P4

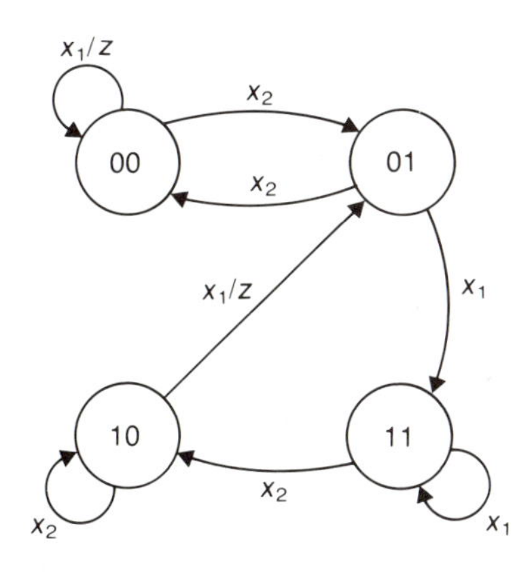

12. For the state diagram that is shown in Figure 8.P4, find the state equations.
13. Construct a pulse-mode state diagram for a device that will recognize five successive x_1 pulses and generate an output pulse at the end of the fifth pulse. An x_2 pulse resets and initializes the counter. Obtain a sequential circuit using only *JK* FFs.
14. Repeat Problem 13 using only *D* FFs.
15. Design a pulse-mode circuit that has two pulse inputs, *A* and *B*, and one pulse output, *z*. The output is to be coincident with the second of the two consecutive *B* pulses immediately following exactly three consecutive *A* pulses.
16. Design a pulse-mode circuit that has two pulse inputs, *A* and *B*, and one pulse output, *z*. The output is to be coincident with the second consecutive *B* pulse immediately following three or more consecutive *A* pulses.

Suggested Readings

Friedman, A. D., and Menon, P. R. "Synthesis of asynchronous sequential circuits with multiple-input changes." *IEEE Trans. Comp.* vol. C-17 (1968): 559.

Kuhl, J. G., and Reddy, S. M. "Multicode single transition time state assignment for asynchronous sequential machines." *IEEE Trans. Comp.* vol. C-27 (1978): 927.

Nagle, H. T., Jr.; Carroll, B. D.; and Irwin, J. D. *Introduction to Computer Logic.* Englewood Cliffs, N.J.: Prentice-Hall, 1975.

Nanya, T., and Tohma, Y. "Universal multicode STT state assignments for asynchronous sequential machines." *IEEE Trans. Comp.* vol. C-28 (1979): 811.

Nordmann, B. J., and McCormick, B. H. "Modular asynchronous control design." *IEEE Trans. Comp.* vol. C-26 (1977): 196.

Rey, C. A., and Vaucher, J. "Self-synchronized asynchronous sequential machines." *IEEE Trans. Comp.* vol. C-23 (1974): 1306.

Rhyne, V. T. *Fundamentals of Digital Systems Design.* Englewood Cliffs, N.J.: Prentice-Hall, 1973.

Sholl, H. A., and Yang, S. C. "Design of asynchronous sequential networks using READ-ONLY memories." *IEEE Trans. Comp.* vol. C-24 (1975): 195.

Design of Fundamental-Mode Circuits

CHAPTER NINE

9.1 Introduction

In the previous two chapters where we examined clocked and pulsed sequential logic, we saw that no action took place until a clock or pulse occurred. The FF inputs could be either stable or varying between clocks, but action occurred only after the arrival of the clock or pulse. In comparison, a system is said to be in *fundamental mode* as long as the external inputs are never allowed to change unless the system has reached a stable state. In fundamental-mode machines the circuit action occurs whenever there is a change in input. These inputs could be pulses or a clock, but for considerations of this chapter we are concerned only with events that occur as a result of a $0 \rightarrow 1$ or $1 \rightarrow 0$ change in the input variable. The circuit state in fundamental-mode circuits is determined by both the memory section outputs and the circuit input variables. *SR* latches are generally used for the memory portion of the circuit.

In a properly designed synchronous circuit, timing problems are eliminated by waiting long enough so that the clock pulse is not introduced until after the external input changes for all FF inputs have reached steady states. In many digital systems, however, it is relatively hard to maintain a fundamental-mode operation. Input signals are derived from different sources and are often random in nature. Consequently, special interference circuits, known as *synchronizers*, usually are used to guarantee the continuation of normal circuit operation.

Fundamental-mode circuit design is the most difficult of the three sequential circuit modes examined, because of the involved timing problems. However, it is also the most powerful since the designer controls every aspect of the circuit action. Each of the FF types considered in Chapter 6 could be easily designed using fundamental-mode design techniques. After mastering the design techniques in this chapter, you should be able to:

- Design and analyze fundamental-mode circuits;
- Understand the concepts of cycles and races;
- Design pulse input circuits with overlapping pulses or an FF with any special characteristics.

9.2 Analysis of Level Sequential Circuits

In the previous two chapters on synchronous and pulsed sequential circuits, we saw that the FF inputs could be stable or varying but no action took place until a clock or a pulse occurred. In fundamental-mode logic the circuit action is initiated by a change in the inputs, and we shall concentrate upon the events that occur as a result of either a $0 \rightarrow 1$ or a $1 \rightarrow 0$ transition in the external input. The circuit state is still determined by both FF outputs and input variables. A general form of the fundamental-mode circuit is shown in Figure 9.1 where the memory elements are shown as delay elements in the system model. The input signals, x_1 through x_p, are referred to as the *external level signals*; the lines z_1 through z_q represent the *primary outputs*; and Q_1 through Q_r represent the *secondary outputs* (state variables) that are fed back to the combinational section after the delay through the *SR* latches.

FIGURE 9.1 **Fundamental-Mode Circuit Model.**

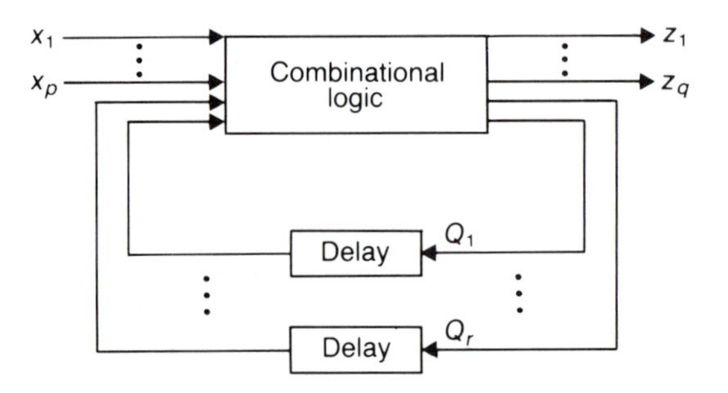

In this particular circuit configuration there exist both stable and transitional states. For a state to be *stable* the combinational portion of the circuit must be generating inputs to the FF portion of the circuit that will not cause the state to change. That is, the present state becomes the next state. To better understand the design procedure of fundamental-mode circuits, the asynchronous delay network of Figure 9.2[*a*] will be considered. Whenever the input variables change, a new state is immediately entered. If this new state requires a change in the FF outputs, the system will be temporarily in a *transitional* state. This transitional state is the result of the nonzero time it takes to set the *SR* latches to their new values. The concept of both stable and transitional states will become more

FIGURE 9.2 **Asynchronous Delay Network: [*a*] Circuit Diagram and [*b*] Timing Diagram.**

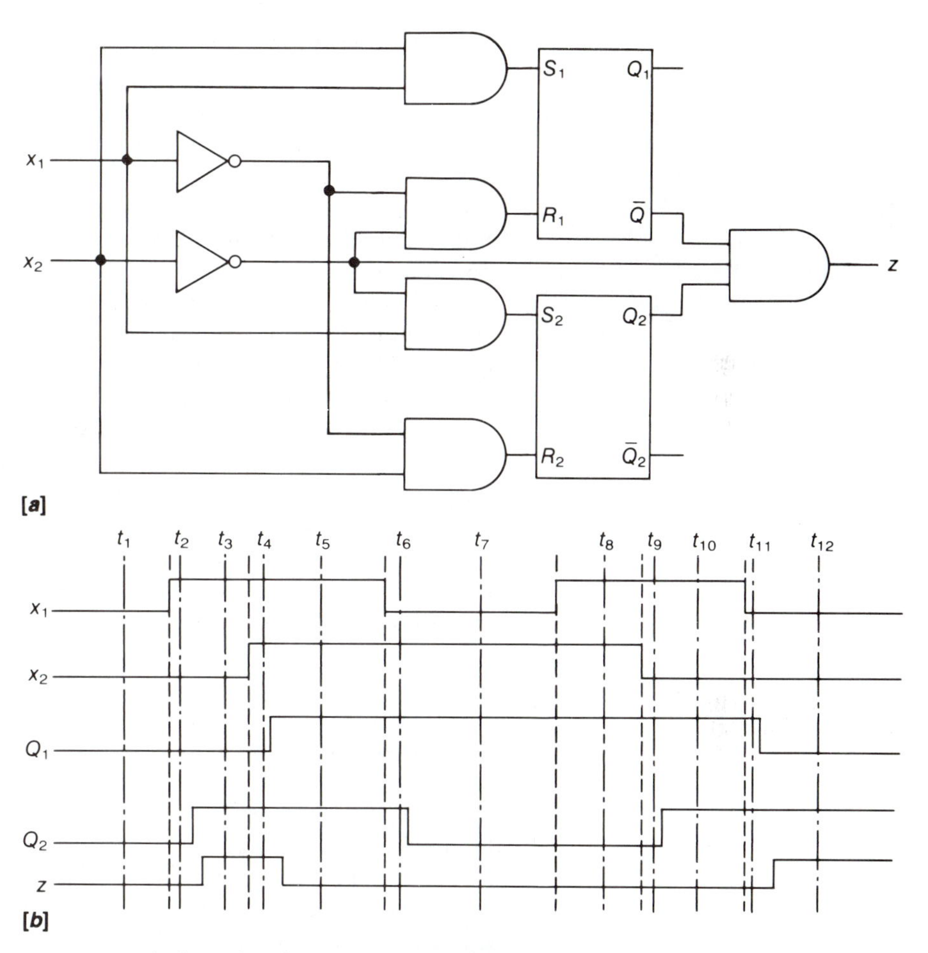

obvious by inspecting the timing diagram of Figure 9.2[*b*]. The state, whether stable or transitional, is identified by the FF values, Q_1Q_2.

At t_4, Q_1Q_2 is only a transitional value corresponding to an x_2 transition, $0 \rightarrow 1$, when $x_1 = 1$, but the present state of Q_1Q_2 soon makes the transition from 01 to the stable state 11. This small time lag is due to the delay element, that is, the gate and latch delays. Similar transitional states are also located at t_2, t_6, t_9, and t_{11}.

The timing diagrams like that in Figure 9.2[*b*] are usually sufficient for the analysis of small circuits only. For larger circuits it may not always be feasible to consider every possible input condition in a timing diagram. A more systematic way is needed to design the circuit. Consequently a tabular form is introduced. First, a transition table is constructed that shows the next states of the latches (referred

FIGURE 9.3 **Flow Table for the Delay Network of Figure 9.2[*a*].**

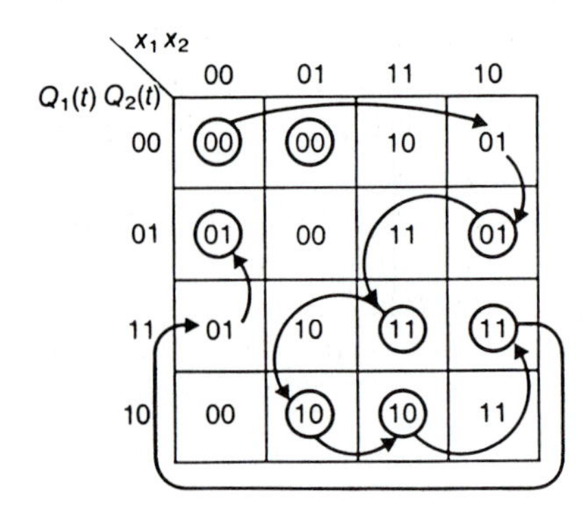

to as FFs from here on) as a function of the present states and inputs. For Figure 9.2[*b*] we can use Equation [6.1] to obtain the next-state equations as follows:

$$Q_1(t + \Delta t) = x_1(t)x_2(t) + [x_1(t) + x_2(t)]Q_1(t)$$
$$Q_2(t + \Delta t) = x_1(t)\bar{x}_2(t) + [x_1(t) + \bar{x}_2(t)]Q_2(t)$$

These equations have been mapped in Figure 9.3. Each column of this figure corresponds to a specific combination of inputs, x_1 and x_2, and each row corresponds to a particular stable present state. The circled states are the stable states and the uncircled states are the transitional states.

We could begin from the present state Q_1Q_2 = (00) when x_1x_2 = 00. For an input transition of x_1x_2 from 00 to 01, a new stable state, (00), is entered. Since both present and next states are the same, the change doesn't have to go through any transitional state. However, if x_1x_2 is changed to 10, the FF state goes to a new stable state (01). Since the two states are different, the system will temporarily be in a transitional state before stabilizing at (01). Since the FFs may not change instantaneously, the transitional state is occupied until the change is complete. This transitional state is given the same designation as that of the stable state toward which the motion has already begun. Consequently the system will go through a transitional state, 01, shown by a transition path. The transitional state is listed at the intersection of the corresponding stable states. Note that in each row the stable state is that for which the next state is the present state. As soon as Q_1Q_2 changes from 00 to 01, a stable state is reached unless the input values of x_1 and x_2 are changed again. The transition flow corresponding to the input sequence of Figure 9.2[*b*] is illustrated by the arrows drawn on the flow table of Figure 9.3. This is not the only possible transition flow of states. Different input sequences would take the circuit through different transition paths. These comments may be initially confusing but are made only to introduce concepts that will be elaborated on in this chapter.

9.3 Flow Table Generation

The basic design algorithm for the fundamental-mode circuits is similar to that for the other sequential machines. First, a flow table is constructed from the problem statement (analogous to the state diagram); then the flow table is reduced to a merged flow table (analogous to the state table with redundant states removed). These steps are then followed by the assignment of states, realization of excitation equations, and, finally, implementation of the logic circuit. The details of each step, however, are different for the fundamental-mode circuits.

The starting point for the fundamental-mode circuit design is a primitive state (or flow) table. The *primitive flow table* is defined as a table that has exactly one stable state in each of its rows. The primitive flow table is eventually reduced to a *merged flow table* that may have fewer rows. Some of the rows of the merged flow table may have more than one stable state and fewer transitional states.

Example 9.1 illustrates the construction of a primitive flow table from a word statement. For this example we shall assume that only one input variable changes at a time and that the input changes are separated in sufficient time to allow the circuit to reach a stable state between changes. This is not to say that two or more input changes are forbidden. In the primitive flow table every change in the input results in a state change, since only one stable state is allowed in each row.

EXAMPLE 9.1

Construct the primitive flow table for a circuit with two inputs, x_1 and x_2, and an output, z. The output z turns on when x_2 makes a $0 \rightarrow 1$ transition if x_1 is a 1 at the time of the transition. However, z turns off when x_1 makes a $1 \rightarrow 0$ transition.

FIGURE 9.4

x_1x_2				
00	01	11	10	z
(1)				0
			(2)	0
		(3)		1
	(4)			0

SOLUTION

The primitive flow table is quite similar to the state diagram explained in earlier chapters. Like state diagrams the primitive flow table lists every possible action of the circuit. It can be begun by assuming an initial state (usually first row, first column) and considering the input transitions that make the output assume a value of 1. Other states are subsequently determined as we attempt to complete the table.

In Figure 9.4 each row represents a state that will eventually be identified with a particular combination of FF outputs. Each row may contain only one stable state. Assume that the circuit is in a stable state (first row) and the inputs are $x_1x_2 = 00$. Note that each of the stable states is numbered and circled. In order to cause an output $z = 1$, the circuit must first enter a stable state by setting $x_1 = 1$. Thereafter the output z would become a 1 when x_2 was changed from a 0 to a 1. A subsequent change of x_1 from a 1 to a 0 would turn off the output. The primitive flow table, as shown in Figure 9.4, initially has four rows corresponding to these four input circumstances. The four states are listed under the columns $x_1x_2 =$ 00, 10, 11, and 01 with outputs of 0, 0, 1, and 0, respectively.

The next step in the design of a flow table is to insert transitional states. When the circuit moves from one stable state to the next, it has to travel through a transitional state. The transitional state has the same designation as the destination state and is listed at the intersection of the same row as the source state and the same column as the destination state. This table is obtained as shown in Figure 9.5. The circled numbers in the flow table represent stable states. The combinational portion of the to-be-designed sequential circuit will have x_1, x_2, and the FF outputs as its inputs. The FF inputs are designed so that no change in the FF outputs occurs if the state is already stable. In other words, a stable condition implies that the next state is the present state itself.

Assume that the circuit is in stable state (1). When the inputs change to 10 the circuit action moves from the beginning stable state (1) to a location under the $x_1x_2 = 10$ column but in the same row. At this time the combinational circuit is developing the FF inputs necessary to cause the FF

FIGURE 9.5

x_1x_2				
00	01	11	10	z
(1)			2	0
		3	(2)	0
	4	(3)		1
1	(4)			0

outputs to change to the binary values corresponding to row 2. This change takes an amount of time corresponding to the gate delays of the gates involved. After this delay the circuit action moves to stable state ②. We shall indicate the transitional state that occurs between stable states ① and ② with an uncircled 2 that indicates that the circuit is in a transitional state but headed toward stable state ②. Similarly, while moving either from state ② to state ③, or from state ③ to state ④, or from state ④ to state ①, the circuit goes through additional transitional states as shown in the flow table of Figure 9.5.

The next step is to define the actions that occur when transitions other than the ones considered occur. For instance, if the circuit is in stable state ① and the inputs change to 01, a stable state must be entered in column 01 with an output of 0. Stable state ④ meets this requirement. This transition is justified by entering transitional state 4 at the intersection of row 1 and column 01. However, if the circuit is in stable state ③ and the inputs x_1x_2 change to 10, the circuit must enter a stable state in column 10 with an output of 1. There is no such state defined as yet in column 10. Stable state ② is present in column 10 but its associated output is 0; so a new stable state must be defined. This new stable state is numbered ⑤ and is justified by entering transitional state 5 at the intersection of row 3 and column 10. Similar arguments are carried out that lead to the completed table shown in Figure 9.6.

FIGURE 9.6

x_1x_2				
00	01	11	10	z
(1)	4	—	2	0
1	—	3	(2)	0
—	4	(3)	5	1
1	(4)	6	—	0
1	—	3	(5)	1
—	4	(6)	2	0

The primitive flow table just completed describes all possible actions. The don't-cares represent the transitions that will not occur due to the restriction that both x_1 and x_2 will not change at the same time.

In actual physical systems it is virtually impossible for two events to occur simultaneously. If the two inputs changed with a time difference less than the minimum response time of the circuit, the input change would appear simultaneous to the circuit. The possibility of simultaneous changes can be handled by defining the desired circuit action when such an event occurs and by adding necessary states. In some of the examples to come, we shall consider possible simultaneous changes in two or more inputs. The next design problem will be carried through to its completion with occasional pauses to look at additional examples related to particular design sequences.

EXAMPLE 9.2

Obtain the merged flow table for a two-button electrical lock that will actuate according to the following sequence:

A: 0 1 1 1 1 0 1

B: 0 0 1 0 1 1 1

z: 0 0 0 0 0 0 1

$z = 1$ opens the lock, *A* and *B* correspond to the two buttons, and *z* corresponds to the circuit output. A switch being depressed corresponds to a 1. Release of both buttons will clear the circuit.

SOLUTION

First, a primitive flow table is constructed that accounts for every possible action of the switches. The completed primitive flow table is shown in Figure 9.7. The circuit starts from the stable state ① and goes through six stable states, ② through ⑦, until the lock opens. This system movement corresponds to the given opening sequence. When the system reaches stable state ⑦, the output becomes high. In addition, there are three states, ⑧, ⑨, and ⑩, placed respectively under columns 11, 01, and 10, to account for all mistaken inputs that might occur. The system stays in one of these three states, ⑧, ⑨, and ⑩, after a mistake is made anywhere. Whenever both of the push buttons are released, the circuit returns to state ①.

In the second step we attempt to eliminate redundant states. Two states are considered equivalent if (a) the states have the same input conditions; (b) the states have the same output values; and (c) for each possible input change there is a transition from these stable states to the same or equivalent stable states. Only stable states in the same column need to be considered. Therefore, the possible equivalencies are as follows:

(6,9), (3,5), (3,8), (5,8), (2,4), (2,10), (4,10), (3,7), (5,7), and (7,8)

The equivalent states could be determined by using an implication table as introduced in Chapter 7. An alternative tabular approach is shown in Figure 9.8.

All possible equivalencies are used as column headings. The states that

FIGURE 9.7

AB				
00	01	11	10	z
①	9	8	2	0
1	9	3	②	0
1	9	③	4	0
1	9	5	④	0
1	6	⑤	10	0
1	⑥	7	10	0
1	9	⑦	10	1
1	9	⑧	10	0
1	⑨	8	10	0
1	9	8	⑩	0

FIGURE 9.8

	6,9	3,5	3,7	3,8	5,7	5,8	7,8	2,4	2,10	4,10
6,9							√			
3,5	√									√
3,8										√
5,8	√									
2,4		√								
2,10				√						
4,10						√				

do not meet the three previously mentioned equivalency requirements are crossed out (with X). No state can be equivalent to state ⑦ since only this state gives the output $z = 1$. The states (3,7), (5,7), and (7,8) are crossed out. Those possible equivalencies that are not crossed out are now listed as row labels. After the implication table is labeled, we place one or more checks (with ✓) in each row and only under those columns whose equivalency is a requirement for the equivalency of the particular row-pairs. As an example, examine the transitional states in the rows containing the stable states ③ and ⑤. For equivalency of two states the corresponding transitional states for all possible input conditions must also be equivalent; that is, the pairs (6,9) and (4,10) must be equivalent for the pair (3,5) to be equivalent. After the table is completed we find that the equivalency of states 6 and 9 depends on the equivalency of states 7 and 8 but (7,8) has already been crossed out. This implies that (6,9) are not equivalent and, therefore, the (6,9) column is crossed out (this time with a slash, /, for illustration purpose). This process leads to the crossing out of two more columns, (3,5) and (5,8). Continue this procedure until no more column headings can be eliminated. Any column heading not eliminated represents an equivalency. In this example, it turns out that there are no equivalencies since none of the column-pairs survived.

FIGURE 9.9

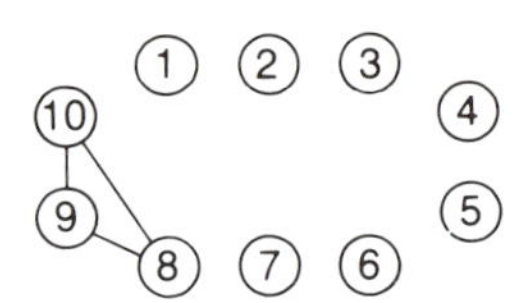

FIGURE 9.10

AB			
00	01	11	10
(1)	9	8	2
1	9	3	(2)
1	9	(3)	4
1	9	5	(4)
1	6	(5)	10
1	(6)	7	10
1	9	(7)	10
1	(9)	(8)	(10)

Next, we shall attempt to merge the primitive flow table. During the construction of the primitive flow table we assigned only one stable state to each row. FF values will eventually be assigned to identify each of these rows. Since the goal is to have a uniquely specified circuit with a minimum number of FFs, we should try to condense the flow table. In the first part of this chapter we pointed out that for fundamental-mode circuits the states were defined by the variable values and the FF states. However, if we can merge more than one stable state in a row, they can still be uniquely identified, using the FF state for the row and the variable values for the column.

The algorithm for constructing the merger diagram is as follows:

a. Display all states.

b. Ignore the output differences.

c. Connect all states that have the same next states. Don't-cares can be interpreted as either 0 or 1.

d. Any group of states for which every possible interconnection exists are called completely connected and may be merged together.

The merger diagram of our flow table, therefore, may be obtained as shown in Figure 9.9.

The interconnections indicate that the states ⑧, ⑨, and ⑩ may be merged to obtain the condensed primitive flow table as shown in Figure 9.10. While merging two or more stable states it should be noted that the entries in a merged row should be such that stable states should dominate over corresponding transitional states, and states of all kind should dominate over don't-cares.

Example 9.2 has illustrated the various steps involved in obtaining a merged flow table. The merged states are said to be *completely connected.* Note that the merged flow table has no output information associated with it. As we shall see later, the output equations are derived by taking output data from the primitive flow table. Note that the reduction of rows and merging of rows do not make the circuit operation more correct but only more economical. At this point we are ready to make state assignments. The state assignments cannot be made arbitrarily as in clocked and pulse-mode sequential circuits. Prior to considering the fundamental-mode state assignment concepts, consider an additional example of reducing primitive flow tables to the corresponding merged flow tables.

EXAMPLE 9.3

For the two primitive flow tables of Figure 9.11, obtain the reduced flow tables.

SOLUTION

FIGURE 9.11

x_1x_2 00	01	11	10
(1)	5	—	2
1	—	3	(2)
—	6	(3)	2
—	6	(4)	2
1	(5)	4	2
1	(6)	4	—

[*a*]

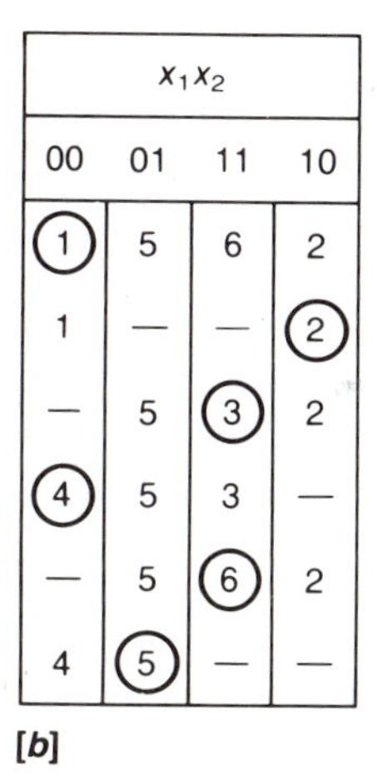

x_1x_2 00	01	11	10
(1)	5	6	2
1	—	—	(2)
—	5	(3)	2
(4)	5	3	—
—	5	(6)	2
4	(5)	—	—

[*b*]

We consider the merger diagram for both as shown in Figures 9.12[*a*–*b*]. In part [*a*], rows (4,6), (1,5), (1,2), and (2,3) may be merged. If 1 and 2 are merged, the only other possible merger could be between 4 and 6. To obtain the most reduced flow table, we would rather merge the pairs (1,5), (2,3), and (4,6). In part [*b*], states (1,2,6) and (3,4,5) may be merged. Note that (2,3,5,6) may not be merged since the pairs (2,5) and (3,6) are not

FIGURE 9.12

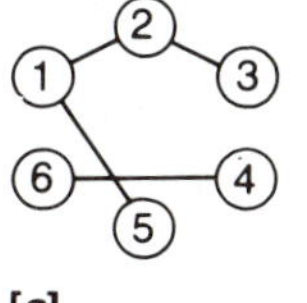

[*a*]

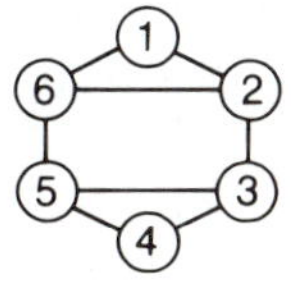

[*b*]

connected. In each of the cases, the don't-cares were used in determining equivalency of states. The resulting merged flow tables are shown in Figures 9.13[*a–b*].

FIGURE 9.13

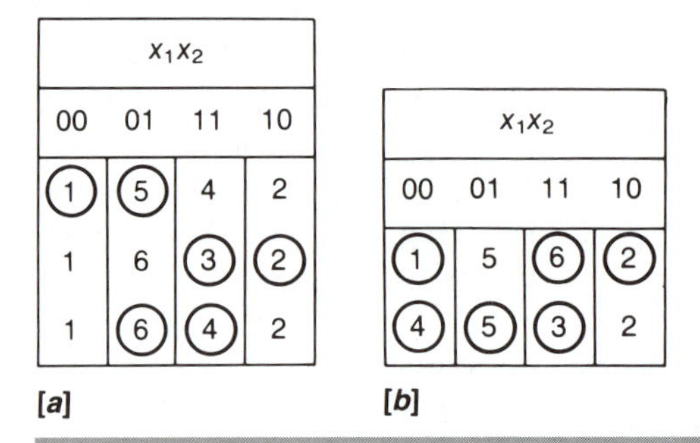

9.4 Cycles and Races

In Chapters 6 and 7 the primary objective of choosing a state assignment for sequential networks was the simplification of logic. However, for the level-mode circuits the primary objective in choosing a particular state assignment is the prevention of critical hazards; simplification of the combinational logic becomes a secondary consideration. In the circuits designed previously, there was ample time for the excitations to settle down before the next pulse arrived. However, the fundamental-mode circuit inputs are not clocked, and consequently unwanted and unpredictable hazards can result. Therefore, it is of extreme importance to make a state assignment such that the circuit is free of critical hazards.

It is important to understand the concepts of cycles and races prior to considering the techniques to avoid circuit hazards. Usually race is the problem and cycle is its solution. Consider the partially complete flow table of Figure 9.14. Imagine yourself at one of the positions in the flow table and follow the circuit action. You may glance to the left and see what values the FFs have at that moment and then look at the code of your position and determine which row you are moving to next. If the code of your position is the code of

FIGURE 9.14 **Representation of Cycles.**

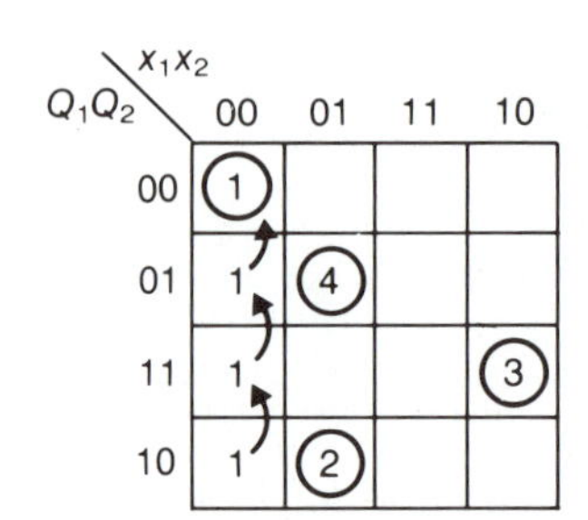

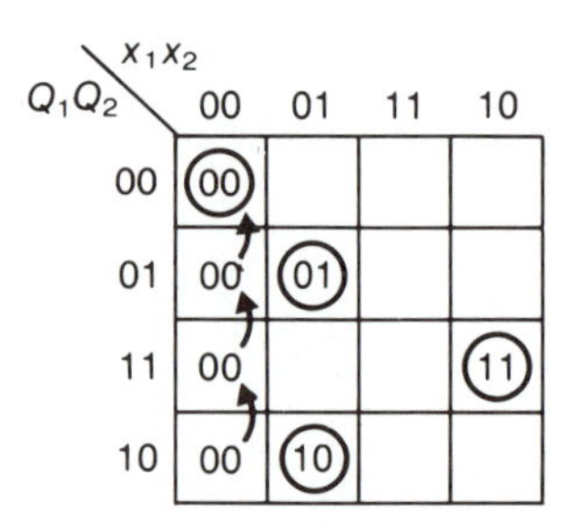

the row you are in, you are already in a stable state and not going anywhere. Notice the arrows in the first column. If the circuit is in states ②, ③, or ④ and the input variables change to $x_1x_2 = 00$, then you are to go to stable state ① in row 1. The occurrence of two or more consecutive unstable states is referred to as a *cycle*. What is demonstrated by the arrows is that the designer has forced the circuit to a cycle. If the circuit is in stable state ② and the input changes to $x_1x_2 = 00$, the circuit moves to the correct column and examines the entry to see where to go next. It is required to move to rows 11, 01, and then to 00 in sequence. The stable state is reached eventually in row 00. Similarly, from stable state ③ there could be a cycle of two successive transitions when there is an input change from $x_1x_2 = 10$ to 00.

The designer could have entered 00 in the intersection of $x_1x_2 = 00$ and $Q_1Q_2 = 10$. In that case the circuit would proceed directly from the stable state ④ to the stable state ① once x_1x_2 is changed to 00. Time is the only difference in the two cases discussed. In the first case it took three times as long to get to stable state ①. The designer may use these cycles to introduce delays in circuitry whenever necessary.

A *race* is said to exist in a circuit when two or more state variables must change values when the circuit is required to make a transition from one stable state to another stable state. A race condition usually results in unexpected and inaccurate performance of the fundamental-mode circuit. There are, however, two major types of races: critical and noncritical. Figure 9.15 shows a noncritical race. If the circuit is in stable state ③ and the inputs x_1x_2 change to 00, the circuit moves to stable state ① without any hindrance. This movement requires both Q_1 and Q_2 to turn off. As mentioned earlier in the chapter it is almost impossible for two events to occur simultaneously. Even if the two FFs are supposedly identical, one would always turn off more quickly than the other. If Q_1 turns off first, the circuit moves in the beginning to row 01 and then to row 00 as Q_2 catches up with Q_1. If Q_2 turns off first, then the circuit goes to row 10 and then to row 00 when Q_1 catches up. The direction the circuit takes to get to stable state ① depends on which FF is faster. The

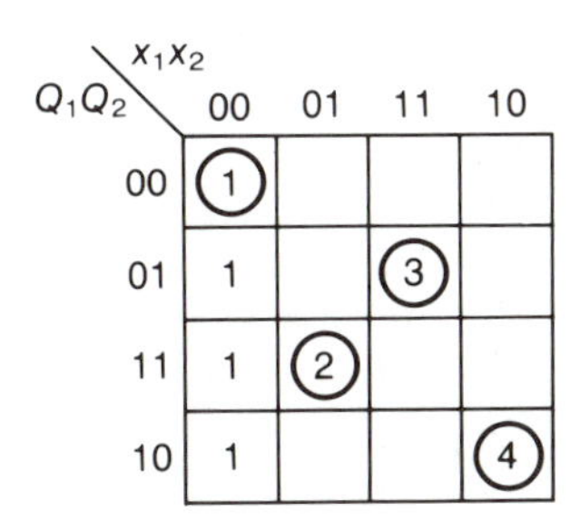

Q_1Q_2 \ x_1x_2	00	01	11	10
00	(00)			
01	00		(01)	
11	00	(11)		
10	00			(10)

FIGURE 9.15 **Example of a Noncritical Race.**

two FFs, therefore, are said to be in a race. The particular race in Figure 9.15, however, is noncritical since the circuit winds up in the proper stable state either way.

Now consider the circuit of Figure 9.16. This is an example of a critical race. If the circuit resides in stable state ③ and x_1x_2 changes to 00, then the circuit will wind up in stable state ② if Q_1 is faster and in stable state ① otherwise. Once the circuit arrives in stable state ②, it is stuck there since the next state and present state are the same. It is always possible that the designer could overlook a race and be fortunate in the choice of FF used in the prototype circuit. During the testing phase the circuit might work. When several circuits are built, however, some would have the race won by one FF, and in the other circuits another FF would win the race. Such anomalous circuit conditions must be strictly avoided. However, one may force the circuit to cycle in the proper direction and thereby eliminate the race problem altogether.

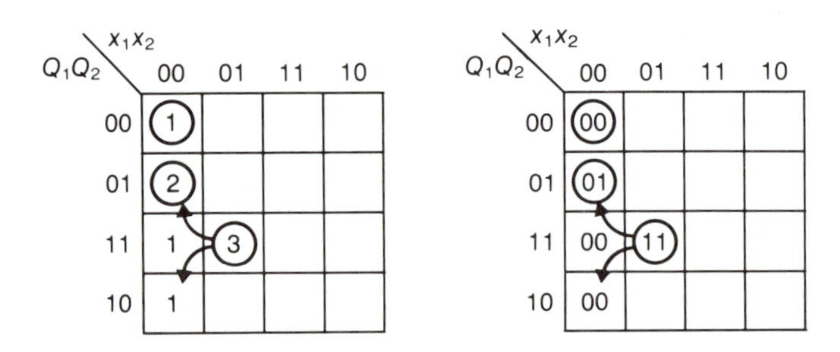

FIGURE 9.16 **Example of a Critical Race.**

Figure 9.17 shows the same situation as in Figure 9.16 but with a cycle to eliminate the critical race. As soon as x_1x_2 changes from 01 to 00 the circuit state traverses from row 11 to row 10 and then to row 00. Note that the transition $11 \rightarrow 10 \rightarrow 00$ does not involve any race since none of these transitions requires more than one FF to change. Potential races exist whenever more than one FF has to change during a state change. Race problems may be eliminated if state assignments are made in such a way as to restrict all transitional movement to adjacent rows only. This restriction is the desired goal of all designers but it is not always possible to achieve.

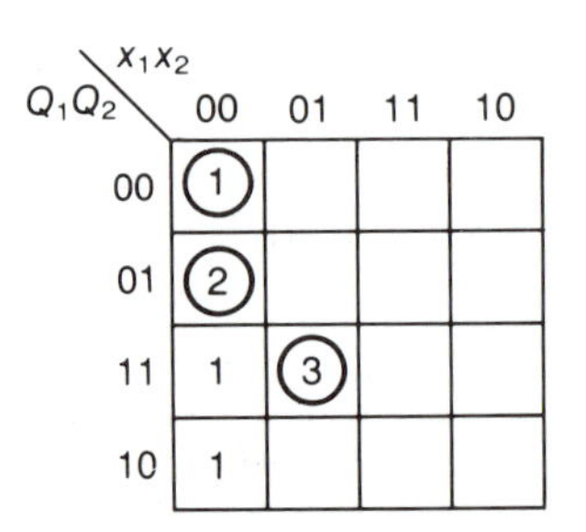

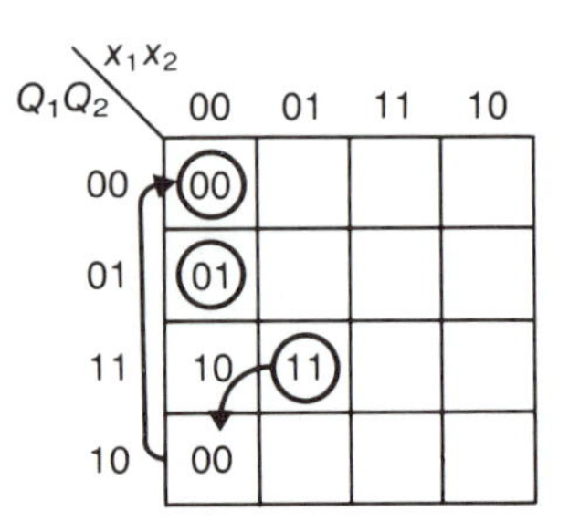

FIGURE 9.17 **Correction to the Race Condition of Figure 9.16.**

In complicated circumstances, cycles are used to design race-free circuits.

As an example, let us try to make the state assignments for the merged flow table of Figure 9.18[*a*]. A letter is initially assigned to each row for identifying the corresponding stable states. Next a list of required row-to-row transitions is made. For instance, if the circuit is in stable state (2) and x_1x_2 changes to 10, the circuit must move to stable state (8). This requires that the circuit move from row a to row d. All other row-to-row transitions are determined similarly. For this merged table the required transitions are as follows:

$$a \to d \quad b \to d \quad c \to b \quad d \to a$$
$$a \to c \quad b \to c \quad c \to a \quad d \to b$$

Using a two-variable assignment map, a, b, c, and d are assigned Q_1Q_2 codes in such a way that the transitions are limited to neighboring rows only. There are four different ways to accomplish this but only one of these assignments is shown in Figure 9.18[*b*].

FIGURE 9.18 [*a*] **Merged Flow Table and** [*b*] **Corresponding State Assignments.**

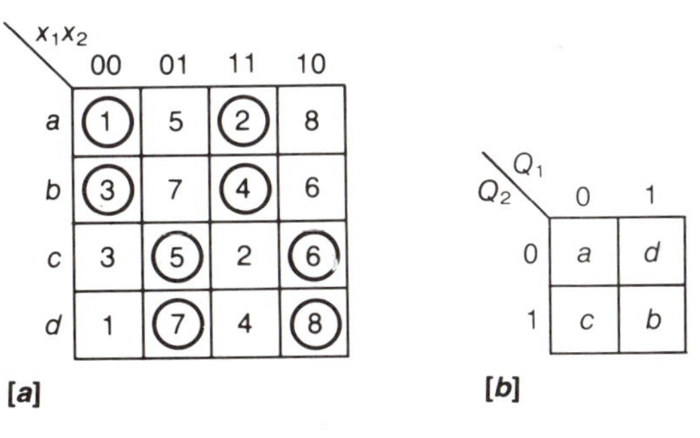

x_1x_2	00	01	11	10
a	(1)	5	(2)	8
b	(3)	7	(4)	6
c	3	(5)	2	(6)
d	1	(7)	4	(8)

[*a*]

Q_2 \ Q_1	0	1
0	a	d
1	c	b

[*b*]

FIGURE 9.19 Merged Flow Table for Example 9.2.

	AB			
	00	01	11	10
a	(1)	9	8	2
b	1	9	3	(2)
c	1	9	(3)	4
d	1	9	5	(4)
e	1	6	(5)	10
f	1	(6)	7	10
g	1	9	(7)	10
h	1	(9)	(8)	(10)

We may next return to our digital lock problem (Example 9.2) and determine the state assignments. The eight-row merged flow table for the problem, as obtained in Example 9.2, requires at least three FFs. The merged flow table is redrawn in Figure 9.19 and each of its rows is given a letter code as shown. To determine the correct assignments to these states a list of all required transitions is made:

$$b \to a \quad g \to a \quad b \to h \quad e \to f$$
$$c \to a \quad h \to a \quad c \to d \quad e \to h$$
$$d \to a \quad a \to h \quad c \to h \quad f \to g$$
$$e \to a \quad a \to b \quad d \to e \quad f \to h$$
$$f \to a \quad b \to c \quad d \to h \quad g \to h$$

Note that there are seven transitions (transitions to a) that are purely noncritical, because all of the states wind up in stable state (1) some time after the input AB = 00 has occurred. There is no other stable state in column 00. Some of the remaining thirteen transitions need extra attention since in the remaining columns there is more than one destination state. Every single row but the last is involved directly in the correct transition toward the opening of the digital lock. This involves the movement $a \rightarrow b \rightarrow c \rightarrow d \rightarrow e \rightarrow f \rightarrow g$. It is quite easy to make assignments so that this movement includes transitions between adjacent rows only. Once this is accomplished the circuit will not be subjected to any race between FFs. Accordingly, the eight states are assigned binary codes as shown in Figure 9.20. The assignments are made in a way so that a is a neighbor of b, b is a neighbor of c, c is a neighbor of d, d is a neighbor of e, e is a neighbor of f, and f is a neighbor of g. The unoccupied cell is assigned to h. Note that this is not the only possible assignment scheme.

FIGURE 9.20 **Arbitrary Assignments of Eight States.**

Q_3 \ Q_1Q_2	00	01	11	10
0	a	b	c	d
1	h	g	f	e

The assignment scheme as proposed in Figure 9.20 takes care of nine different transitions of the aforementioned thirteen transitions. That leaves four remaining transitions, which are as follows:

$b \rightarrow h$

$c \rightarrow h$

$d \rightarrow h$

$f \rightarrow h$

In order to guarantee race-free operation, these four transitions could be made to cycle through one or several other states. In many problems it will be necessary to add another FF. The additional unused states are then used as paths through which to cycle to avoid ending up in an unwanted state due to the critical race. In our digital lock example, with an additional FF we would have twice as many cells in the map of Figure 9.20. Fortunately, in this design example we can make proper assignments with only three FFs. Let us try to explore the possibilities now.

The possible ways to cycle these four transitions are many, but we shall list only those that would require the least amount of propagation delay. They are determined using Figure 9.20 and are listed as follows:

$b \rightarrow h$: $b \rightarrow a \rightarrow h$

$b \rightarrow g \rightarrow h$

$c \rightarrow h$: $c \rightarrow d \rightarrow e \rightarrow h$

$c \rightarrow f \rightarrow e \rightarrow h$

$c \rightarrow f \rightarrow g \rightarrow h$

$c \rightarrow b \rightarrow a \rightarrow h$

$c \rightarrow d \rightarrow a \rightarrow h$

$d \rightarrow h$: $d \rightarrow a \rightarrow h$

$d \rightarrow e \rightarrow h$

$f \rightarrow h$: $f \rightarrow g \rightarrow h$

$f \rightarrow e \rightarrow h$

Before the final choice is made, these cycles are checked with the merged flow table about their final consequences. This comparison would reveal that two of the listed transitions are very undesirable. They are $c \rightarrow f \rightarrow e \rightarrow h$ and $c \rightarrow f \rightarrow g \rightarrow h$. These cycles cannot be used since both of them would end up at the stable state ⑥. Also note that the inclusion of $c \rightarrow b \rightarrow a \rightarrow h$ would solve two of the races, $b \rightarrow h$ and $c \rightarrow h$, at the same time. There are two reasonable choices for column $AB = 01$; either $d \rightarrow c \rightarrow b \rightarrow a \rightarrow h$ or $c \rightarrow b \rightarrow a \rightarrow h$ and $d \rightarrow a \rightarrow h$. Both of these cycles are illustrated in Figure 9.21[*a*], which has been rearranged to conform to a reflected binary code. It is interesting to note that the $d \rightarrow c \rightarrow b \rightarrow a \rightarrow h$ cycle solves three of the races: $d \rightarrow h$, $c \rightarrow h$, and $b \rightarrow h$. This leaves only the $f \rightarrow h$ race. The two choices $f \rightarrow g \rightarrow h$ and $f \rightarrow e \rightarrow h$ for solving the race $f \rightarrow h$ are also shown in the primitive flow table.

Finally, Figure 9.21[*b*] is obtained by considering only the $d \rightarrow c \rightarrow b \rightarrow a \rightarrow h$ and $f \rightarrow g \rightarrow h$ cycles. Note that in the case of a noncritical race the transitional state is given the code of the steady state toward which the circuit is headed. If cycles are used for timing purposes, the transitional state is given the code of the next row in the cycle.

In the next step the excitation table corresponding to the *SR* latches is realized using the FF control conditions. The excitation

FIGURE 9.21 **Primitive Flow Table with Proper State Assignments.**

$Q_1Q_2Q_3$	AB 00	01	11	10	
000	①	9	8	2	*a*
001	1	⑨	⑧	⑩	*h*
011	1	9	⑦	10	*g*
010	1	9	3	②	*b*
110	1	9	③	4	*c*
111	1	⑥	7	10	*f*
101	1	6	⑤	10	*e*
100	1	9	5	④	*d*

[*a*]

$Q_1Q_2Q_3$	AB 00	01	11	10
000	000	001	001	010
001	000	001	001	001
011	000	001	011	001
010	000	000	110	010
110	000	010	110	100
111	000	111	011	011
101	000	111	101	001
100	000	110	101	100

[*b*]

maps are constructed the same as for clocked and pulsed sequential circuits. Figure 9.22 shows the excitation maps for only S_1 and R_1 inputs of Q_1. The S_1 and R_1 Boolean equations are obtained as follows:

$$S_1 = ABQ_2\overline{Q}_3$$
$$R_1 = \overline{A}\overline{B} + \overline{A}Q_2\overline{Q}_3 + \overline{B}Q_3 + AQ_2Q_3$$

FIGURE 9.22 **S_1 and R_1 Excitation Maps for the Circuit Described by Figure 9.21[*b*].**

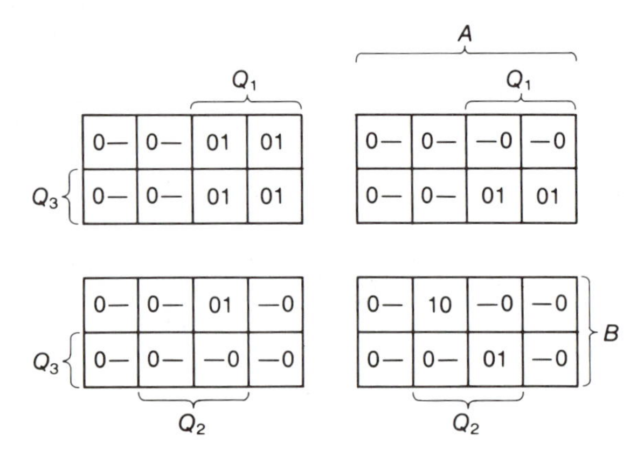

The derivation of the Boolean equations for the remaining FF inputs are left as an exercise (see Problem 1 in this chapter).

The output of the circuit is to be a 1 when the circuit has reached stable state ⑦. Referring to Figure 9.21, it can be seen that in the state ⑦, AB = 11 and $Q_1Q_2Q_3$ = 011 and, therefore, $z = AB\overline{Q}_1Q_2Q_3$. The circuit for only Q_1 and the output is shown in Figure 9.23.

FIGURE 9.23 **Q_1 Excitation Circuit for Example 9.2.**

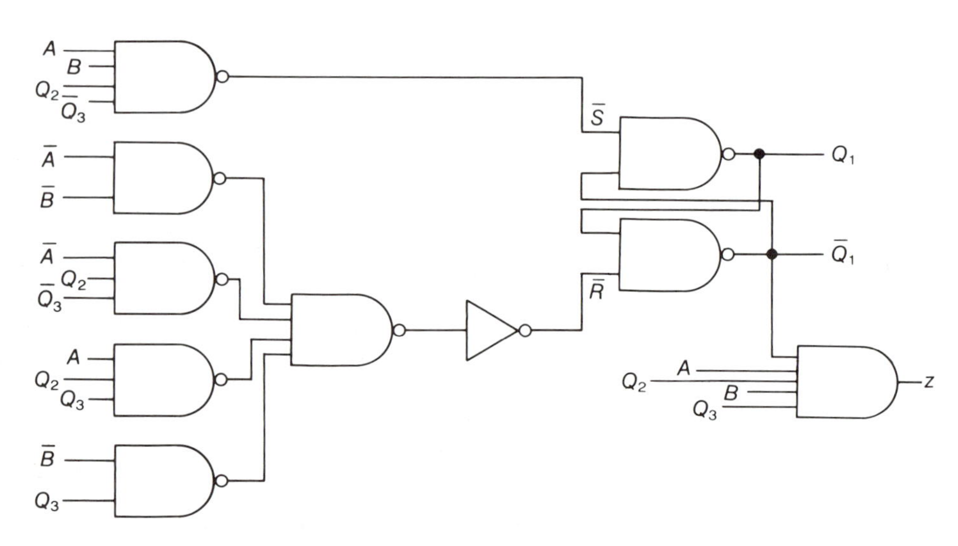

The technique used to obtain a race-free fundamental-mode machine from a merged flow table may be summarized as follows:

Step 1. The state assignments are made so that as many transitions as possible are made between adjacent rows.

Step 2. The remaining transitions are satisfied by means of cycles if sufficient cycle paths are available. Otherwise, additional states are created even at the cost of additional FFs. The cycles are then implemented to satisfy the remaining transitions.

Making the state assignments is a critical part of the fundamental-mode design process. Several different examples will now be given.

EXAMPLE 9.4

Make a race-free state assignment to the merged flow table of Figure 9.24.

SOLUTION

FIGURE 9.24

x_1x_2			
00	01	11	10
(1)	2	4	(5)
3	(2)	(7)	(8)
(3)	(6)	(4)	5

If the merged flow table rows are labeled, respectively, a, b, and c, the transitions in question are

$a \longrightarrow b$

$a \longrightarrow c$

$b \longrightarrow c$

$c \longrightarrow a$

In this example no matter how the assignments are made, a transition is required that results in a race situation. Since there are only three states, the additional state can be used without adding an additional FF. Let us make the assignments arbitrarily: $a = 00$, $b = 01$, $c = 10$, and $d = 11$. This assignment scheme takes care of three of the four transitions. The remaining transition, $b \longrightarrow c$, requires both of its FFs to change. However, this may now be handled by means of the cycle $b \longrightarrow d \longrightarrow c$, which takes the system through the unused state, d. Consequently, all of the race conditions are avoided. The flow table with this assignment may now be obtained as shown in Figure 9.25.

FIGURE 9.25

x_1x_2			
00	01	11	10
(1)	2	4	(5)
3 ↓	(2)	(7)	(8)
3 ↓			
(3)	(6)	(4)	5

x_1x_2			
00	01	11	10
00	01	10	00
11	01	01	01
10	—	—	—
10	10	10	00

EXAMPLE 9.5

Obtain a race-free assignment for the merged flow table of Figure 9.26, using as few additional states as possible.

SOLUTION

FIGURE 9.26

x_1x_2			
00	01	11	10
(1)	(2)	3	4
7	(6)	(3)	5
(7)	6	(8)	(4)
7	2	3	(5)

The stable states of rows 1 through 4 are denoted, respectively, by a through d. The necessary transitions of the system are obtained as follows:

$$\begin{array}{llll} a \rightarrow b & b \rightarrow c & c \rightarrow b & d \rightarrow a \\ a \rightarrow c & b \rightarrow d & d \rightarrow b & d \rightarrow c \end{array}$$

A critical transition diagram may now be drawn as shown in Figure 9.27. From the necessary transition list, note that b occurs in the most transitions (five times). One way to solve this problem is to create cycles between a and c, between c and d, and between a and d. This choice does not involve state b at all. The choice leads to the introduction of three cycle states, a', c', and d', as shown in Figure 9.28. These cycle states require the addition of another FF. The primitive flow table with the added cycle states is obtained as shown in Figure 9.29.

An adjacency map is now prepared for satisfying the adjacencies, (ab), (bc), (bd), (da'), $(a'c)$, (ad'), $(d'c)$, (ac'), and $(c'd)$, of the states as shown in Figure 9.30. Note that this is only one of the ways the adjacency map could be completed. The map is filled by trial and error. Using the assignments from the adjacency map of Figure 9.30, the transition table may now be

FIGURE 9.27

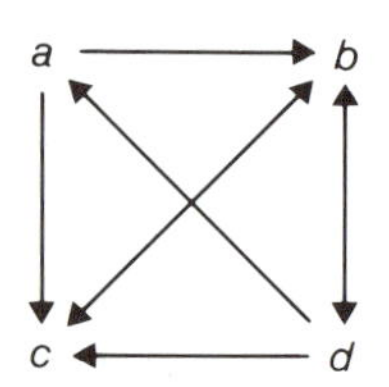

FIGURE 9.28

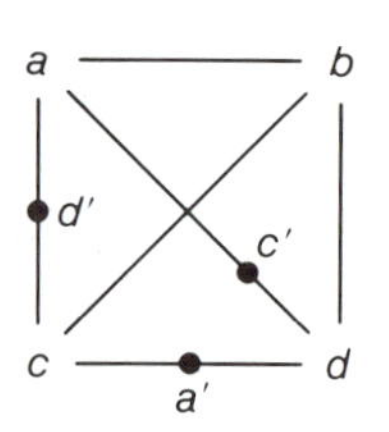

FIGURE 9.29

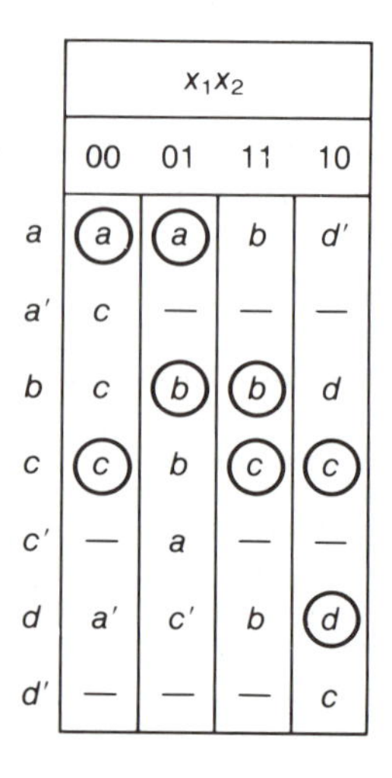

	x_1x_2			
	00	01	11	10
a	(a)	(a)	b	d'
a'	c	—	—	—
b	c	(b)	(b)	d
c	(c)	b	(c)	(c)
c'	—	a	—	—
d	a'	c'	b	(d)
d'	—	—	—	c

FIGURE 9.30

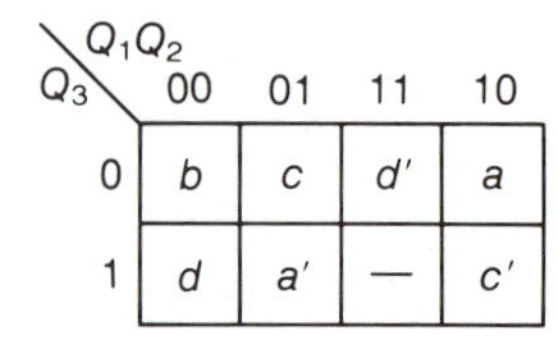

$Q_3 \backslash Q_1Q_2$	00	01	11	10
0	b	c	d'	a
1	d	a'	—	c'

obtained. The table with rows rearranged is obtained as shown in Figure 9.31.

FIGURE 9.31

$Q_1Q_2Q_3$ \ x_1x_2		00	01	11	10
b	000	010	000	000	001
d	001	011	101	000	001
a′	011	010	---	---	---
c	010	010	000	010	010
d′	110	---	---	---	010
—	111	---	---	---	---
c′	101	---	100	---	---
a	100	100	100	000	110

In the last two examples the race problem was avoided either by making use of the available unused state (Example 9.4) or by introducing an additional FF (Example 9.5). The critical transitions were diverted by forcing the system to go through one or several cycle states. However, there is another valid method that could be used as well for obtaining a race-free circuit. This technique involves replication of states, whereby each of the states is given several binary codes so that it is a neighbor of all other states. This replication method usually is expensive because it requires additional FFs, but it is more straightforward than other methods. What this procedure requires is to increase the number of FFs such that the available states would be enough to replicate each of the original states. The number of replications and the way these states are assigned binary values are critical. The assignments should be such that each of the states becomes a neighbor of each other state at least at one location.

EXAMPLE 9.6

Repeat Example 9.5 using replication of states.

SOLUTION

There could be several ways to accomplish replication of states. One of the schemes is shown in Figure 9.32 where each of the states is a neighbor of the others. The scheme requires three FFs. Note the relationships of the states. For example, state a is a neighbor of b since $Q_1Q_2Q_3 = 000$ and $Q_1Q_2Q_3 = 010$ are adjacent; it is also a neighbor of c since $Q_1Q_2Q_3 = 001$

FIGURE 9.32

Q_3 \ Q_1Q_2	00	01	11	10
0	a	b	b	d
1	a	c	c	d

FIGURE 9.33

	$Q_1Q_2Q_3$ \ x_1x_2	00	01	11	10
a	000	000	000	010	001
a	001	001	001	000	011
b	010	011	010	010	110
b	110	111	110	110	100
c	011	011	010	011	011
c	111	111	110	111	111
d	100	101	000	110	100
d	101	111	001	100	101

and $Q_1Q_2Q_3 = 011$ are adjacent; and, finally, it is also a neighbor of d since $Q_1Q_2Q_3 = 000$ and $Q_1Q_2Q_3 = 100$ are adjacent. The same argument may be repeated for the remaining states.

Using these state assignments, the excitation table can be completed as shown in Figure 9.33. Consider the $a \rightarrow b$ transition. The state a is represented by both $Q_1Q_2Q_3 = 000$ and $Q_1Q_2Q_3 = 001$. Consequently, when moving from $Q_1Q_2Q_3 = 000$ to state b, the system would move through the transitional state $Q_1Q_2Q_3 = 010$. However, when moving from $Q_1Q_2Q_3 = 001$ to state b, the best cycle path to follow would be $001 \rightarrow 000 \rightarrow 010$. The system at first moves internally from 001 to 000 (during all of this time the system resides in state a) and then moves on to 010. To accomplish both of these transitions, 010 and 000 are entered under column 11, next to rows 000 and 001, respectively. Let us consider one more case: the $a \rightarrow c$ transition. If the system begins from 001, it could go through the transitional state 011. And if it starts from 000, the system goes through the cycle $000 \rightarrow 001 \rightarrow 011$. Consequently, 001 and 011 are entered under column 10, next to rows 000 and 001, respectively. Similar reasonings were used in the case of other states in completing the excitation table of Figure 9.33.

The replication technique could be used to accommodate any number of states. As the number of states increases, the number of FFs necessary to perform replication also increases. For example, the two adjacency maps in Figures 9.34[*a*–*b*] can be used for handling the requirements of six and eight states, respectively.

FIGURE 9.34 **Replicating Adjacency Maps: [*a*] for Six States and [*b*] for Eight States.**

Q_3Q_4 \ Q_1Q_2	00	01	11	10
00	a	a	c	d
01	b	b	c	d
11	e	f	e	e
10	e	f	f	f

[*a*]

Q_4Q_5 \ $Q_1Q_2Q_3$	000	001	011	010	110	111	101	100
00	a	b	b	b	c	d	e	e
01	a	b	a	a	c	d	f	f
11	g	g	h	g	d	d	f	e
10	h	h	h	g	c	c	f	e

[*b*]

Comparison of the techniques used in Examples 9.4, 9.5, and 9.6 demonstrates that the nonreplicating state assignments would result in a simpler circuit. In the nonreplicating cases there is the possiblility of having many don't-cares in the corresponding transition table. This configuration results in simpler excitation equations. The designer must decide whether the time spent in the more difficult process is justified by the savings in circuit components.

9.5 Fundamental-Mode Output Maps

The previous sections dealt with the steps necessary for obtaining reliable and efficient excitation equations in a fundamental-mode circuit. Only output equations remain to be explored in more detail. The output maps for the fundamental-mode circuits are constructed according to the following procedures:

1. The output value for each of the merged stable states is the same as that shown in the primitive flow table.
2. The output values corresponding to the transitional states are determined by the outputs of the source and destination stable states. The four possibilities are shown in Figure 9.35.

FIGURE 9.35 **Fundamental-Mode Output.**

Case	Output Value of		Output Value of Transitional State
	Source	Destination	
I	0	0	0
II	0	1	—
III	1	0	—
IV	1	1	1

If the outputs in both source and destination stable states are 0 (case I), 0 must be chosen for the output of the transitional state to avoid a momentary output of 1 in the transitional state. The same reasoning gives a value of 1 for the output of the transitional state in case IV. In case II there will be no unwanted momentary output pulse if a 0 or a 1 is assumed for the transitional state. The only difference is that with a 1 in the transitional state, the output turns on sooner than with a 0 in the transitional state. The circuit requirements may dictate that the output should turn on as soon as possible, in which case a 1 should be placed in the transitional state. The argument for case III is the same as that for case II.

If more than one stable state passes through a transitional state, then the output value must be chosen to satisfy each of the transitions to eliminate the possiblity of a *glitch* (momentary 1) or *drop-out* (momentary 0). The next few examples illustrate some of the ideas discussed in this and the previous sections.

EXAMPLE 9.7

Obtain the output equations corresponding to the merged flow table and the respective stable state outputs shown in Figure 9.36.

SOLUTION

FIGURE 9.36

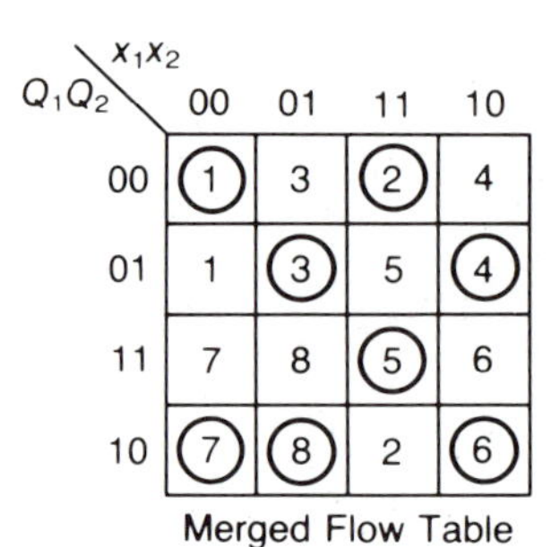

Merged Flow Table

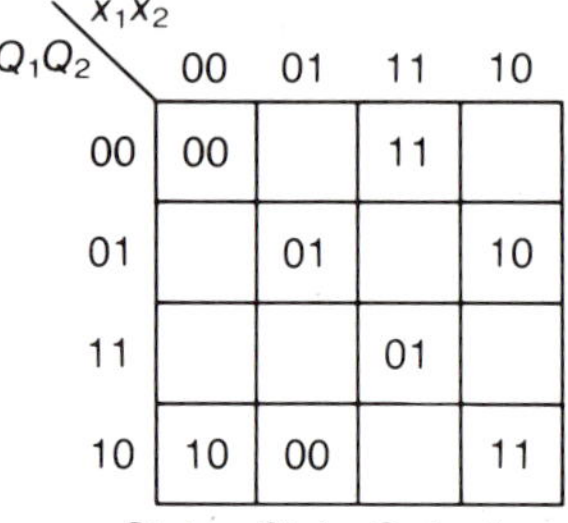

Stable State Outputs z_1z_2

The resulting output map, as shown in Figure 9.37, is obtained following the rules discussed previously. The output of the transitional states are determined and then entered in the output map of Figure 9.37. Consider the transitional state at the intersection of column 01 and row 00. This is encountered when moving from either ① or ② to ③. The z_1z_2 values of ①, ②, and ③ are, respectively, 00, 11, and 01. The value of z_1 is changing either from 0 to 0 or from 1 to 0. In the first case, z_1 for the transitional state should be 0, and in the second case it should be —. Consequently, z_1 is set at 0 as the best compromise. Similarly, z_2 is determined to be 1. Finally, 01 is entered at the corresponding location of the output map. This line of reasoning is carried out until the map of Figure 9.37 is completed. The output equations are then determined by the usual method as

$$z_1 = x_1\overline{Q}_2 + x_1\bar{x}_2 + \bar{x}_2Q_1$$
$$z_2 = x_1Q_1 + x_2\overline{Q}_1 + x_1x_2$$

FIGURE 9.37

Q_1Q_2 \ x_1x_2	00	01	11	10
00	00	01	11	10
01	00	01	01	10
11	--	0—	01	—1
10	10	00	11	11

z_1z_2

EXAMPLE 9.8

Obtain the primitive flow for a negative edge-triggered T FF. Consider T as its regular input, C as its clock input, and Q as its output.

SOLUTION

We know of the control conditions of a T FF from Figure 6.39. The primitive flow table can be obtained, therefore, as shown in Figure 9.38. Note that the output Q changes only when the inputs, CT, change from 11 to 01. This characteristic corresponds to a negative edge-triggered FF.

FIGURE 9.38

CT				
00	01	11	10	Q
(1)	3	—	2	0
1	—	4	(2)	0
1	(3)	4	—	0
—	5	(4)	2	0
6	(5)	7	—	1
(6)	5	—	8	1
—	3	(7)	8	1
6	—	7	(8)	1

EXAMPLE 9.9

Design a circuit that outputs a short pulse each time an input variable x makes a $0 \rightarrow 1$ to $1 \rightarrow 0$ transition.

SOLUTION

In the solution we will make use of the time the circuit spends in the transitional states. The ouput will be 1 during transitions. The length of the output pulse will be controlled solely by cycles. The state table for the single-input problem is obtained as shown in Figure 9.39. The transition table of Figure 9.40 is then obtained by making proper state assignments. The resultant SR equations are

$$S_1 = xQ_2$$
$$R_1 = \bar{x}Q_2$$
$$S_2 = \bar{R}_2 = Q_1 \oplus x$$

FIGURE 9.39

x		
0	1	z
(1)	2	0
1	2	1
1	2	1
1	(2)	0

From the output map, as shown in Figure 9.41, the resultant output equation is obtained as follows:

$$z = Q_2 + Q_1 \oplus x$$

The output z will be high for three transition times. The pulse width could

FIGURE 9.40

Q_1Q_2 \ x	0	1
00	00	01
01	00	11
11	01	10
10	11	10

FIGURE 9.41

Q_1Q_2 \ x	0	1
00	0	1
01	1	1
11	1	1
10	1	0

be shortened to two transition times by letting $z = Q_2$. The pulse could be lengthened by adding another FF and having the circuit cycle move through the additional rows.

9.6 Summary

In this chapter the concepts of races and cycles were developed along with the working principles and design techniques of a fundamental-mode circuit. The fundamental mode is a very powerful design tool. A designer would probably use clocked and pulsed sequential design almost exclusively, but in extremely difficult timing situations the technique presented in this chapter would prove to be the most valuable.

Problems

1. Obtain S_2, S_3, R_2, and R_3 equations for the problem of Figure 9.21[*b*]. Also obtain the Q_2 and Q_3 excitation circuits.
2. For the merged table of Figure 9.P1, complete the state assignments for a race-free situation. Find the Boolean expression and draw the circuit.

FIGURE 9.P1

State Assignments (Q_1Q_2)	x = 0	x = 1	z
	(1)	2	0
00	(3)	4	1
01	3	(2)	0
	1	(4)	1

3. Find the merger diagram for the primitive flow table of Figure 9.P2 and rewrite the merged flow table.

FIGURE 9.P2

x_1x_2				
00	01	11	10	z
(1)	6	5	2	0
1	4	3	(2)	0
8	4	(3)	7	1
8	(4)	3	7	1
1	6	(5)	2	0
1	(6)	5	2	0
8	6	5	(7)	1
(8)	4	5	7	1

4. Determine the SR equations and the output equation for the circuit described by the information given in Figure 9.P3. Draw the circuit for a race-free system. The optimum circuit will be determined by the sum of gates and inputs. The output is z.

FIGURE 9.P3

State Assignments (Q_1Q_2)	AB 00	01	11	10
00	(1)	(7)	8	2
	(3)	(5)	6	4
01	3	7	(8)	(2)
	3	7	(6)	(4)

Merged Flow Table

AB 00	01	11	10
1	0		
0	1		
		1	1
		0	1

z

5. Complete the output table and write the equation for the output, using Figure 9.P4. Use don't-cares wherever necessary.

FIGURE 9.P4

x_1x_2				
00	01	11	10	z
(1)	2	—	3	0
4	(2)	5	—	1
1	—	6	(3)	0
(4)	7	—	8	1
—	2	(5)	3	1
—	7	(6)	8	0
1	(7)	6	—	0
4	—	5	(8)	1

Flow Table

Q_1Q_2	x_1x_2 00	01	11	10
00	(1)	2	(5)	3
01	1	(7)	6	(3)
11	(4)	7	(6)	8
10	4	(2)	5	(8)

Merged Flow Table

6. Design a fundamental-mode circuit where the output, z, equals 0 when the two inputs, x_1 and x_2, are equal. $z = 1$ results when $x_1 = 0$ and x_2 changes from 0 to 1, and when $x_1 = 1$ and x_2 changes from 1 to 0. No other input change causes any output change.

7. Find a two-level NAND realization of the sequential circuit with two inputs, A and B, and one output, z, under the following conditions:
   a. $z = 0$ when $B = 1$,
   b. z changes to 1 when $B = 0$ and A changes from 0 to 1,
   c. z remains at 1 until B goes to 1 and forces z back to 0.
8. Repeat Problem 7 with a two-level NOR format.
9. Solve the cycle problem for the fundamental-mode circuit of Figure 9.P5 using as few additional states as possible.

FIGURE 9.P5

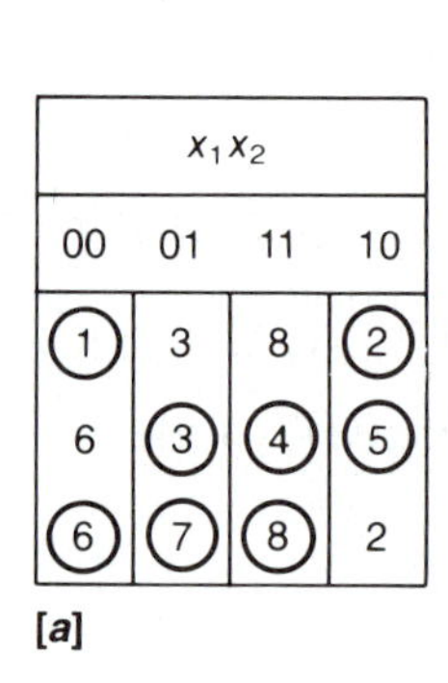

x_1x_2			
00	01	11	10
(1)	3	8	(2)
6	(3)	(4)	(5)
(6)	(7)	(8)	2

[*a*]

x_1x_2			
00	01	11	10
(1)	2	8	6
5	(2)	(3)	(4)
(5)	7	3	(6)
1	(7)	(8)	4

[*b*]

x_1x_2			
00	10	11	10
(1)	2	6	10
1	(2)	(3)	(4)
1	(5)	(6)	(7)
1	(8)	(9)	(10)

[*c*]

10. Solve Problem 9 using replication of states.
11. Obtain the fundamental-mode circuit for the merged flow table of Figure 9.P6. For the cycle use (a) few additional states and (b) replication of states.

FIGURE 9.P6

x_1x_2			
00	01	11	10
(1),0	2	3	(4),1
(5),1	6	(3),0	7
5	(2),1	(8),1	7
1	(6),0	3	(7),0

12. Obtain the circuit for the problem of Example 9.8.
13. Design the circuit for a negative edge-triggered clock D FF using the fundamental-mode principles.

Suggested Readings

Friedman, A. D., and Menon, P. R. "Synthesis of asynchronous sequential circuits with multiple-input changes." *IEEE Trans. Comp.* vol. C-17 (1968): 559.

Kuhl, J. G., and Reddy, S. M. "Multicode single transition time state assignment for asynchronous sequential machines." *IEEE Trans. Comp.* vol. C-27 (1978): 927.

Nagle, H. T., Jr.; Carroll, B. D.; and Irwin, J. D. *An Introduction to Computer Logic*. Englewood Cliffs, N.J.: Prentice-Hall, 1975.

Nanya, T., and Tohma, Y. "Universal multicode STT state assignments for asynchronous sequential machines." *IEEE Trans. Comp.* vol. C-28 (1979): 811.

Nordmann, B. J., and McCormick, B. H. "Modular asynchronous control design." *IEEE Trans. Comp.* vol. C-26 (1977): 196.

Rey, C. A., and Vaucher, J. "Self-synchronized asynchronous sequential machines." *IEEE Trans. Comp.* vol. C-23 (1974): 278.

Rhyne, V. T. *Fundamentals of Digital Systems Design*. Englewood Cliffs, N.J.: Prentice-Hall, 1973.

Sholl, H. A., and Yang, S. C. "Design of asynchronous sequential network using READ-ONLY memories." *IEEE Trans. Comp.* vol. C-24 (1975): 195.

Introduction to Counters, Registers, and Register Transfer Language

CHAPTER TEN

10.1 Introduction

With the study of flip-flops and sequential circuits behind us, the study of counters and registers will be a natural and straightforward extension. Counters and registers are essential to the design of advanced circuits found in digital computers. *Counters* are employed to keep track of a sequence of events, and *registers* are used to store and manipulate data that contribute to all or many of these events. In other words, most of the robust digital systems would have two major functional units: a unit where the manipulations are conducted and a unit that is used for regulating the events of the first unit. Registers and associated logic subunits help to make the first unit, and counters could be used for running the second unit. Therefore, without an understanding of flip-flops, counters, and registers, design of digital systems would be impossible.

Counters are particularly common in the control and arithmetic units of processors, where they are used to keep track of the sequence of instructions in a program, to distribute the sequence of timing signals, for frequency division for causing time delays, for counting, and a host of other similar operations. Counters may count in binary or in nonbinary fashion. They are commercially available in a large variety of medium-scale integrated devices. The basic operational characteristic of a counter is sequential; for every present state there is a well-defined next state. The design of a counter involves designing combinational logic that decodes the present state and enables entry into the next state of the counting sequence. Counters are generally classified into two groups: synchronous and asynchronous. A *synchronous counter* has all FFs change state synchronously with the clock input whether a periodic clock or an aperiodic pulse occurs. An *asynchronous* (or *ripple*) *counter* is made up of FFs that do not change simultaneously with the clock input.

Another application for FFs is for storing bits of information. When FFs are configured to store multi-bit information, they are referred to as *registers*. Registers are classified according to the way information bits are stored and retrieved. If data are stored and removed at either end of a multi-bit register, one bit at a time, the register is referred to as a *serial* or *shift register*. However, if all bits of the word are stored or retrieved simultaneously, the register is referred to as a *parallel register*.

Another area that needs to be investigated is how a digital system, however small it may be, is built, integrated, and run. The control unit of a system can be designed using the methods that were developed in the last three chapters. However, these sequential design techniques are inadequate for the representation of subsystems that are used strictly in the manipulation of data. The tool that has been found useful in accomplishing this representation is known as the *register transfer language* (RTL). RTL helps to translate a specification mechanically into its hardware realization.

The beginning of this chapter is devoted to the development and study of various counters and registers. This discussion is then gradually expanded to include the basics of RTL. Finally, RTL is used in the design of complete digital systems. After studying this chapter, you should be able to:

- Design and analyze both synchronous and asynchronous counters;
- Design and analyze serial, parallel, and hybrid registers;
- Design and analyze systems that have counters and registers;
- Translate complex operations into equivalent RTL sequences;
- Use RTL in the design of data and control units.

10.2 Synchronous Binary Counters

Synchronous counters are distinguished from asynchronous (or ripple) counters in that the clock pulses in synchronous counters initiate changes in the FFs used in the counter. The simplest possible counter is a single-bit counter that alternates between two states, 0 and 1. A toggle FF using a single JK FF, with both inputs tied to 1 ($J = K = 1$), will function as a single-bit counter alternating between the two states with the occurrence of each clock. The output of the FF has a frequency that is one-half the clock frequency.

A two-bit binary up-counter with four states was already designed in Example 7.5. Such a counter consists of two JK FFs whose states Q_2Q_1 could be assumed to move in sequence through 00, 01, 10, 11, 00, 01, and so on. The corresponding J and K inputs of the two FFs are given by

$$J_1 = K_1 = 1$$

$$J_2 = K_2 = Q_1$$

Note that these equations are slightly different from those given in Example 7.5. The positions of the FFs have been reversed and output equations are abandoned altogether. We can take the outputs directly from the FFs.

We shall now attempt to synthesize a three-bit binary up-counter of the nonterminal type. With every clock input the counter moves to the next higher state. Consider the FF outputs to correspond to the present state. The first step in the design sequence of a sequential circuit is to begin with a state diagram and a state table followed by the assignment of states. The state diagram, the state table, and excitation maps of a three-bit counter are shown in Figure 10.1.

The excitation maps of Figure 10.1[*c*] may be used to obtain the *JK* equations as follows:

$$J_1 = K_1 = 1$$
$$J_2 = K_2 = Q_1$$
$$J_3 = K_3 = Q_1Q_2$$

The resulting circuit diagram is shown in Figure 10.2.

We learned in Chapter 5 that many of the complex combinational designs are realized using heuristic techniques. This simplification is more true when modularity is evident in the system. Quite similarly, many sequential design problems can be accomplished without going beyond state tables. A close examination of the state table of Figure 10.1[*b*] reveals the presence of a certain degree of regularity. Note that Q_1 changes with every clock pulse and, in general, Q_i changes state if all less significant bits are 1. Similar conclusions about the regularity of counter design can be made by inspecting the respective *JK* equations of the three counters that we have considered thus far. An inspection of the *J* and *K* equations leads us to the conclusion that, based on a regular pattern, these equations can be extended to *J* and *K* equations for the *n*th bit of a multi-bit up-counter as follows:

$$J_n = Q_{n-1}Q_{n-2} \cdots Q_3Q_2Q_1 = Q_{n-1}J_{n-1}$$
$$K_n = Q_{n-1}Q_{n-2} \cdots Q_3Q_2Q_1 = Q_{n-1}K_{n-1}$$

There are two different ways of connecting the inputs to successive FFs based on the two forms of the equation for the *n*th term. Both are illustrated in Figure 10.3. Part [*a*] of the figure shows a configuration where the FF outputs are combined in parallel. The propagation delay at the input of each FF is the same for all stages. However, the fan-in to the AND gate and the fan-out of each FF increase as the number of counter stages is increased. Figure 10.3[*b*] shows an equivalent configuration using the second form of the J_n and K_n equations. The fan-in of the AND gates is always two; how-

FIGURE 10.1 **Three-Bit Binary Up-Counter: [*a*] State Diagram, [*b*] State Table, and [*c*] Excitation Maps.**

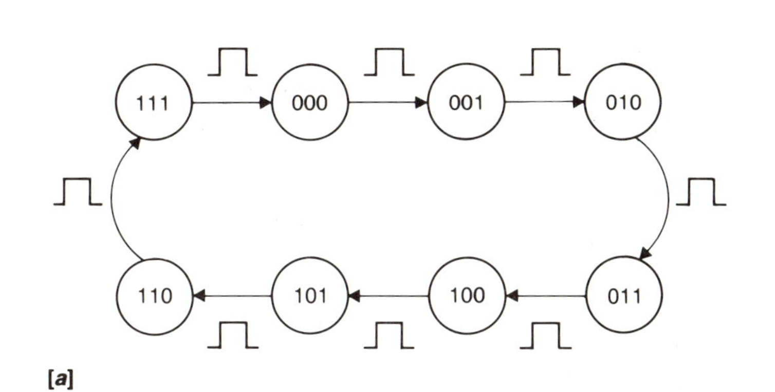

[*a*]

PS	
$Q_3Q_2Q_1$	NS
000	001
001	010
010	011
011	100
100	101
101	110
110	111
111	000

[*b*]

J_1 (Q_3 rows; Q_2 over the right two columns)

1	—	—	1
1	—	—	1

J_2

0	1	—	—
0	1	—	—

J_3

0	0	1	0
—	—	—	—

K_1 (Q_3 rows; Q_1 under the middle two columns)

—	1	1	—
—	1	1	—

K_2

—	—	1	0
—	—	1	0

K_3

—	—	—	—
0	0	1	0

[*c*]

FIGURE 10.2 **Logic Circuit of a Three-Bit Up-Counter.**

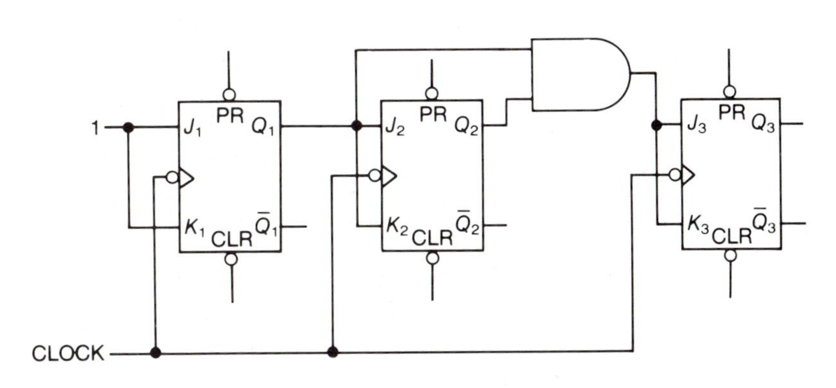

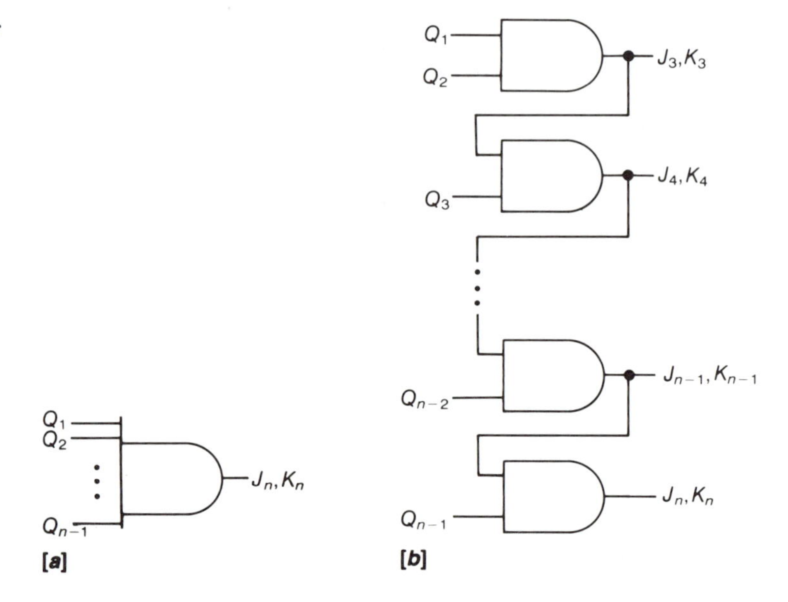

FIGURE 10.3 **Configurations for the *J* and *K* Inputs of an *n*-Bit Counter: [*a*] Parallel and [*b*] Serial.**

ever, the propagation delay to the *n*th FF increases as the number of counter stages is increased. Note that the first method allows faster clocking and counting but results in fan-in and fan-out problems for large counters.

Figure 10.4 shows a synchronous *n*-bit binary up-counter using *JK* FFs and two-input AND gates. Two control signals are added in this circuit, CLEAR and COUNT. A high on the CLEAR input resets the counter; the counter remains reset until the CLEAR signal is withdrawn. The COUNT signal is used to disable the clock pulses. The designer can use this to block the clock input and hold any nonzero count state. If the preset (PR) inputs are effectively

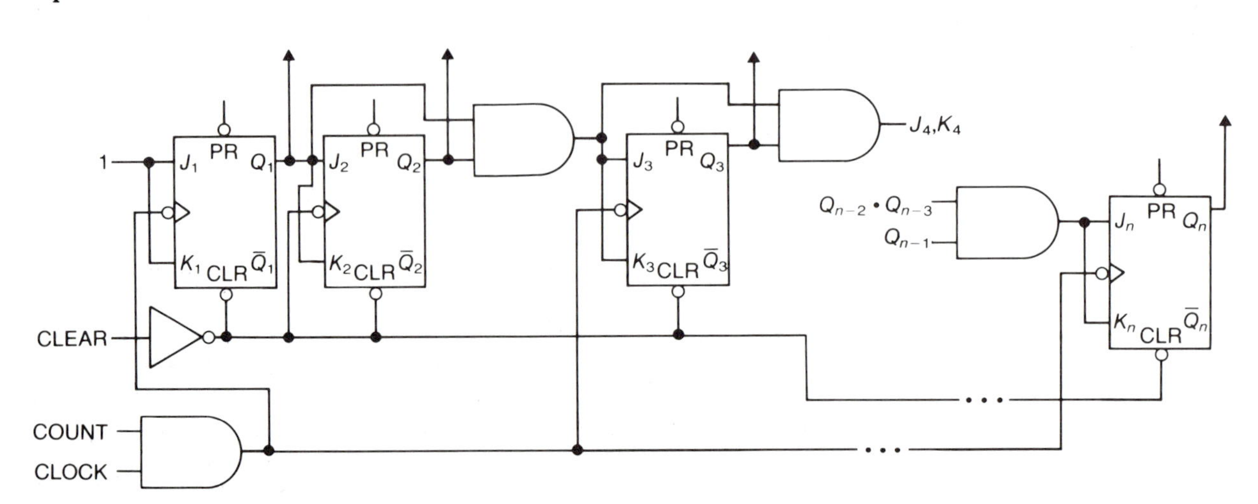

FIGURE 10.4 **n-Bit Synchronous Up-Counter.**

used, one may even be able to set the counter to its maximum count state.

There are many occasions in a digital system when a down-counter is required. A binary number is set into the counter that then counts toward zero as the clock pluses occur. These counters can be designed in the same way as up-counters. The equations for the down-counters are also seen to have regularities. The J and K equations for an n-bit binary down-counter are obtained as follows:

$$J_1 = K_1 = 1$$
$$J_2 = K_2 = \overline{Q}_1$$
$$J_3 = K_3 = \overline{Q}_1\overline{Q}_2 = J_2\overline{Q}_2 = K_2\overline{Q}_2$$
$$\cdots\cdots\cdots\cdots\cdots\cdots\cdots\cdots\cdots\cdots\cdots\cdots$$
$$J_n = K_n = \overline{Q}_{n-l}\,\overline{Q}_{n-2}\ldots\overline{Q}_3\overline{Q}_2\overline{Q}_1 = J_{n-1}\overline{Q}_{n-1} = K_{n-1}\overline{Q}_{n-1}$$

By comparing these equations with those of the up-counters, we note that the resulting circuit is similar in nature. The J_n and K_n inputs are taken from the two-input AND gate, whose inputs are $\overline{Q}_{n-1}$ and K_{n-1}. Note in the case of the up-counters the corresponding AND inputs were Q_{n-1} and J_{n-1}.

For some of the applications a counter may be required to count up or down. One such application could be a counter device that keeps track of total cars inside a parking garage. As each car enters the garage, the counter counts up; as each car leaves, the counter counts down. This is a more complex design than that of all counters considered so far since it requires at least one control signal, E, to determine the direction of the count. We may assume that when $E = 1$ the circuit counts up, and when $E = 0$ the circuit counts down. One way to synthesize such a circuit would be to follow the standard steps for the design of sequential circuits (see Chapter 7, Problem 6, for example). However, we can combine the equations for up- and down-counters to derive the respective J and K equations of an n-bit, up-down counter as follows:

$$J_1 = K_1 = 1$$
$$J_2 = K_2 = EQ_1 + \overline{E}\overline{Q}_1$$
$$J_3 = K_3 = EQ_1Q_2 + \overline{E}\overline{Q}_1\overline{Q}_2$$
$$\cdots\cdots\cdots\cdots\cdots\cdots\cdots\cdots\cdots\cdots$$
$$J_n = K_n = EQ_{n-1}Q_{n-2}\ldots Q_2Q_1 + \overline{E}\overline{Q}_{n-1}\overline{Q}_{n-2}\ldots\overline{Q}_2\overline{Q}_1$$

The excitation function of the up-counter is ANDed with E, and that of the down-counter is ANDed with $\overline{E}$, and, finally, the two corresponding composite functions are ORed to obtain the J and K equations. Consequently, when $E = 1$ these equations reduce to those of an up-counter, and when $E = 0$ the equations reduce to

FIGURE 10.5 **Four-Bit Binary Up-Down Counter.**

those of a down-counter. The implementation of a four-bit, up-down counter is shown in Figure 10.5.

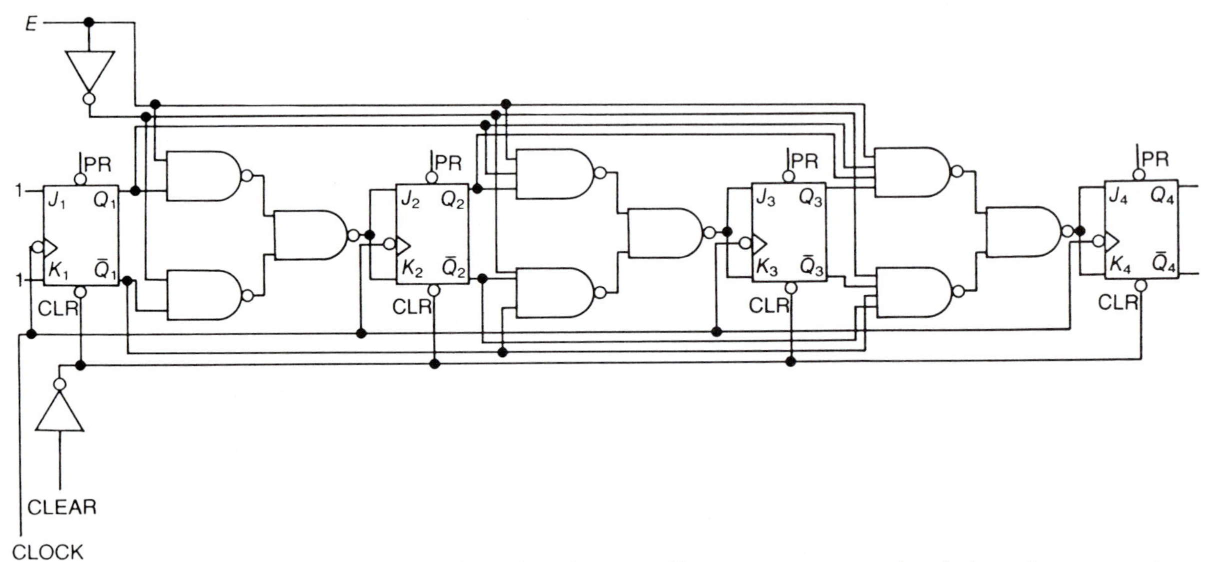

As pointed out earlier, a counter can be designed to count in a nonbinary manner as well. Two examples of nonbinary counters are a BCD decade counter and a Gray code counter. In the former, the counter counts 0000 through 1001 and then resets back to 0000. A four-bit Gray code counter, on the other hand, counts 0000, 0001, 0011, 0010, 0110, 0111, 0101, 0100, 1100, 1101, 1111, 1110, 1010, 1011, 1001, and 1000 in that order and then resets to 0000 before resuming count-up operation again. Example 10.1 illustrates the design of a BCD decade counter.

EXAMPLE 10.1

Obtain the J and K equations for a BCD up-counter.

SOLUTION

The design steps, followed in the usual sequence, consist of the state diagram (Figure 10.6), the state and transition table (Figure 10.7), and excita-

FIGURE 10.6

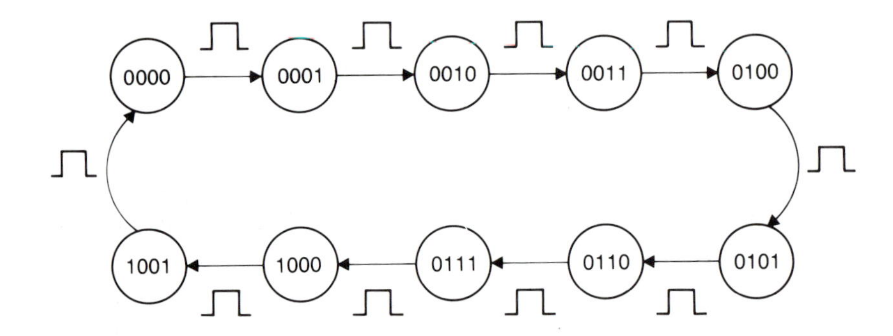

FIGURE 10.7

PS $Q_4Q_3Q_2Q_1$	NS	J_4K_4	J_3K_3	J_2K_2	J_1K_1
0000	0001	0—	0—	0—	1—
0001	0010	0—	0—	1—	—1
0010	0011	0—	0—	—0	1—
0011	0100	0—	1—	—1	—1
0100	0101	0—	—0	0—	1—
0101	0110	0—	—0	1—	—1
0110	0111	0—	—0	—0	1—
0111	1000	1—	—1	—1	—1
1000	1001	—0	0—	0—	1—
1001	0000	—1	0—	0—	—1

tion maps (Figure 10.8). The resulting J and K equations are obtained from Figure 10.8 as follows:

$$J_1 = K_1 = 1$$
$$J_2 = Q_1\overline{Q}_4$$
$$K_2 = Q_1$$
$$J_3 = K_3 = Q_1Q_2$$
$$J_4 = Q_1Q_2Q_3$$
$$K_4 = Q_1$$

FIGURE 10.8

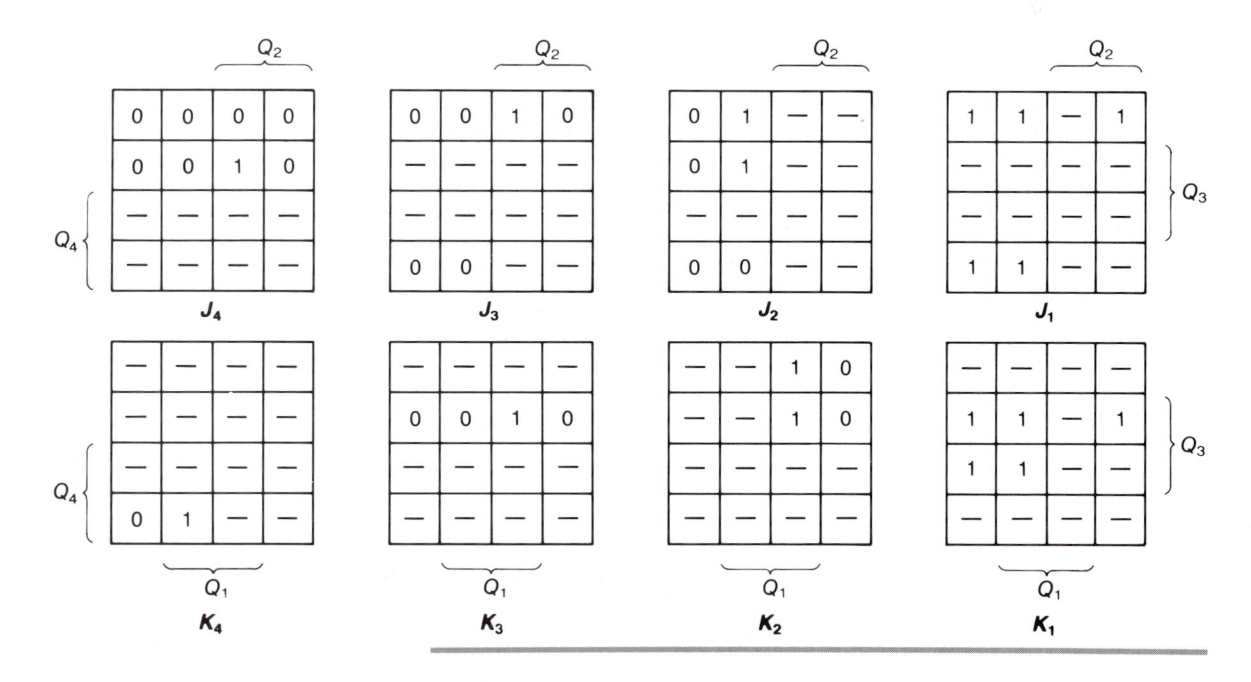

Steps similar to those in Example 10.1 can be used to determine the circuit of a BCD down-counter. Such an exercise gives the J and K equations for a BCD down-counter as follows:

$$J_1 = K_1 = 1$$
$$J_2 = \bar{Q}_1(Q_3 + Q_4)$$
$$K_2 = \bar{Q}_1$$
$$J_3 = \bar{Q}_1 Q_4$$
$$K_3 = \bar{Q}_1 \bar{Q}_2$$
$$J_4 = K_4 = \bar{Q}_1 \bar{Q}_2 \bar{Q}_3$$

The design of the BCD counter considered in Example 10.1 is a classical one. It is not necessary for a designer to go through all of these cumbersome steps if an already working circuit can be modified to suit the requirement of the new design. For example, consider a four-bit binary up-counter, also known as the *modulo-16* (or divide-by-16) counter. A synchronous BCD counter works exactly as the already designed four-bit, modulo-16 counter until state 9 ($Q_4Q_3Q_2Q_1 = 1001$) is reached. The four-bit counter advances to 1010, while the BCD counter starts all over at 0000. Assuming that a four-bit counter is to be modified to a BCD counter, it is necessary to reset FFs when they are about ready to go to 1010 otherwise. The modification process requires circuit changes to satisfy the following conditions when the four-bit up-counter reaches count 1001:

Q_1, the LSB, should become a 0,
Q_4, the MSB, should become a 0,
Q_2 should be prevented from becoming a 1.

An inspection of the circuit of Figure 10.4 reveals that the least significant FF is always in a toggle mode, which implies that Q_1 will change to a 0 by itself. Note also that the MSB, Q_4, of the four-bit counter changes only when the count reaches either 0111 or 1111. This change happens because the inputs J_4 and K_4 are both held to a 0 during the other counts. Consequently, it is necessary to supply K_4 with a 1 instead of a 0 when the count reaches 1001. This can be accomplished if Q_1 is fed directly to the K_4 input and K_4 is disconnected from the J_4 input. The next consideration is to prevent Q_2 from switching back to 1 once the count reaches 1001. This action is accomplished by supplying J_2 and K_2 with the ANDed output of Q_1 and $\bar{Q}_4$. This modification does not cause any problem because $\bar{Q}_4$ is a 1 until the count reaches 1000. And once the count reaches 1000, Q_2 need not be turned on in a BCD counter. In summary,

Q_1 should be fed to K_4,
K_4 and J_4 should not be connected,
$Q_1\overline{Q}_4$ should be fed to J_2 and K_2.

Consequently, the BCD counter may be obtained by modifying the modulo-16 counter circuit, as shown in Figure 10.9.

FIGURE 10.9 **Synchronous BCD Up-Counter.**

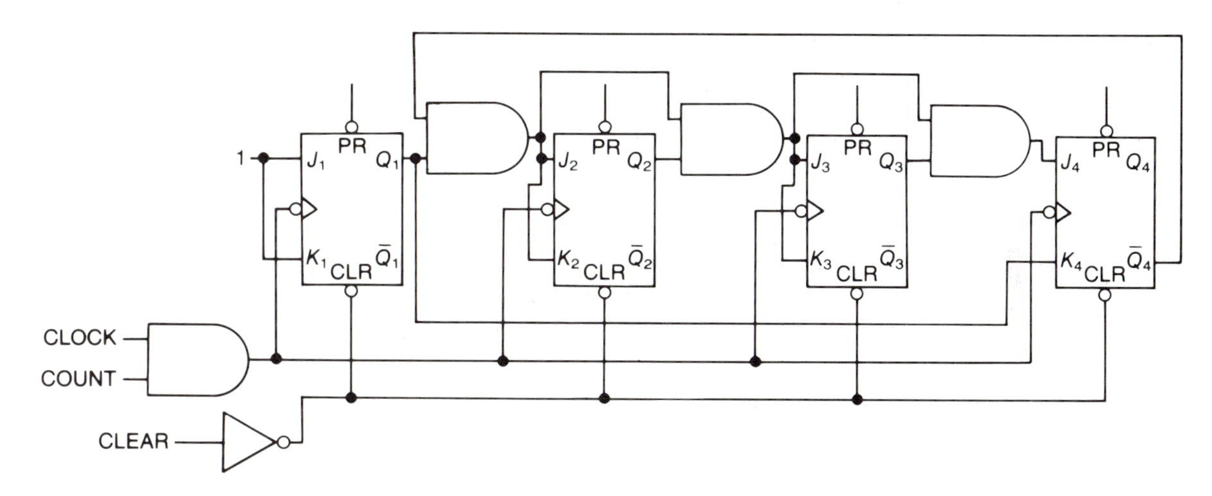

Synchronous circuits requiring irregular count sequences, such as the Gray counter mentioned earlier, are usually designed using the familiar step-by-step technique. A count sequence is said to be *irregular* if it is not magnitude ordered. In general, it is harder to come up with an irregular counter by modifying a regular counter. Examples 10.2 and 10.3 relate to counters with irregular count sequences.

EXAMPLE 10.2

Obtain the state table for the counter shown in Figure 10.10, starting from the count 000. The MSB of the count is at Q_3.

SOLUTION

FIGURE 10.10

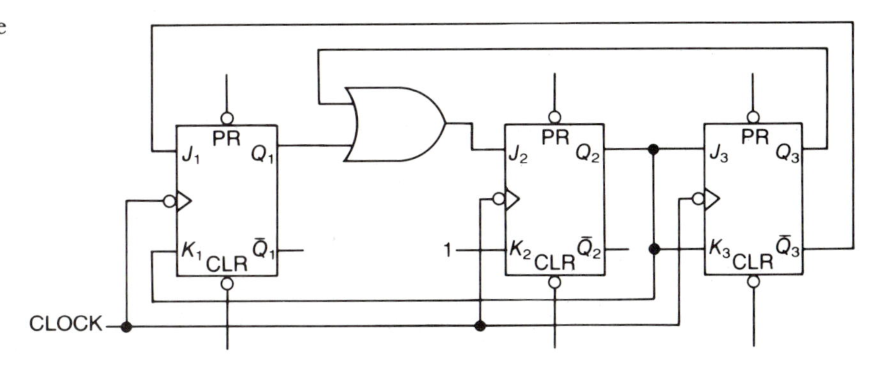

This circuit can be analyzed by assuming a present state and using our knowledge of JK FF operation. Corresponding to each of the present states, the next state is found by determining the corresponding J and K values of each FF. When $JK = 00$, the corresponding Q remains unchanged; when $JK = 01$, Q is reset; when $JK = 10$, Q is set; and when $JK = 11$, Q toggles. The resulting state table is obtained as shown in Figure 10.11. The circuit has an irregular state sequence 0, 1, 3, 4, 6, and repeat. Consequently this circuit is an irregular counter.

FIGURE 10.11

PS	J_3K_3	J_2K_2	J_1K_1	NS
000	00	01	10	001
001	00	11	10	011
011	11	11	11	100
100	00	11	00	110
110	11	11	01	000

EXAMPLE 10.3

Obtain a synchronous counter that produces the count sequence 0, 2, 4, 3, 6, 7, 0, ...

SOLUTION

The state table and the JK excitations are obtained as shown in Figure 10.12. The easiest way to solve this problem would be to consider the present and corresponding next states. The JK excitations necessary for the corresponding transitions are then determined for each of the cases. Note that states 1 and 5 don't occur in the count sequence. Consequently, we assume the corresponding next states and the JK excitations to be don't-cares. This assumption should help in the minimization step. The J and K Boolean equations are obtained using K-maps. They are as follows:

FIGURE 10.12

PS	NS			
$Q_3Q_2Q_1$	$Q_3Q_2Q_1$	J_3K_3	J_2K_2	J_1K_1
000	010	0—	1—	0—
001	——	——	——	——
010	100	1—	—1	0—
011	110	1—	—0	—1
100	011	—1	1—	1—
101	———	——	——	——
110	111	—0	—0	1—
111	000	—1	—1	—1

$$J_1 = Q_3$$
$$K_1 = 1$$
$$J_2 = 1$$
$$K_2 = \overline{Q_1 \oplus Q_3}$$
$$J_3 = Q_2$$
$$K_3 = Q_1 + \overline{Q}_2$$

The resulting counter is obtained as shown in Figure 10.13

FIGURE 10.13

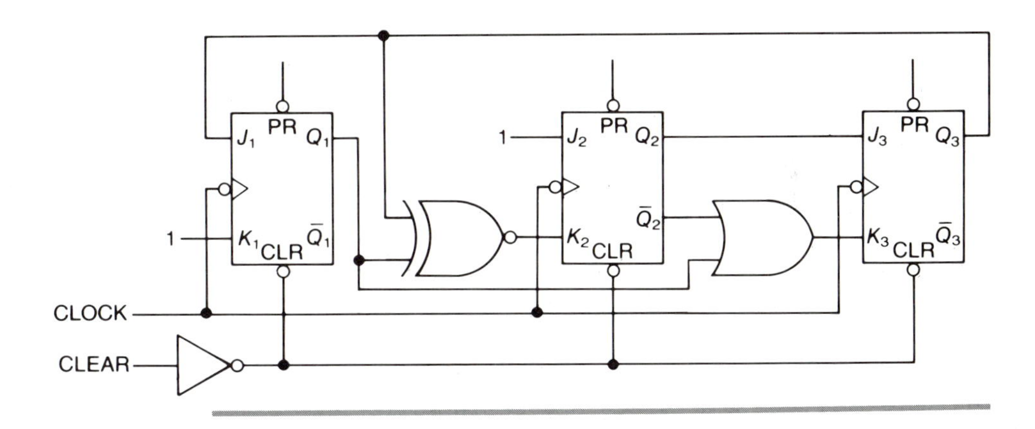

It is appropriate at this point to offer a word of caution. Since there is are various cascadable multi-mode binary and nonbinary counters available, a designer might be tempted to use them in his or her design as a starting chip. Before doing so, the designer must thoroughly examine the circuit specifications and diagrams. If *JK* FFs are used, the familiar 1's-catching problem (see Section 6.5 for details) may crop up here. More specifically, if the clock input is not at the proper level, erroneous operation might result if a particular mode is changed. There may be times, therefore, when a counter may have to be designed from scratch.

10.3 Asynchronous Binary Counters

All of the counters considered thus far employed synchronous circuits, that is, the FF actions were synchronized with the clock pulse. The advantage of a synchronous counter is that all bits of a count change simultaneously except for slight differences in FF delays. If a specific count is being decoded, all bits are available at the same time, eliminating momentary errors at the decode output. We shall now introduce asynchronous counters, also known as ripple counters. The FF clock inputs in a ripple counter are not tied together.

In fact, the clock inputs are cascaded from output to input (almost like the rippling of carries in a ripple adder). They are used to reduce the amount of control logic required to construct a binary counting sequence. Asynchronous counters have limitations but also provide less expensive counter options for those cases where their limitations will not affect the circuit. The AND gates between FFs in the synchronous binary counter design may be eliminated by observing the counter state table. The LSB needs to be changed in every present-state to next-state transition. In all bit locations, Q_i changes each time Q_{i-1} makes a transition from a 1 to a 0.

A negative edge-triggered *JK* FF in a toggle mode changes state each time the signal connected to the clock input makes a $1 \rightarrow 0$ transition. An asynchronous counter using *T*-configured *JK* FFs has its least significant FF activated by the system clock. The $1 \rightarrow 0$ transitions of that FF may be used as the trigger (clock input) signal for the next most significant FF. This process of triggering is continued for as many bits as desired.

The logic circuit of a four-bit ripple counter is shown in Figure 10.14[*a*], where four *T* FFs are cascaded together. The output of

FIGURE 10.14 **Four-Bit Ripple Counter: [*a*] Circuit and [*b*] Timing Diagrams.**

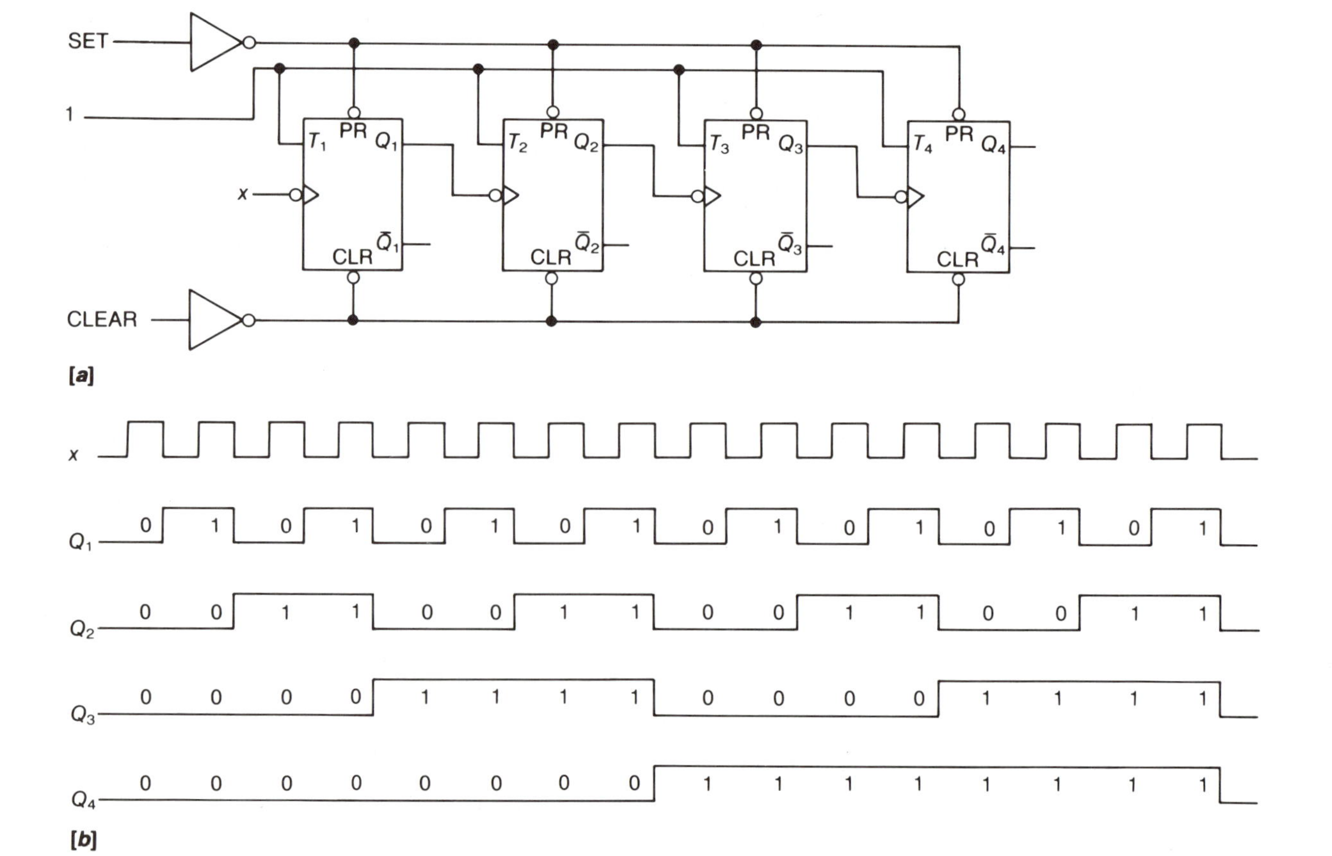

each FF provides the clock signal for the next FF. The timing diagram without delays for the four-bit ripple counter is shown in Figure 10.14[*b*]. Examination of the timing diagrams shows that the frequency of the Q_4 pulse is one-sixteenth of the frequency of the input pulse x. Each stage of the counter divides the frequency of the preceding stage by two.

Counters like the one shown in Figure 10.14[*a*] are simple in concept, but have at least two disadvantages: a forced regular binary sequence and speed. The first disadvantage is not so much of a serious handicap, but the speed is. In reality, the rippling effect through the FFs causes delay between each count that is proportional to the number of FFs in the counting chain. Consider the timing diagram of Figure 10.15 that shows the situation existing in the counter when the count is 1111. Q_1 does not change to 0 coincident with the trailing edge of the sixteenth x-pulse until time t_f, the propagation delay of each FF. The same is generally true for the synchronous counter, but for the asynchronous counter Q_2, Q_3, and Q_4 change at times $2t_f$, $3t_f$, and $4t_f$, respectively, beyond the negative edge of the sixteenth x-pulse. For an n-bit ripple counter to reach a valid count before the next clock pulse, $T > nt_f$, where T is the period of the input pulse. If the final count is all that is of interest, the condition $T > t_f$ is all that must be met. In this situation, changes in the LSB are constantly rippling to higher-order bits. After the last pulse is input, it will be nt_f before the final count can be correctly read. Note that in Figure 10.15 the counter does not go through the transition 1111 → 0000. Instead the counter passes through the state transition sequence 1111 → 1110 → 1100 → 1000 → 0000. These transitions occur in rapid succession but they result in undesired transient conditions that might cause further problems

FIGURE 10.15 **Timing Consequences of a Four-Bit Asynchronous Counter During 1111 → 0000 Transition.**

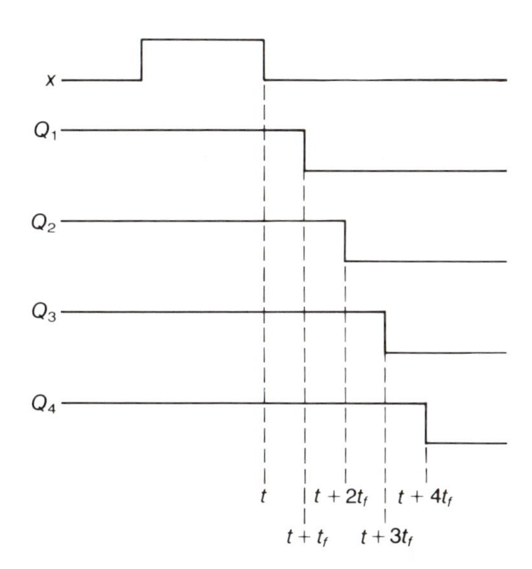

in a circuit that is being driven by such a counter. This must be borne in mind when such counters are used in generating a controller of a complex system.

If the circuit of Figure 10.14[*a*] is modified so that $\overline{Q}_1$, $\overline{Q}_2$, and $\overline{Q}_3$, are used respectively as the clock inputs for the second, third, and fourth T FFs, then the circuit will operate as a four-bit ripple down-counter. The ripple counter of Figure 10.14[*a*] may also be modified to generate a resettable counter. Say, for example, we are interested in generating only BCD counts. Such a counter is shown in Figure 10.16. The circuit operates as a modulo-16 ripple counter, but when the state 1010 is reached the counter resets immediately. This resetting is accomplished by means of a combinational decoding scheme that is tied with the reset inputs of every FF. The counter stays at the count of 1010 for the decoding gate delay plus the reset input-to-output delay.

The ripple counter of Figure 10.16 exhibits other transient

FIGURE 10.16 **Asynchronous BCD Up-Counter: [*a*] Circuit and [*b*] Timing Diagram.**

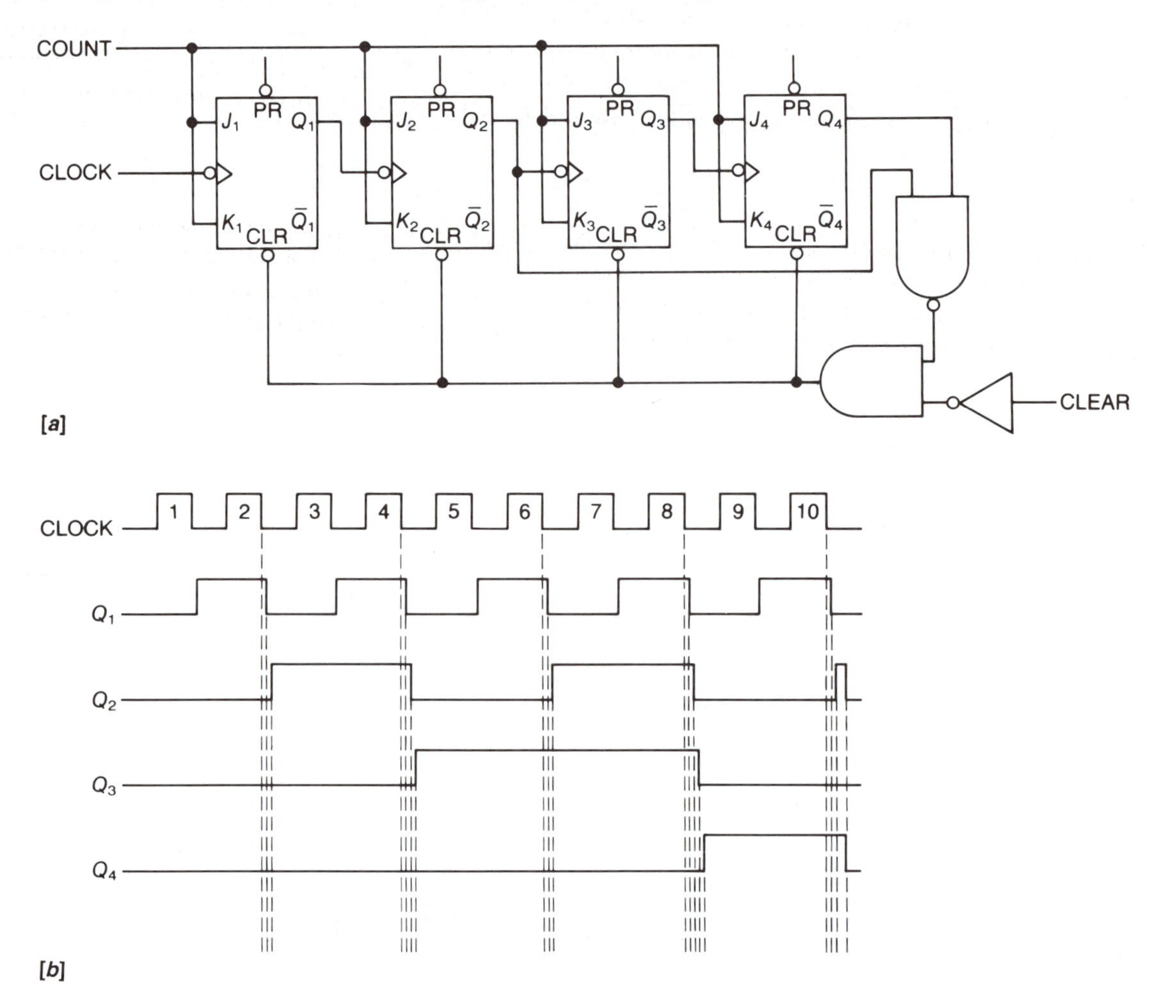

behavior that needs to be closely examined. The transient behavior of the counter is listed as follows:

0001 → 0010 (ideal), 0001 → 0000 → 0010 (actual),
0011 → 0100 (ideal), 0011 → 0010 → 0000 → 0100 (actual),
0101 → 0110 (ideal), 0101 → 0100 → 0110 (actual),
0111 → 1000 (ideal), 0111 → 0110 → 0100 → 0000 → 1000 (actual),
1001 → 0000 (ideal), 1001 → 1000 → 1010 → 0000 (actual).

The worst-case transient occurs during the transition 0111 → 1000. Since an intermediate count is to be decoded, the clock period must be longer than the 0111 → 1000 propagation delay.

Another alternative to the BCD counter design involves feeding the clock input of a modulo-5 counter with the output of a single T FF. The combination of a modulo-2 and modulo-5 counter results in a modulo-10 counter. Such cascading of one counter with another should be pursued whenever possible. Some of the counter designs that we have considered thus far have demonstrated how to reset a counter once a specific count has been reached. There are cases where a different approach may be necessary. It is always possible to preset a counter to a specific count by means of a *jam-entry* scheme, as shown in Figure 10.17. The bit that is to replace the old value at the Q_i location is fed directly to the corresponding entry point, X_i. The new bit is transferred to the FF output when the clock pulse occurs. The counter will begin counting from the preset count

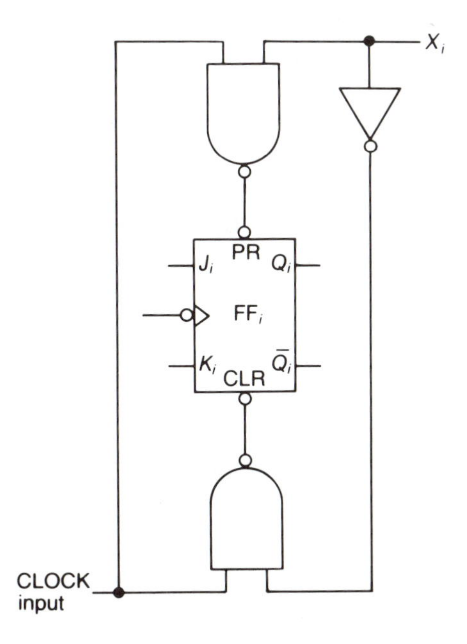

FIGURE 10.17 **Jam-Entry Scheme for Presetting the *i*th FF.**

as further clock pulses arrive at the input. This is possible only in counters made up of independent FFs.

Because of the problems of asynchronous counters, they should be used with a certain degree of caution. The designer must be aware of their limitations. The next section introduces the characteristics of IC counters. In a subsequent section (Section 10.9), more counter configurations will be introduced that exhibit several advantages over the synchronous and asynchronous counters.

10.4 IC Counters

In the first three sections of this chapter we have used both classical and heuristic design techiques to design counters. Similar multi-bit counters are available in IC form. A typical four-bit binary counter module with inputs and outputs is shown in Figure 10.18. We are already familiar with most of its features. The different inputs are described as follows:

A, B, C, D: These inputs are used for presetting the counter to an initial value. It has been assumed that A is the LSB and D is the MSB.

Q_A, Q_B, Q_C, Q_D: These are the FF outputs of the counter. Q_A corresponds to the LSB and Q_D corresponds to the MSB.

CARRY-OUT (CO): This output becomes a 1 when the count equals 1111 and the ENABLE control input is a 1. CARRY-OUT is equivalent to $Q_A Q_B Q_C Q_D \cdot$ ENABLE.

LOAD: This control input is used to load *A, B, C,* and *D* values into the counter. When the LOAD input is a low (0), the values are either loaded immediately if loading is asynchronous or loaded at the next clock pulse in the synchronously operated counter.

CLEAR (CLR): This control input when set to a 0 causes the counter to be cleared. The counter is cleared immediately if it has an asynchronous clear. In a counter with synchronous clear the output changes coincident with the next clock pulse.

ENABLE (E): This input must be high for the counter to count.

CLOCK (CK): The 1 → 0 transitions of this input are counted by the counter when the ENABLE input is a 1.

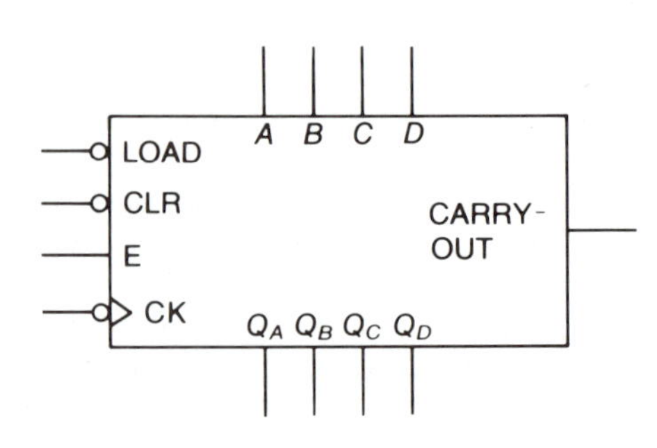

FIGURE 10.18 **Four-Bit Binary Counter Module.**

The four-bit IC counters may be cascaded together, as shown in Figure 10.19, to form counters of any bit length. When counter #1 has a count of 1111 and its ENABLE is a 1, the CARRY-OUT is a 1, which activates the second counter. When the next clock occurs, counter #1 resets to 0000 and counter #2 counts up to 0001 and the CARRY-OUT of counter #1 is reset to a 0. The eight-bit counter output then becomes 0001 0000. During the next 15 pulses, counter #2 would remain disabled and counter #1 would keep on counting upward. Again when counter #1 reaches count 1111, counter #2 counts up to 0010. And this process continues as more inputs appear at counter #1. Each time the setup of Figure 10.19 receives a total of 16 clock pulses, the #2 counter counts up by 1.

FIGURE 10.19 **Eight-Bit IC Counter.**

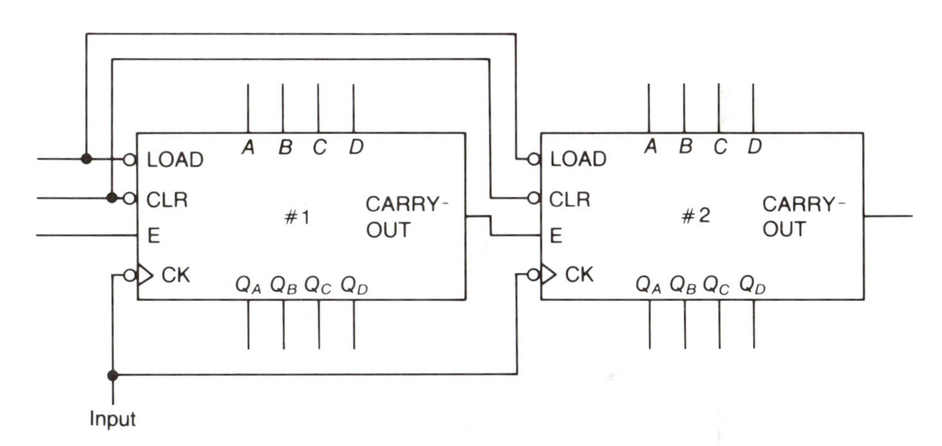

The LOAD and CLR controls can be manipulated in many different ways to obtain many count variations. Example 10.4 illustrates such manipulations of a binary IC counter mode.

EXAMPLE 10.4

Design a counter using the module of Figure 10.18 that outputs a 1 each time six counts have been received.

SOLUTION

One of the ways to accomplish this design is to make use of the CARRY-OUT output, which is 1 when the count reaches 1111. We can load the counter with an initial value that will cause an output to occur five pulses later. The CARRY-OUT is then used to reload the initial value. We begin from 1010, and when the counter reaches 1111 the CARRY-OUT would give a 1. Note that the first output upon turning the power on may not even be 1010. The CARRY-OUT either may become 1 before six pulses have occurred or may require up to 15 pulses. The number of pulses is dependent on what value the counter assumes upon power-up. The state diagram of the required sequence is shown in Figure 10.20.

FIGURE 10.20

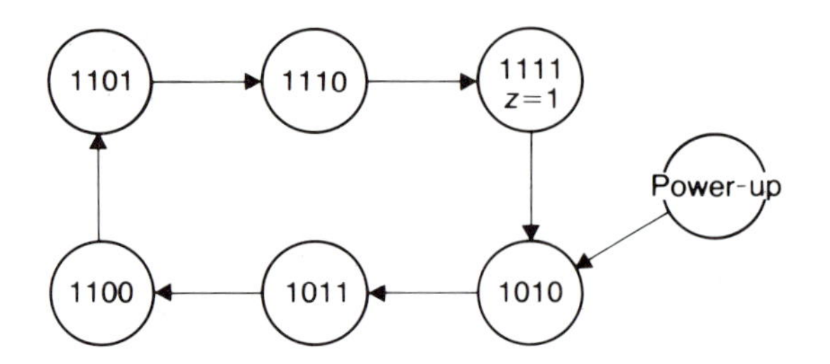

Successive outputs should occur after every six pulses. The circuit configuration using an IC counter with a synchronous load is obtained as shown in Figure 10.21. When the counter reaches 1111 the CARRY-OUT becomes a 1, causing a low at the LOAD input. This low forces the counter to begin again from the 1010 state.

FIGURE 10.21

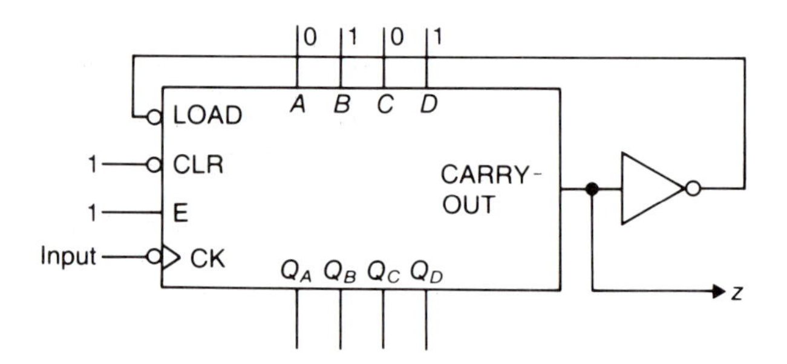

An alternate solution to this problem may be obtained by making use of the CLR input. In this case we begin from 0000, and once the count reaches 0101 an output is made to occur, and at the same time the counter is reset. A count of 0101 is decoded using external logic, and the counter is then made to reset using the synchronous CLR. The first pulse may require as many as seven pulses or none, depending on what the counter state is after power-up. In the worst case the beginning state could be 0110 and the counter would be reset once the count reaches 1101. The state diagram and the corresponding circuit configuration are obtained as shown in Figure 10.22. Note that the output *z* will be a 1 whenever the count is 0101 or 1101. The count will be 1101 only if the counter shows a count larger than 0101 at power-up.

FIGURE 10.22[a]

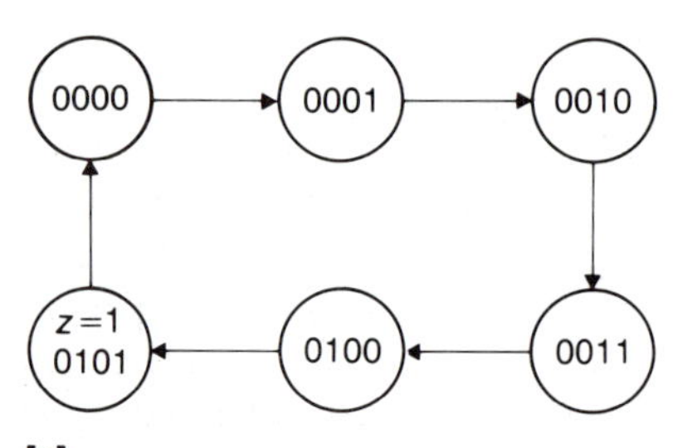

[a]

FIGURE 10.22[b]

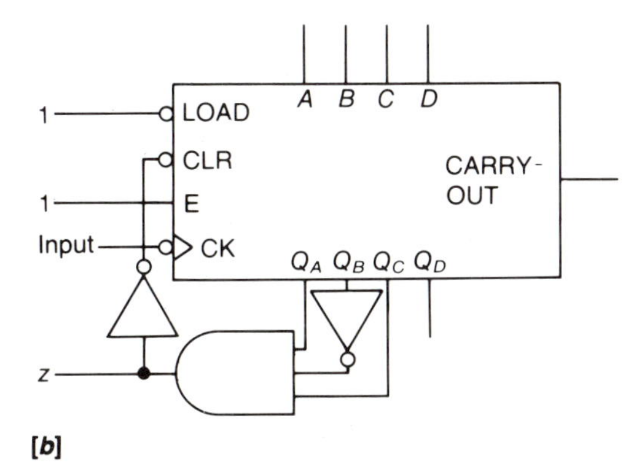

Given a modulo-m counter to count n pulses, where $n \leq m$, and then to restart the count, it is thus advisable to follow either of the following schemes:

a. Use the CARRY-OUT and load $m - n$ into the preset inputs. The CARRY-OUT will be 1 every n pulses.

b. Decode a count of $n - 1$ and use the output of the decode gate to clear the counter. The decode gate will be active every n clock pulses.

10.5 Basic Serial Shift Registers

The shift register is one of the most extensively used functional devices in digital systems. A *shift register* consists of a group of FFs connected so that each FF transfers its bit of information to the adjacent FF coincident with with each clock pulse. In other words, shift registers store bits of information, behaving like temporary memory, and upon external command shift those information bits one position to either right or left, depending on the design of the device.

The action of a right-shift register whose shift-right serial input is tied to a 1 is illustrated in Figure 10.23. The bits shifting out of the right-most FF are lost. With each clock input the bits move one position to the right while a 1 moves in at the MSB. After 11 clock pulses, the data in the register prior to shifting are replaced by a string of 1s. A quite useful application of shift-right registers requires a connection of the right-most FF output to the input of the left-most FF. In that case the LSB is not lost but appears at the MSB. Consequently, after two clock pulses, for example, 00101100101 would be replaced by 01001011001. Such a shift-right register is known as a *circulate-right* register. In the event these same data were stored in a shift-left register and the MSB output was connected to the LSB input, the data would be 10110010100 after two clock pulses. This latter type of register is known as a *circulate-left* register.

FIGURE 10.23 **Shift-Right Register Action.**

After Clock	Bit Pattern
0	00101100101
1	10010110010
2	11001011001
3	11100101100
4	11110010110
5	11111001011
⋮	⋮
10	11111111110
11	11111111111

A typical stage, n, in a multi-bit serial shift-right register can be designed using the design procedures described in the earlier chapters. Figure 10.24 shows the state diagram of the FF concerned. The FF state reflects the current content of that position of the shift register. If the current state of the nth bit of the shift register is a 0 and that of the $(n + 1)$th bit is a 0, the nth bit remains unchanged when the clock occurs. Similarly, Q_n does not change if the present states of both Q_n and Q_{n+1} are 1. For the other cases, Q_n changes and takes the value of Q_{n+1}. Figures 10.23 and 10.24 illustrate the action of serial shift-right registers and make it obvious that the function of each of the FFs is governed by the same next-state equations. This observation reduces the design of a multiple-bit shift register to the design of a single stage. For n bits, n such stages are cascaded.

The circuit action described by the state diagram in Figure 10.24 is that of a D FF. This can be verified by comparing the state diagram of Figure 10.24 with that of Figure 6.40[c]. Multi-bit, shift-right registers can be built using edge-triggered D FFs or JK FFs connected as D FFs. Such shift-right registers are shown in Figure 10.25 where SRI is the entry point for *shift-right input* and SRO is the exit point for *shift-right output*. While using D FFs, the Q output of each stage is connected to the D input of the succeeding stage. If JK FFs are used, the Q and $\overline{Q}$ outputs of each stage are connected to

FIGURE 10.24 **State Diagram for the Q_n Bit of a Multi-Bit, Shift-Right Register.**

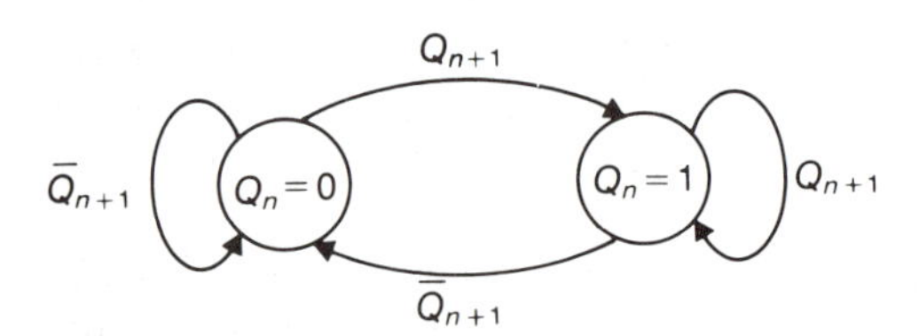

FIGURE 10.25 **Four-Bit, Shift-Right Register: [*a*] Using *D* FFs and [*b*] Using *JK* FFs.**

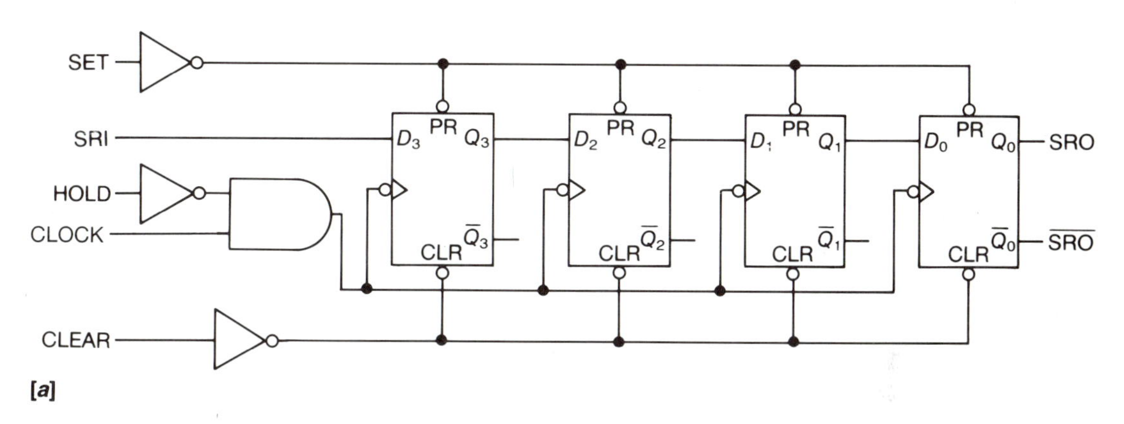

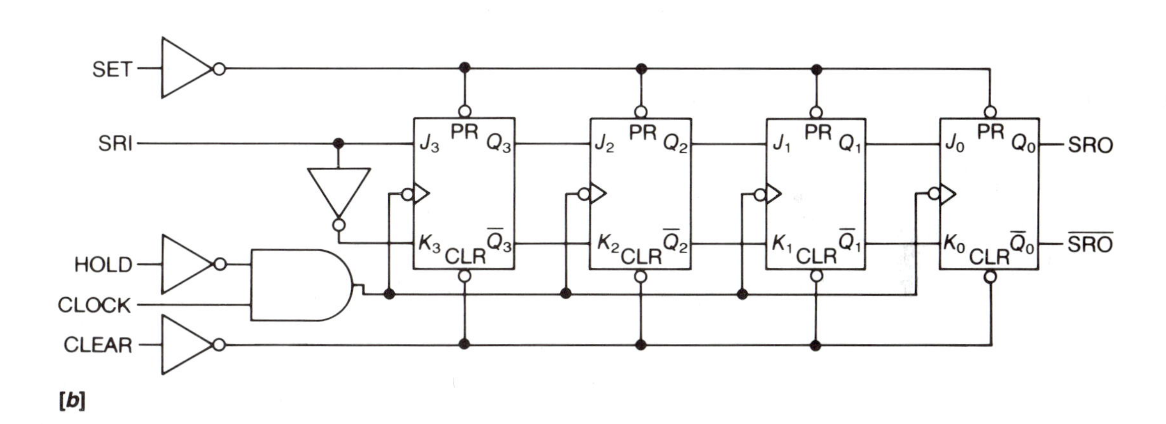

the J and K inputs of the next stage. The first JK FF is modified to that of a D FF by supplying the data directly to the J input and the complement of data to the corresponding K input. Note that the remaining FFs do not have an *inverter* between their J and K inputs since each of the J inputs is already a complement of the corresponding K input. A high at the CLR input would reset the register FFs, and similarly a high at the SET input would place a 1 at each of the FFs. A high at the HOLD input would disable the FFs and this could be used for storing the bits indefinitely. SET, CLR, and HOLD inputs must be fed with 0 for operating the register in serial mode.

The shift registers of Figure 10.25 are classed as serial-in, serial-out, shift-right registers. Similar design techinques may be used to obtain a shift-left register. In fact, both shift-right and shift-left capabilities may be combined to obtain a bidirectional serial-in, serial-out shift register. A controlled shift-left register is shown in Figure 10.26 that has an additional control input SLE that determines what it does on the next clock pulse. The SLI is the entry point for *shift-left input* and SLO is the exit point for *shift-left output*. When the *shift-left enable*, SLE, is low, the FF output is fed back to its data input. In this way, digital information bits may be restored indefinitely. It is interesting to see how the HOLD input has been eliminated in this case. The clock inputs are still allowed to excite the FFs. However, in the previous case, as shown in Figure 10.25, the clock input was not allowed into the FFs. Again when the SLE control is high, the serial input sets up the right-most FF, Q_0 sets up the second FF, Q_1 sets up the third FF, and so on. With SLE high, the circuit functions as a shift-left register. The serial output is obtained out of the left-most FF.

FIGURE 10.26 **Controlled Shift-Left Register.**

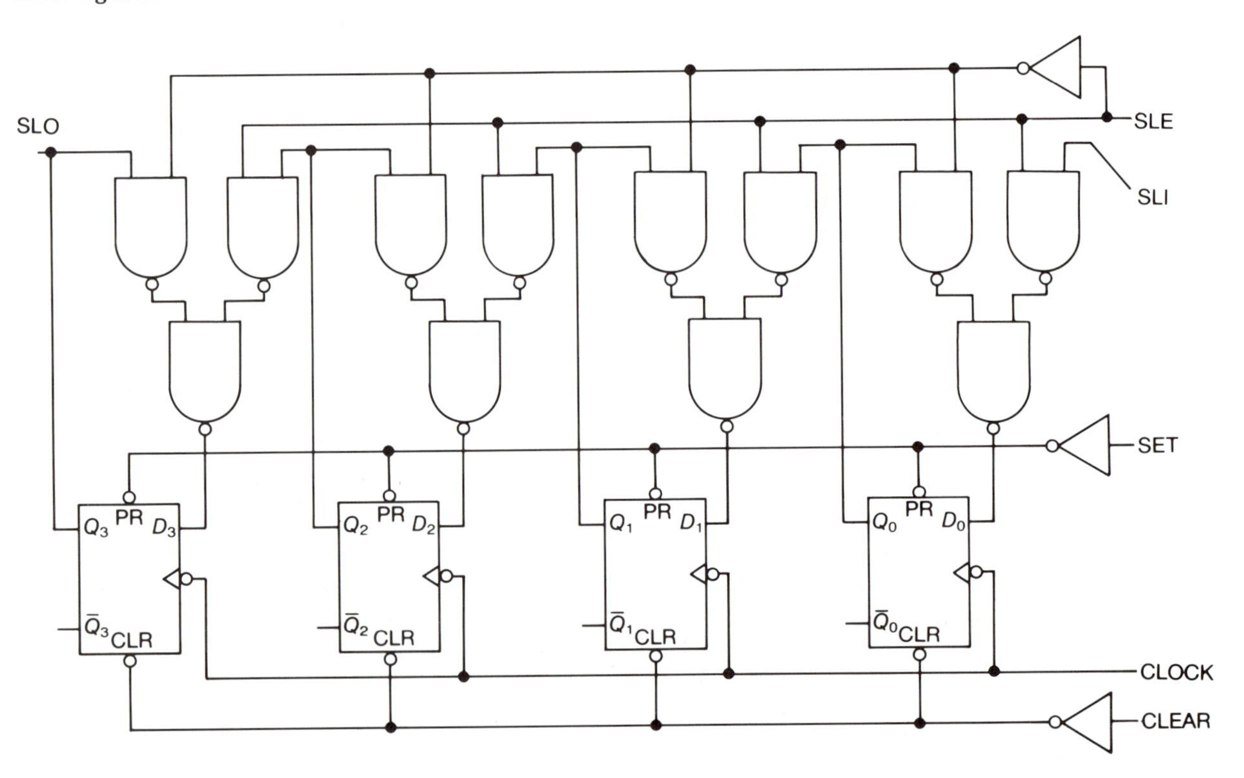

The shift-left and shift-right registers can be combined to obtain the bidirectional shift register of Figure 10.27. This register is similar in design to that of an up-down counter where an up-counter is combined with a down-counter. The direction of shift is controlled by the inputs SLE and SRE. SLE and SRE control inputs cannot be 1 simultaneously, except when HOLD = 1, so they are mutu-

FIGURE 10.27 **Bidirectional Shift Register.**

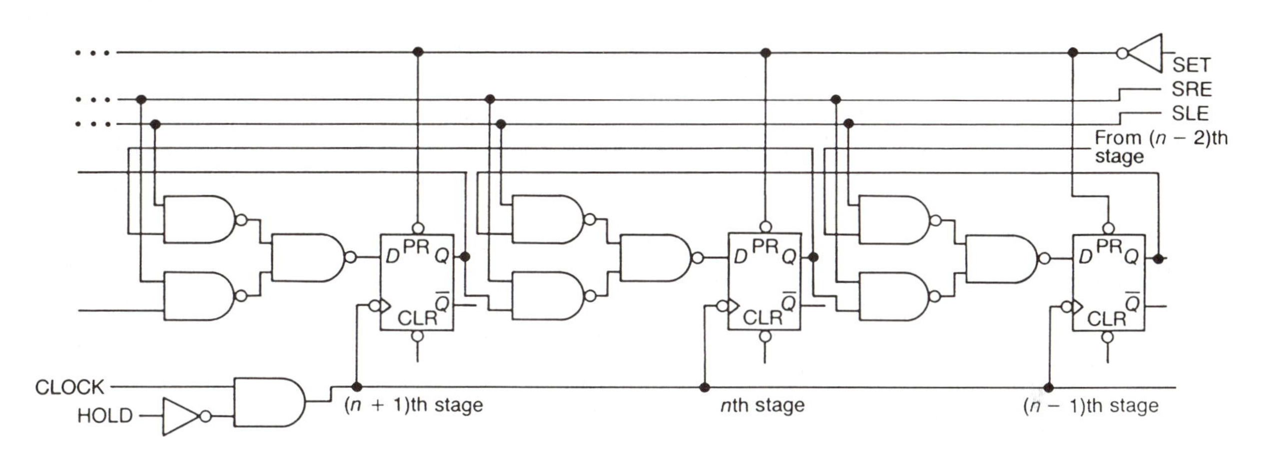

ally exclusive. When the SRE input is high, each of the FFs loads the respective Q output of the FF on its immediate left. When the SLE input is high, each of the FFs loads the Q output of the FF on its immediate right. When both SLE and SRE are 0, the FFs are all reset. Note this time how the CLR control has been eliminated. When HOLD = 0 the register functions in its serial mode, and when HOLD = 1 the old data bits are restored. We might decide to eliminate one of the two shift controls. In that event we may decide to keep SRE only by making sure that SLE has been replaced with the complement of SRE. The register would function as a shift-right type when SRE = 1 and as a shift-left type when SRE = 0. But this arrangement causes a problem if we need to reset the FFs at any time. This problem could be solved, however, by feeding the complement of a CLR control to each of the FF resets.

An important application of shift registers is their role in arithmetic operations. A binary number can be multiplied by 2 by shifting the number one bit to the left and divided by 2 by shifting the register content one bit to the right. As we will see later, the bits shifted in at one end and out at the other end are not unimportant; they are used in arithmetic operations in many instances.

10.6 Parallel-Load Shift Registers

An n-bit, serial-load shift register requires n clock pulses to load an n-bit word. A *parallel-load shift register*, in comparison, loads all information bits simultaneously. Both serial-in and parallel-load shift registers have specific applications in digital systems. A parallel-in, serial-out shift register using master-slave *SR* flip-flops is shown in Figure 10.28. The parallel data are loaded using the jam-entry

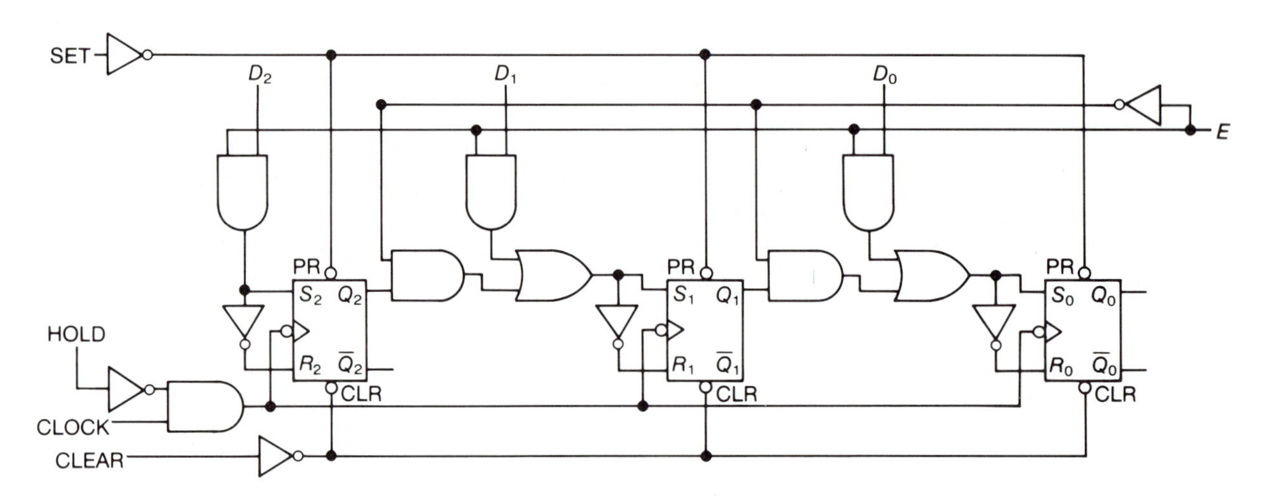

FIGURE 10.28 **Three-Bit, Parallel-In, Serial-Out Shift Register.**

scheme that was discussed in Section 10.3. When the enable signal E is high, the data are loaded into the register in parallel. Again, if E is low, the Q output of the FF of every stage is shifted to the right by means of the combinational gates. In either case the HOLD control must be held low. Parallel-in, serial-out shift registers allow accepting data n bits at a time on n lines and then sending them one bit after another on one line. This is a standard mode of communication between digital systems.

At the receiving end of two digital systems communicating over a single data line, it is necessary to collect n bits and then transfer them in parallel to the receiving system that is designed to handle n bits simultaneously. Figure 10.29 shows the logic circuit of such a

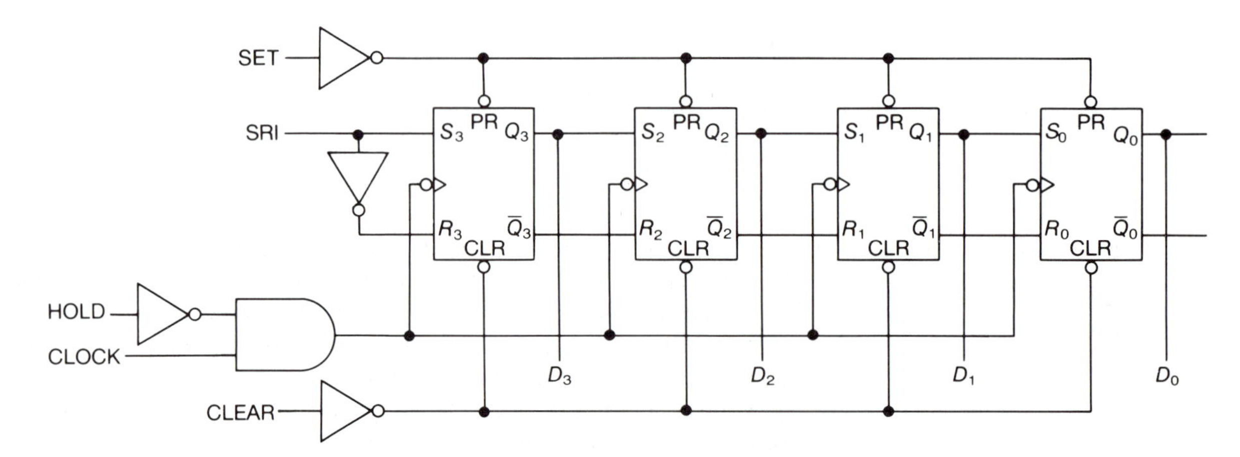

FIGURE 10.29 **Four-Bit, Serial-In, Parallel-Out Shift Register.**

serial-in, parallel-out register. The serial data are entered to the S input of the left-most FF while the data are transferred in parallel from the Q outputs. The register is organized in exactly the same way as that of Figure 10.25[*b*]; however, all of its Q outputs are available, which is not the case for all shift registers. Both parallel loading access features are included in the four-bit, parallel-in, parallel-out register shown in Figure 10.30. The CLR, SET, and HOLD inputs are set low for normal operation. When the LOAD is low, the shift register performs the shift-right operation. When the LOAD is high, the inputs I_3, I_2, I_1, and I_0 are loaded in parallel into the register coincident with the next clock pulse. The outputs O_3, O_2, O_1, and O_0 are available in parallel from the Q output of the FFs.

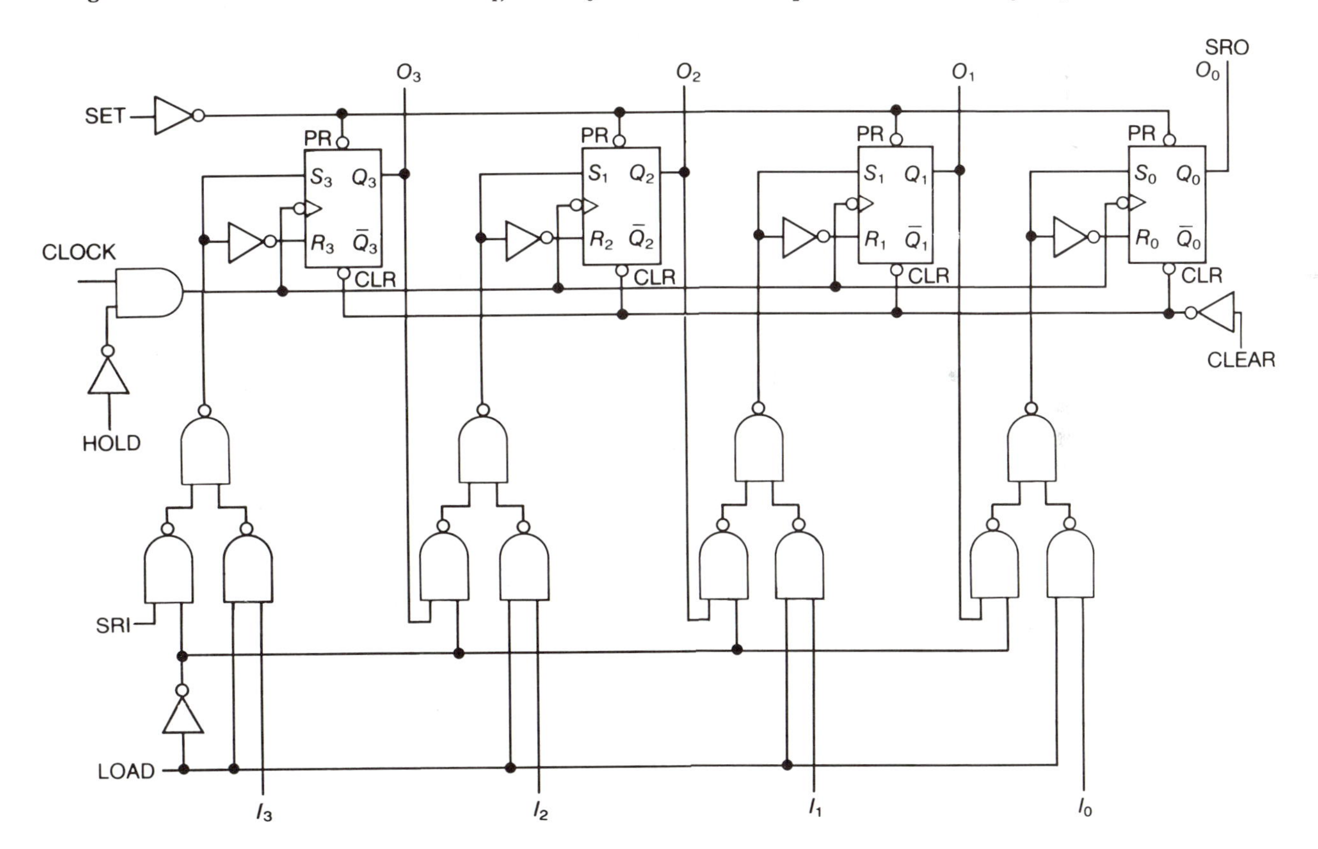

FIGURE 10.30 **Four-Bit, Parallel-In, Parallel-Out Shift Register.**

10.7 Universal-Shift Registers

A *universal-shift register* is a versatile shift register that has capabilities for parallel loading, parallel outputs, bidirectional shifting, and bidirectional serial input and output. In other words, it is capable of operating in all of the register modes described previously. There could be two different ways to realize a universal-shift register: either by modifying a parallel-in, parallel-out shift register or by building one from scratch.

The circuit of Figure 10.31[*a*] shows the logic diagram of a four-bit, parallel-in, parallel-out shift-right register. Its internal circuitry and functions are exactly like those of the circuit of Figure 10.30. It has five entry points for the data inputs—SRI, I_3, I_2, I_1, and I_0—four outputs—O_3 through O_0—and five control inputs—SET, HOLD, CLR, CK, and LOAD. The register is capable of either serial or parallel entry and serial and parallel output. Note that in the circuit of Figure 10.30 the normal serial mode is only shift-right. Such devices are also available commercially. When LOAD is low the register works as a shift-right register, and when LOAD is high it works as a parallel-in register.

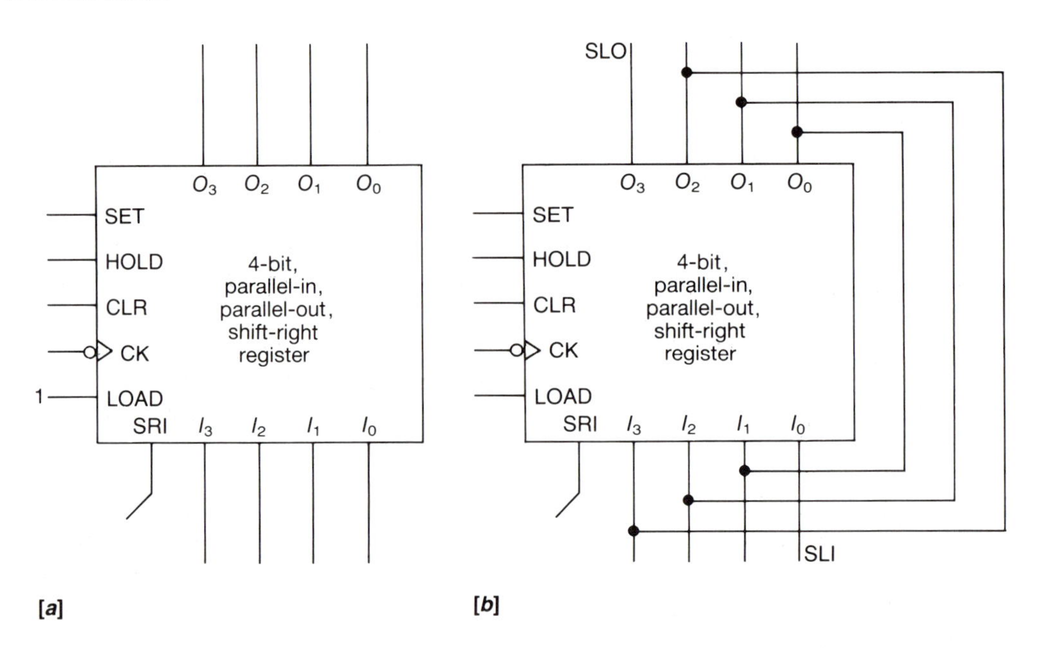

FIGURE 10.31 **Block Diagram of Figure 10.30: [*a*] in Shift-Right Mode and [*b*] in Shift-Left Mode.**

The circuit of Figure 10.30 can be externally modified to perform a left-shift operation, as shown in Figure 10.31[*b*]. This modification is accomplished by selecting the parallel-mode, LOAD = 1, and by connecting each parallel input to the output of the FF on the immediate right. This form of shifting is called a *wired shift* since the shifting action is accomplished by external connections.

An n-bit, universal-shift register can be designed by means of n D FFs and n 1-of-4 multiplexers. The ith bit of such a universal register is shown in Figure 10.32. Now all that remains is to cascade n such units. When $AB = 00$ the register performs a parallel-in operation, when $AB = 01$ the register restores the old value, when $AB = 10$ the register performs a right-shift operation, and when $AB = 11$ the register performs a left-shift operation.

FIGURE 10.32 *i*th Bit of a Universal-Shift Register.

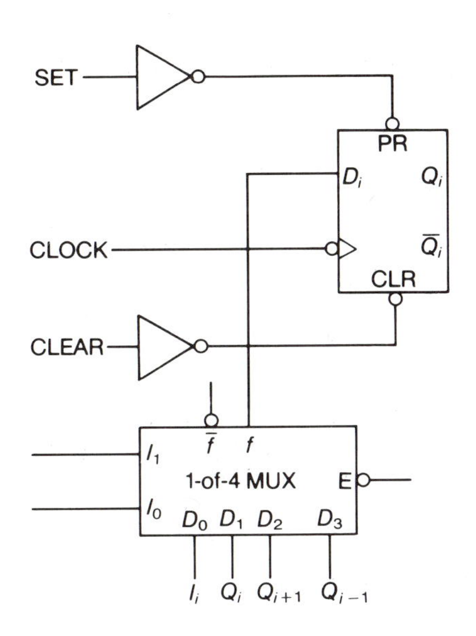

10.8 Shift Registers as Counters

Shift register ICs are used at times to generate counts or controlled sequences. As a result registers are used extensively in multiple address coding, parity bit generators, and random bit generators. The output of each stage and its complement must be accessible for these applications. These register outputs are used to drive combinational feedback logic, as shown in Figure 10.33. The feedback logic determines the next state of Q_n. In the case of a bidirectional register, the feedback logic controls shift-left and shift-right signals and sets up a 1 or 0 to the appropriate SLI and SRI input.

The state diagram of a four-bit shift register with the J_3 input of the input FF available is shown in Figure 10.34. If the shift register is initially in the state $Q_3Q_2Q_1Q_0 = 1001$, then there are two possi-

FIGURE 10.33 Feedback Shift-Right Register Configuration.

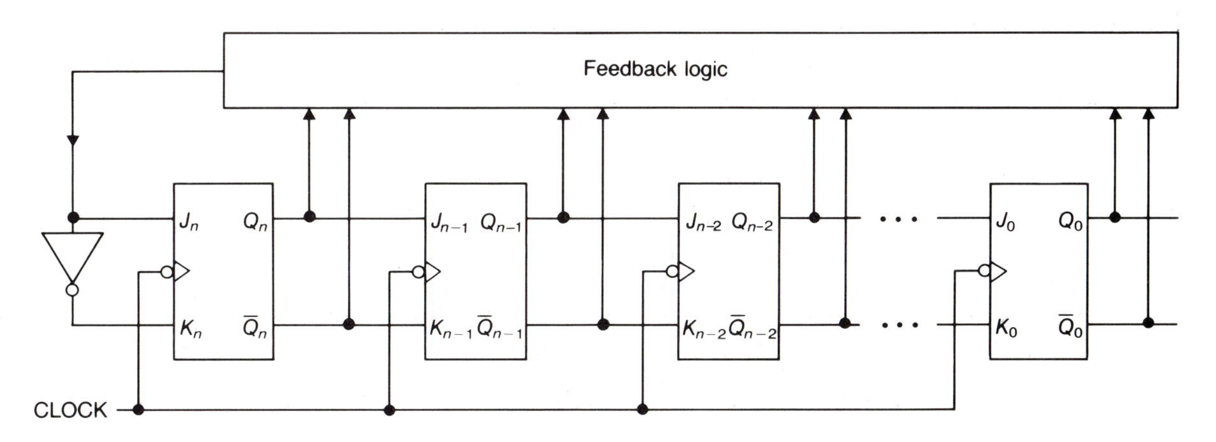

FIGURE 10.34 **Universal State Diagram for a Four-Bit Feedback Register.**

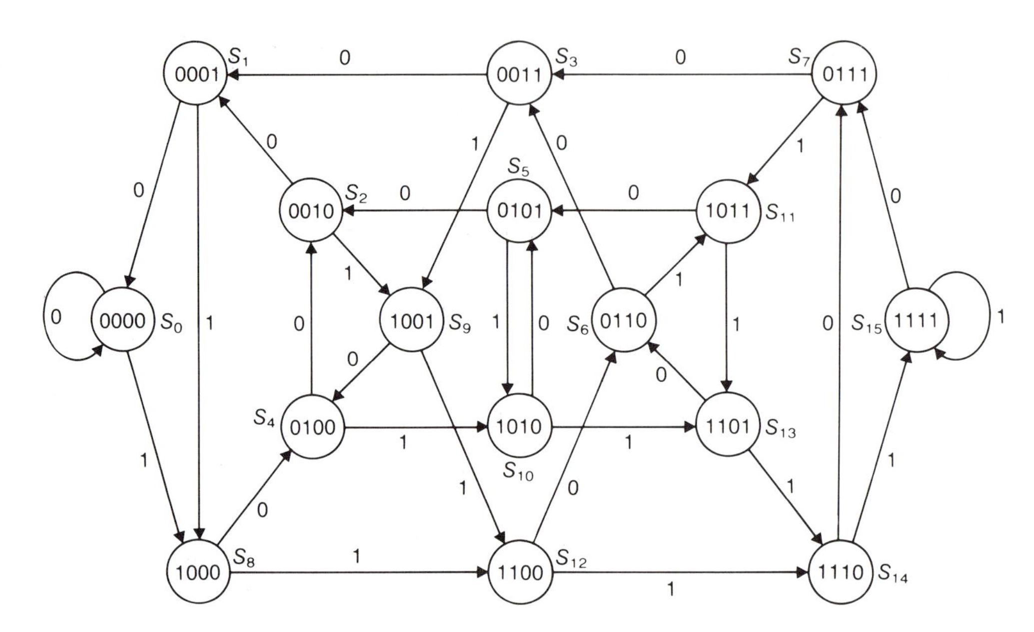

ble next states. These are 0100 if the J_3 input is a 0, or 1100 if the J_3 input is a 1. These values correspond to a shift-right operation. All possible internal states of the register and all possible transitions between the states are considered in this state diagram.

In order to design a counter or sequence generator, the designer selects the desired sequence of states on the universal state diagram. Based on the desired sequence the feedback logic is designed so that the register will cycle through the selected sequence of states. Example 10.5 illustrates the technique.

EXAMPLE 10.5

Design a decade sequence generator using a shift register that follows the sequence

$S_0 \rightarrow S_8 \rightarrow S_{12} \rightarrow S_{14} \rightarrow S_{15} \rightarrow S_7 \rightarrow S_{11} \rightarrow S_5 \rightarrow S_2 \rightarrow S_1 \rightarrow S_0$

SOLUTION

The state transitions and the corresponding feedback logic condition are obtained as shown in Figure 10.35. The resulting K-map for the feedback function J_3 is obtained as shown in Figure 10.36. The feedback function is

$$J_3 = \bar{Q}_1\bar{Q}_0 + Q_2\bar{Q}_0 + \bar{Q}_3Q_2Q_1$$

The feedback logic is realized using combinational logic. It is then fed to the J_3 input of the shift-right register, as shown in Figure 10.37, to obtain the desired sequence generator. Note that the nonsequence states lead eventually to the sequence as follows:

$$S_3 \rightarrow S_1, \quad S_4 \rightarrow S_{10} \rightarrow S_5, \quad S_9 \rightarrow S_4, \quad \text{and} \quad S_{13} \rightarrow S_6 \rightarrow S_{11}$$

FIGURE 10.35

Transitions PS →	NS	J_3
0000	1000	1
1000	1100	1
1100	1110	1
1110	1111	1
1111	0111	0
0111	1011	1
1011	0101	0
0101	0010	0
0010	0001	0
0001	0000	0

FIGURE 10.36

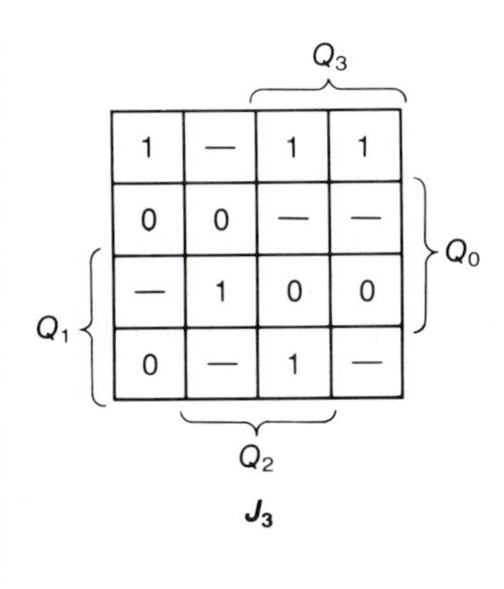

FIGURE 10.37

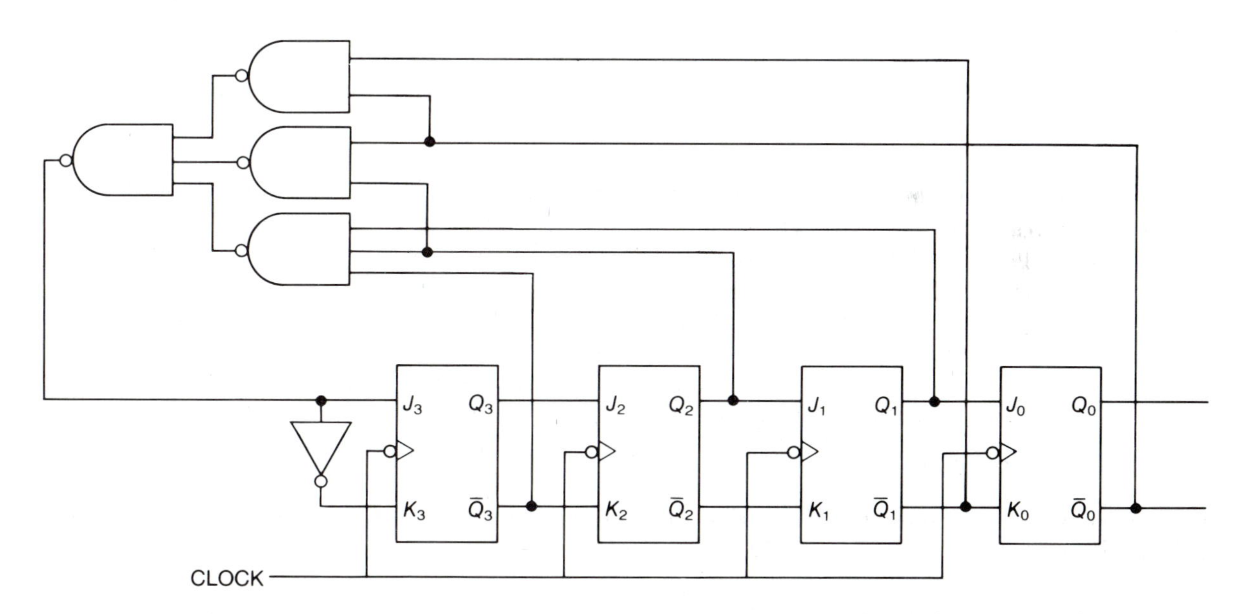

Consider, for example, that J_3 is a 0 whenever the present register state is at S_3 (= 0011). Therefore, the system would shift from state S_3 to S_1 (= 0001) as per the state diagram of Figure 10.34. If the sequence generator in this example enters an unused state due to a glitch or on power-up, it will return to the decade sequence after a maximum of two clock pulses.

10.9 Counter and Register Applications

There are hundreds of applications of counters and shift registers. They are used extensively in computers. In general, digital computers process numbers by repeated arithmetic and logic operations. Execution of a specific instruction usually involves moving the instruction and data between registers. The data are operated on by the ALU as they are transferred between registers. These transfer sequences are in turn controlled by sequential circuits. In particular, registers provide the means for the storage of bits as they are being processed. On the other hand, counters keep track of the next memory location and count the intervals in the sequences that control these complex operations. In this section we shall consider only a few of their many important applications.

The operations in digital computers are performed in parallel in most cases since this is a faster mode of operation. In comparison, serial operations are slower but require less complicated and less expensive circuits. Consider the *add* function. In Chapter 5 the design of parallel addition circuits was examined in detail. The techniques developed were reasonably fast but they involved very complex circuitry. Frequently, the designer must make a trade-off between time and the number of components.

The add operation can also be performed by loading the addend and augend into two serial shift registers and shifting one bit from each register into a single-bit FA, as shown in the block diagram of Figure 10.38. The carry-out of the FA is stored in a D FF and fed back as the carry-in to the FA to be added to the next pair of signifi-

FIGURE 10.38 **Serial Adder Configuration.**

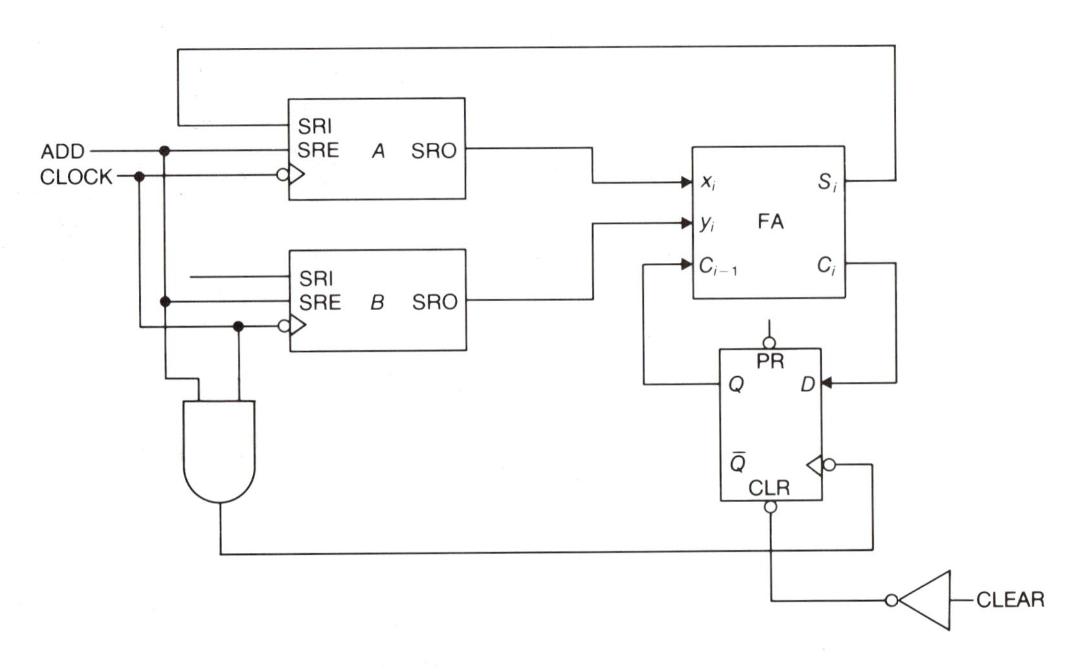

cant bits from the shift registers. The sum bit is shifted into the shift register containing the augend as the augend bits are continually being shifted out to the right.

Initially the shift registers *A* and *B* hold the augend and addend and the *D* FF is cleared. The summation is achieved by connecting each pair of bits, through shifting, together with the previous carry-out into the FA circuit and by transferring the sum bits serially, into the register *A*. The ADD command starts and stops the operation. When ADD is high the registers perform a shift-right operation at each clock, and when ADD is low the registers maintain a hold mode. In the next chapter we shall consider every aspect of how to design such a serial adder circuit. In the meantime we will develop other relevant concepts.

Operations in digital systems are controlled by a sequence of timing pulses. The control unit in a serial computer must generate a signal that remains high for a number of pulses equal to the number of bits in the shift registers. For example, the serial adder system of Figure 10.38 requires a control signal, ADD, for its operation. Figure 10.39[*a*] shows a control circuit that generates a signal that remains high for a period of 16 clock periods. The four-bit IC counter and the *SR* FF are initially CLEARed. The BEGIN signal

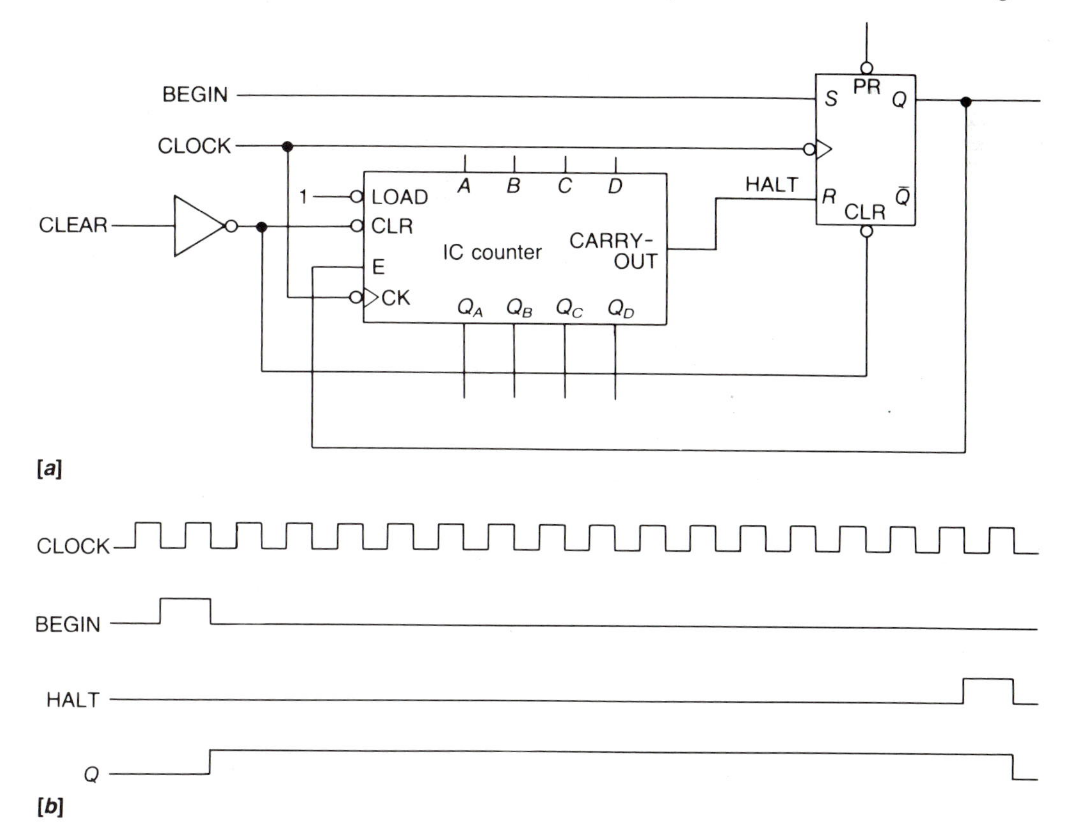

FIGURE 10.39 **Generation of Timing Sequences: [*a*] Circuit and [*b*] Timing Diagram.**

sets the *SR* FF, which in turn enables the counter. The FF output Q remains high for 16 pulses, as shown in the accompanying timing diagram of Figure 10.39[*b*]. When the counter reaches count 1111, HALT is activated, which in turn resets the FF. The BEGIN signal is synchronized with the clock and is made to stay on for one clock period. It could be made to stay on for a longer period; however, if it is made to last for more than 15 clock periods, the circuit will not function as expected. This HALT signal might be used in another similar circuit to generate a BEGIN pulse.

In a parallel mode of operation, a single pulse is generally used to specify the time at which an operation should be executed. Shift registers may be used to realize such a timing circuit when connected as a *ring counter.* A shift-right register used as a ring counter is shown in Figure 10.40. A feedback path is provided from the serial output to the serial input of the shift register. A shift register con-

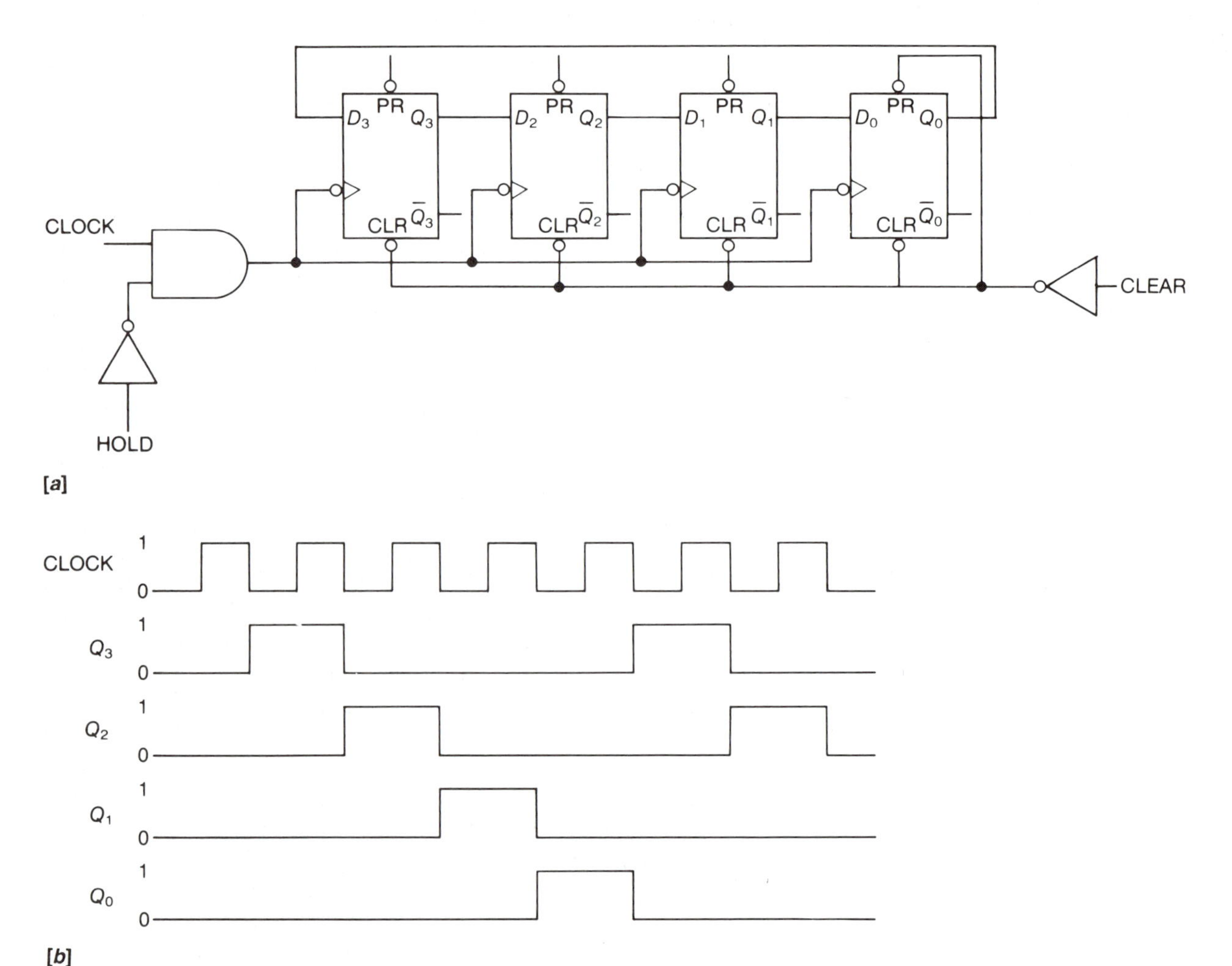

FIGURE 10.40 **Four-Bit Ring Counter: [*a*] Circuit and [*b*] Timing Diagram.**

nected as such circulates the register contents. Initially the CLEAR input is set to a 1. This presets the right-most FF to a 1 and clears the others. The starting output word, therefore, is 0001, and as the clock pulses occur the output word becomes 1000, 0100, 0010, 0001, and so on. Only four unique states are possible. This circuit would allow timing four sequential operations by tying each of the four operation's initiating lines to one bit of the ring counter. Each operation would then be active one-fourth of the time.

A four-bit binary counter has 16 counting states, whereas the ring counter has only four states. Thus the ring counter makes an extremely uneconomical use of FFs, which is more true for systems requiring a large number of timing signals. An excellent alternative to using an uneconomical ring counter is to use an n-bit binary counter and an n-to-2^n line decoder. This combination is often referred to as a *Moebius* or *Johnson counter* (no relation to the coauthor). It also involves a shift-right register like that of a ring counter, but it is connected in a *switch-tail* configuration.

As stated earlier, an n-bit ring counter provides only n distinguishable states. The number of states can be doubled if the shift register is connected in a switch-tail configuration, as shown in the four-bit Johnson counter of Figure 10.41. Here $\overline{Q}_0$ instead of Q_0 is fed back as the D_3 input. The register shifts one bit to the right with every clock pulse, and at the same time the complement value of the fourth FF is transferred to the left-most FF. Consequently, this would result in eight different counting states: 0000, 1000, 1100, 1110, 1111, 0111, 0011, and 0001. These eight states must be decoded to give eight distinct timing sequences. Unlike the previous circuit, the least significant FF need not be preset. For minimum chip count, a ring counter or a Johnson counter is the best choice for a timing circuit.

We have emphasized that counters are often used for sequencing various arithmetic and/or logic operations. And in most instances these operations are executed only if certain conditions are met. Figure 10.42 shows a four-state controller that can control four distinct operations. The sequencer circuit consists of a two-bit synchronous counter, a 1-of-4 MUX, and a 2-4 line decoder. Each of these four operations is activated by a low on the respective sequencer output. The sequencer output, D_n, will become low during its allocated time only if C_n (representing the corresponding condition) is a 1. The counter is first CLEARed. Consequently, it STARTs at the 00 address when the function F_0 is performed provided the condition C_0 is met. The function corresponds to a specific operation, and the condition C_0 may be the result of one or several test results. When the counter reaches the 01 value the function F_1 is executed, provided that the condition C_1 is met. The synchronous count allows the operations to be executed only at regular intervals. Once the sequencer is STARTed, the test conditions determine whether or

FIGURE 10.41 **Four-Bit Johnson Counter: [*a*] Circuit and [*b*] Timing Diagram.**

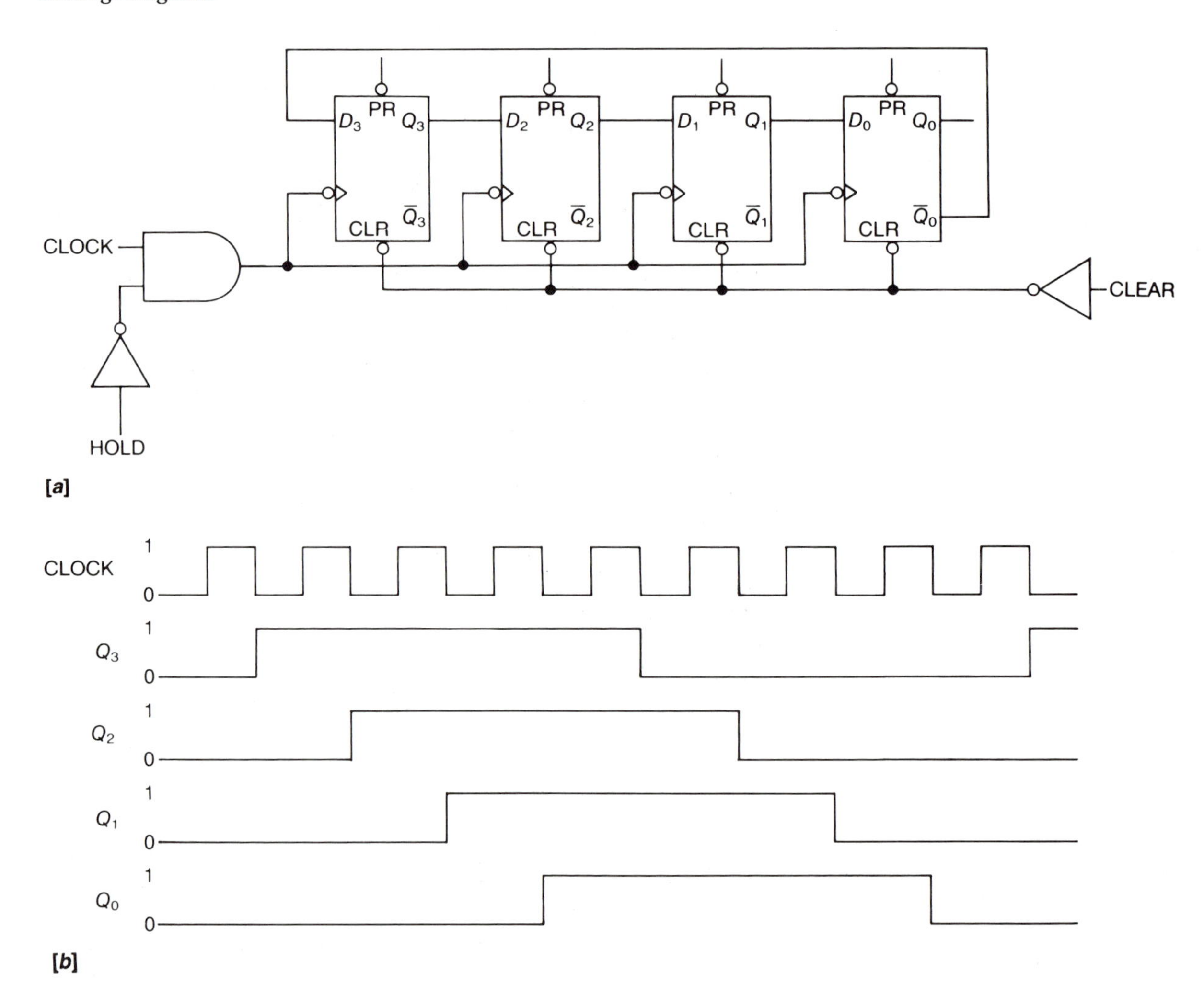

not the function is selected by the decoder. If the condition is not met, the decoder remains disabled, and as a result the corresponding function is not executed. The significance of a sequencer circuit, such as that of Figure 10.42, becomes meaningful only when it is allowed to control a complex digital system. Prior to the sequencer implementation, concepts of RTL (Register Transfer Language) will be introduced in Section 10.11.

FIGURE 10.42 **Two-Bit Operation Sequencer.**

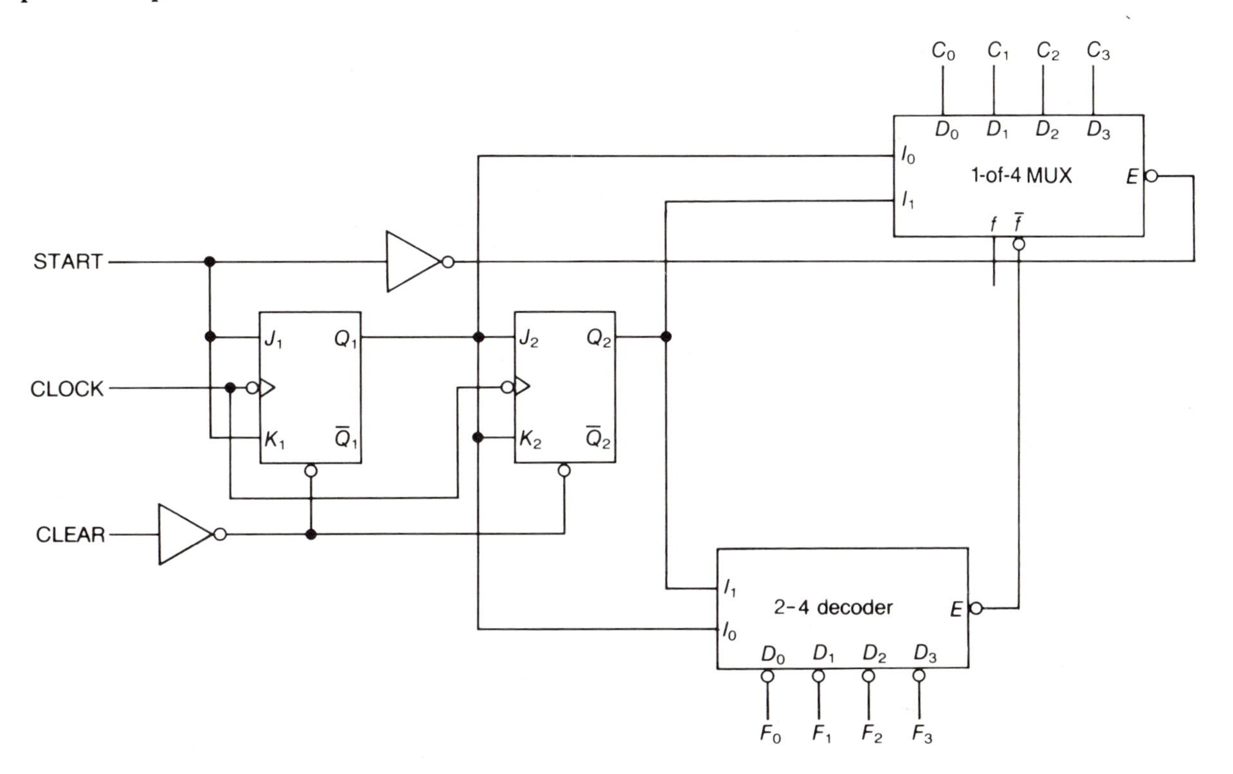

10.10 Bus Concept

Digital systems have large numbers of registers, and as part of the computational process it is often required to transfer data from one register to another. Consider the case of m registers with n bits in each. To allow direct inter-register transfer it would be necessary to have a total of $(m!)n$ data paths, which would require an awesome number of wires running between registers. Such a data transfer problem is solved by a group of wires called a *bus*. The bus concept is analogous to a mass transport system where each commuter waits in line until the transport becomes available (in this analogy the bus can carry only one passenger). For a parallel transfer of n bits the bus consists of only n lines, as shown in Figure 10.43[a].

Each of the 1-of-8 MUXs is equipped with eight input lines, three select lines, and one output line. The least significant inputs of each MUX are connected to the respective FFs of register A. The FF A_7 is connected to the first MUX, A_6 is connected to the second MUX, A_5 to the third MUX, and so on. The next significant input of each MUX is connected to the respective FFs of register B. This process is continued until all eight registers are connected. Figure

FIGURE 10.43 [*a*] Register-to-Bus Transfer Circuit for Eight Registers, [*b*] Block Diagram of Figure 10.43[*a*], and [*c*] Bus-to-Register Transfer Circuit for Eight Registers.

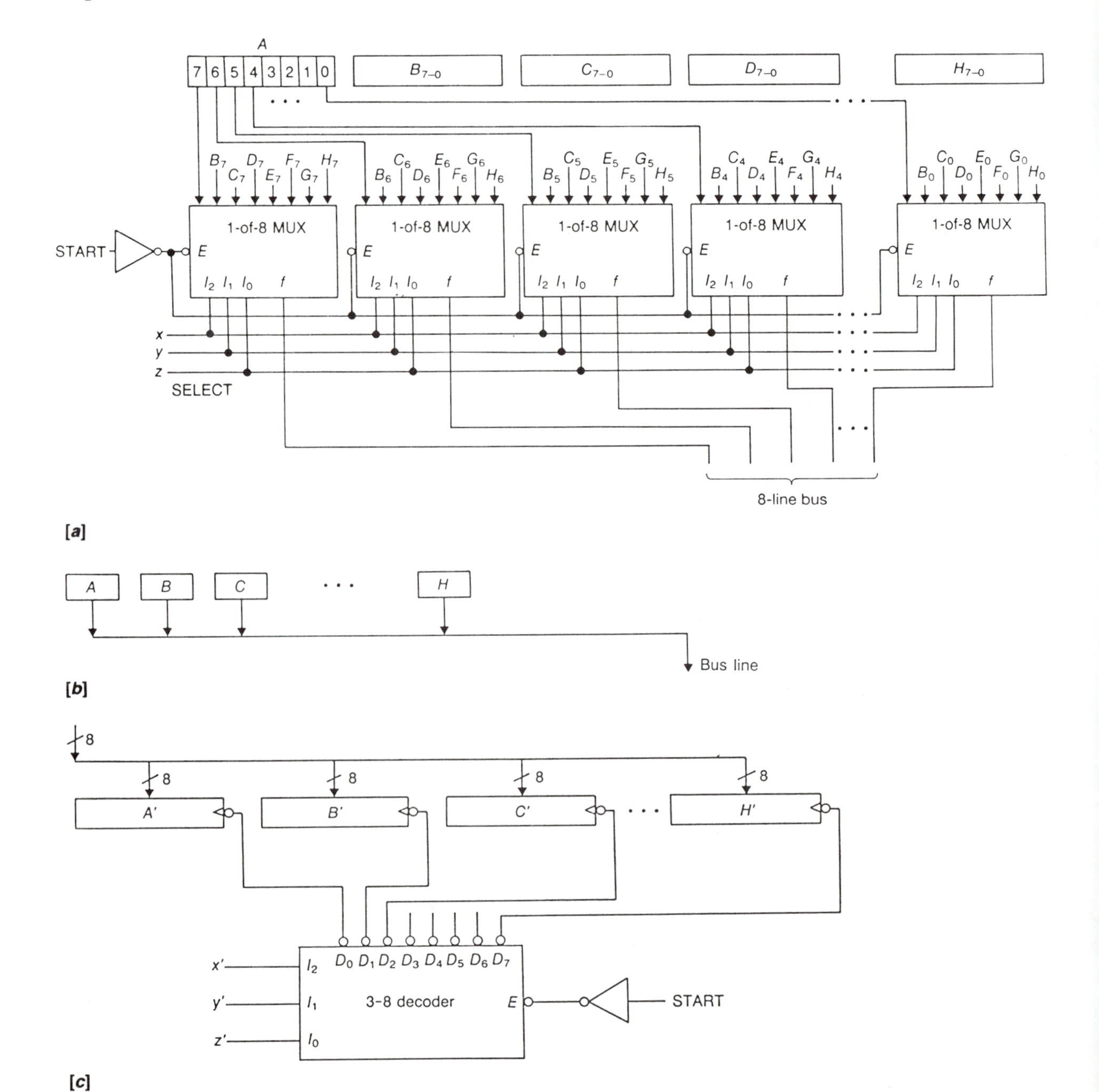

10.43[*a*] shows the necessary connections. When the select lines, x, y, and z, are all low the least significant input to each of the MUXs is selected, and consequently the bus is loaded with the contents of register A. Each combination of the x, y, and z inputs selects the contents of a particular register and the contents are then attached to the bus. The simplified block diagram is shown in Figure 10.43[*b*]. It is possible to design a bus system without the MUXs if the registers have tri-state outputs or are connected to the bus using tri-state buffers. Finally, the contents of the bus are required to reach a certain destination register. This requirement is accomplished by the circuit arrangement of Figure 10.43[*c*] where the select lines x', y', and z' determine the particular destination register by means of a 3-8 line decoder. Upon receiving the proper select inputs, the contents of the bus are loaded into the selected register.

Another example of the use of a bus involves a simple memory device. Registers often are assembled together to form a larger storage array. This arrangement of registers is referred to as a *scratch-pad* memory. Figure 10.44[*a*] shows an arrangement of registers that can store up to four four-bit words. The device consists of four registers that in turn have four FFs each. When the WRITE ENABLE (WE) is low, the four data inputs, I_0, I_1, I_2, and I_3, are routed to a particular location of each register as specified by the entries in the WRITE SELECT (WS) lines. A 00 on the WS lines will store the input bits in the respective 0th cell of the registers. Similarly, 01, 10, and 11 applied at the WS lines would respectively select the first, second, and third bit of each register. Stored data from the scratch-pad memory may be retrieved through the four output lines by applying a low to the READ ENABLE (RE) and necessary address bits to the READ SELECT (RS) lines. Figure 10.44[*b*] shows the logic diagram of the memory formed using gated D latches.

Scratch-pad memory, although very fast, is not extensively used. It is not particularly suitable for LSI because too many pin-outs

FIGURE 10.44 **16-Bit Scratch-Pad Memory: [*a*] Block Diagram**

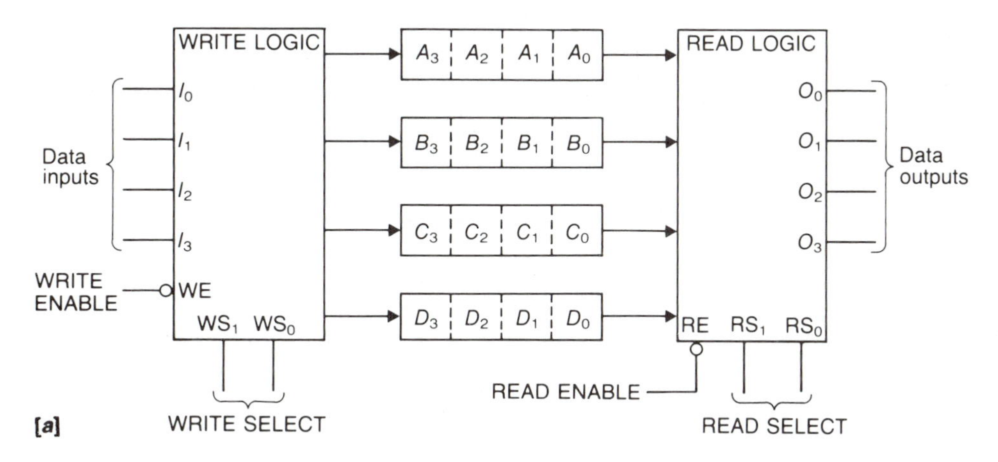

FIGURE 10.44 (Continued) **16-Bit Scratch-Pad Memory: [*b*] Circuit Showing *D* Latch at the *i*th Row and *j*th Column and the *i*th Output.**

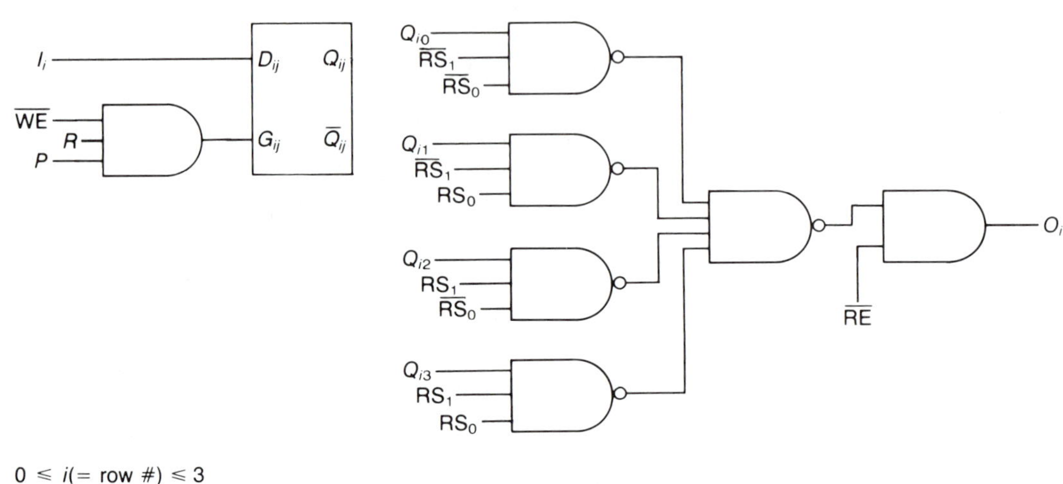

$0 \leqslant i(= \text{row } \#) \leqslant 3$

$0 \leqslant j(= \text{column } \#) \leqslant 3$

$$R = \begin{cases} \overline{WS_1} & \text{for } j = 0,1 \\ WS_1 & \text{for } j = 2,3 \end{cases}$$

$$P = \begin{cases} \overline{WS_0} & \text{for } j = 0,2 \\ WS_0 & \text{for } j = 1,3 \end{cases}$$

RS lines: RS_1, RS_0
WS lines: WS_1, WS_0

[*b*]

make such a chip economically unattractive. Again, being fabricated from several gates/FFs, scratch-pad uses significant power and chip area. We shall look at some of the alternative memory sources in Chapter 12.

The applications of shift registers and counters are limited only by the imagination of the designer. Their importance will become more obvious as the design of digital systems continues in this and subsequent chapters.

EXAMPLE 10.6

Design a four-bit register capable of performing the 2's complement operation on its content.

SOLUTION

One of the procedures for performing the 2's complement operation (see Section 1.4 for details) requires that all bits on the left of the least significant nonzero bit be complemented. Using this scheme, a register can be built for loading the numbers in parallel and then performing the 2's complement operation using additional logic.

Define a control input, LOAD, to be used for selecting the parallel load operation, and its complement, $\overline{\text{LOAD}}$, for selecting the 2's complement operation. The loading excitation corresponding to a register consisting of only T FFs is given by

$$T_i = \text{LOAD} \cdot I_i + \overline{\text{LOAD}} \cdot (Q_{i-1} + Q_{i-2} + \cdots + Q_0) \quad \text{for } i > 0$$

where I_i is the parallel data input to the ith FF. The first term corresponds to loading the parallel input and the remaining terms correspond to performing a complement operation if at least one of the less significant bits is a 1. The equation is valid as long as the FFs have been CLEARed initially.

Note that in this complementing scheme Q_0 remains unchanged. Accordingly, the excitation equation for the least significant FF is given by

$$T_0 = \text{LOAD} \cdot I_0 + \overline{\text{LOAD}} \cdot (Q_0)$$

The overall excitation equation for the ith FF is now obtained as follows:

$$T_i = \text{LOAD} \cdot I_i + \overline{\text{LOAD}} \cdot (Q_{i-1} + (Q_{i-2} + (Q_{i-3} + (\cdots + (Q_0 + 0) \ldots))))$$

Consequently, if at least one of the bits on the right is a 1, the bit in question is complemented. The equation may be used to obtain the register circuit shown in Figure 10.45. The register should be LOADed with the data

FIGURE 10.45

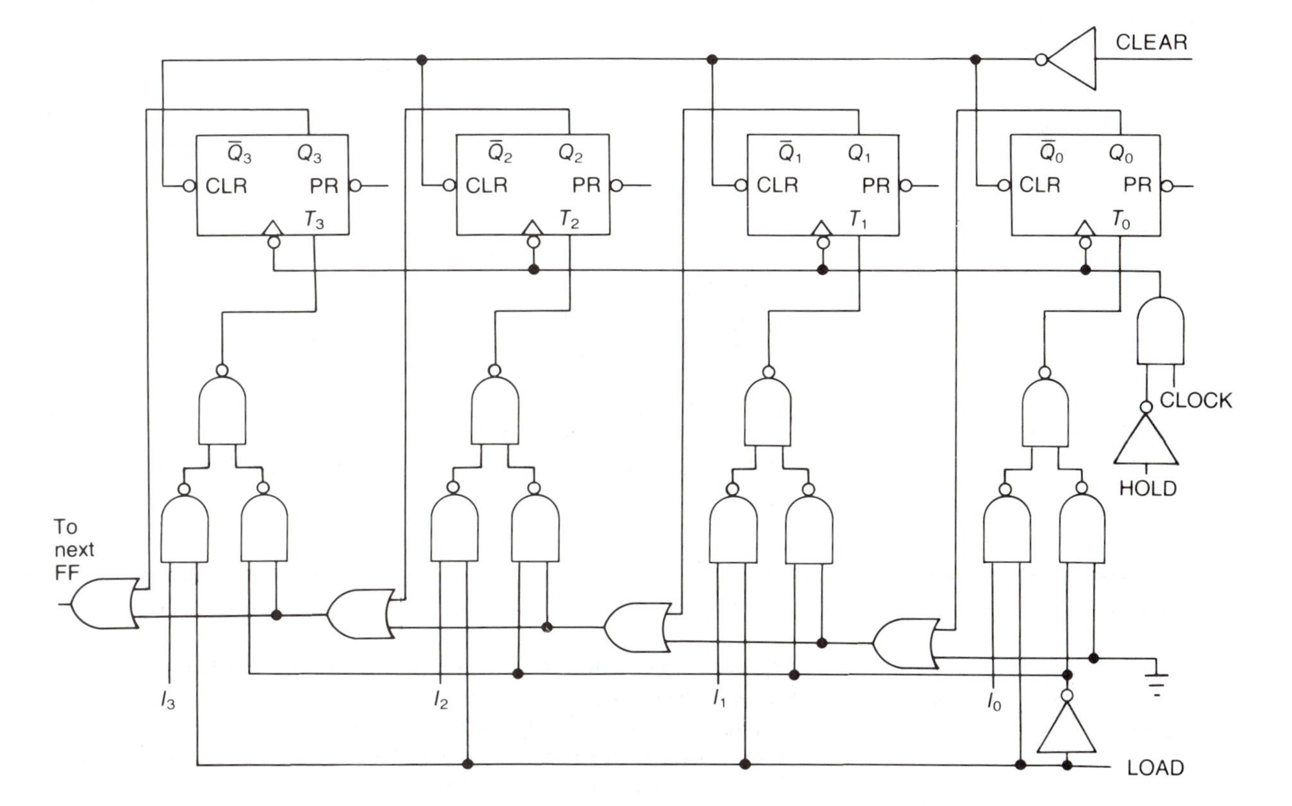

only after the FFs have been CLEARed. The corresponding 2's complement is then obtained by supplying a low LOAD input. Note that a high HOLD could be used at any time to keep the register content unchanged indefinitely.

EXAMPLE 10.7

Design a digital wristwatch that displays the month, day, and time accurate up to 1 second using BCD–to–seven-segment LED display devices. All display except for seconds should be adjustable by the corresponding external ADJUST switches. Assume that the ADJUST switches can be used only if the external DISABLE switch has been turned on. The external BEGIN switch may be used to resume the operation. Assume further that the MINUTE (MIN) ADJUST automatically resets the seconds and you have a 60 Hz quartz crystal to run your device.

SOLUTION

A detailed examination of the problem reveals that this device can be made to function in the prescribed way in several steps:

Step 1. Sixty of the 60 Hz clock pulses may be counted in sequence to indicate 1 second.

Step 2. Sixty seconds may be counted in sequence to indicate 1 minute.

Step 3. Sixty minutes may be counted in sequence to indicate 1 hour.

Step 4. Twelve hours may be counted in sequence to indicate one-half day.

Step 5. Two sequences of 12 hours may be counted to indicate 1 day.

Step 6. Days are counted, and once the count equals the maximum number of days in a given month, the month count should be made to go up by 1 and the day count should be set to 1.

Step 7. At the completion of 12 months, the month count should be set to 1.

From the steps listed it appears that this problem can be solved by interconnecting several counters and decoders. A list of the components needed for the wristwatch includes the following:

scale-of-sixty counters,

up-counter that counts 1 through 12,

binary-to-BCD converters,

decoding circuit to determine the last day of a month,

multiplexers,

T FF,

seven-segment display devices.

Steps 1–3 can be realized using three modulo-60 counters. Two modulo-12 counters could be used to implement Steps 4–7. Steps 5 and 6 can be implemented, respectively, using a T FF and last-day decoder circuit. BCD converters and seven-segment displays will be used for the purpose of display, and MUXs will be used for routing the data.

The IC counter module of Figure 10.18 will be used as the basic unit for producing the counter modules a and b as shown, respectively, in Figures 10.46 and 10.47.

Internally, module a consists of two four-bit IC counters cascaded together. The carry-out of the first is made to enable the second counter (see Figure 10.19 for a similar circuit). When the count reaches 59

FIGURE 10.46

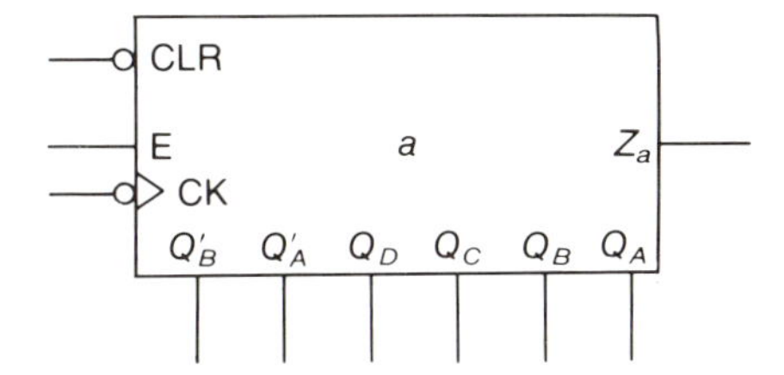

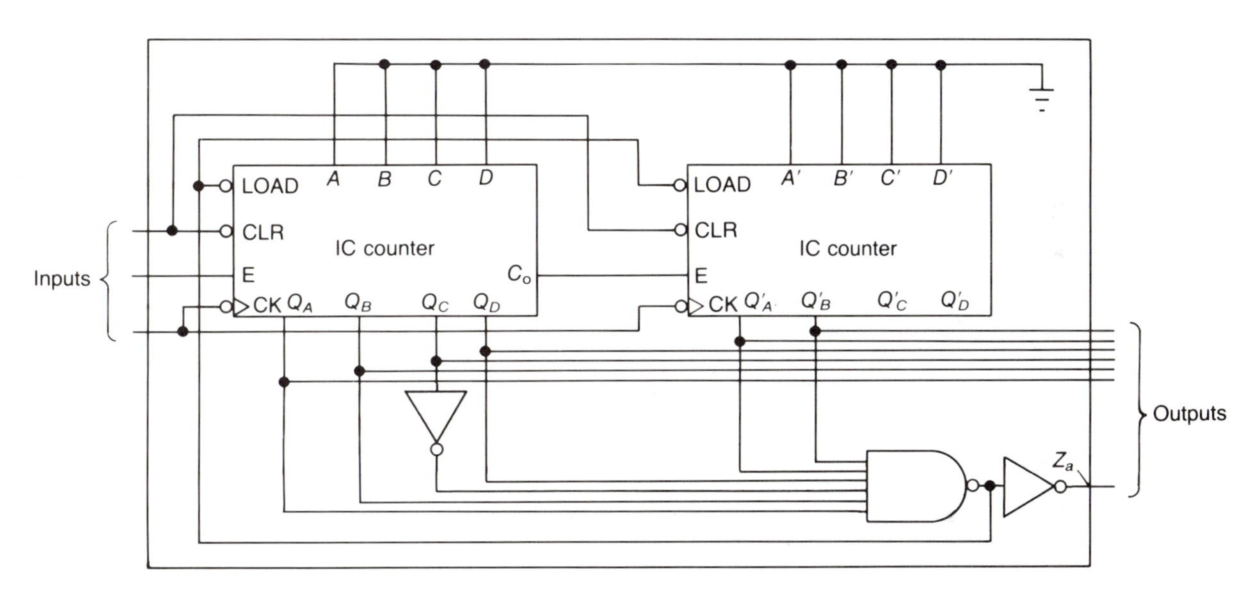

(111011), zero count is LOADed. However, module *b* consists of a single four-bit IC counter that is made to LOAD 0001 every time the count reaches 1100.

Module *c* was designed already in Example 5.7 (Figure 5.24). The only other module that remains to be designed is the decoding circuit for identifying the last day of a month. In terms of their lengths, the months may be classifed into four groups, respectively having a total of 28, 29, 30, and 31 days. The month of February has 28 days normally but has 29 days in a leap year. The module *d* may be designed accordingly, as shown in Figure 10.48, using a 1-of-4 MUX and two four-bit IC counters. The end-of-month is decided upon by the four values of *AB*. The MUX output becomes a 1 at the end of 28, 29, 30, and 31 input pulses when *AB* is 00, 01, 10, and 11, respectively. The decoding circuit is set to 1 each time an end-of-month has been located.

The modules may now be assembled to yield the wrist watch using the steps described earlier. In Step 1 the 60 Hz oscillator output is introduced into module *a*, as shown in Figure 10.49. A high SEC output indicates that an integral multiple of 60 input pulses has been counted. This is followed by Step 2, as shown in Figure 10.50, where a BEGIN input would allow another module *a* to count SEC pulses. For every integral multiple of 60

FIGURE 10.47

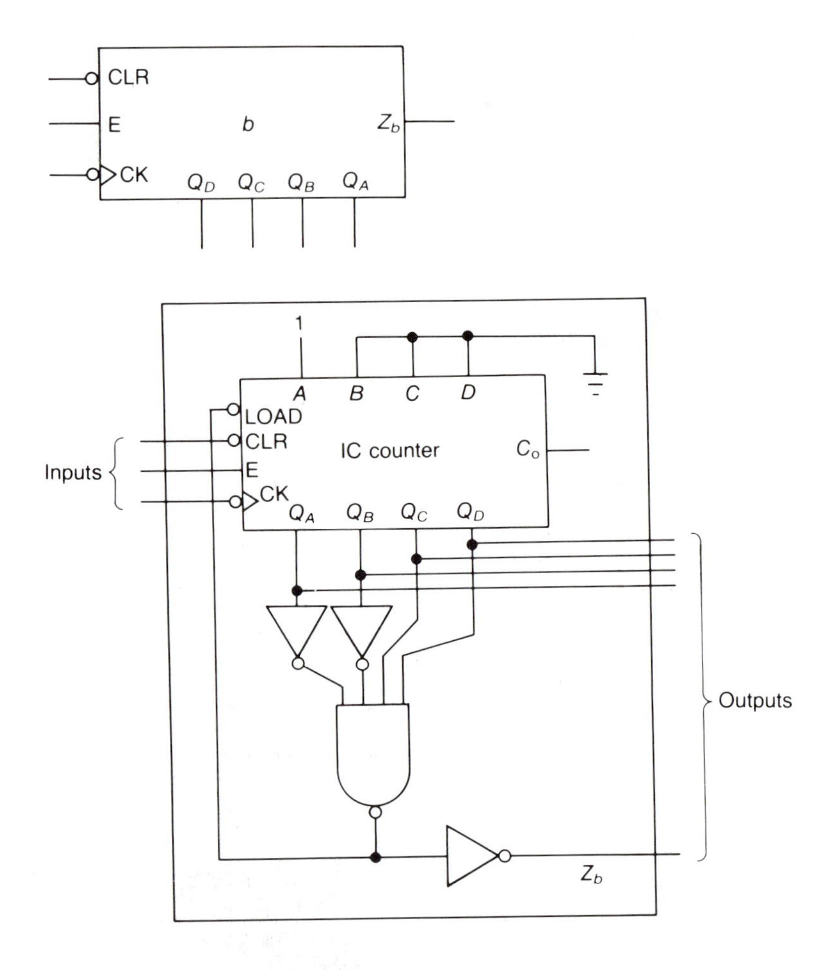

SEC pulses, a high will be generated at the minute output, MIN. The outputs of this module *a* are then converted to equivalent BCD numbers by means of three binary-to-BCD converter modules as shown in the circuit. For an explanation of this multi-bit, binary-to-BCD conversion scheme, review Example 5.8.

The BCD output is displayed by means of BCD–to–seven-segment LED display devices. The display will become 00 whenever the MIN ADJUST input is activated. This allows for the fact that the second display is cleared automatically whenever the minute display needs to be adjusted. Note, however, that the DISABLE input must always be activated prior to MIN ADJUST operation. The resulting low at *M* output would be used for all of the remaining ADJUST operations.

In Step 3, as shown by the circuit of Figure 10.51, the output of Figure 10.50 is fed into another module *a*. This configuration is made possible by means of a 1-of-2 MUX selectable by *M*. A high *M* would cause the circuit to count MIN pulses. In comparison, a low *M* would allow the adjustments of MIN counts. The HOUR output becomes a 1 corresponding to every integral multiple of 60 MIN pulses.

FIGURE 10.48

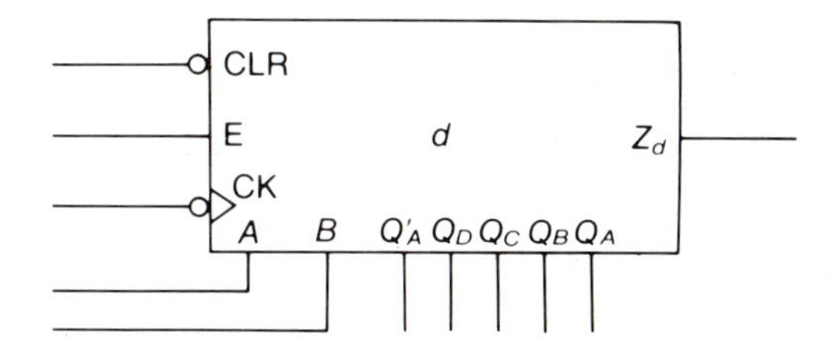

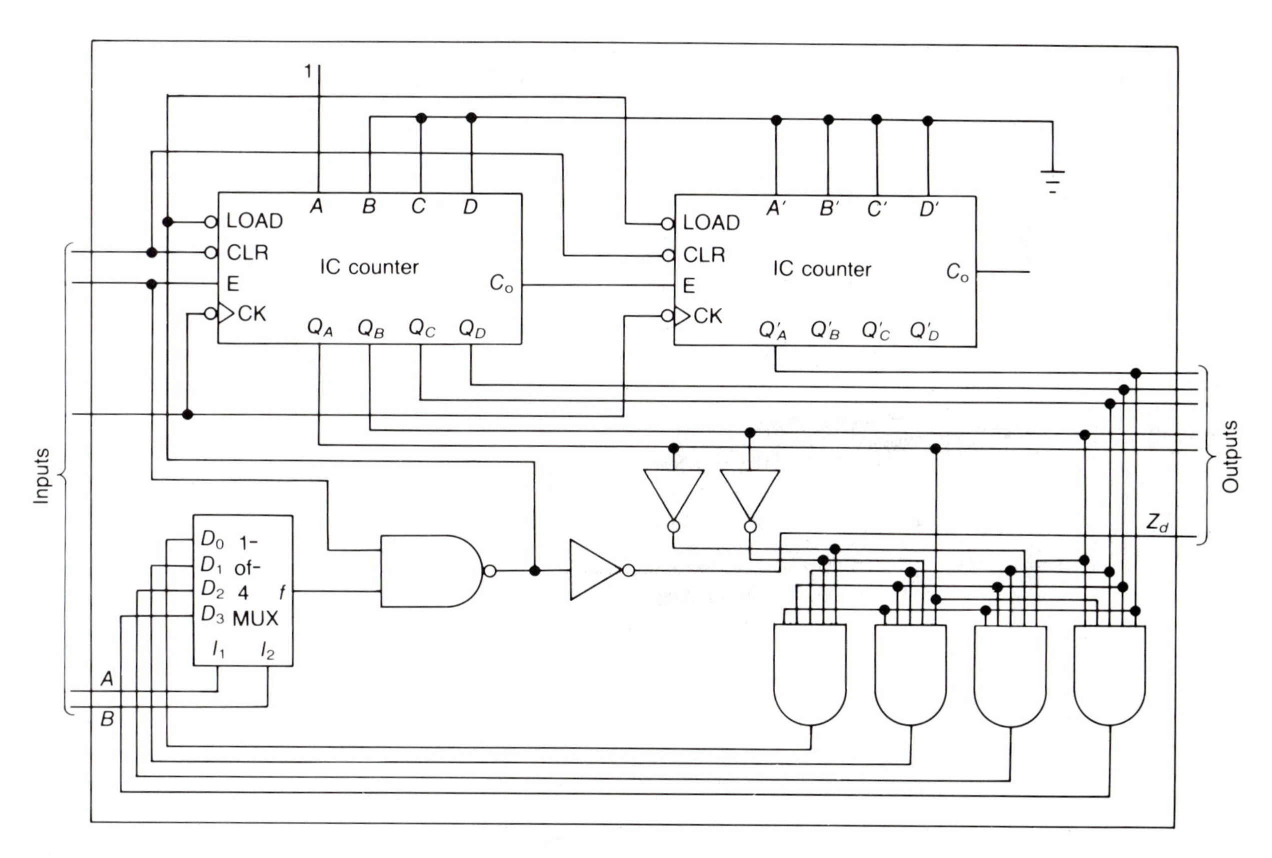

FIGURE 10.49

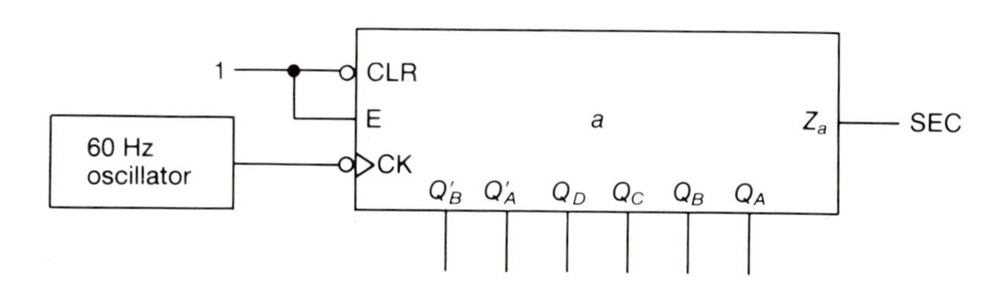

FIGURE 10.50

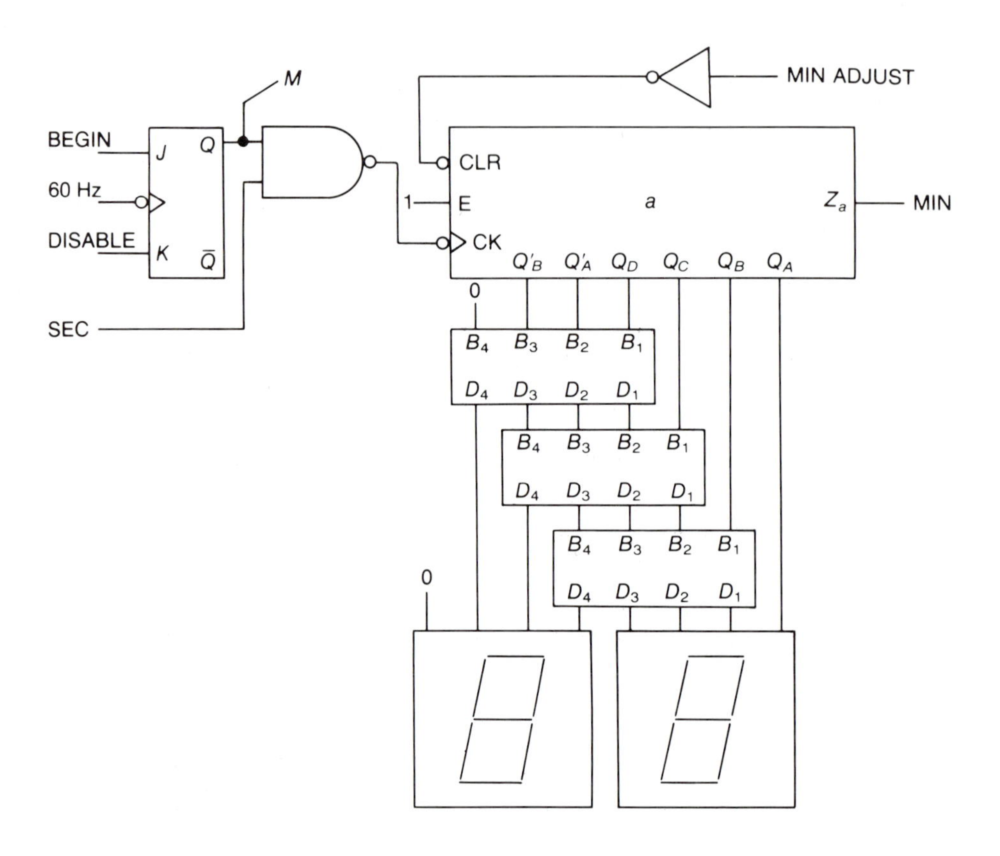

FIGURE 10.51

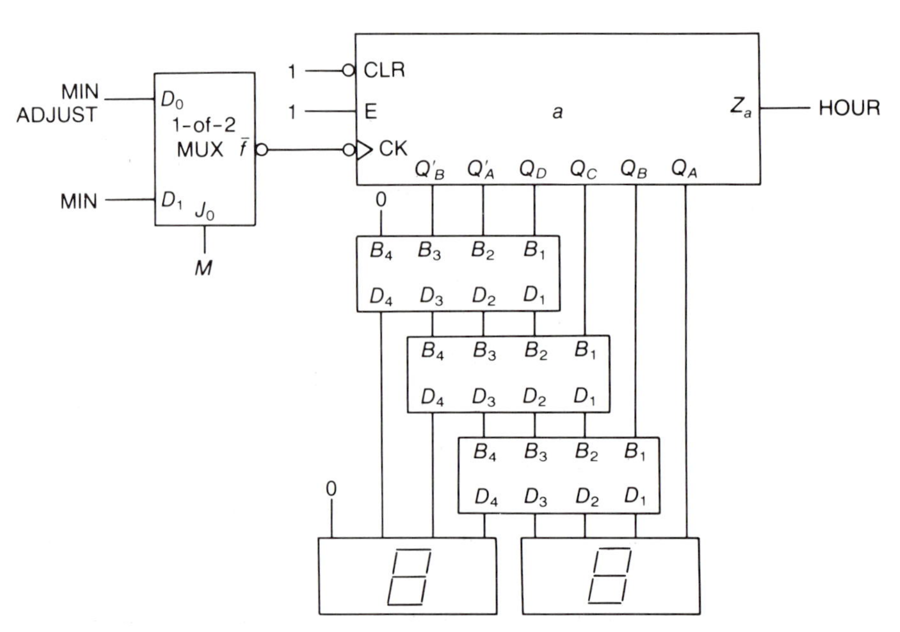

The circuitry in Step 4 is constructed using a module b, as shown in Figure 10.52. Its working principle is very similar to that of Figure 10.51. The 1/2-DAY output becomes a 1 corresponding to every integral multiple of 12 HOUR pulses.

The circuit of Figure 10.53 corresponds to both Steps 5 and 6. The counting circuit goes up by a 1 corresponding to every other 1/2-DAY pulse. Every time the end-of-month is located by means of AB inputs, the circuit of module d is set to day 1. The MONTH output becomes high each time the end-of-month is identified.

FIGURE 10.52

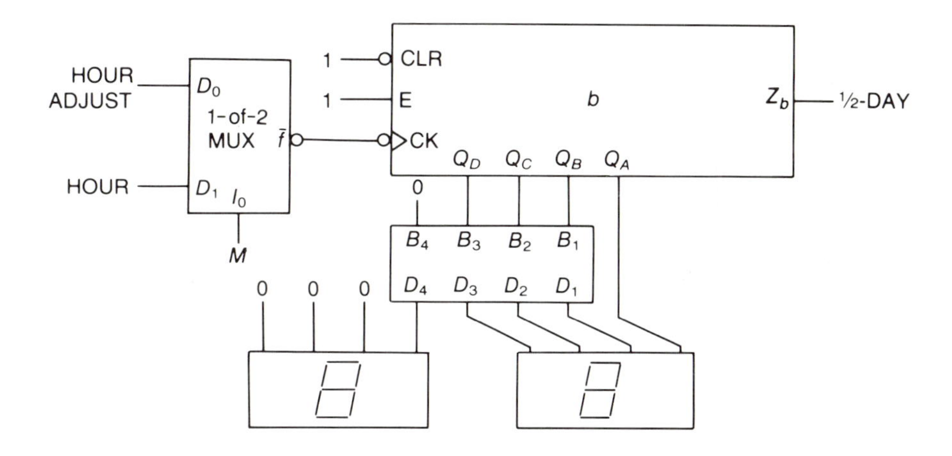

FIGURE 10.53

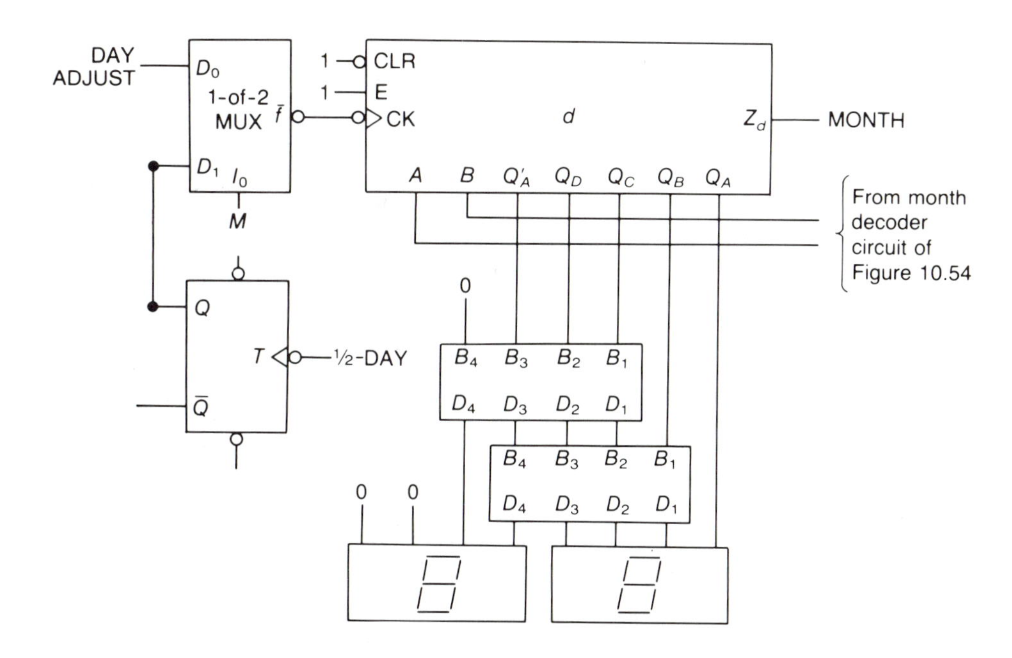

In the last step, as shown in Figure 10.54, the MONTH output is fed into a module *b*. The FF outputs, Q_D, Q_C, Q_B, and Q_A, and the external input, LEAP, are decoded to generate A and B. These outputs, A and B, are fed into module *d* of Figure 10.53. Note that LEAP is the extra input required for indicating that the current year is a leap year. This input needs to be entered prior to February 29. The month decoder circuit of Figure 10.54 examines the four FF outputs of module *b*. For the months January, March, May, July, August, October, and December, both A and B are set high. For all other months, except February, $A = 1$ and $B = 0$. In February $A = B = 0$ if LEAP = 0, and $A = 0$ and $B = 1$ if LEAP = 1. Consequently, the decoder has the following Boolean equations:

$$A = \overline{Q_D + Q_C + Q_A}$$
$$B = Q_D \oplus Q_A + \overline{Q}_D\overline{Q}_C \cdot \text{LEAP}$$

FIGURE 10.54

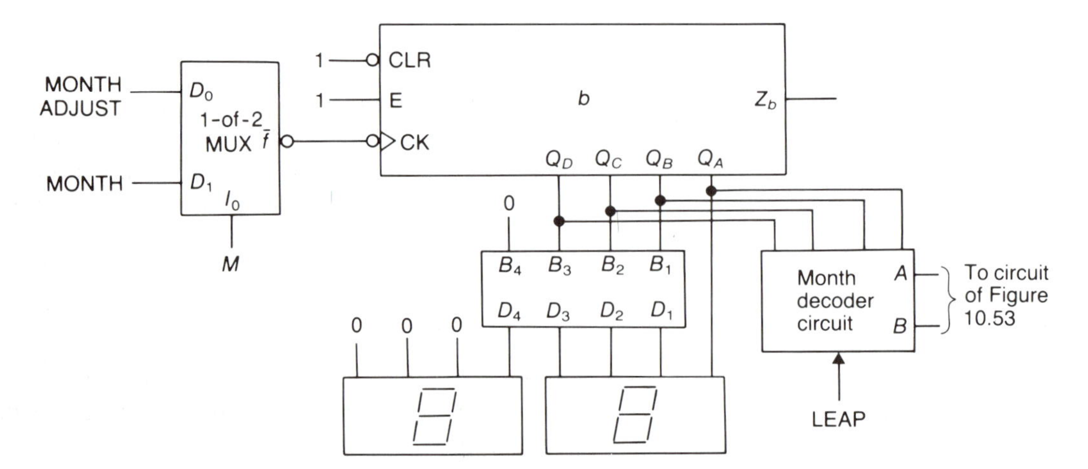

The design is now complete. After counting the chips involved, let's call this a table clock. Note that BCD counters instead of binary counters could have been used in this design, which would have reduced the total chip count. However, the incentives behind this particular design were to demonstrate the use of binary counters, to demonstrate the use of binary-to-BCD converter modules, and to demonstrate the thought processes involved in digital circuit designs.

10.11 Register Transfer Language Operations

The diversity of register types and applications points to the need for concise language to describe the flow of information (and processing enroute) of bits between registers. The most commonly used way to attain this goal is by an informal scheme called *register transfer language,* RTL, which was introduced first by I. S. Reed. A thorough investigation would reveal that a complete register transfer description is made up to two units: data and control. The *data unit* consists of registers, data paths, and logic necessary to implement a set of register transfers. The *control unit*, on the other hand, generates necessary signals in a specific sequence to regulate the register transfers within the data unit. The RTL scheme has the ability to specify the hardware involved in both units.

The simplest of all RTL operations is represented by $P \leftarrow Q$, which indicates that the data in register P are replaced by the data in register Q. It is also understood that both of these registers have the same number of bits. Such an operation can be completed during a single clock period, and thus it corresponds to a single-state transition of a sequential machine. One clock period, referred to as the *cycle time*, may be taken as the basic unit of time at the MSI complexity level. For uniformity, the following standardization is essential:

1. The contents of the registers should be denoted by one or more letters with the first always in uppercase. A concatenated register is the result of joining two or more registers in a string (represented by the register symbols separated only by commas) so that the LSB of the first-mentioned register is one bit to the left of the MSB of the second one, and so on.
2. Transfer between the registers will be considered parallel. In other words, all of the bits will transfer at the same instant of time.
3. The bits of each register shall be numbered from right to left. A_0 and A_{n-1} respectively represent the LSB and the MSB of an n-bit register named A. Note also that $A_{1,4}$ represents the bits 1 and 4 of register A, A_{1-4} represents bits 1 through 4 of register A, and A_M represents the subset of bits, M, of the register A.

Table 10.1 lists some of the most important RTL examples that include arithmetic, bit-by-bit logic, shift, rotate, scale, and conditional operations. In order to differentiate between the arithmetic and the logic operations, the following convention is maintained. The arithmetic addition is represented by a $+$ symbol, the logical OR operation by a $\vee$ symbol, and the logical AND operation by a $\wedge$ symbol. The shift, rotate, and scale operations are generally represented by two lowercase letters. For shift and rotate, the first letter indicates the type of operation (r for rotate and s for shift) and the second letter indicates the particular direction (r for right and l for left). Furthermore, in the rotate operation the LSB and the MSB are considered to be adjacent. For all shift operations a 0 will be assumed to occupy the vacant bit. For the scale operations scl indicates scale left and scr indicates scale right.

TABLE 10.1 **Examples of RTL**

Type of Operation	Meaning	Register Bits after Operation
General		
$A_3 \leftarrow A_2$	Bit 2 of A to bit 3 of A	A = 11110
$A_3 \leftarrow B_4$	Bit 4 of B to bit 3 of A	A = 11110
$A_{1\text{-}3} \leftarrow B_{1\text{-}3}$	Bits 1 through 3 of B to bits 1 through 3 of A	A = 11000
$A_{1,4} \leftarrow B_{1,4}$	Bits 1 and 4 of B to bits 1 and 4 of A	A = 10100
$A_{1\text{-}3} \leftarrow B_z$	Groups of bit Z of B to bits 1 through 3 of A	A = 11000
Arithmetic		
$B \leftarrow 0$	Clear B	B = 00000
$A \leftarrow B + C$	Sum of B and C to A	A = 11001
$A \leftarrow B - C$	Difference $B - C$ to A	A = 10111
$C \leftarrow C + 1$	Increment C by 1	C = 00010
Logic		
$A \leftarrow B \wedge C$	Bit-by-bit AND result of B and C to A	A = 00000
$A \leftarrow B \vee C_4$	OR operation result of B with bit 4 of C to A	A = 11000
$C \leftarrow \overline{C}$	Complement C	C = 11110
$B \leftarrow \overline{B} + 1$	2's complement of B	B = 01000
$B \leftarrow A \oplus C$	X-OR operation result of A and C to B	B = 10111
Serial		
$B \leftarrow$ sr B	Shift right B one bit	B = 01100
$B \leftarrow$ sl B	Shift left B one bit	B = 10000
$B \leftarrow$ sr2 B	Shift right B two bits	B = 00110
$B \leftarrow$ rr B	Rotate right B one bit	B = 01100
$B \leftarrow$ rl2 B	Rotate left two bits	B = 00011
$B \leftarrow$ scr B	Scale B one bit (shift right with sign bit unchanged)	B = 11100
$B \leftarrow$ scl B	Scale B one bit (shift left with sign bit unchanged)	B = 10000
$B,C \leftarrow$ sr2 B,C	Shift right concatenated B and C two bits	B,C = 0011000000
Conditional		
IF ($B_4 = 1$) $C \leftarrow 0$	If bit 4 of B is a 1, then C is cleared	C = 00000
IF ($B \geq C$) $B \leftarrow 0$, $C_1 \leftarrow 1$	If B is greater than or equal to C, then B is cleared and C is set to 1	B = 00000 C = 00011

Initial values: A = 10110, $B = 1\underbrace{100}_{z}0$ and C = 00001.

The operations as described in Table 10.1 may now be combined to write complex functions or a sequence of operations. The control conditions are included along with the operation state to distinguish one set of executions from another. Recall from the circuits of Figure 10.43 that the execution of an operation and the transfer of data are usually regulated by one or several control conditions. For example, the loading of the contents of A into the bus may be expressed as follows:

$$\bar{x} \cdot \bar{y} \cdot \bar{z}\text{: bus} \leftarrow A\ ;$$

The control condition $\bar{x}\bar{y}\bar{z}$ is separated from the corresponding operation by the ":" sign, while the sign ";" indicates the end of the operation. This RTL statement indicates that when $\bar{x}\bar{y}\bar{z} = 1$, the content of register A should be transferred to the bus. When more than one operation, $A \leftarrow B$ and $B \leftarrow \text{sr } B$, are to be performed under the same control condition $S = 1$, the operation is expressed as follows:

$$S\text{: } A \leftarrow B\ ;\ B \leftarrow \text{sr } B\ ;$$

The operations that are performed simultaneously follow the same control condition. Accordingly, when S is true, B is transferred to A and is also restored after being shifted right one bit. Similarly, the RTL operations describing the function of the circuit of Example 10.6 may be summarized as follows:

$$\text{CLEAR: } R \leftarrow 0\ ;$$

$$\text{LOAD: } R \leftarrow I\ ;$$

$$\overline{\text{LOAD}}\text{: } R \leftarrow \bar{R} + 1\ ;$$

where R is the register receiving the input bits from register I. These three RTL statements describe the action of a significant amount of hardware. In this and the next chapter we shall be discussing several of the complex circuits and their correlation with the corresponding RTL statements. Such one-to-one correspondence will make us appreciate the simplification that results from the use of RTL.

Consider the circuit shown in Figure 10.55. Here we have two registers, A and B, having four D FFs each. For simplicity the FFs

FIGURE 10.55 **Realization of a Register Copying Operation.**

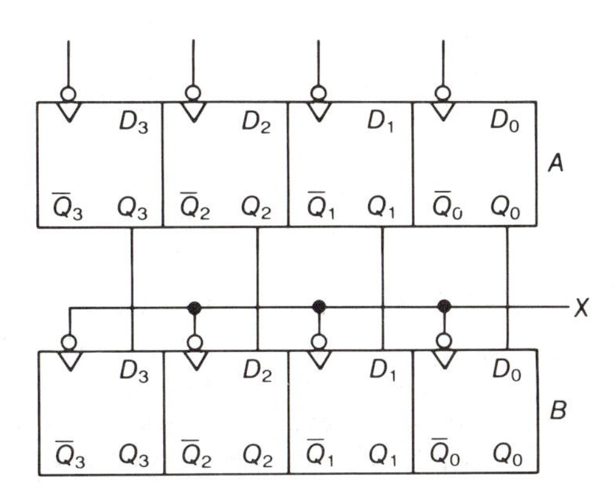

within the registers are not internally cascaded together. The FF outputs of register A are connected to the respective D inputs of register B. Corresponding to the trailing edge of X input, the contents of register A are loaded into register B. This hardware operation will be represented by the following RTL statement:

$$X\colon B \leftarrow A\ ;$$

This could also be written as follows:

$$X\colon B_3 \leftarrow A_3,\ B_2 \leftarrow A_2,\ B_1 \leftarrow A_1,\ B_0 \leftarrow A_0\ ;$$

Both of these operations are equivalent; but we would prefer to use the first form since it is more concise.

Note that a transfer operation is really a copying operation where the contents of the source register remain unaltered. This type of RTL operation is the most common, but there are other possibilities. Consider the four situations of Figure 10.56. In each of these cases each of the registers is of four-bit length.

In Figure 10.56[*a*] the $\overline{Q}$ outputs of register A are connected to the respective D inputs of register B. A clock input P would result in the following transfer:

$$P\colon B \leftarrow \overline{A}\ ;$$

Figure 10.56[*b*] shows register A where its MSB is connected to a 0. Each of the Q_n outputs of register A, except for the LSB, is fed to the D_{n-1} input of the same register. A pulse at R would cause the following transfer:

$$R\colon A \leftarrow \text{sr}\ A\ ;$$

FIGURE 10.56 **Realization of: [*a*] a Complement Transfer Operation, [*b*] a Shift-Right Operation.**

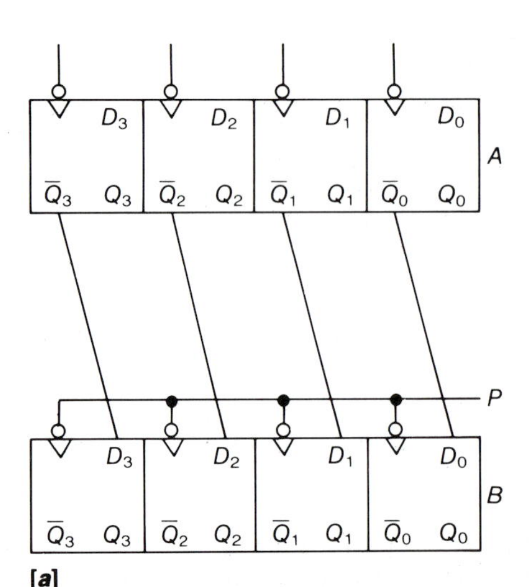

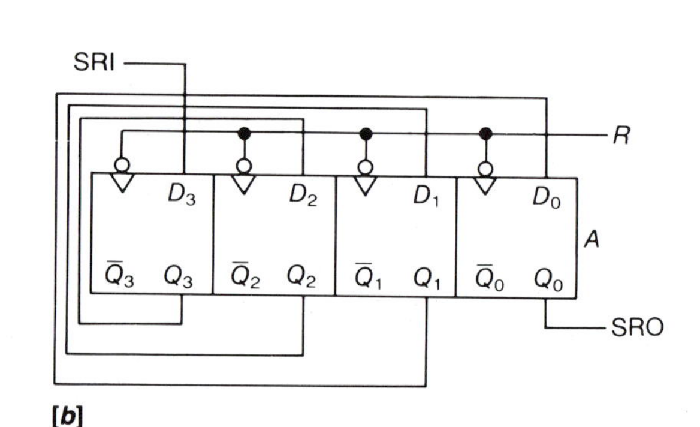

FIGURE 10.56 (*Continued*)
Realization of: [*c*] a Rotate-Right Operation, and [*d*] a Logical Operation and Transfer.

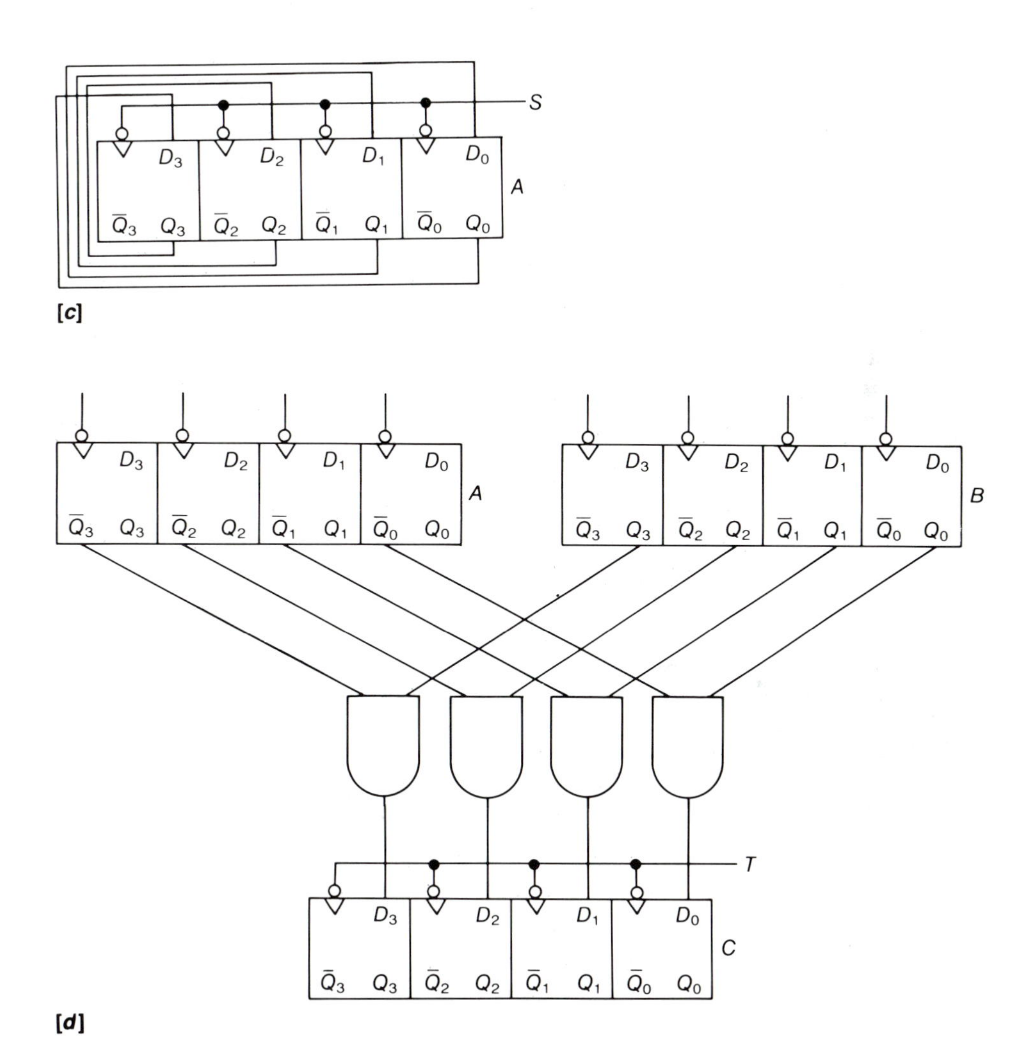

Similarly, Figures 10.56[*c–d*] respectively correspond to the following register transfers:

S: $A \leftarrow \text{rr } A$;

T: $C \leftarrow \overline{A} \wedge B$;

Note that in all of these valid RTL expressions, the source and destination registers have the same number of bits.

EXAMPLE 10.8

Design a typical stage for performing the following operations:

T_1: $A \leftarrow 0$;

T_2: $A \leftarrow A \vee B$;

T_3: $A \leftarrow A \wedge B$;

T_4: $A \leftarrow \overline{A}$;

where A and B are two multi-bit registers of equal bit size.

SOLUTION

You might decide to use JK FFs for the design of register A. Register B doesn't need to be designed because it is no different than a regular register with parallel outputs. The JK excitations necessary to turn on the ith FF for performing the required operations are obtained as follows.

$A \leftarrow 0$ is possible when $K_i = T_1$

$A \leftarrow A \vee B$ is possible when $J_i = T_2 \cdot B_i$ and $K_i = 0$

$A \leftarrow A \wedge B$ is possible when $J_i = 0$ and $K_i = T_3 \cdot \overline{B}_i$

$A \leftarrow \overline{A}$ is possible when $J_i = K_i = T_4$

The corresponding circuit is obtained as shown in Figure 10.57.

FIGURE 10.57

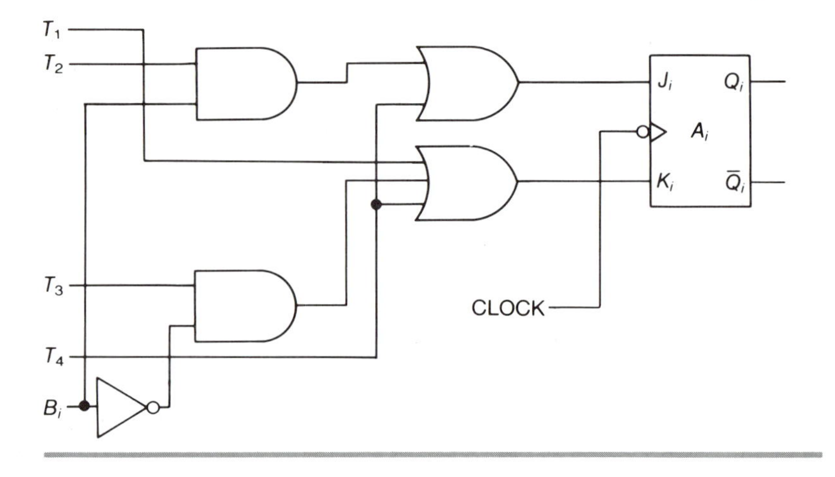

10.12 RTL Applications

The importance of RTL to describe the internal operations of a digital system is primarily due to the flexibility with which a design can be described and the direct way the data and control circuitry can be realized from RTL statements. Consider, for example, a system with a four-bit input, I, a four-bit output, O, and three four-bit registers, X, Y, and Z, in which the following algorithm is to be implemented:

A: $X \leftarrow I$;

B: $Z \leftarrow \text{sl2}\ X$;

C: $Y \leftarrow \overline{X}$;

D: $Y \leftarrow Y \wedge Z$;

E: IF $(Y_3 = 1)$ $O \leftarrow 0$ ELSE $O \leftarrow Y$;

The algorithm begins at state *A* and continues through state *E*. The input, *I*, is shifted left two bits and stored in *Z*, and the complemented input is stored in *Y*. This operation is followed by a bit-by-bit AND operation between the contents of *Y* and *Z*, and the result is stored subsequently in *Y*. If $Y_3 = 1$, only 0000 is placed on the output lines; otherwise, the contents of *Y* are placed on the output lines. In brief, this particular algorithm includes various logic, shift, and conditional transfers.

We shall now attempt to develop the hardware necessary from this RTL algorithm. Each of these RTL statements has two significant parts. In each statement, the right side of the $\leftarrow$ sign specifies the signals that must be generated for either storage in registers or transfer to output lines. Correspondingly, the left side of $\leftarrow$ specifies the destination registers or output lines. For this example $\overline{X}$ can be obtained by taking lines from the $\overline{Q}$ outputs of register *X*. No logic gates would be necessary for the shifting operation, as evident from Figure 10.56[*b*]. However, four AND gates would be necessary in state *D*. Figure 10.58 shows the connections necessary to develop the required signals, *I*, $\overline{X}$, sl2 *X*, *Y*, $Y \wedge Z$, and 0.

The next step is to feed the generated signals to the respective destination register or output lines. Figure 10.59 shows the involved connections. The complete RTL algorithm requires five clock periods. Corresponding to the first RTL statement, the inputs, *I*, are loaded to the *D* inputs of register *X* at the trailing edge of clock *A*, a transfer pulse fed to the clock input of register *X* during the allocated time for the first RTL statement. Similarly, at the trailing edge of clock *B*, $A_1A_0$00 are loaded into register *Z*. This is equivalent to the transfer of *A* after it has been shifted left twice. The third RTL statement requires that $\overline{X}$ be transferred to *Y*. This could have been possible simply by loading $\overline{X}$ to register *Y* at the trailing edge of clock *C*. However, it can be seen that the fourth RTL statement also includes a load operation involving register *Y*. To allow for these load operations, two separate ports of AND gates are used to select the inputs to *Y*, and a port of OR gates is used to combine them. The $\overline{Q}$ outputs of register *X* are transferred through one of the AND ports by *CL*, a level signal that is high during the period available for the third RTL statement. Similarly, the corresponding outputs of both *Y* and *Z* are transferred through the other AND port by *DL*. This AND port functions as the bit-by-bit AND logic circuit and also as the select port for register *Y*. The outputs of these two AND ports are loaded into register *Y* at the trailing edge of clocks *C* and *D*, respectively. The conditional transfer of the fifth RTL statement is realized when *EL* is high by ANDing each of the *Y* bits with $\overline{Y}_3$ and *EL*. It is appropriate to point out that when data are loaded into a register, they remain there until the power is turned off or other data are loaded. However, when data are placed

FIGURE 10.58 **Preliminary Step for the Development of Signals: [*a*] *I*, [*b*] $\overline{X}$ and sl2 *X*, [*c*] *Y* and $Y \wedge Z$, and [*d*] 0.**

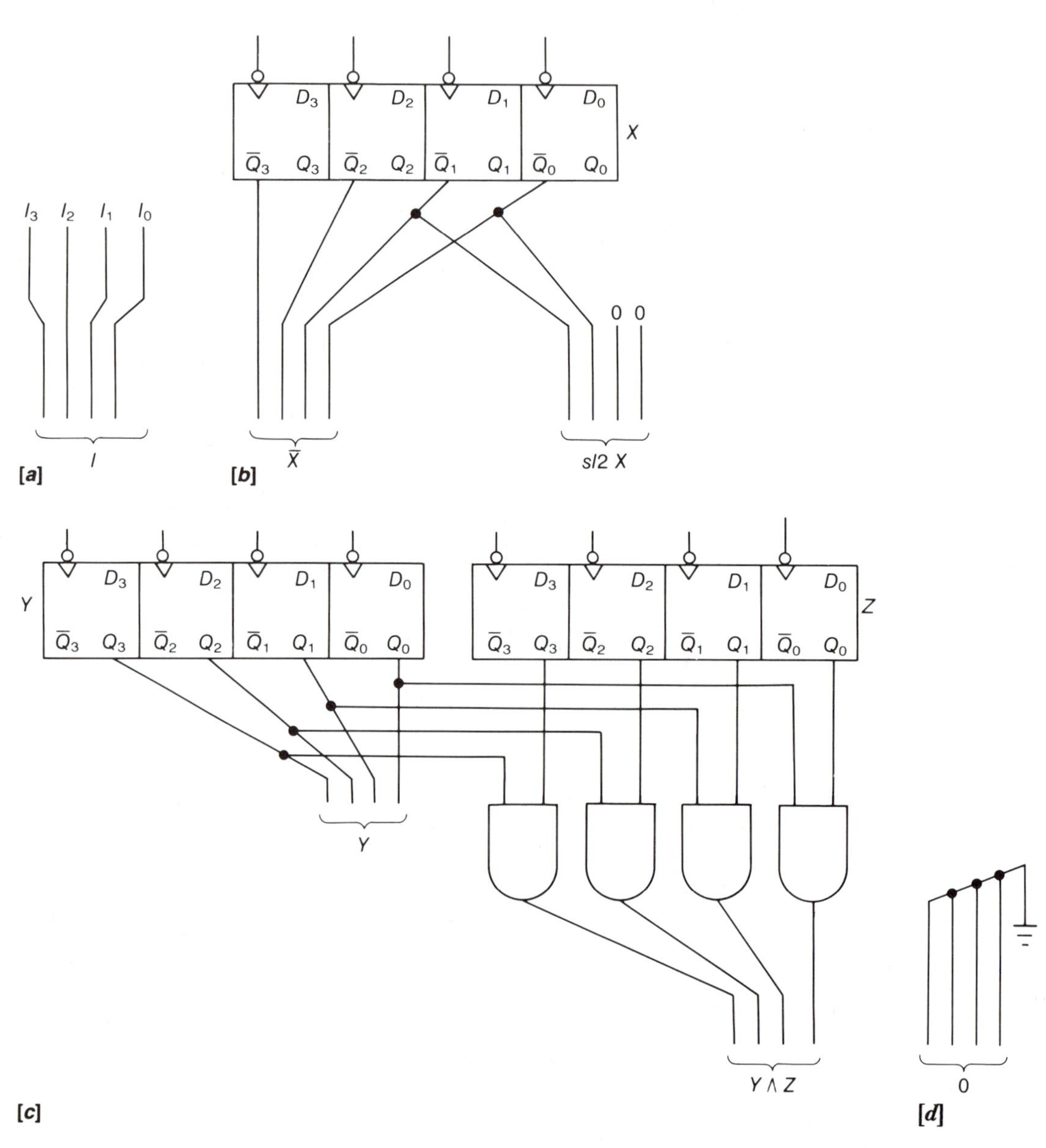

on the output lines, they remain there only during the steps for which they are valid.

The next step in the design is to regulate this algorithm by generating *A, B, CL, C, DL, D,* and *EL* signals in the proper order as specified by the algorithm. To run the system, a synchronizing sys-

FIGURE 10.59 **Complete Data Unit.**

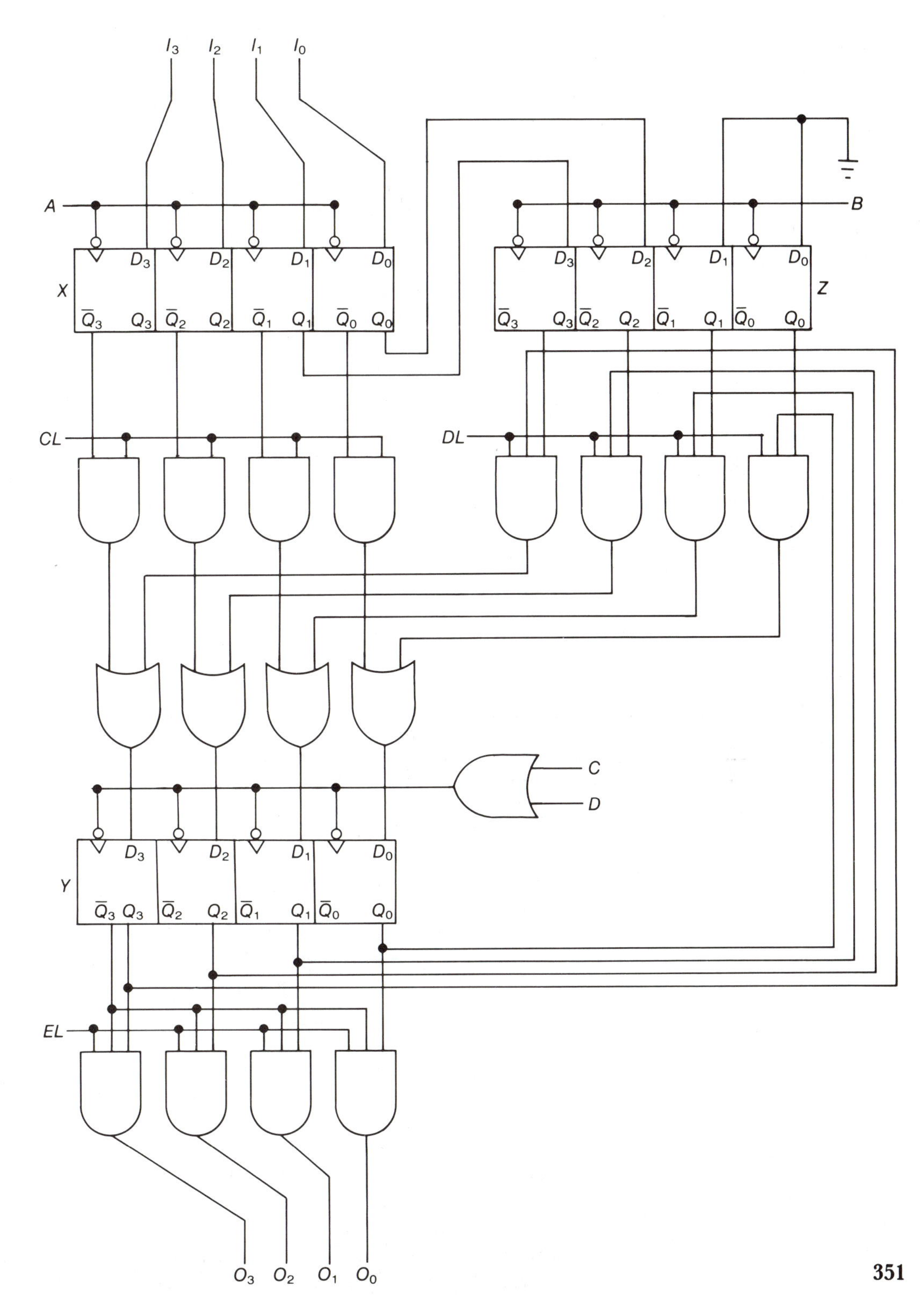

tem clock signal is necessary, as shown by the timing diagram of Figure 10.60. The time during which each step is valid is indicated by the level signals. When a level signal is ANDed with the system clock, the corresponding transfer clock pulse is generated. Even though AL and BL are not used for any of the operations in Figure 10.59, they are considered necessary in the control unit since they are used to generate A and B clocks, respectively. However, there is no need to generate the E clock signal.

We have already seen in Section 10.9 how various register organizations can be used to produce an operation sequencer. This particular algorithm requires only a five-bit ring counter, as shown in Figure 10.61. Using CLR input, the LSB of this ring counter is set and the remaining FFs are reset asynchronously. Consequently the FF outputs start generating the respective level signals in the correct order. The corresponding transfer clock pulses are realized by ANDing the level signals with the system clock.

It is now appropriate to consider several practical aspects about the functioning of this control unit. In order to correctly change Q_5 from a 0 to a 1, the CLR operation must be done in such a way as to assure a full clock period for the first RTL statement. The circuit might be expected to HALT this algorithm at the end of the algo-

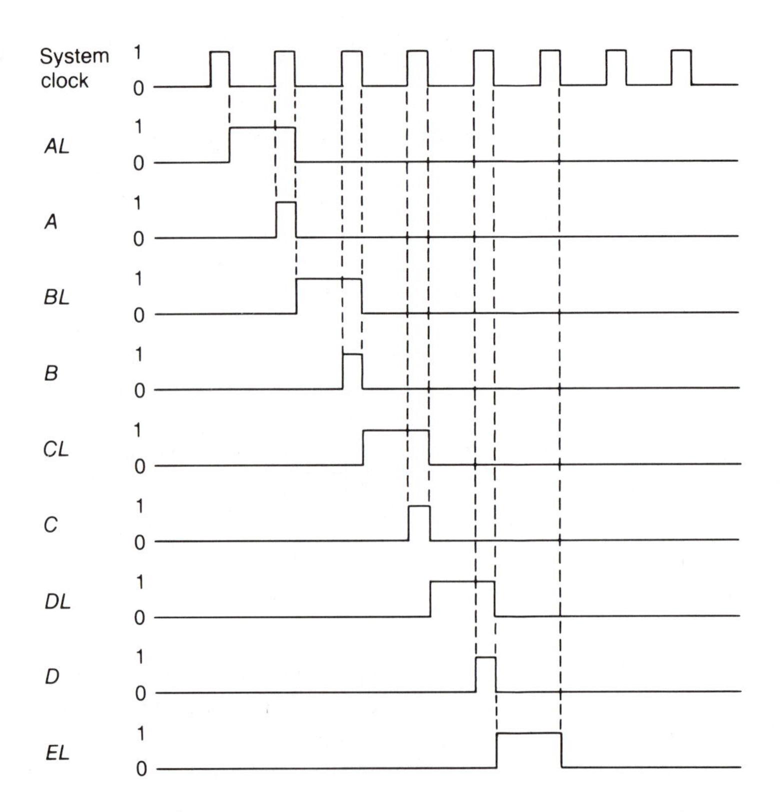

FIGURE 10.60 **Timing Diagram of the Control Unit.**

FIGURE 10.61 **Controller for the Data Unit of Figure 10.59.**

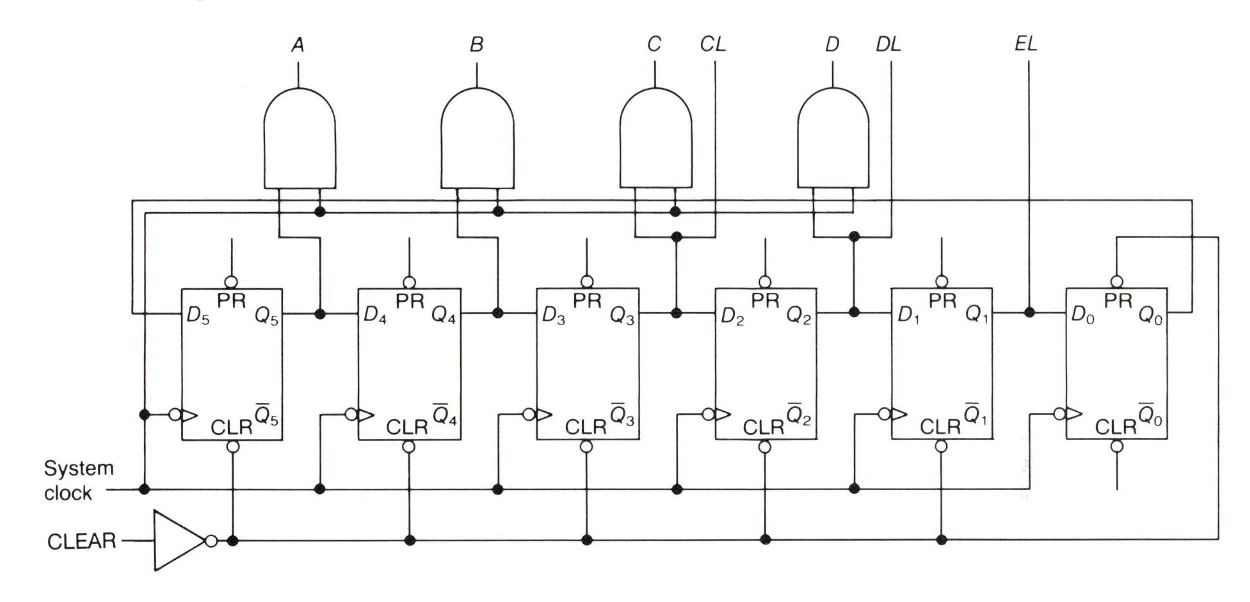

rithm, which could be accomplished by ANDing the system clock with $\overline{Q}_1$ prior to supplying it to various points of the circuit. This arrangement will disable the system clock when $Q_1 = 1$. *EL* becomes high and stops the control circuit from repeating the algorithm. Consequently the outputs would be available indefinitely. In many of the arithmetic circuits of the next chapter we will consider the HALT operation in more detail.

Throughout this section we have preferred to use negative edge-triggered FFs, primarily because transfer clock pulses are easily derived by ANDing the corresponding level signal with the system clock. Generating the transfer clock pulses would not be as easy if positive edge-triggered devices were used. Thus in the event a designer is faced with using a positive edge-triggered device, several modifications are necessary. These modifications involve inverting the system clock before feeding it to the clock input of positive edge-triggered devices. Furthermore, NAND gates rather than AND gates should be used for generating the transfer clock pulses.

10.13 Summary

In covering the application of what is known as the traditional sequential machines, our studies moved through synchronous and asynchronous counters; serial, parallel, and mixed-mode registers; and operation sequencers. An understanding of these functional units was subsequently applied to the development of RTL (register transfer language). This unique tool was then implemented in the design of both data and controller units. An understanding of the concepts presented is necessary, for much of the remaining material in this book is dependent on understanding this chapter.

Problems

1. Determine the state diagram of the three-bit programmed counter whose excitation equations are given as follows:

$$D_1 = Q_3Q_1 + Q_2\bar{Q}_1$$
$$D_2 = \bar{Q}_2(Q_1 + Q_3)$$
$$D_3 = \bar{Q}_3$$

2. Design a three-bit, Gray-code counter using (a) D FFs and (b) JK FFs.
3. Obtain a four-bit presettable asynchronous counter. Discuss its operation and significant characteristics.
4. Design a modulo-12 counter using a modulo-3 and a modulo-4 counter. Discuss its functions and characteristics.
5. The registers of Figure 10.38 store six bits each. A holds 010101 and B holds 001010. List the binary values in A and Q after each shift. Assume that Q was cleared initially.
6. You are given D FFs and several 1-of-4 MUXs and nothing else. Obtain the logic diagram for a three-bit register that is able to do the following: hold the present data, shift right, shift left, and load new data in parallel.
7. Design a four-bit circulate-right shift register using (a) D FFs and (b) JK FFs.
8. Verify the equations of the four-bit down-counter given in Section 10.2.
9. Design a four-bit shift register using JK FFs and a minimum number of assorted gates that performs the shift-left operation.
10. Explain how the unused sequences of Example 10.5 are obtained.
11. Design a divide-by-2048 counter using specific four-bit binary counters. The counter consists of a single FF, Q_A, followed by three cascaded FFs that form a divide-by-8 counter. It has two inputs, A and B, and four standard FF outputs, Q_A through Q_D. The two reset inputs, R_1 and R_2, clear the FFs when both are high. The counter counts in sequence when at least one of the reset inputs is low.
12. Obtain a circuit of as few FFs as possible to sequence 16 different operations.
13. Consider the truth table for a four-bit, Gray code–to–four-bit

binary equivalent conversion. It can be seen that if the MSB of the codes is disregarded, the numbers 8 through 15 will be found to be mirror images of the numbers 0 through 7. Design a parallel Gray code–to–parallel and serial binary converter using a four-bit shift register and only one FF. Explain the detailed functioning of the circuit.

14. Draw the logic diagram of a four-bit register with clocked *JK* FFs having control inputs for the increment, complement, and parallel transfer micro-operations. Show how the 2's complement can be implemented in this register.

15. Design a two-bit counter that counts up when control variable C is a 1 and counts down when C is a 0. No counting occurs when the control variable D is a 0.

16. Using a four-bit binary counter with synchronous clear and asynchronous load and clock action on the leading edge, complete the necessary circuit to make a two-digit BCD counter. Make it as hardware-efficient as possible.

17. You have available a four-bit adder and a large assortment of gates, counters, multiplexers, and decoders. Design and draw the circuit for a device that will multiply a two-bit (plus sign) sign-magnitude quantity by 3. *Note*: You don't have to use an adder.

18. Using the four-bit adder and any additional circuits you want, design a circuit that will multiply a number $X_2X_1X_0$ (assume signed magnitude with the sign bit handled elsewhere) by 2.5. The result is to be rounded to the next highest integer if the product results in a fractional part.

19. Design the circuit whose block diagram is shown in Figure 10.P1, using decoders, counters, MUXs, FFs, and so on. The circuit has as inputs a clock and four select lines. The select lines determine the number by which the input clock frequency is divided. The output is a clock divided by the select line value. If F is the frequency of the input clock and D is the binary value of the select lines, the output frequency, f, is $F/(D + 1)$.

FIGURE 10.P1

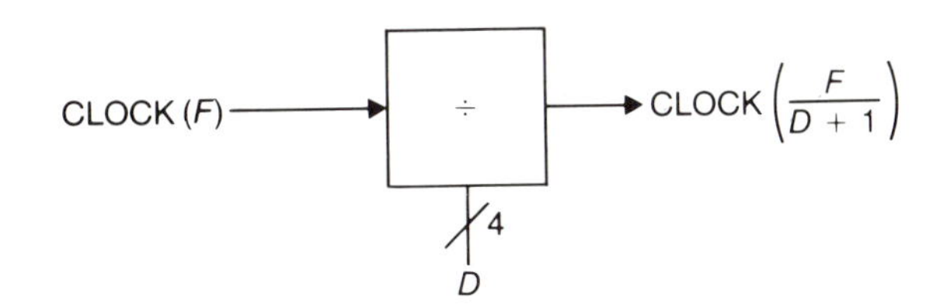

20. Design a two-bit counter that counts up one count at a time when $C = 0$ and counts up two counts per clock when $C = 1$.

21. Using a four-bit universal-shift register, design a 12-bit, serial-in, parallel-out, left-shift register.
22. Using a four-bit universal-shift register, design a 12-bit, parallel-in, serial-out, right-shift register.
23. Using a four-bit universal-shift register, design a 16-bit, universal-shift register.
24. Use four-bit binary counters in parallel and assorted gates to realize a divide-by-39 counter. Show the circuit configuration.
25. Design a typical stage that implements the following logic microoperations:

    a. Q_1: $A \leftarrow A \vee \overline{B}$
       Q_2: $A \leftarrow \overline{A} \wedge B$
       Q_3: $A \leftarrow \overline{A \vee B}$
       Q_4: $A \leftarrow \overline{A \wedge B}$

    b. P_1: $A \leftarrow \overline{A} \vee B$
       P_2: $A \leftarrow A \wedge \overline{B}$
       P_3: $A \leftarrow \overline{A \vee \overline{B}}$
       P_4: $A \leftarrow \overline{\overline{A} \wedge B}$
       P_5: $A \leftarrow A \oplus B$

26. Describe the function and characteristics of the counter circuit of Figure 10.P2.

FIGURE 10.P2

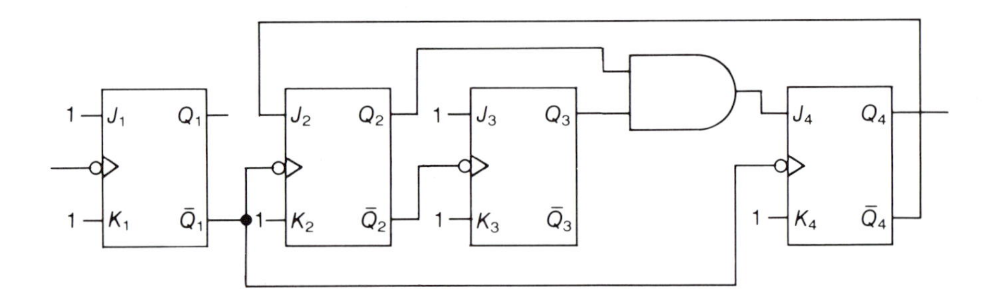

27. Obtain the universal state diagram for a four-bit, shift-left feedback register.
28. Obtain the universal state diagram for a four-bit bidirectional feedback register.
29. Design the additional circuitry that would be necessary in the table clock design of Example 10.7 so that the external LEAP input would not be needed. The table clock circuitry would have to take into consideration the consequence of a leap year on its own.
30. Design a wristwatch using BCD counters instead of binary counters. Comment on this design after comparing it with that of Example 10.7.
31. Design the complete data and control units for the algorithm of Problem 25a.
32. Design the complete data and control units for the algorithm of Problem 25b.

33. Using a seven-segment decoder and a display driver, design an electronic die as illustrated in Figure 10.P3. With equal probability the die should display the digits 1 to 6 on the seven-segment display when a switch is depressed and released. You may use a 1 MHz oscillator and a make/break push-button switch.

FIGURE 10.P3

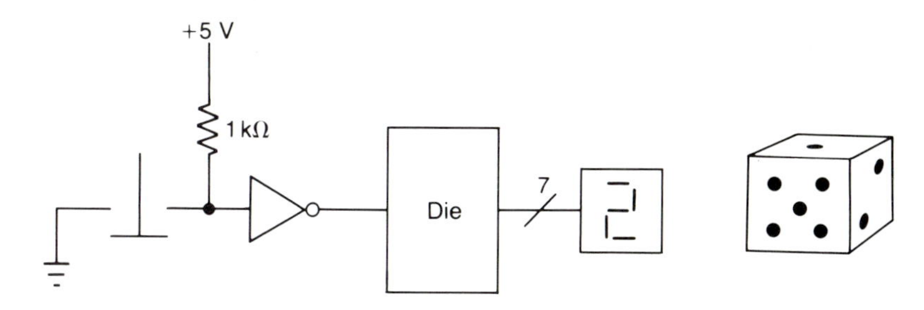

Suggested Readings

Awwal, A. A. S., and Karim, M. A. "A digital pseudo-random number generator scheme." Paper presented at the Proceedings of the Twenty-Eighth Midwest Symposium for Circuits and Systems, Louisville, Kentucky, August 1985.

Barbacci, M. R. "A comparison of register transfer languages for describing computers and digital systems." *IEEE Trans. Comp.* vol. C-24 (1975): 137.

Bell, C. G.; Eggert, J. L.; Garson, J.; and Williams, P. "The description and use of register transfer modules (RTM's)." *IEEE Trans. Comp.* vol. C-21 (1972): 495.

Berndt, H. "Functional microprogramming as a logic design aid." *IEEE Trans. Comp.* vol. C-19 (1970): 902.

Chambers, W. G., and Jennings, S. M. "Linear equivalence of certain BRM shift-register sequences." *Elect. Lett.* vol. 20 (1984): 1018.

Dadda, L. "Composite parallel counters." *IEEE Trans. Comp.* vol. C-29 (1980): 942.

David, R. "Testing by feedback shift register." *IEEE Trans. Comp.* vol. C-29 (1980): 668.

Divan, D. M.; Hancock, G. C.; Hope, G. S.; and Burton, T. H. "Microprogrammable sequential controller." *IEE Proc. E., Comp. & Dig. Tech.* vol. 131 (1984): 201.

Dormido, S., and Canto, M. A. "Synthesis of generalized parallel counters." *IEEE Trans. Comp.* vol. C-30 (1981): 699.

Dormido, S., and Canto, M. A. "An upper bound for the synthesis of generalized parallel counters." *IEEE Trans. Comp.* vol. C-31 (1982): 802.

Downing, C. P. "Proposal for a digital pseudorandom number generator." *Elect. Lett.* vol. 20 (1984): 435.

Hayes, J. P. *Digital System Design and Microprocessors.* New York: McGraw-Hill, 1984.

Hill, F. J., and Peterson, G. R. *Digital Systems: Hardware Organization and Design.* 2d Ed. New York: Wiley, 1978.

Hill, F. J., and Peterson, G. R. *Digital Logic and Microprocessors*. New York: Wiley, 1984.

Kline, R. *Structured Digital Design Including MSI/LSI Components and Microprocessors*. Englewood Cliffs, N.J.: Prentice-Hall, 1983.

Lo, H. Y.; Lu, J. H.; and Aoki, Y. "Programmable variable-rate up/down counter for generating binary logarithms." *IEE Proc. E., Comp. & Dig. Tech.* vol. 131 (1984): 125.

Manning, F. B., and Fenichel, R. R. "Synchronous counters constructed entirely of J-K flip-flops." *IEEE Trans. Comp.* vol. C-25 (1976): 300.

Rhyne, V. T. "Serial binary-to-decimal and decimal-to-binary conversion." *IEEE Trans. Comp.* vol. C-19 (1970): 808.

Swartzlander, E. E. "Parallel counters." *IEEE Trans. Comp.* vol. C-22 (1973): 1021.

Tien, P. S. "Sequential counter design techniques." *Comp. Des.* vol. 10 (1971): 49.

Vasanthavada, N. S. "Group parity prediction scheme for concurrent testing of linear feedback shift registers." *Elect. Lett.* vol. 21 (1985): 67.

Winkel, D., and Prosser, F. *The Art of Digital Design—An Introduction to Top-Down Design*. Englewood Cliffs, N.J.: Prentice-Hall, 1980.

Design of Advanced Arithmetic Circuits

CHAPTER ELEVEN

11.1 Introduction

In previous chapters various techniques for digital subsystem design have been introduced. Many of these digital subsystems have been refined and exist as ICs, such as multi-bit adders, shift registers, counters, decoders, multiplexers, comparators, and many other MSI devices. MSI devices have already been used for realizing various complex operations. We have also seen how the various elements can be connected to form still more complex systems, described by RTL.

Our ultimate goal is to understand sufficiently how complex digital systems, of which the computer is one important example, work in order to design the different functional components that when interconnected form the complete system architecture. Whenever an arithemtic or logic operation is carried out on a digital computer, the operation generally involves multiple movement of multi-bit data from register to register. These register transfers are intimately linked with the control unit that coordinates the transfers in a certain specific order. The RTL algorithm of the complex operation provides the necessary clues for the design of both the data and the corresponding control units. This chapter presents the RTL techniques for designing different computer circuits that are useful in a host of different arithmetic manipulations. After studying this chapter, you should be able to:

- Design data and control units for addition and subtraction circuits;
- Design data and control units for multiplication and division circuits;
- Write RTL algorithms;
- Implement RTL algorithms.

11.2 Binary Serial Adder/Subtracter

The possible circuit for a binary serial adder was introduced in Section 10.9. Starting from the LSB, one bit of addend and one bit of augend are shifted right into a single-bit FA. The resultant sum bit is stored at the MSB of either the addend or augend register. The carry-out of the add operation, however, is fed back as carry-in to the same FA along with the next shifted addend and augend bits. This shift-add scheme would be continued until all bits of the addend and augend were processed. In this section 2's complement capability will be incorporated into the proposed adder design to allow it to be used as a serial subtracter also. For simplicity of understanding we shall limit the design to only four bits. The control strategies necessary for operating the circuit will also be developed.

When you analyze the operation of such a unit, it becomes obvious that the unit must be provided with several inputs to control the operation. As obvious from the circuit of Figure 10.38, the two data inputs are either a four-bit addend and a four-bit augend or a four-bit minuend and a four-bit subtrahend. At least one control input to the unit is necessary to indicate whether the intended operation is addition or subtraction. In addition, a system clock pulse must be provided so that sequencing of the different operations may be synchronized and controlled. Examples of such control sequencing were introduced in Figure 10.39 and Figure 10.42.

The first step in our design process is to split this data unit into several subunits and design each of these units separately. The identifiable parts are: four-bit registers for storing the two four-bit data, one single-bit FA for addition, and a flip-flop for storing the carry-out. This process of identifying the subunits follows directly from our previous discussion of Figure 10.38. The circuit of Figure 5.12 should suggest ideas about how to convert an adder to an adder/subtracter unit. You will see that an X-OR gate is necessary to complement the subtrahend, and a combinational unit must be designed and incorporated for indicating an overflow condition. Finally for the system to work correctly, a control circuit needs to be designed for controlling the different sequential operations. The serial adder/subtracter circuit consequently requires the following subunits:

an augend/minuend register,

an addend/subtrahend register,

a single-bit FA,

a carry/borrow FF,

a two-input X-OR gate,

an overflow circuit,

a control circuit.

Figure 11.1 shows a simplified block diagram for such an arithmetic unit.

FIGURE 11.1 **Block Diagram of a Serial Adder/Subtracter.**

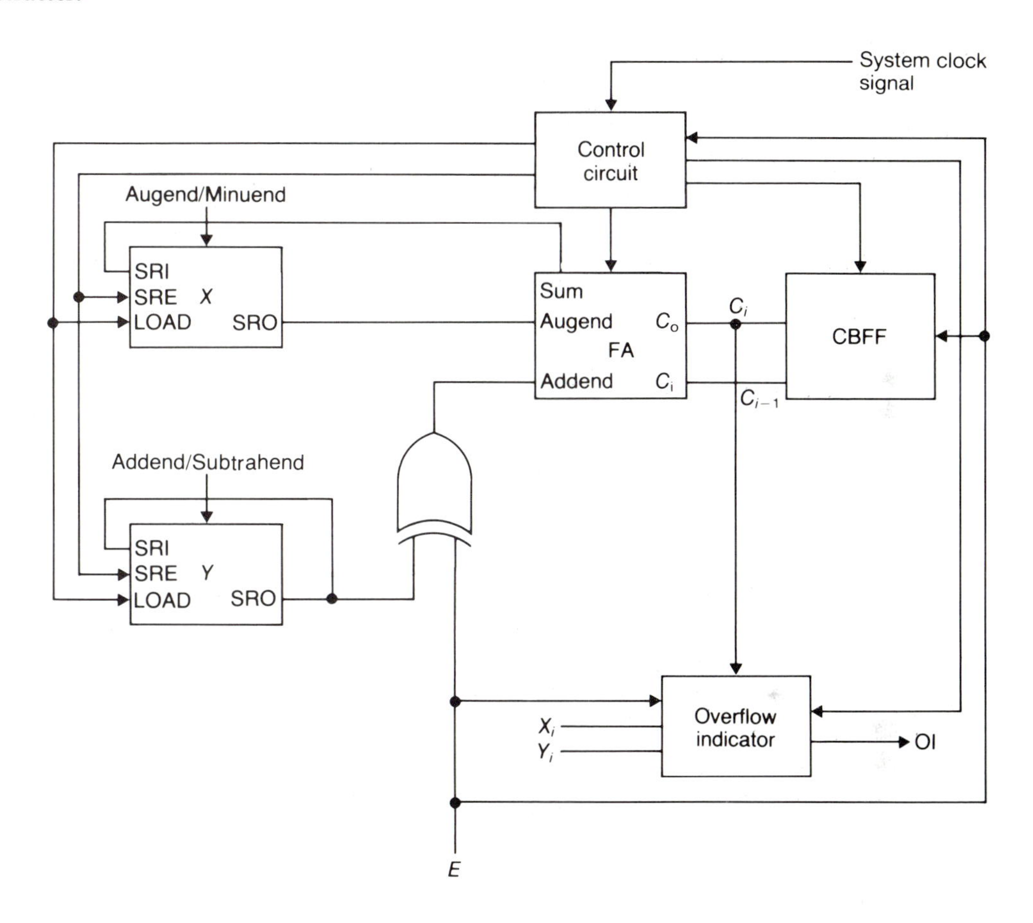

The subtraction operation is best performed using 2's complement arithmetic. This method involves taking the 1's complement of each bit of register Y as it enters the FA and setting the initial content of the carry/borrow storage FF, CBFF, to a 1. The control variable, E, selects an addition when equal to 0, and subtraction when equal to 1. The two values to be operated on are to be loaded into the registers in parallel. However, when $E = 0$ the initial value of the CBFF should be reset to 0.

The fundamentals of RTL were introduced in detail in Chapter 10. The specifications of advanced digital systems often take the form of one or more algorithms that must be implemented in hardware. Algorithms of this kind are often most conveniently expressed by means of a set of RTL statements. The RTL algorithm of the n-bit adder/subtracter unit can be expressed as follows:

Algorithm 1

$S = 1$: $X \leftarrow$ (augend/minuend) ; $Y \leftarrow$ (addend/subtrahend) ;
$\mathrm{CBFF}_0 \leftarrow E$; $\mathrm{OI} \leftarrow 0$; $N \leftarrow 0$;

$S = 2$: $X_{n-1} \leftarrow X_0 + (Y_0 \oplus E) + \mathrm{CBFF}_0$; $Y \leftarrow \mathrm{rr}\ Y$;
$\mathrm{CBFF}_0 \leftarrow$ (carry-out) ; $X_{0-(n-2)} \leftarrow X_{1-(n-1)}$;
$N \leftarrow N + 1$;

$S = 3$: IF $(N < n - 1)$ $S \leftarrow 2$ ELSE IF $(N = n - 1)$ $S \leftarrow 4$
ELSE $S \leftarrow 5$;

$S = 4$: IF $(X_0[Y_0 \oplus E]\overline{\mathrm{CBFF}_0} + \overline{X}_0\overline{[Y_0 \oplus E]}\mathrm{CBFF}_0 = 1)$
$\mathrm{OI} \leftarrow 1$; $S \leftarrow 2$;

$S = 5$: STOP ;

It can be seen that even after the detection of an overflow, the algorithm goes on to determine a sum that is wrong. This may be considered a waste of cycle time (see Problem 3). However, from an examination of the RTL program, the necessary control sequences for the correct operation of the circuit can be determined.

The process STARTs from the step $S = 1$. The registers are loaded with the n-bit words and the CBFF is initially loaded with E. When $E = 0$ the initial carry-in is 0, that is, CBFF $\leftarrow 0$. When $E = 1$ a 1 is stored at the CBFF. The addition of this 1 to the LSB completes the 2's complement requirement that 1 be added to the 1's complement that is generated as the subtrahend is serially passed through the X-OR gate. The variable N keeps count of the number of add operations performed. A combinational comparator could be used to determine if the current count is less than, equal to, or greater than $n - 1$. The overflow indicator, OI, is turned on (set equal to 1) when there is an overflow. An overflow cannot be determined until there is only one more addition. The RTL expression corresponding to the step $S = 4$ determines whether or not an overflow would occur. If two numbers are added, there is always the possibility of an overflow if the word length of the result is constrained to be the same as that of the incoming operands. In order for the result to be valid the carry into and out of the sign bit must be the same. At step $S = 4$ the sign bits of the two values are in X_0 and Y_0. The four cases for which an overflow would result are

$$X_0 = Y_0 = 1, \quad E = 0, \quad \text{and} \quad \mathrm{CBFF}_0 = 0$$

$$X_0 = Y_0 = 0, \quad E = 0, \quad \text{and} \quad \mathrm{CBFF}_0 = 1$$

$$X_0 = 1, \quad Y_0 = 0, \quad E = 1, \quad \text{and} \quad \mathrm{CBFF}_0 = 0$$

$$X_0 = 0, \quad Y_0 = 1, \quad E = 1, \quad \text{and} \quad \mathrm{CBFF}_0 = 1$$

Note that the carry-in into the sign bit is available at $CBFF_0$ during step $S = 4$.

The two registers involved in storing the augend/minuend and addend/subtrahend have almost identical requirements. Both of the registers must be capable of shifting right and loading data in parallel. The register X stores the sum bit in its MSB while the register Y must be able to rotate right. Figure 11.2 shows a configuration for the four-bit register X using D FFs, where I_3–I_0 are the augend/minuend input data. The Y register is very similar except for the external connections. The rotate-right capability of Y is achieved simply by connnecting Q_0 to the shift-right input, SRI. A high at the LOAD input causes the loading of parallel input while a high at the shift-right enable, SRE, line causes the shift register to move its bits to the right.

FIGURE 11.2 **Augend/Minuend Register.**

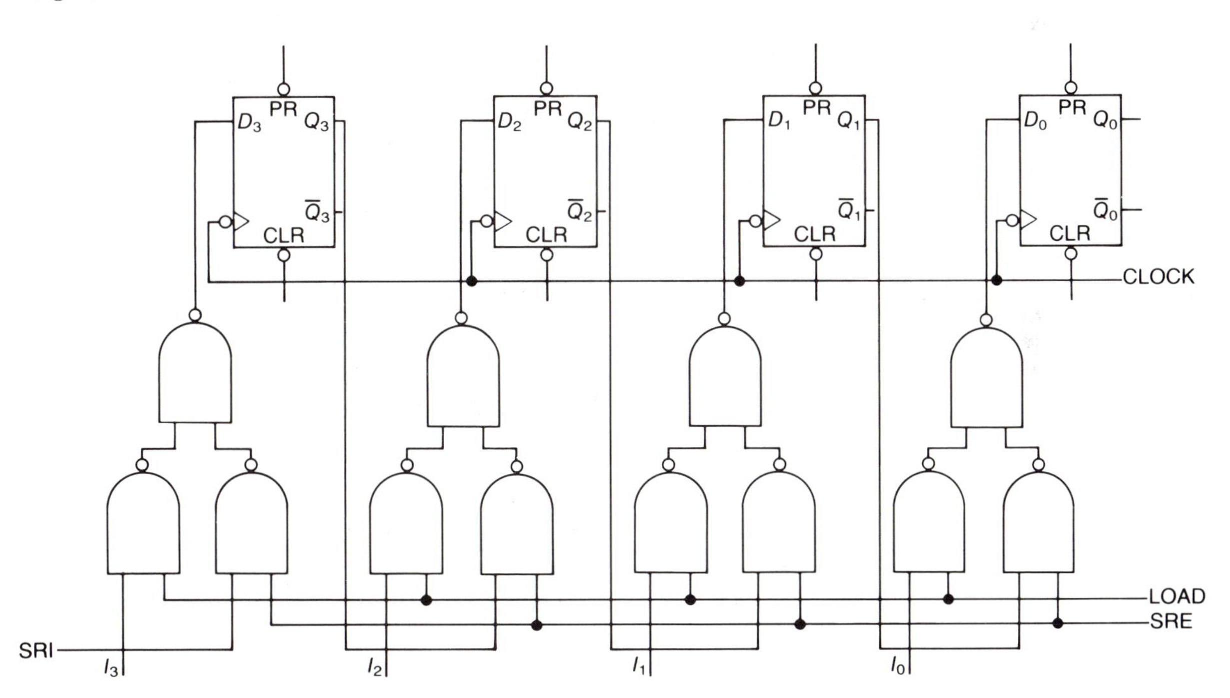

The single-bit FA circuit necessary for the serial adder/subtracter is no different than the one designed in Section 5.2. There is no need to consider a CLA design since only a single-bit FA is required. Equations [5.7] and [5.8] may now be modified to express the sum bit, S_i, and the carry-out, C_i, as follows:

$$S_i = X_0 \oplus (Y_0 \oplus E) \oplus C_{i-1} \qquad [11.1]$$

and

$$\begin{aligned} C_i &= X_0(Y_0 \oplus E) + C_{i-1}[X_0 \oplus (Y_0 \oplus E)] \\ &= X_0(Y_0 \oplus E) + C_{i-1}[\overline{X}_0\overline{Y}_0E + X_0 + \overline{X}_0Y_0\overline{E}] \\ &= X_0(Y_0 \oplus E) + C_{i-1}(Y_0 \oplus E) + X_0C_{i-1} \end{aligned} \qquad [11.2]$$

The inputs at the CBFF are controlled by a variable SRE such that when $S = 1$, E should be loaded as the initial carry, and when $S \neq 1$ the carry-out should be shifted in. The input to the CBFF (which could be built using a D FF), therefore, is obtained as follows:

$$\begin{aligned} D = {} & \text{SRE} \cdot [X_0(Y_0 \oplus E) + C_{i-1}(Y_0 \oplus E) + X_0C_{i-1}] \\ & + \overline{\text{SRE}} \cdot E \end{aligned} \qquad [11.3]$$

where SRE is a control signal variable that is on during all of the steps except when $S = 1$.

The overflow condition is detected whenever the carry-out into the sign bit is either a 1 corresponding to a positive result, or a 0 corresponding to a negative result. Consequently, the overflow indicating function is not evaluated until the step $S = 4$ has occurred. When $S = 4$, an *overflow control signal*, OCS, may be generated by the control unit to monitor this operation. The overflow indicating function, therefore, is given by

$$\text{OI} = \text{OCS} \cdot [X_0(Y_0 \oplus E)\overline{\text{CBFF}}_0 + \overline{X}_0\overline{(Y_0 \oplus E)}\text{CBFF}_0] \qquad [11.4]$$

Figure 11.3 shows the realization of the sum, carry-out, and overflow indicating function. The initial enable, E, is stored in an unclocked SR FF whose set input is connected to the SUBTRACT command and reset input is connected to the ADD command. The JK FF is used for indicating the overflow. The K input of this FF is turned on during the step $S = 1$ that clears the overflow indicator. This is accomplished by making sure that the control unit generates a *clear overflow signal*, COS, when $S = 1$.

The only other part of the circuit that is not designed as yet is the control circuit. This unit must generate the clock signal, CK, and the control signals, SRE, OCS, and COS. It must also generate the signals identifying the steps in the process. Upon re-examination of the RTL statements it should be apparent that in steps $S = 3$ and $S = 4$ there are no actions that cannot be in step $S = 2$. In fact these operations are also realizable along with those in step $S = 2$ by making careful use of the count, N, to determine whether or not they should occur. Therefore, steps $S = 2$ through $S = 5$ can be combined as follows to give a two-step RTL sequence:

FIGURE 11.3 **Adder/Subtracter and Overflow Logic.**

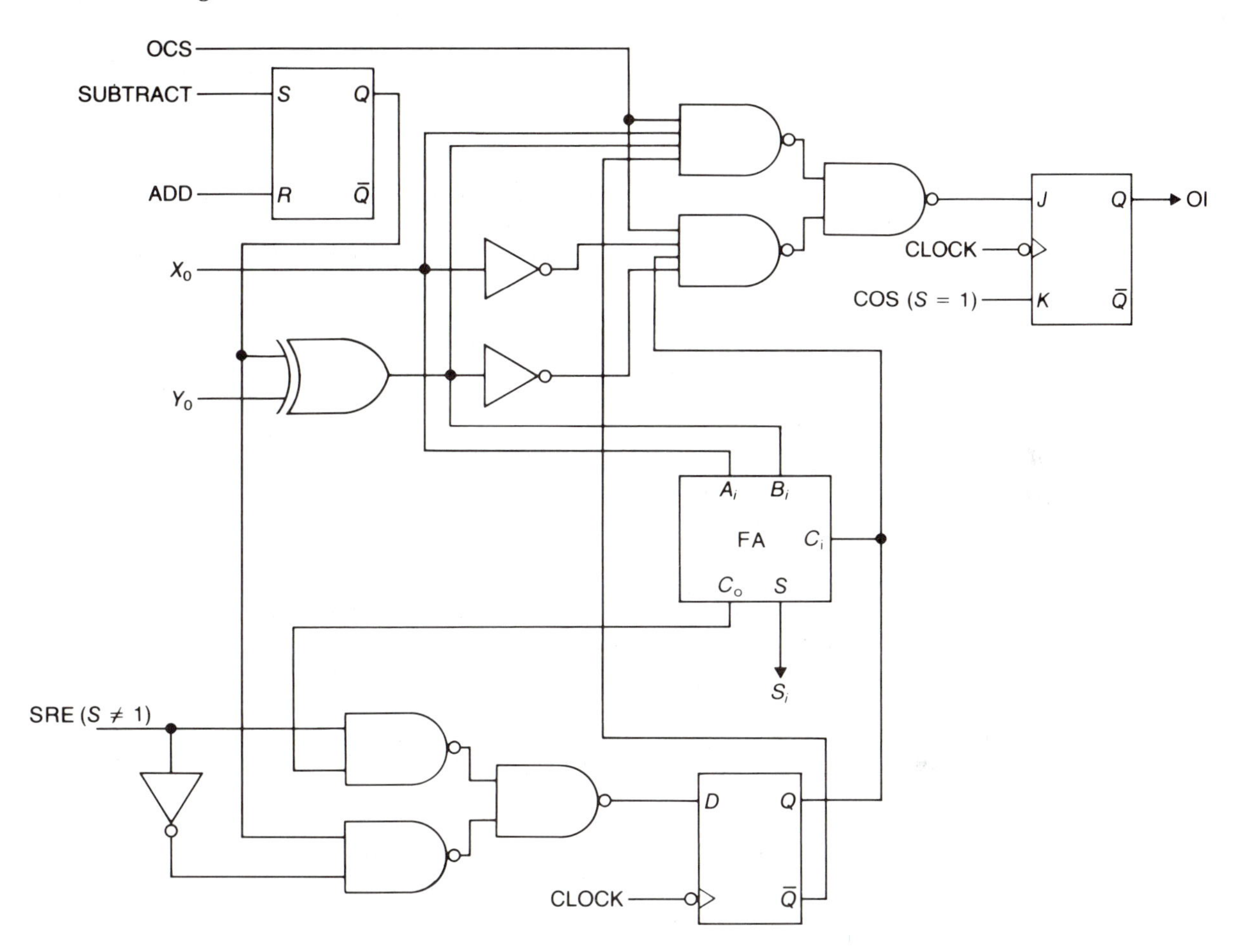

$S = 2$: $X_{n-1} \leftarrow X_0 + (Y_0 \oplus E) + \text{CBFF}_0$; $Y \leftarrow \text{rr } Y$;
$\text{CBFF}_0 \leftarrow$ (carry-out) ; $X \leftarrow \text{sr } X$; $N \leftarrow N + 1$;
IF $(N = n - 1)$ OI $\leftarrow X_0(Y_0 \oplus E)\overline{\text{CBFF}_0}$
$+ \overline{X_0}\overline{(Y_0 \oplus E)}\text{CBFF}_0$;
IF $(N = n)$ $S \leftarrow 3$ ELSE $S \leftarrow 2$;

$S = 3$: STOP ;

Since in this case $n = 4$, a combinational comparator is necessary that would test if the present count is either less than, or greater than, or equal to 3. Such a circuit could be designed and built quite easily (see Chapter 5, Problem 31). If the comparator was designed for testing four-bit numbers (a two-bit comparator would have been okay in our case), it would have two four-bit input ports, A and B, and three output lines, LT (less than), GT (greater than), and EQ (equal to). When A is less than B, the output LT would be high; when A is greater than B, the output GT would be high; and when A and B are equal, the output EQ would become high.

In order to design the circuit the designer should accurately estimate the time needed by the system to go through each of the sequencing cycles. The estimation should account for the worst-case propagation delay for each of the gates, FA, shift register FFs, and comparator. It is always a good practice to include a safety margin to account for component aging and wiring delays.

Assume that the propagation delay for each cycle, including the safety margin, is at maximum about 250 ns. To operate the circuit at maximum speed, a clock is required with a frequency of 4 MHz (= 1/250 ns). The control circuit for this serial adder/subtracter is shown in Figure 11.4[*a*]. It consists of a one-shot, a 4 MHz system clock, a four-bit binary counter module, a four-bit comparator, two

FIGURE 11.4 **Control Unit for the Four-Bit Adder/Subtracter: [*a*] Circuit.**

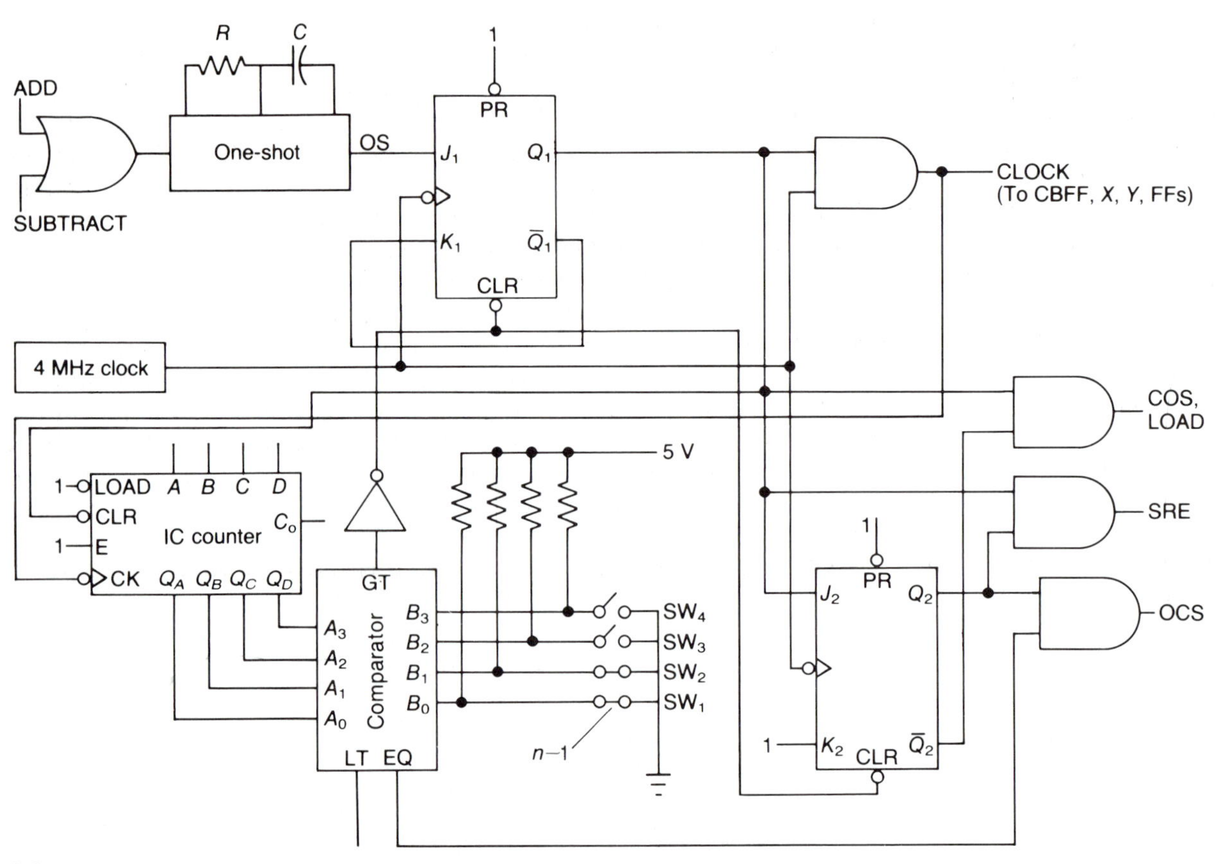

[*a*]

FIGURE 11.4 (Continued) **Control Unit for the Four-Bit Adder/Subtracter: [b] State Diagram.**

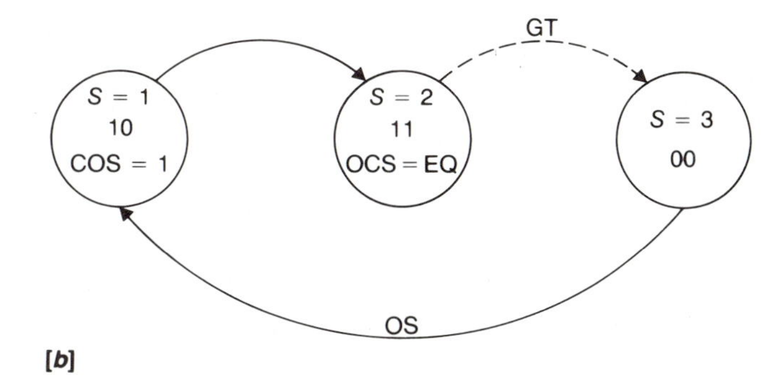

state FFs, and a few assorted gates. The broken line in the state diagram of the control circuit shown in Figure 11.4[*b*], connecting $S = 2$ and $S = 3$, indicates that the state change occurs asynchronously using the clear or preset inputs of the FFs. In this example when the count is greater than the $n - 1$ value (i.e., GT = 1), the circuit enters the $S = 3$, STOP, state. In $S = 3$, the clock to the arithmetic circuit is inhibited until another ADD or SUBTRACT signal is received.

As soon as the *one-shot output*, OS, becomes high, corresponding to either an ADD or a SUBTRACT input, a 4 MHz CK output is generated and is fed to CBFF, X, Y, various FFs, and counter. This also allows the signal COS to clear the overflow signal and LOAD signal to load the two registers, X and Y, in parallel. Starting with the next clock input, the counter would begin counting upward and the data bits in the registers would begin shifting to the right. In addition, when the count reaches $n - 1$, the overflow control signal, OCS, would also be turned on to check for possible overflow. Note that in this case the count is being compared with 0011 (B input). The size of the comparator and that of the counter, as shown in Figure 11.4[*a*], would allow this control unit to be used for controlling up to 15 bits of data. In such a case all one needs to do is to set the appropriate switches (SW). Example 11.1 gives the values for a four-bit add operation. Each new line in the example implies the occurrence of a clock pulse.

EXAMPLE 11.1

Add 0001 to 0110.

SOLUTION

S	X	Y	$CBFF_0$	OI	N
1	0110	0001	0	0	0
2	1011	1000	0	0	1
2	1101	0100	0	0	2
2	1110	0010	0	0	3
2	0111	0001	0	0	4
3	0111	0001	0	0	4

The sum is 0111 with no overflow.

It may be appropriate to point out several things about the control circuit of Figure 11.4[*a*]. A one-shot has been used here for starting the control operation. As discussed earlier in Section 6.7, one-shots are not to be used in cases where an alternative solution is available. This is not to say that there is no alternative solution to the one-shot used in our control unit. The one-shot has been used here strictly to demonstrate one of its many applications. Again, in Chapter 10 we learned several ways to use a register for counting sequences. A multi-bit ring counter could even be used (see Figure 10.40) for that purpose. The number of counter bits would depend on the number of bits present in the data. The MSB of the counter could be used for providing LOAD and COS signals while the outputs of the next four FFs could be ORed to provide the SRE signal. The other control signals, like that of OCS and the one necessary to turn off the system clock output, could also be generated by decoding appropriate FFs.

11.3 Serial-Parallel Multiplication Schemes

Perhaps the easiest multiplication technique is the one that involves signed magnitude numbers. In that case the data consist of the absolute value and a sign bit indicating whether or not the number is positive. The sign of the product is determined by separately operating on the signs of the multiplier and the multiplicand. The product sign is positive if the signs of the two operands are similar, and negative otherwise. The multiplication of two multi-bit binary numbers can be performed using the rules developed for multiplication in grade school. Consider, for example, the multiplication of 14 and 13 in both decimal and in binary.

$$
\begin{array}{r}
14_{10} \\
\times\ 13_{10} \\
\hline
42 \\
14 \\
\hline
182_{10}
\end{array}
\qquad\qquad
\begin{array}{r}
1110_2 \\
\times\ 1101_2 \\
\hline
1110 \\
0000 \\
1110 \\
1110 \\
\hline
10110110_2
\end{array}
$$

And, indeed, $182_{10} = 10110110_2$. Note that in both decimal and binary multiplications there is a possibility of having twice as many digits in the product as are in the multiplier. For a digital multiplication of two N-bit words, a register of size $2N$ bits is required for storing the product.

After examining this multiplication process, the sequence of repetitive shift and add operations becomes obvious. Using the symbolic representation of numbers, the algorithm for signed magnitude multiplication can be easily developed. Any binary value may be represented as a fraction with an appropriate power of two as a scale factor, that is, $0.0111 \times 2^4 = 111$. Consequently for the development of the following algorithm, the numbers are all assumed to be fractions. In general, the magnitude of a fractional binary number may be represented as

$$X = \sum_{i=1}^{N} X_i 2^{-i}$$

where X_i is either a 1 or a 0. The magnitude of binary number 0.01101, for example, may be expressed as

$$\frac{X_1}{2} + \frac{X_2}{4} + \frac{X_3}{8} + \frac{X_4}{16} + \frac{X_5}{32}$$

where $X_1 = X_4 = 0$ and $X_2 = X_3 = X_5 = 1$. If the product of X and Y is desired, it may be expressed as

$$\begin{aligned} X \cdot Y &= 2^{-1}X_1 \cdot Y + 2^{-2}X_2 \cdot Y + 2^{-3}X_3 \cdot Y + \cdots \\ &\quad + 2^{-N}X_N \cdot Y \\ &= 2^{-1}(X_1 \cdot Y + 2^{-1}(X_2 \cdot Y + 2^{-1}(X_3 \cdot Y + \cdots \\ &\quad + 2^{-1}(X_N \cdot Y) \ldots))) \end{aligned}$$

Note that "multiplying by 2^{-1}" amounts to a right shift by one bit. The multiplication rule can now be summarized as follows:

Step 1. Initialize the product subtotal to 0.

Step 2. Start scanning from the right-most bit of the multiplier, X. If $X_n = 1$, Y is added to the subtotal and then the sum is shifted right one bit. However, if $X_n = 0$, a 0 is added to the subtotal and then the sum is shifted right one bit.

Step 3. Step 2 is repeated n times for an n-bit word.

EXAMPLE 11.2

Multiply 0.0110_2 and 0.1001_2.

SOLUTION

Y = multiplicand = 0.0110_2 (3/8 in base 10)

X = multiplier = 0.1001_2 (9/16 in base 10)

$X_4 = 1$: the product subtotal is 0.0110_2

$X_3 = 0$: the product subtotal is 0.000110_2

$X_2 = 0$: the product subtotal is 0.0000110_2

$X_1 = 1$: the product subtotal is 0.00110110_2 (27/128 in base 10)

The product, therefore, is 0.00110110_2.

There are many different ways to perform multiplication. In this section we shall study several of these. The simplest of all cases occurs when both of the numbers are positive. Consider multiplying two eight-bit positive numbers, which results in a 16-bit product. The multiplier circuit would require two shift registers, X and Y, for storing the two numbers; a 16-bit register, P, for storing the product; and a 16-bit FA. The register that stores the multiplicand must also have 16 bits since the multiplicand needs to be shifted left repetitively. The RTL program for the serial-parallel multiplication scheme can be summarized as follows:

Algorithm 2

$S = 1$: $P \leftarrow 0$; $N \leftarrow 1$; $X \leftarrow$ (multiplicand) ;
$Y \leftarrow$ (multiplier) ;

$S = 2$: IF $(Y_0 = 1)$ $P \leftarrow X + P$; $N \leftarrow N + 1$;

$S = 3$: $Y \leftarrow \text{sr } Y$; $X \leftarrow \text{sl } X$; IF $(N \neq 9)$ $S \leftarrow 2$;

$S = 4$: STOP ;

where N is the current count. Figure 11.5 shows the corresponding multiplication block diagram. The multiplicand is loaded into the

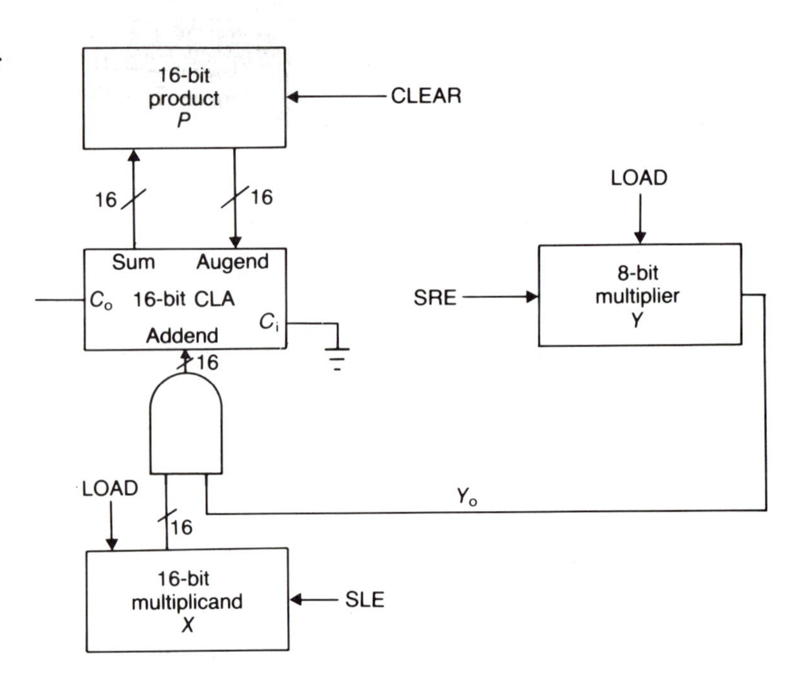

FIGURE 11.5 **Block Diagram for Realizing Algorithm 2.**

eight right-most FFs of the register X. The multiplier is loaded into the eight-bit register Y. Since at no time would the carry-out of the CLA adder be a 1, no attempt is made to store it. The control unit running this circuit needs to generate at least four control signals: CLEAR, LOAD, SRE, and SLE. The CLEAR and LOAD are used to set the registers for the operation. The test performed at $S = 2$ guarantees that the value of X is added to the partial product only if the LSB of Y is a 1. This conditional add can be performed with hardware by ANDing each bit of the multiplicand, X, with Y_0.

A closer examination of Algorithm 2 and the block diagram of Figure 11.5 reveals that as the multiplier is being shifted right, the register Y has unused FF bits on the extreme left. Again, the register P uses only half of its full capability in the very beginning and does not use its total capacity until the very end. In fact, as register P starts making use of one more FF on its left, register Y starts vacating one more FF. These observations allow an improved version of the RTL program that is summarized as follows:

Algorithm 3

$S = 1$: $P \leftarrow 0$; $N \leftarrow 0$; $X \leftarrow$ (multiplicand) ;
$Y \leftarrow$ (multiplier) ; $C \leftarrow 0$;

$S = 2$: IF ($Y_0 = 1$) $C,P \leftarrow X + P$; IF ($N = 7$) $S \leftarrow 4$;

$S = 3$: $C,P,Y \leftarrow$ sr C,P,Y ; $N \leftarrow N + 1$; $S \leftarrow 2$;

$S = 4$: STOP ;

A single FF register, C, is placed on the immediate left of register P such that it is able to store the carry-out of the CLA adder. In this algorithm the register Y initially stores the eight-bit multiplier and eventually stores the least significant eight bits of the product. The registers C, P, and Y are physically concatenated such that the shift-right operation during the step $S = 3$ involves all of these registers. The bit from C shifts into the MSB of P, and the LSB of P occupies the MSB of Y. Figure 11.6 shows the block diagram for the improved design. The bits in register X do not need to be shifted left any more since the product bits are being continually shifted to the right. However, X is still being added to the more significant eight bits of the product. This algorithm also eliminates the need of having an SLE signal.

If we must work with signed complement numbers, the problem becomes more complex because we have to consider four different cases. One possibility is to consider the signs separately, take the absolute value of both the multiplicand and the multiplier, and use a multiplication technique for positive numbers. The final product is then complemented or not, according to the sign of the product.

FIGURE 11.6 **Block Diagram for Realizing Algorithm 3.**

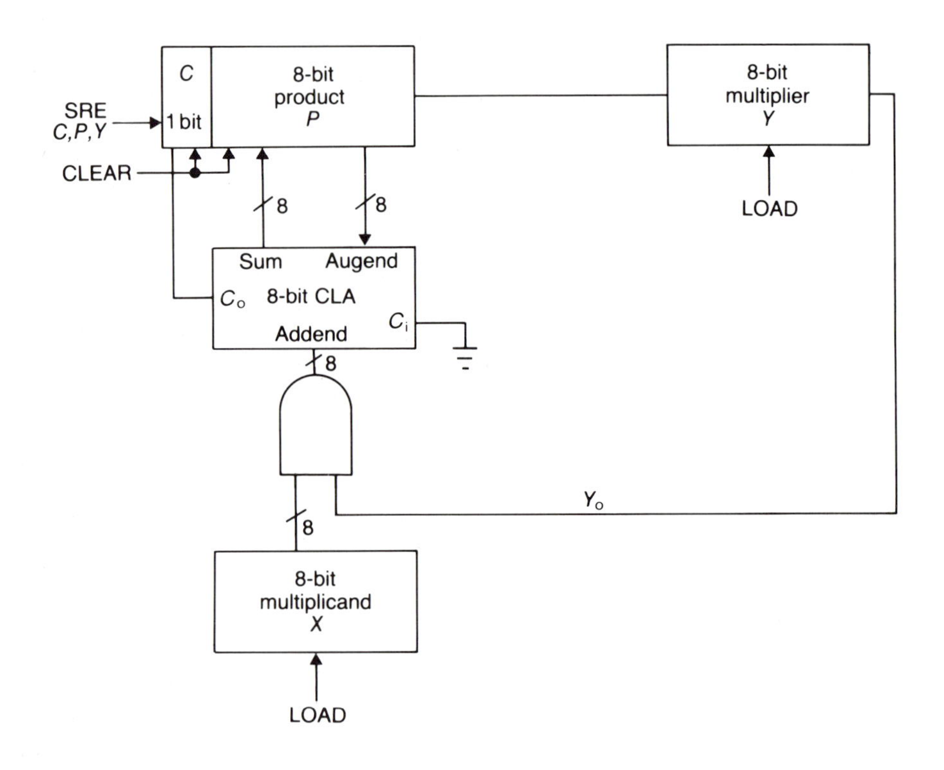

The RTL program for the 2's complement sign-and-magnitude, serial-parallel, n-bit multiplication using this method is obtained as follows:

Algorithm 4

$S = 1$: $P \leftarrow 0$; $N \leftarrow 1$; $X \leftarrow$ (multiplicand) ;
$Y \leftarrow$ (multiplier) ; $C \leftarrow 0$;

$S = 2$: IF $[(\overline{X}_{n-1} \wedge \overline{Y}_{n-1}) = 1]$ $M \leftarrow 0$;
IF $[(\overline{X}_{n-1} \wedge Y_{n-1}) = 1]$ $M \leftarrow 1, Y \leftarrow \overline{Y} + 1$;
IF $[(X_{n-1} \wedge \overline{Y}_{n-1}) = 1]$ $M \leftarrow 1, X \leftarrow \overline{X} + 1$;
IF $[(X_{n-1} \wedge Y_{n-1}) = 1]$ $M \leftarrow 0, X \leftarrow \overline{X} + 1$,
$Y \leftarrow \overline{Y} + 1$;

$S = 3$: $C,P \leftarrow P + (X \wedge Y_0)$; $N \leftarrow N + 1$;

$S = 4$: $C,P,Y \leftarrow \text{sr } C,P,Y$; IF $(N \neq n + 1)$ $S \leftarrow 3$;

$S = 5$: IF $(M = 1)$ $P,Y \leftarrow \overline{P,Y} + 1$;

$S = 6$: STOP ;

At $S = 2$ the test is made for the various combinations of sign bits. If the signs are equal, M is reset to a 0; otherwise it is set to a 1. The sign of the product is eventually restored at $S = 5$.

Figure 11.7 shows the block diagram of the circuit corresponding to the multiplication Algorithm 4. This diagram is the same as the one in the previous figure except for the inclusion of the subunits needed for performing the 2's complement conversion. Note that the **ENABLE** inputs are not all activated at the same instant. A shift register that is able to perform the 2's complement operation was already designed in Example 10.6. The 2's complement operation could also be done using an adder/subtracter such as the one

FIGURE 11.7 **Block Diagram for Realizing Algorithm 4.**

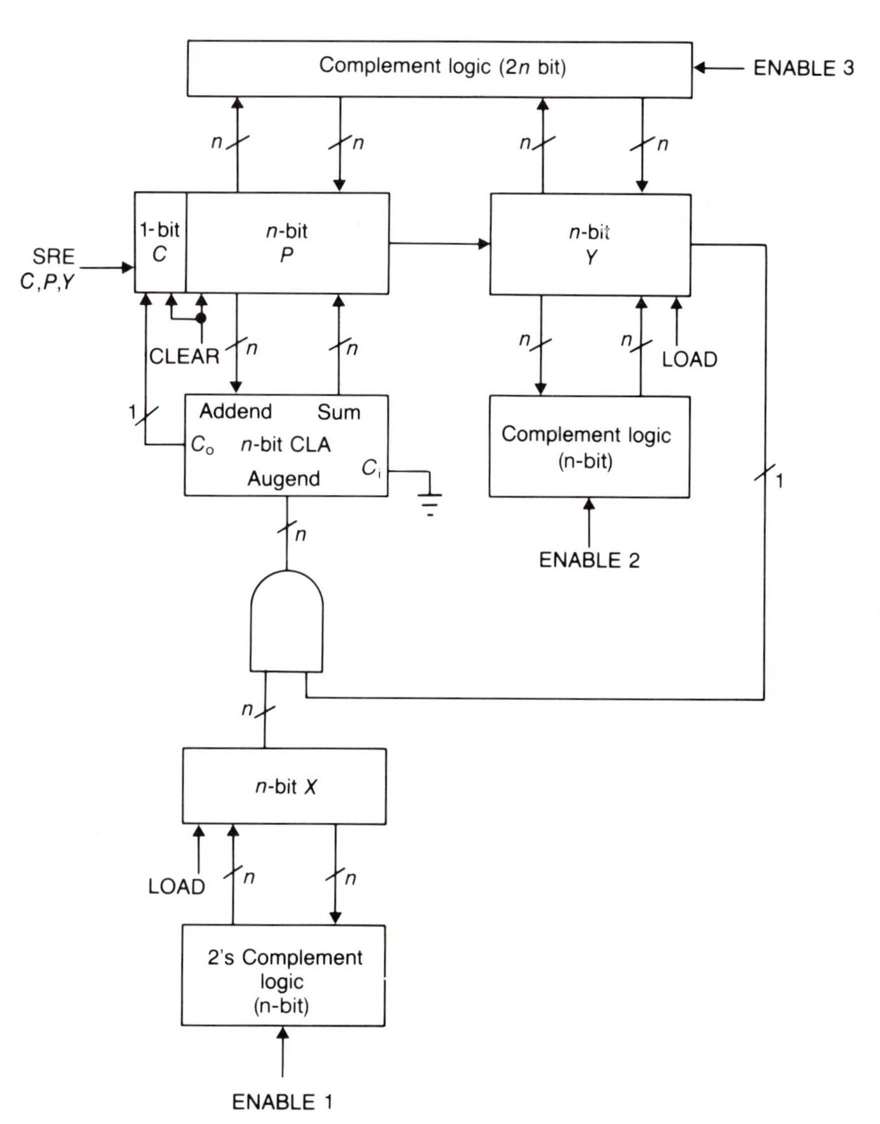

shown in Figure 5.12. In that case the to-be-complemented number is introduced at the B inputs, and all of the A inputs are tied to a 0.

The multiplication of 2's complement numbers becomes very complicated if we do not take the complement of negative quantities and hope to be able to adjust the sign after multiplication is complete. Consider the multiplication process of two 2's complement numbers. If the multiplier is positive and the multiplicand is negative, the multiplication does not pose too much of a problem; however, the simple addition of the multiplicand to the partial product is not sufficient. Consider Example 11.3 where the multiplicand is negative.

EXAMPLE 11.3

Work out the multiplication $(5)_{10} \times (-7)_{10}$

SOLUTION

$-7_{10} = 1\ 001_2$ and $5_{10} = 0\ 101_2$

$$
\begin{array}{r}
1001_2 \\
\times\ 0101_2 \\
\hline
11111001 \\
+\ 00000000 \\
+\ 11100100 \\
+\ 00000000 \\
\hline
111011101_2
\end{array}
$$

The overflow indicates a negative product. Consequently the 2's complement is performed to obtain $-(00100011)_2 = -(35)_{10}$.

When the multiplicand is added, it is necessary for a string of ones to be cascaded on the left of partial products. The number of ones would be such that each of the partial products consists of a total of eight bits. In Example 11.3 if the first partial product were 00001001, it would have simply implied the addition of 9 and not of -7. The eight-bit representation of the negative multiplicand, therefore, becomes necessary.

Next consider the case where the multiplier is a 2's complement negative number. This multiplication could be accomplished as follows:

Step 1. Scan the 2's complement of the multiplier from the right to the left.

Step 2. For each 1 found, add the 2's complement of the multiplicand to the partial product. Otherwise add a 0.

Step 3. Continue repeating Step 2 until all bits of the multiplier have been considered.

The general RTL program can be expressed as follows:

Algorithm 5

$S = 1$: $P \leftarrow 0$; $N \leftarrow n$; X $\leftarrow$ (multiplicand) ; $Y \leftarrow$ (multiplier) ;

$S = 2$: IF $(Y_{n-1} = 0)$ $S \leftarrow 7$;

$S = 3$: IF $(Y_0 = 0)$ $P,Y \leftarrow$ sr P,Y , $S \leftarrow 3$; $N \leftarrow N - 1$;

$S = 4$: $P \leftarrow P + \overline{X} + 1$; IF $(N = 0)$ $S \leftarrow 9$;

$S = 5$: $P,Y \leftarrow$ scr P,Y ; IF $(N = 0)$ $S \leftarrow 9$;

$S = 6$: IF $(Y_0 = 1)$ $P \leftarrow P + \overline{X} + 1$; $N \leftarrow N - 1$; $S \leftarrow 5$;

$S = 7$: $P \leftarrow P + (X \wedge Y_0)$; $N \leftarrow N - 1$;

$S = 8$: $P,Y \leftarrow$ scr P,Y ; IF $(N = 0)$ $S \leftarrow 9$ ELSE $S \leftarrow 7$;

$S = 9$: STOP ;

Whenever necessary to cascade strings of ones to the left of the partial product, the scale operation is found to be useful. An scr (scale-right operation) is very similar to the sr (shift-right operation) except that the MSB retains its old value. To handle situations where an overflow might occur, as illustrated by Example 11.3, the registers P and Y are usually provided with an additional bit on the left.

We shall now consider an algorithm similar to Algorithm 3 for the multiplication of unsigned positive numbers and accordingly attempt to obtain its circuit. The registers X, Y, and P require n bits to store the multiplicand, multiplier, and partial product, respectively. P and Y will display the product. In addition, a single FF, R, is used for temporarily storing the carry-out bit. This FF is unnecessary if the shift/add operation is done in the same clock period. The RTL instructions for the multiplication scheme are obtained as follows:

Algorithm 6

$S = 1$: $P \leftarrow 0$; $N \leftarrow 0$; $X \leftarrow$ (multiplicand) ;
$Y \leftarrow$ (multiplier) ;

$S = 2$: $R,P \leftarrow P + (Y_0 \wedge X)$; $R,P \leftarrow$ rr R,P ;

$S = 3$: $R,Y \leftarrow$ rr R,Y ; $N \leftarrow N + 1$; IF $(N < n - 1)$ $S \leftarrow 2$;

$S = 4$: STOP ;

The R, P, X, and Y registers may all be constructed with D FFs, using the circuit configuration of Figure 10.30. To add flexibility in loading these registers, MUXs could be used. Figure 11.8 shows a representative hardware block diagram of the proposed multiplication data unit. In reality, each of the MUXs, except one, is equiva-

FIGURE 11.8 **Block Diagram of the Multiplication Unit for Algorithm 6.**

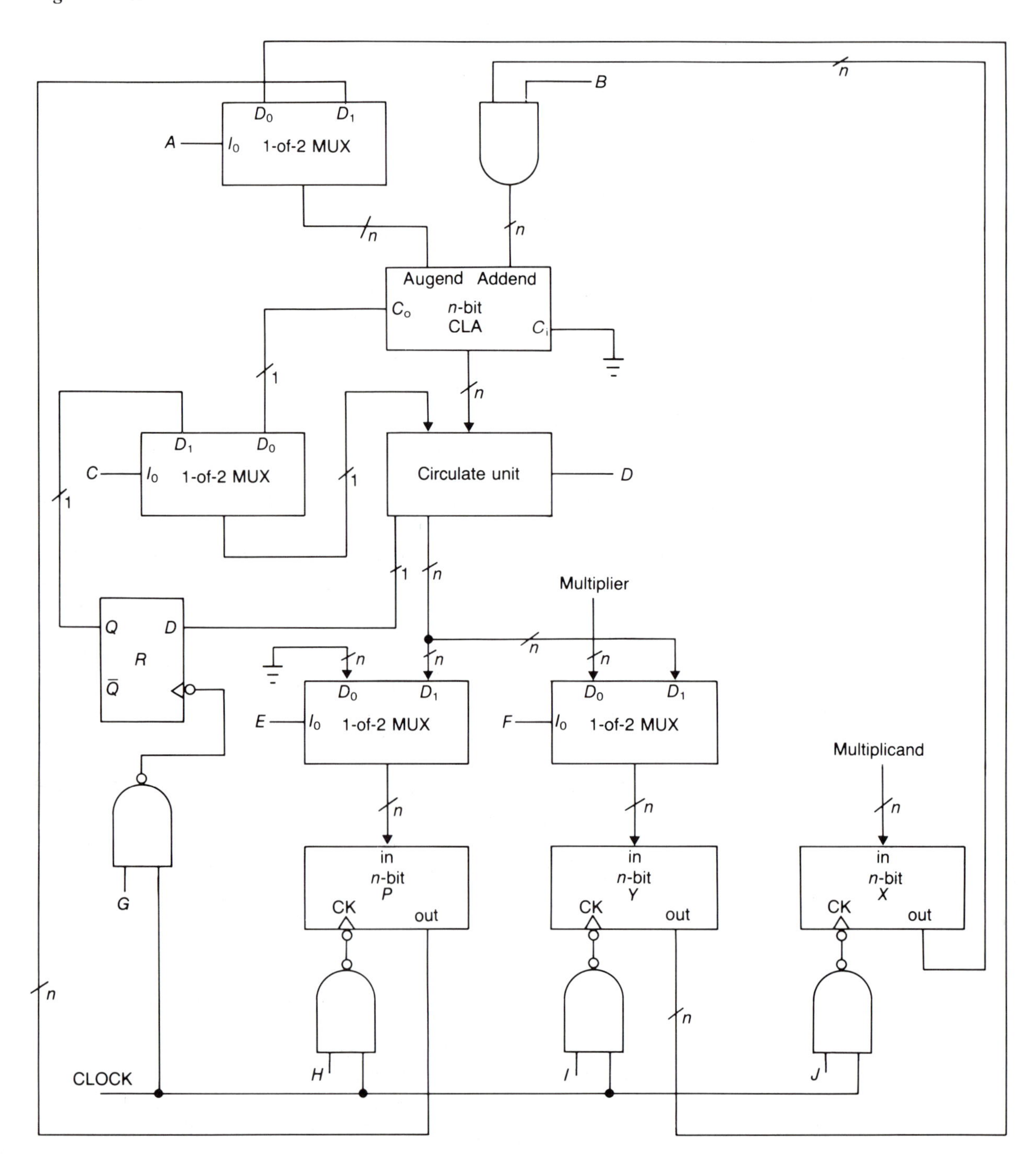

lent to n 1-of-2 MUXs. Similarly the AND gate at the CLA input port represents n two-input AND gates.

Ten of the control lines, A through J, in Figure 11.8 are used basically for loading and rotating operations. The complete list of the various actions caused by these ten control inputs are described as follows:

A selects either P or Y,

B selects either 0 or X,

C selects either the current carry-out or R output,

D performs rotation when equal to 1 and selects uncirculated output otherwise,

E selects either 0 or the circulated output,

F selects either the multiplier or the circulated output,

G, H, I, J enables R, P, Y, and X registers, respectively, for loading the input data.

The mechanics of the initialization step ($S = 1$) are quite straightforward. The addition operation is performed at $S = 2$ by proper manipulations of A and B controls. For the rotation operation (at $S = 2$ and $S = 3$), the content of R is loaded into the circulate unit by setting $C = 1$ while the other operand is introduced through the CLA adder. For example, P contents could be introduced by making sure that $A = 1$ and $B = 0$. The actual rotation, however, is performed by setting $D = 1$.

In order to control the multiplication unit of Figure 11.8, a sequencer needs to be designed so that it may generate the aforementioned control signals. An analysis of the requirements indicates that four additional subunits are necessary to monitor and run the operation: an event FF, a sequencer, a counter, and a sequence decoder. The event FF is necessary to decide when to BEGIN and when to STOP the algorithm. A sequencer is required to generate the four state sequences, $S = 1$ through $S = 4$, and the counter is needed to count the number of times the operation has gone through the steps $S = 2$ and $S = 3$. Finally, the sequencer and counter signals are to be decoded to provide the proper control signals.

Since there are only four sequences, we could use a five-stage, self-starting ring counter. Such sequencers were introduced in Section 10.9. However, a modification must be made so that when the test "Is $N < n - 1$?" yields a negative answer, the control could be switched to step $S = 2$. The control unit accordingly has the block diagram of Figure 11.9. For its internal consumption, the control unit would require three additional signals: K, L, and M. The

FIGURE 11.9 **Control Unit for the Circuit of Figure 11.8.**

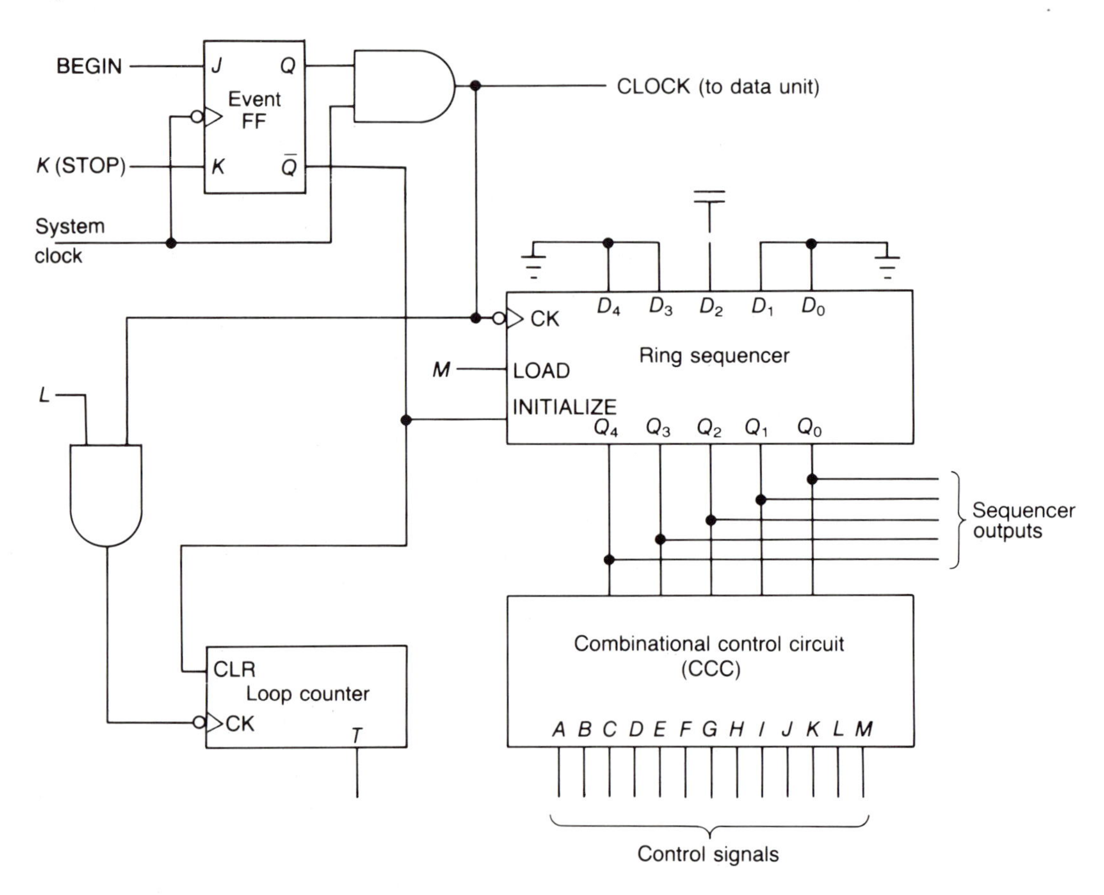

function of the three control signals necessary to monitor the control unit are listed as follows:

> K stops the event FF,
>
> L enables the counter to count up,
>
> M loads the inputs, D_4 through D_0, into the sequencer. This action overrides the normal sequencing operation.

The ring sequencer is INITIALIZEd to state 10000 once the event FF is set by the BEGIN command. With subsequent clock input the sequencer goes through the states 01000, 00100, 00010, and 00001. These sequencer states are used for generating the 13 control signals, A through M, by means of the combinational control circuit, CCC. The loop counter is CLEARed along with the initialization of the sequencer. When the loop counter produces a high at T, it implies that the loop count has already exceeded $n - 2$.

Once the J input of the event FF goes high, the clock signal enables the sequencer and the loop counter. The occurrences of the four timing cycles (corresponding to the steps $S = 1$ through $S = 4$) are respectively indicated by a high at Q_3, Q_2, Q_1, and Q_0. These timing ouputs are also used to generate the 13 control signals to direct the algorithm. When the control counter is in state 01000, the accumulator must be cleared and the registers Y and X should be loaded with the multiplier and the multiplicand. This is achieved by turning on three lines, H, I, and J, when all other control lines are maintained low. The algorithm step $S = 2$ is performed at the counter state 00100 and is achieved by setting $A = 1$, $B = Y_0$, $D = 1$, $E = 1$, and $G = H = 1$. As a result the content of the register X is added to the contents of the accumulator only when the LSB of the register Y is a 1; the result is stored in R and P after being rotated right one bit. Then during the step $S = 3$ the rotation between R and Y contents and the loop increment are performed. This rotation is achieved by setting C, D, F, G, I, and L to 1. Furthermore, a test is performed at the $S = 3$ state to determine the next state. If $T = 0$, the next step should be $S = 2$; otherwise the algorithm must stop. This additional task is accomplished by providing T to K and $\overline{T}$ to D_2 and M. Thus if the count is less than $n - 1$, 00100 ($S = 2$) is loaded in the sequencer. However, if the count equals $n - 1$, K becomes high and, consequently, the event FF is turned off.

The control operations just described are summarized in the table of Figure 11.10. This table is called a *control matrix*. It is used frequently to derive the control equations for the design of the CCC subunit. The CCC equations, therefore, are as follows:

$$A = E = Q3$$

$$B = Y_0 \cdot Q3$$

$$C = F = L = Q4$$

$$D = G = Q3 + Q4$$

$$H = Q2 + Q3$$

$$I = Q2 + Q4$$

$$J = Q2$$

$$K = T$$

$$M = \overline{T}$$

FIGURE 11.10 **Control Matrix Table for the Circuits of Figures 11.8 and 11.9.**

State	Control signals												
	A	*B*	*C*	*D*	*E*	*F*	*G*	*H*	*I*	*J*	*K*	*L*	*M*
Q1 ($S = 0$)													
Q2 ($S = 1$)								1	1	1			
Q3 ($S = 2$)	1	Y_0		1	1		1	1					
Q4 ($S = 3$)			1	1		1	1		1		T	1	$\overline{T}$

The design of the serial-parallel multiplication circuit using Algorithm 6 is essentially complete.

In general, the serial-parallel multiplication schemes are sequential in nature. The algorithms discussed in this section, therefore, are much slower than the schemes where all partial products could be generated simultaneously and added. In the next section we shall examine some of the highly parallel multiplication circuits.

11.4 Combinational Array Multiplier Logic

Several multiplication schemes were introduced in the last section. These were all serial, involving a single partial product generation at a time. In comparison, multiplication using combinational array logic takes very little time. In this technique the product is generated almost in the same way as in the manual multiplication process.

The unsigned parallel binary product of two binary numbers,

$$A = \sum_{i=0}^{m-1} a_i 2^i$$

and

$$B = \sum_{j=0}^{n-1} b_j 2_j$$

is given by

$$A \times B = \sum_{i=0}^{m-1} \sum_{j=0}^{n-1} a_i b_j 2^{i+j}$$

The manual multiplication process for $m = n = 4$, for example, yields the array as shown in Figure 11.11. The process begins by finding a total of 16 partial products. The partial products are subsequently added as shown in the illustration to give an eight-bit

FIGURE 11.11 **Manual Multiplication of Two Four-Bit Numbers.**

				a_3	a_2	a_1	a_0
				b_3	b_2	b_1	b_0
				a_3b_0	a_2b_0	a_1b_0	a_0b_0
			a_3b_1	a_2b_1	a_1b_1	a_0b_1	
		a_3b_2	a_2b_2	a_1b_2	a_0b_2		
	a_3b_3	a_2b_3	a_1b_3	a_0b_3			
P_7	P_6	P_5	P_4	P_3	P_2	P_1	P_0

product. Each of the partial products, a_ib_j, may be formed from the individual multiplicand, a_i, and multiplier, b_j, bits by means of a simple AND gate. The partial products are then fed into an array of FAs to generate the products bits, P_k. Note that in this example there could be up to four single-bit numbers in a column. The addition of more than three single-bit numbers was already considered in Example 5.1. Using similar ideas, the FAs are arranged as shown in Figure 11.12. The sum output of one row appears as an input to the adder directly below it, and the carry-out bit is usually shifted into an adder on the immediate left. Finally the last rows of the FAs are organized in the same form as a ripple adder.

FIGURE 11.12 4 × 4 **Combinational Multiplier Array.**

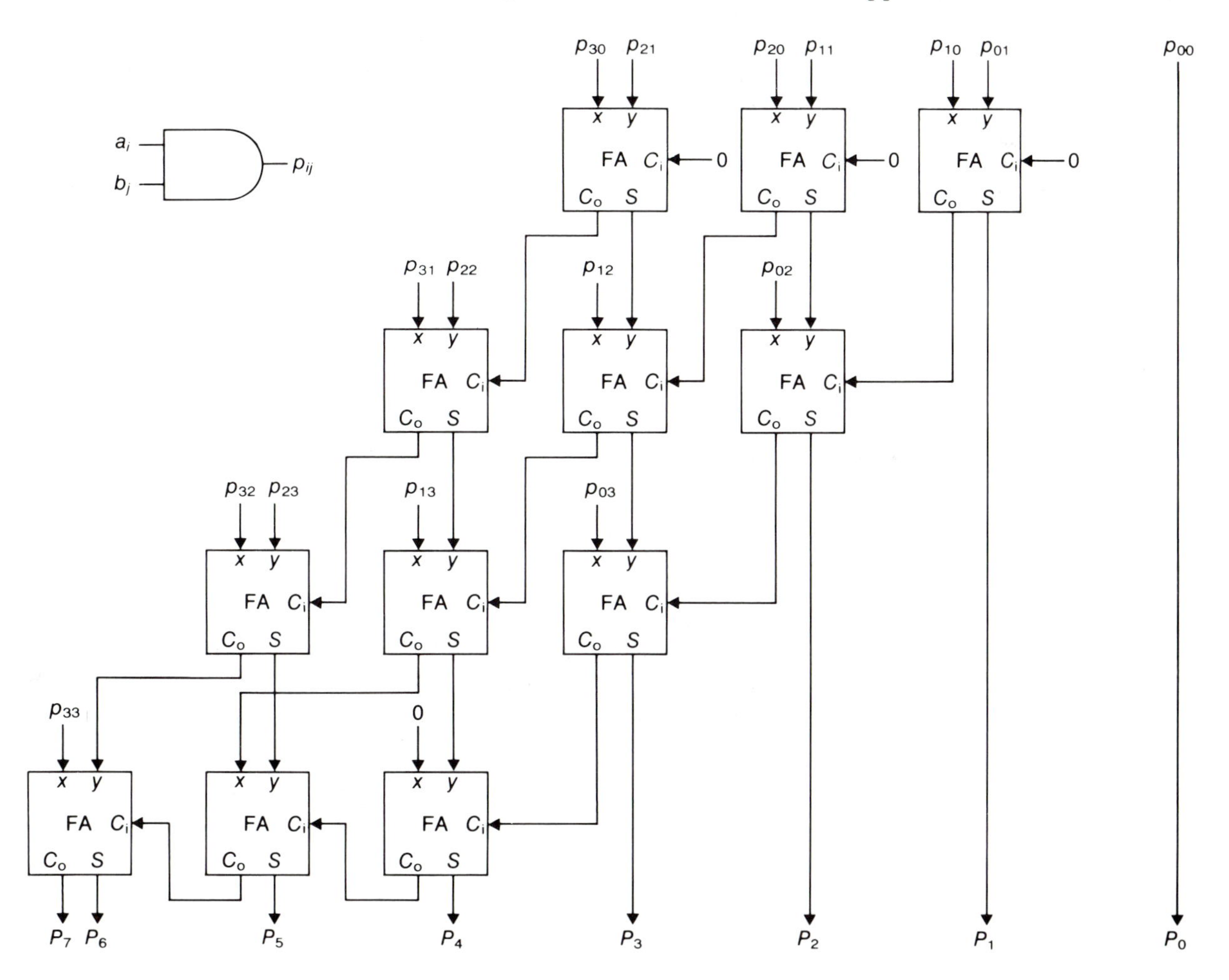

An additional diagonal of FAs will be required for each additional bit of A, and an additional row of FAs will be required for each additional bit of B. Therefore a total of mn AND gates and $(m - 1)n$ FAs are necessary for an m-bit × n-bit multiplication. The multiplication time still depends on the choice of the FA circuit where the worst-case propagation delay will be along the right-most diagonal and the last row. The regularity of the circuit pattern in Figure 11.12 has allowed the implementation of such an array in LSI form for providing high-speed multiplication.

EXAMPLE 11.4

Obtain a 4 × 2 generalized multiplication module, as shown in Figure 11.13, such that the module output

$p = (a \times b) + c$

where a and c are four-bit inputs and b is a two-bit input. Use two four-bit FAs and AND gates for the design.

SOLUTION

FIGURE 11.13

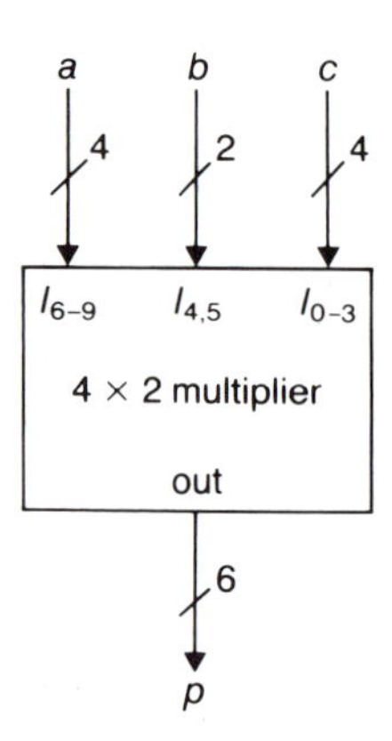

We already know from earlier discussions that the product, p_{ij}, of two bits, a_i and b_j, can be obtained simply by means of a two-input AND gate. Since b is a two-bit input, p can be written as

$$\begin{aligned} p &= (a \times b) + c \\ &= (a \times b_1)2^1 + (a \times b_0)2^0 + (c)2^0 \end{aligned}$$

This is equivalent to saying that p could be achieved by following these steps:

Step 1. Add $(a \times b_0)$ and c.

Step 2. Shift-left $(a \times b_1)$ one bit and then add it to the sum obtained in Step 1.

Accordingly, the circuit for the module is obtained as shown in Figure 11.14.

FIGURE 11.14

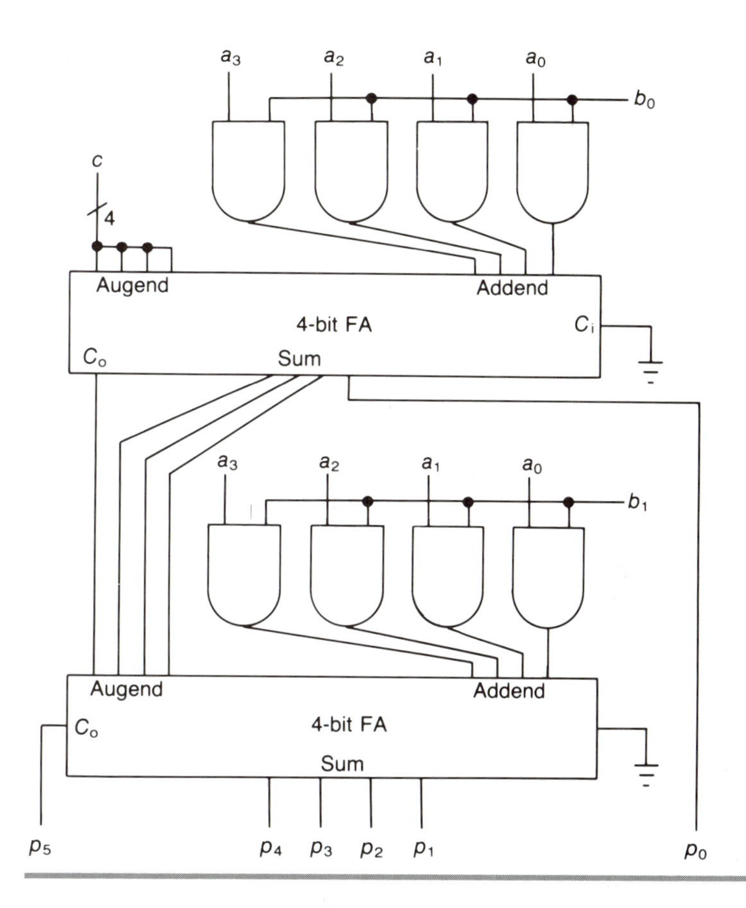

EXAMPLE 11.5

Obtain a generalized 4 × 4 multiplication module such that

$p = (a \times b) + c$

where a, b, and c are four-bit inputs and p is an eight-bit output.

SOLUTION

When both a and b are four-bits each, the output of the multiplication module should correspond to

$$\begin{aligned} p &= (a \times b_3)2^3 + (a \times b_2)2^2 + (a \times b_1)2^1 + (a \times b_0)2^0 + c \\ &= a(2b_3 + b_2)2^2 + a(2b_1 + b_0)2^0 + c \end{aligned}$$

Note that multiplying by 2^2 corresponds to a left shift of two bits. Hence p could be obtained as follows:

Step 1. Feed a and the least significant two bits of b and c to a generalized 4 × 2 multiplication module.

Step 2. Feed a, the remaining bits of b, and all but the two LSBs of the result obtained in Step 1 to another generalized 4 × 2 multiplication module.

Consequently, the generalized 4 × 4 multiplication module is obtained using two generalized 4 × 2 multiplication modules as shown in Figure 11.15.

FIGURE 11.15

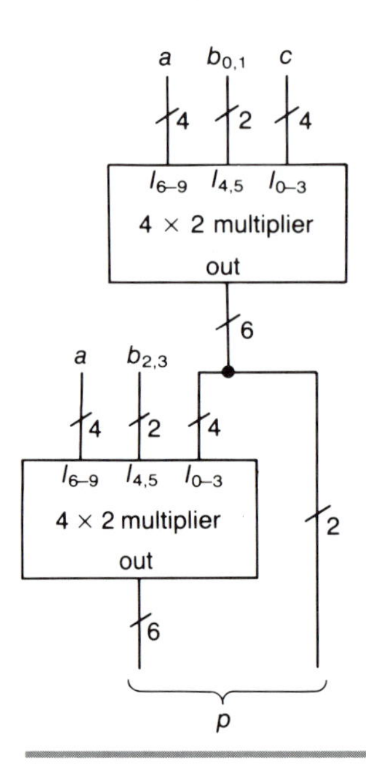

11.5 Carry-Save Multiplication

We have seen that the combinational multiplier array logic is much faster than the serial multiplication schemes of Section 11.3. However, there are other schemes used to save time. It was pointed out in the last section that a major amount of multiplication time is lost by the carries while they ripple through the adders. One of the ways to overcome this problem is to use the CLA adders in place of the FAs. There is, however, a sequential solution to this purely combinational shortcoming. Provided that none of the carries in transit up the adder array is lost at each stage, it is very much in order to synchronize the propagation of carries with the generation of subtotals. In this scheme, known commonly as the *carry-save multiplication* technique, the carries from the addition of the partial products are initially stored in a set of *D* FFs and used as one of the adder inputs during the next addition. This storage prevents the carries from rippling through different adders.

The scheme consists of two distinct phases: *carry-save* (CS) and *carry-propagate* (CP). The CS phase involves the usual add-and-shift operation where the carry is temporarily stored in a *D* FF. The carry is not allowed to go from one FA column to the next. The true product, however, is obtained in the final CP phase by allowing the propagation of carry-outs. The overall RTL multiplication algorithm is detailed as follows:

Algorithm 7

$S = 1$: $X \leftarrow$ (multiplicand) ; $Y \leftarrow$ (multiplier) ; CP $\leftarrow 0$;
CS $\leftarrow 1$; $P \leftarrow 0$; $C \leftarrow 0$; $N \leftarrow n$;

$S = 2$: $P_i \leftarrow P_i \oplus C_i \oplus (X_i \wedge Y_0)$; $C_i \leftarrow (\text{carry-out})_i$;
$N \leftarrow N - 1$;

$S = 3$: $P,Y \leftarrow \text{sr } P,Y$; IF ($N \neq 0$) $S \leftarrow 2$ ELSE CS $\leftarrow 0$,
CP $\leftarrow 1$;

$S = 4$: $P \leftarrow P + C$;

$S = 5$: STOP ;

where n is the number of bits in the multiplier.

The step at $S = 2$ is equivalent to bit-by-bit addition where the carry-outs are stored in the register C. The register C could consist of n D FFs interconnected in a special way. This CS phase is continued until the process of add-and-shift has been conducted n times. Finally the step $S = 4$ includes the conventional summation involving carry-propagation. Note that the CS and the CP phases are mutually exclusive. Figure 11.16 shows the circuit of the carry-save adder that is suitable for both CS and CP operations.

The ENABLE control signal is turned on when the step $S = 2$ is reached. As long as the CS signal and Y_0 bit is ON, X is added bit-by-bit to the partial product. The carry-outs and the sum bits are stored at the corresponding D FFs of registers C and P, respectively. Note that the registers P and Y are both shift-right type and are concatenated to each other. When Y_0 is a 0 during the step $S = 2$, a 0 is added to each of the P-bits. However, when CP is high each of the carry-outs moves to the next FA on the left as one of the adder inputs. Consequently, the last step is equivalent to a straightforward ripple addition. The more significant half of the final product would be available in register P and the remaining half in register Y.

11.6 Booth's Multi-Bit Algorithm

The scheme for signed magnitude multiplication has several shortcomings. The absolute values of the inputs must be generated from the signed complement data. In addition, the signs of the operands should be handled separately. Finally, the resultant product has to be transformed back to its correct signed complement form. In this section we introduce a new variation of the standard shift-and-add scheme, called *Booth's algorithm,* that is equally applicable to both positive and negative operands. It accommodates working with 2's complement numbers without pre- or postadjustments. It is also useful in fixed-point calculations that tend to have multiplier distributions that peak close to zero. Such numbers generally have long chains of ones and zeros at the most significant end.

For a radix complement of n-bit numbers. $A = r^n - a$ and $B =$

FIGURE 11.16 **n-Bit, Carry-Save Adder Circuit.**

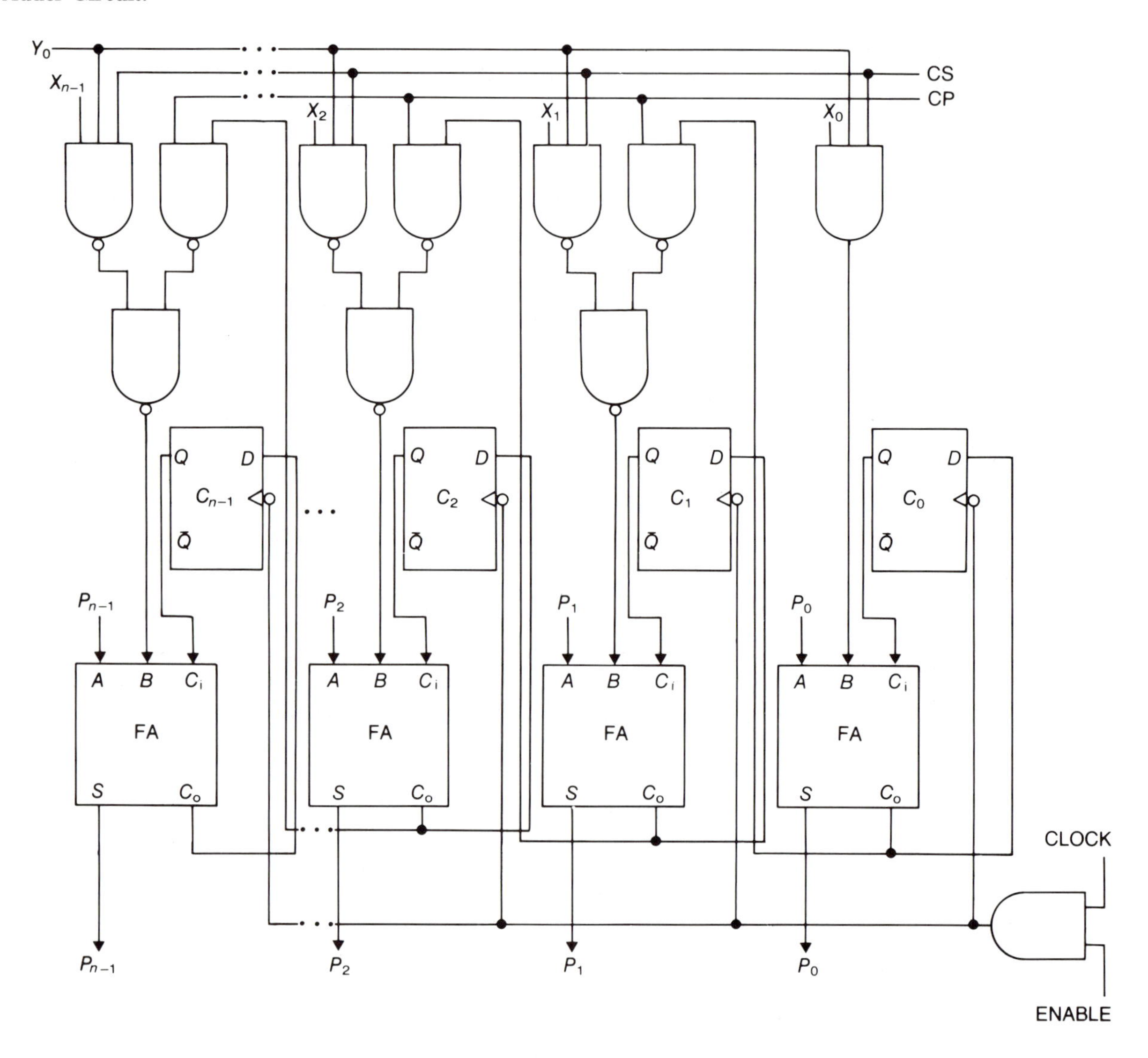

$r^n - b$, the product AB is equal to $r^{2n} - ab$ where r is the radix and a and b are the magnitudes of A and B, respectively. A problem occurs if one of the numbers (b, for example) is positive while the other is negative. This results in a significant error given by

$$\begin{aligned} \text{error} &= [r^{2n} - ab] - AB \\ &= [r^{2n} - ab] - b(r^n - a) \\ &= r^n(r^n - b) \end{aligned}$$

Therefore, if the radix complement numbers are multiplied directly, a messy correction factor must be added. Booth's algorithm eliminates this correction phase completely. The presence of a multiplier having m consecutive ones implies m additions and m shifts. Again since m consecutive ones from bit weights 2^q to 2^p (where $p > q$) are equivalent to $2^{p+1} - 2^q$, the corresponding operation is equivalent to a total sequence of one subtraction in the beginning, m subsequent shifts, and a final addition step. For example, a multiplier $(00111100110)_2$ would generally require six add-and-shift operations. However, it would require only four if the multiplier were rewritten as $(01000001000 - 00000100010)_2$. In effect, groups of zeros are ignored. Groups of ones are replaced by a single subtraction of the multiplicand at the position of the least significant 1 and an addition at one bit to the left of the most significant 1.

In order to determine the test criterion for detection of a string of ones, an extra bit, Y_{-1}, is added next to the least significant Y bit. The test condition may be decided upon by observing simultaneously Y_0 and Y_{-1} as the bits in Y are being continually shifted to the right. Thus $Y_0Y_{-1} = 00$ indicates the absence of a string in this location; $Y_0Y_{-1} = 01$ indicates the detection of the left end of the string; $Y_0Y_{-1} = 10$ indicates the detection of the right end of the string; and $Y_0Y_{-1} = 11$ indicates the continuation of the string. The multiplicand is subtracted from the partial product once the right end of the string is detected. Correspondingly the multiplicand is added to the partial product upon encountering the left end of the string. Otherwise, the partial product is only shifted.

The RTL program for Booth's multi-bit scheme can be summarized as follows:

Algorithm 8

$S = 1$: $X \leftarrow$ (multiplicand) ; $Y \leftarrow$ (multiplier) ; $P \leftarrow 0$;
$N \leftarrow n$; $Y_{-1} \leftarrow 0$;

$S = 2$: IF $((Y_0,Y_{-1} = 0{,}0$ OR $1{,}1)$ AND $N = 0)$ $S \leftarrow 5$;
IF $(Y_0,Y_{-1} = 0{,}0$ OR $1{,}1)$ $S \leftarrow 3$;
IF $(Y_0,Y_{-1} = 0{,}1)$ $P \leftarrow P + X$; $S \leftarrow 3$;
IF $(Y_0,Y_{-1} = 1{,}0)$ $P \leftarrow P + \overline{X} + 1$;

$S = 3$: $P,Y,Y_{-1} \leftarrow$ scr P,Y,Y_{-1} ; $N \leftarrow N - 1$;

$S = 4$: IF $(N \neq 0)$ $S \leftarrow 2$;

$S = 5$: STOP ;

The basic block diagram for this multiplication circuit is shown in Figure 11.17. It is quite similar to those that were examined previously except for an additional test circuit and a dual adder/subtracter unit. The left part of the final result is contained in register P

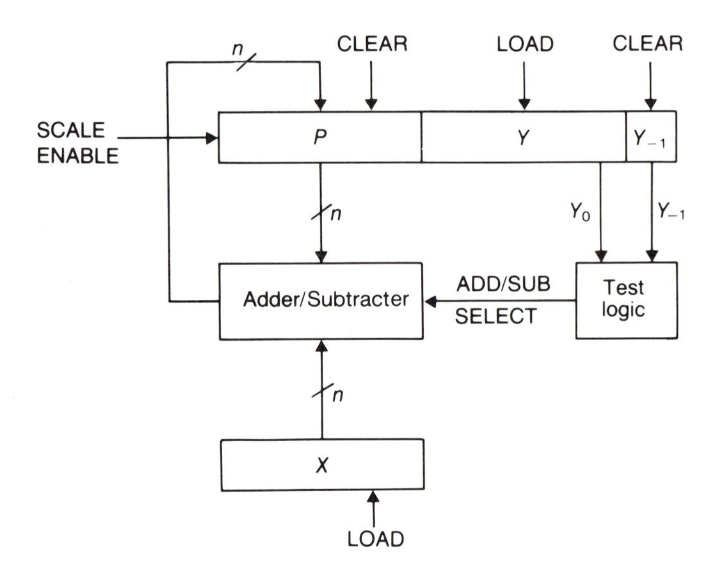

FIGURE 11.17 **Block Diagram of an *n*-Bit Booth Multiplier.**

while the right part of the product is contained in register Y. It is important to note that the LSB of Y is not considered as part of the result. For two n-bit numbers, only $n + 1$ tests and n scale-right operations are necessary.

11.7 Techniques for Division

The previous algorithms for different addition, subtraction, and multiplication schemes are quite sufficient for devising means for division processes. Division is frequently carried out as a reverse process of multiplication. In fact, many of the multiplication techniques are directly applicable to the division algorithms. The manual division process with which we are familiar is nothing but repeated comparison, subtraction, and shift operations. In comparison, binary division is simpler than the decimal division since it is not necessary to determine how many times the dividend or the partial remainder fits within the divisor. If the divisor does not go into the partial remainder, the divisor is moved to the right one bit before the next comparison is made. This right-shift operation may also be accomplished by shifting the dividend to the left while the divisor is kept fixed. Such a division technique, known as the *comparison method*, is realizable by the block diagram of Figure 11.18. The subtraction is achieved by adding the 2's complement of the divisor to the partial remainder.

The n-bit divisor is loaded into register D, and the $2n$-bit dividend is stored in registers X and Y. The dividend is shifted to the left, and the divisor is subtracted from it by the 2's complement mechanism. The relative comparison test is performed by the comparator unit (CU) that determines whether or not the subtraction will be performed. When $X \geq D$ the YES/$\overline{\text{NO}}$ output is a 1. The quotient bit 1 is loaded in Y_0, and the partial remainder is shifted to

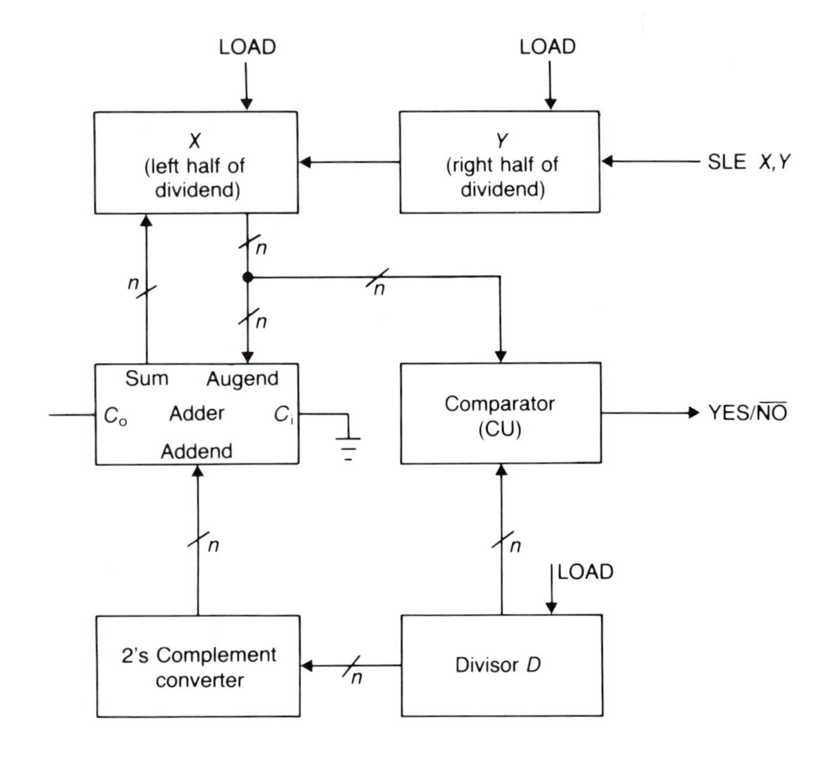

FIGURE 11.18 **Block Diagram of a Comparison Division.**

the left to repeat the cycle. If YES/$\overline{\text{NO}}$ is a 0, it implies that $X < D$ so that the new quotient bit is a 0. In that event the subtraction is not performed and the partial remainder is shifted to the left only. Finally, after n iterations the register X contains only the remainder, and the register Y contains the quotient.

The RTL algorithm of this division scheme, corresponding to only positive numbers, is obtained as follows:

Algorithm 9

$S = 1$: $X \leftarrow$ (left-half of dividend) ; $Y \leftarrow$ (right-half of dividend) ; $D \leftarrow$ (divisor) ; $N \leftarrow 0$; $E \leftarrow 0$;

$S = 2$: IF ($D_{n-1} \neq 1$ OR $D = 0$) $E \leftarrow 1, S \leftarrow 10$;

$S = 3$: IF ($X_{n-1} \neq 1$) $X,Y \leftarrow$ sl X,Y , $Y_0 \leftarrow 0, N \leftarrow N + 1,$ $S \leftarrow 3$;

$S = 4$: IF (YES/$\overline{\text{NO}} \neq 1$ OR $N = n$) $S \leftarrow 10$;

$S = 5$: pause ;

$S = 6$: $X \leftarrow X + \overline{D} + 1$; $Y_0 \leftarrow 1$;

$S = 7$: $X,Y \leftarrow$ sl X,Y ; $N \leftarrow N + 1$;

$S = 8$: IF ($N = n$) $S \leftarrow 10$;

$S = 9$: IF (YES/$\overline{\text{NO}} = 1$) $S \leftarrow 6$ ELSE $Y_0 \leftarrow 0, S \leftarrow 7$;

$S = 10$: STOP ;

The test at step $S = 2$ investigates the validity of the division and its compatibility with the hardware. The two reasons for which the error FF, E, would be turned on are either division by a zero, or the final quotient requiring more than n bits. The divide-overflow must be avoided in normal computer operations because the entire quotient will be too long for transfer into other units that have words of standard length. When the dividend is twice as long as the divisor, a divide-overflow condition would occur if the high-order half bits of the dividend constitute a number greater than or equal to the divisor. The comparison test of step $S = 4$ finds out if there would be a divide-overflow. In either step the operation is terminated prematurely.

The CU for this algorithm is the most crucial circuit. A single-bit CU may be designed readily from the corresponding truth table. Several such units may then be combined to come up with the desired multi-bit comparator circuit. The truth table, as shown in Figure 11.19[*a*], relates the two input bits, X_i and D_i, to the two outputs, M_i and C_i. The match output, M_i, indicates that the divisor is smaller than the dividend, and the compare output, C_i, indicates that at least one more test is necessary at the next significant bit position before a match could be discovered. It follows, therefore, that if an undecided condition occurs, the bit on the immediate right needs to be considered. And if a match has already been located at any bit position, no further tests on the right are necessary.

Figure 11.19[*b*] shows the circuit for a single-bit CU and Figure 11.19[*c*] shows a combination of these units to form an n-bit comparator module. The right-most AND output indicates whether or not the contents of the two registers are equal. A high YES/$\overline{\text{NO}}$ output indicates that $X \geq D$, and a low YES/$\overline{\text{NO}}$ indicates that $X < D$.

A second type of division algorithm is known as the restoration method, where the divisor is subtracted from the partial remainder to determine YES/$\overline{\text{NO}}$ status. When the difference turns out to be negative, the old remainder is restored before continuing any further. The best accuracy, however, is obtained by normalizing both the numerator and the denominator; that is, both of these are shifted to the left until the MSB has a 1 in it. The sign bits of both the numerator and the denominator are handled separately to determine the sign of the quotient. The basic block diagram is shown in Figure 11.20, and the details of the RTL algorithm for the corresponding sign-and-magnitude division is obtained as follows:

Algorithm 10

$S = 1$: $X \leftarrow$ (left-half of dividend) ; $Y \leftarrow$ (right-half of dividend) ; $N \leftarrow n - 1$; $D \leftarrow$ (divisor) ;

$S = 2$: $Q_n \leftarrow X_n \oplus D_n$; $E,X \leftarrow X + \overline{D} + 1$;

$S = 3$: IF $(E = 1)$ $E,X \leftarrow X + D$, $OFFF_0 \leftarrow 1$, $S \leftarrow 12$;

$S = 4$: $E,X \leftarrow X + D$, $OFFF_0 \leftarrow 0$;

$S = 5$: $E,X,Y \leftarrow$ sl E,X,Y ;

$S = 6$: IF $(E = 0)$ $E,X \leftarrow X + \overline{D} + 1$ ELSE $S \leftarrow 8$;

$S = 7$: IF $(E = 0)$ $E,X \leftarrow X + D$, $S \leftarrow 10$ ELSE $S \leftarrow 9$;

$S = 8$: $X \leftarrow X + \overline{D} + 1$;

$S = 9$: $Y_0 \leftarrow 1$;

$S = 10$: $N \leftarrow N - 1$;

$S = 11$: IF $(N \neq 0)$ $S \leftarrow 5$;

$S = 12$: STOP ;

Besides the standard registers, X, Y, and D, five additional FFs are required in the restoration scheme. The FFs, X_n and D_n, store the signs of the dividend and the divisor. The Q_n FF indicates the sign of the quotient, and the overflow flip-flop, OFFF, is turned on whenever there is a divide-overflow. The FF E stores the carry-out of all addition operations. It has been assumed for Algorithm 10 that the registers X, Y, and D each contain n bits. The final remainder will appear at the register X while the quotient will be stored in Y.

Besides these two division schemes, there is one more variation of the division operation, known as the *nonrestoration method.* When the divisor subtraction produces a negative result, the old remainder is not restored, but an alternative quicker and simpler process is followed. In the restoring scheme either X would have to be restored by discarding $X - D$, or D would have to be added to $X - D$ to regain X. However, if we think in terms of binary, then the next trial subtraction will be with $D/2$ to yield $X - (D/2)$. Undoubtedly it is quicker to form $X - (D/2)$ by adding $D/2$ to the result of the previous trial difference, $X - D$, rather than by restoring X. The algorithm for this nonrestoring scheme, therefore, is obtained as follows:

Step 1. Subtract the divisor from the dividend. For a positive result, the quotient bit is a 1; otherwise the quotient bit is a 0.

Step 2. If the result in Step 1 is positive/negative, subtract/add the right-shifted divisor from/to the partial remainder. For the new positive result the quotient bit is a 1; otherwise the quotient bit is a 0. Repeat Step 2.

The nonrestoration scheme works well in the case of fractions. The process is continued until sufficient accuracy is obtained. For integer operation, however, it can produce unreasonable results since the last trial addition/subtraction may or may not be successful. In any event this scheme decreases the number of steps neces-

FIGURE 11.19 **Comparator Design: [*a*] Truth Table for Each Module, [*b*] Circuit for Each Module, and [*c*] Connections for *n*-Bit Comparison.**

Inputs		Outputs	
X_i	D_i	C_i	M_i
0	0	1	0
0	1	0	0
1	0	0	1
1	1	1	0

[*a*]

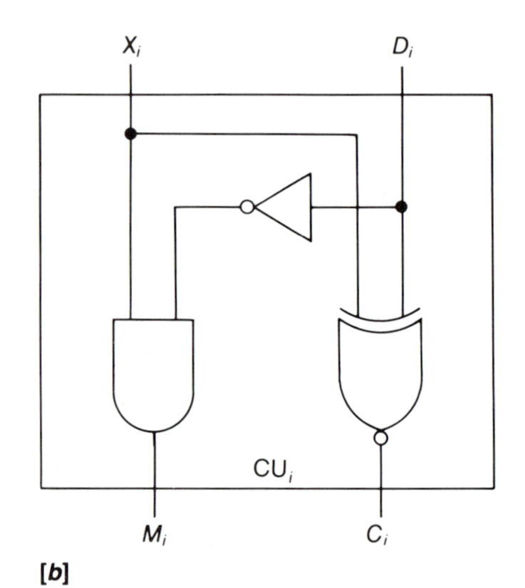

[*b*]

[*c*]

FIGURE 11.20 **Restoration Divider.**

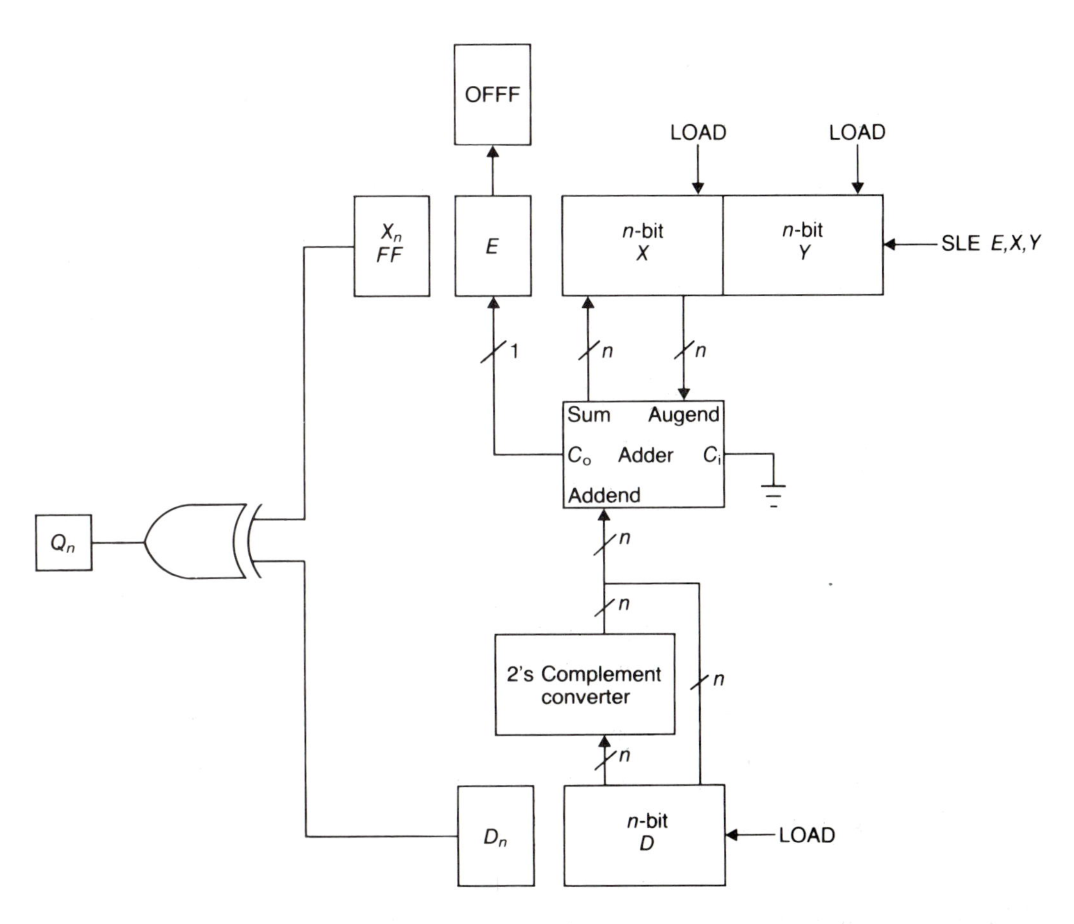

sary in Algorithm 10, but requires a control circuit that must remember the previous result.

As a last example we will present a challenging problem of finding the square root of a given number. The reason for considering this particular application is that it involves arithmetic operations of various types, but it is an iterative process terminated only by a test of convergence. The algorithm computes the square root of a number, A, to within an accuracy specified by a constant, C, as follows:

Step 1. Approximate the square root of A as $B_1 = A/2$ as the first approximation.

Step 2. Approximate the square root of A as $B_n = [B_{n-1} + (A/B_{n-1})]/2$ as the nth approximation for all $n > 1$.

Step 3. Continue repeating Step 2 until two successive approximations differ by less than C.

The success of these steps could be verified easily by actually considering a positive number for A and then running through these steps. The RTL program for the square root scheme may be written as follows:

Algorithm 11

$S = 1$: $X \leftarrow A$; $Y \leftarrow C$;

$S = 2$: $Z \leftarrow$ sr A ;

$S = 3$: $Q \leftarrow X/Z$;

$S = 4$: $Q \leftarrow Q + Z$;

$S = 5$: $Q \leftarrow$ sr Q ;

$S = 6$: $D \leftarrow Z - Q$;

$S = 7$: IF $(D > C)$ THEN $Z \leftarrow Q$, S $\leftarrow 3$;

$S = 8$: STOP ;

Algorithms like this one may be developed for different functions like cube root, square, sine, cosine, tangent, and so on. The basic operations such as addition, subtraction, multiplication, division, and their combinations may be used for generating almost any mathematical function. The steps toward achieving such a goal consist of

understanding the characteristics of the function,

organizing the realization scheme into several RTL statements,

developing the block diagram of the system,

designing the individual data modules,

developing the control matrix,

designing the control unit for controlling the RTL sequence.

11.8 Summary

In this chapter various schemes for performing different arithmetic operations were introduced. In each of these cases the design schemes were translated into RTL algorithms. Most of the corresponding circuits were decomposed naturally into two parts: a data processing unit and a control unit. The RTL algorithms served as the real bridges between the designs of either of these two units.

Problems

1. Given two numbers x and y stored in the registers A and B, respectively, design a serial arithmetic unit that computes $|x|$

$-\ |y|$. You should assume that the numbers are in signed magnitude notation.

2. Repeat Problem 1 where the arithmetic unit utilizes a parallel adder and the numbers are in 1's complement.
3. Design a control circuit for Algorithm 1 such that the sum would not be calculated any further once an overflow has been detected. Discuss the functions of all individual components.
4. Design a binary multiplier when the numbers are represented in 1's complement.
5. Design a working circuit for Algorithm 2.
6. Design a working circuit for Algorithm 3.
7. Design a working circuit for Algorithm 4.
8. Design a working circuit for Algorithm 5.
9. Design a working circuit for Algorithm 7.
10. Design a working circuit for Algorithm 8.
11. Design a working circuit for Algorithm 9.
12. Design a working circuit for Algorithm 10.
13. Design a working circuit for Algorithm 11.
14. Design a circuit that multiplies two four-bit numbers. The first bit indicates the sign. The result should be a seven-bit number.
15. Design a combinational array multiplier logic using CLA adders. Compare your design with that of Figure 11.12.
16. Design the RTL program for the nonrestoration division process.
17. Design (a) the sequencer, (b) the loop counter, and (c) the CCC of Figure 11.9.
18. What corrections are needed in the circuits of Figures 11.8 and 11.9 if consideration is given to the fact that the ring counter might end up in one or more unused states? Explain.
19. Replace the ring counter of the sequencer in Figure 11.9 with a regular binary up-counter. Discuss the modifications necessary. Why? Where?
20. Design a multiplication scheme by modifying Algorithm 5 so that 1's complement numbers may be handled.
21. Design a working circuit for the scheme of Problem 19.
22. Design a combinational circuit for multiplying two eight-bit numbers using (a) 8×2 multiplication units and (b) 8×4 multiplication units. (See Examples 5.2, 11.3, and 11.4 for reference.)

Suggested Readings

Agrawal, D. P. "Negabinary carry look ahead adder and fast multiplier." *Elect. Lett.* vol. 10 (1974): 312.

Agrawal, D. P., and Rao, T. R. N. "On multiple operand addition of signed binary adders." *IEEE Trans. Comp.* vol. C-27 (1978): 1068.

Ahmad, M. "Iterative schemes for high speed division." *Comp. J.* vol. 15 (1972): 333.

Ambikairajah, E., and Carey, M. J. "Technique for performing multiplication on a 16-bit microprocessor using an extension of Booth's algorithm." *Elect. Lett.* vol. 16 (1980): 53.

Basu, D., and Jayashree, T. "On a simple post correction for nonrestoring division." *IEEE Trans. Comp.* vol. C-24 (1975): 1019.

Booth, A. D. "A signed binary multiplication technique." *Quarterly J. of Mech. & Appl. Math.* vol. 4. pt. 2 (1951): 236.

Bywater, R. E. H. *Hardware/Software Design of Digital Systems.* Englewood Cliffs, N.J.: Prentice-Hall, 1981.

Chen, I. N., and Willoner, R. "An $0(n)$ parallel multiplier with bit-sequential input and output." *IEEE Trans. Comp.* vol. C-28 (1979): 721.

Corsini, P., and Forsini, G. "Uniform shift multiplication algorithms without overflow." *IEEE Trans. Comp.* vol. C-27 (1978): 256.

Dao, T. T., David, M., and Gossart, C. "Complex number arithmetic with odd-valued logic." *IEEE Trans. Comp.* vol. C-29 (1980): 604.

Daws, D. C., and Jones, E. V. 'Hardware-efficient bit sequential adders and multipliers using mode-controlled logic." *Elect. Lett.* vol. 16 (1980): 434.

Deegan, I. D. "Cellular multiplier for signed binary numbers." *Elect. Lett.* vol. 7 (1971): 436.

Deverell, J. "Pipeline intérative arithmetic array." *IEEE Trans. Comp.* vol. C-24 (1975): 317.

Dhurkadas, A. "Faster parallel multiplier." *Proc. IEEE.* vol. 72 (1984): 134.

Ercegovac, M. D., and Long, T. *Digital Systems and Hardware/Firmware Algorithms.* New York: Wiley, 1985.

Gardiner, A. B., and Hont, J. "Cellular array arithmetic unit with multiply and divide." *Proc. IEEE.* vol. 119 (1972): 659.

Gibson, J. A., and Gibbard, R. W. "Synthesis and comparison of two's complement parallel multipliers." *IEEE Trans. Comp.* vol. C-24 (1975): 1020.

Gnanasekaran, R. "On a bit-serial input and bit-serial output multiplier." *IEEE Trans. Comp.* vol. C-32 (1983): 878.

Goodman, R., and Feldstein, A. "Round of errors in products." *Comput.* vol. 15 (1975): 263.

Guild, H. H. "Fully interative fast array for binary multiplication and addition." *Elect. Lett.* vol. 5 (1969): 263.

Habibi, A., and Wintz, P. A. "Fast multipliers." *IEEE Trans. Comp.* vol. C-19 (1970): 153.

Hoekstra, J. "Systolic multiplier." *Elect. Lett.* vol. 20 (1984): 995.

Hwang, K. *Computer Arithmetic.* New York: Wiley, 1979.

Johnson, E. L. "A digital quarter square multiplier." *IEEE Trans. Comp.* vol. C-29 (1980): 258.

Johnson, E. L., and Jong, M. T. "An implicit division technique using a digital multiplier." *IEEE Circuits & Syst.* vol. 4 (1982): 16.

Jong, M. T. "Binary fraction multiplication algorithms." *IEEE Circuits & Syst.* vol. 1 (1979): 8.

Kingsbury, N. G. "High speed binary multiplier." *Elect. Lett.* vol. 7 (1971): 277.

Kline, R. *Structured Digital Design Including MSI/LSI Components and Microprocessors.* Englewood Cliffs, N.J.: Prentice-Hall, 1983.

Lai, H. C., and Muroga, S. "Logic networks of carry-save adders." *IEEE Trans. Comp.* vol. C-31 (1982): 870.

Ling, H. "High-speed binary adder." *IBM J. Res. & Dev.* vol. 25 (1981): 156.

Lo, H. Y.; Ju, J. H.; and Aoki, Y. "Programmable variable-rate up/down counter for generating binary logarithms." *IEE Proc. E., Comp. & Dig. Tech.* vol. 131 (1984): 125.

Lyon, R. F. "Two's complement pipeline multiplier." *IEEE Trans. Commun.* vol. COM-12 (1976): 418.

MacSorley, O. L. "High-speed arithmetic in binary computers." *Proc. IRE.* vol. 49 (1961): 67.

McCanny, J. V., and McWhirter, J. G. "Completely iterative, pipe-lined multiplier array suitable for VLSI." *IEE Proc. G., Elect. Cir. & Syst.* vol. 129 (1982): 40.

McDonald, T. G., and Guha, R. K. "The two's complement quasi-serial multiplier." *IEEE Trans. Comp.* vol. C-24 (1975): 1233.

Meggitt, J. E. "Pseudo-division and pseudo-multiplication process." *IBM J. Res. & Dev.* vol. 6 (1962): 210.

Mori, R. D. "Suggestion for an IC fast parallel multiplier." *Elect. Lett.* vol. 5 (1969): 50.

Ninke, W. H., and Ritchie, G. R. "Shift register binary rate multipliers." *IEEE Trans. Comp.* vol. C-26 (1977): 276.

Oklobdzija, V. G., and Ercegovac, M. D. "An on-line square-root algorithm." *IEEE Trans. Comp.* vol. C-31 (1982): 70.

Petry, F. E. "Two's complement extension of a parallel binary division by ten." *Elect. Lett.* vol. 19 (1983): 718.

Pezaris, S. D. "A 40 ns 17 bit by 17 bit array multiplier." *IEEE Trans. Comp.* vol. C-20 (1971): 422.

Quatse, J. T., and Keir, R. A. "A parallel accumulator for a general-purpose-computer." *IEEE Trans. Elect. Comp.* vol. EC-16 (1971): 165.

Rubinfield, L. P. "A proof of the modified Booth's algorithm for multiplication." *IEEE Trans. Comp.* vol. C-24 (1975): 1014.

Singh, S., and Waxman, R. "Multiple operand addition and multiplication." *IEEE Trans. Comp.* vol. C-22 (1973): 113.

Stenzel, W. J.; Kubitz, W. J.; and Garcia, G. H. "A compact high-speed multiplication scheme." *IEEE Trans. Comp.* vol. C-26 (1977): 948.

Strader, N. R., and Rhyne, V. T. "A canonical bit-sequential multiplier." *IEEE Trans. Comp.* vol. C-31 (1982): 791.

Swartzlander, E. E. "The quasi-serial multiplier." *IEEE Trans. Comp.* vol. C-22 (1973): 317.

Swartzlander, E. E. "Merged arithmetic." *IEEE Trans. Comp.* vol. C-29 (1980): 946.

Taylor, F. J., and Huang, C. H. "An autoscale residue multiplier." *IEEE Trans. Comp.* vol. C-31 (1982): 321.

Trivedi, K. S. "On the use of continued fractions for digital computer arithmetic." *IEEE Trans. Comp.* vol. C-26 (1977): 700.

Trivedi, K. S., and Ercegovac, M. D. "On line algorithms for division and multiplication." *IEEE Trans. Comp.* vol. C-26 (1977): 681.

Verber, C. M. "Integrated optical architectures for matrix multiplication." *Opt. Engn.* vol. 24 (1985): 74.

Wallace, C. S. "A suggestion for a fast multiplier." *IEEE Trans. Comp.* vol. C-13 (1964): 14.

Zohar, S. "Rounding and truncation in radix (−2) systems." *IEEE Trans. Comp.* vol. C-25 (1976): 464.

Memory Devices

CHAPTER TWELVE

12.1 Introduction

We are familiar with several types of memory from our applications of them in earlier chapters, including ROM and PLA that were used in Chapter 4 for realizing combinational logic. The bit patterns that make up the instructions and data of a computer program are stored in memory so that the computer can perform its prescribed function. In Chapter 6 it was shown that a flip-flop is able to store one bit of information, and in Chapter 10 it was shown how an array of flip-flops, that is, a register, may store multiple bits of information. In particular, a scratch-pad memory was developed as an example of register applications.

Memory is made up of two state devices in which the bits are stored. The bit combinations at a given memory location are referred to as a word. The number of bits in a word is used as an important parameter of the computer. As the technologies make transitions from one generation to the next, computer memory also continues to change in size, price, ease of use, and bit density. This chapter introduces several types of memory, their characteristics, and functions with the intention of providing a flavor of this most important computer component. After studying this chapter, you should be able to:

- Understand the interrelationship among memory, memory address register, and memory buffer register;
- Understand READ and WRITE operations;
- Understand and use magnetic, semiconductor, and optical memories;
- Assemble several smaller memory units to make a larger memory.

12.2 Memory Basics

A memory is organized in groups of bits called *words*. The bits forming a word are such that they are involved in the same transfer of data and are stored in only one memory register. A combination of eight bits is commonly called a *byte*. The word is a binary number that represents either an instruction, an address, or data.

Each word in a memory has two specific attributes: its *address*, which is used to locate the word within the memory, and the data bits stored at that address. External data are generally written into the memory one word at a time and can be recalled in a subsequent step as data or as an instruction. The external world communicates with the memory by means of address lines, control lines, and data input/output (I/O) lines. A block diagram of a memory with the attributes just described is shown in Figure 12.1. The control lines specify the direction of data transfer. A WRITE control directs data to be stored in a certain location of the memory while a READ control permits the transfer of already-stored memory data to the outside. The SELECT line is used to select a particular memory device such as this. It is of extreme importance when the system has two or more memory units.

FIGURE 12.1 **Block Diagram of a Memory Unit.**

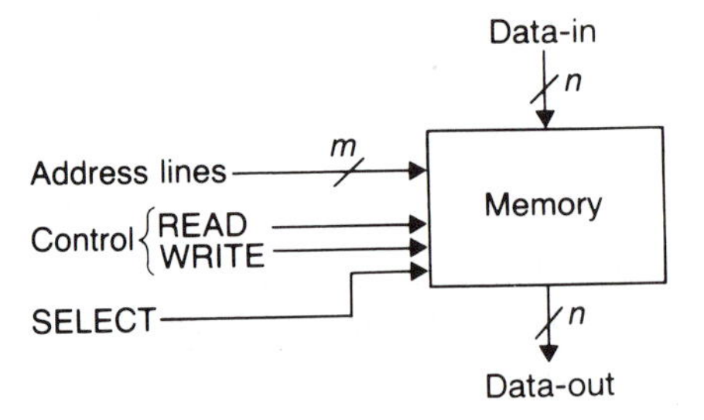

From our previous encounters with register files, ROMs, and PLAs, we know that m address lines can select up to 2^m different memory words. A decoder inside the memory decodes the m address inputs and selects one of the 2^m memory words. For example, 10 address lines may select up to $2^{10} = 1024$ words. Memory size is usually given as integer values of 1K words; 1024 words is commonly referred to as 1K of words. Likewise, 16 address lines would allow selection of 64K words ($64 \times 1024 = 65{,}536$ active locations). In the particular memory of Figure 12.1, each word consists of n bits. The memory is referred to as having $2^m \times n$-bit words.

Since address and data may come from many sources and data read may go to several destinations, the application and timing of memory can be quite complex. The memory model, therefore, is simplified if we associate two external registers with the memory. One of these two registers is used to hold the address and is called the *memory address register* (MAR). The other register, called the *memory buffer register* (MBR) or *memory data register* (MDR), holds the data to be written into the memory or the data from the last memory read. In some cases the MAR and MBR are considered a part of the memory itself. The address lines of the memory unit are permanently connected to the outputs of the MAR. The MBR, on the other hand, serves as a temporary storage register (called a *buffer*) for transferring data to and from the memory. Figure 12.2 shows the block diagram of the MBR, MAR, and an associated 1K × 8-bit memory. This 1K × 8-bit memory contains a total of 2^{10} words times 2^3 bits/words = 2^{13} bits, or 8192 bits.

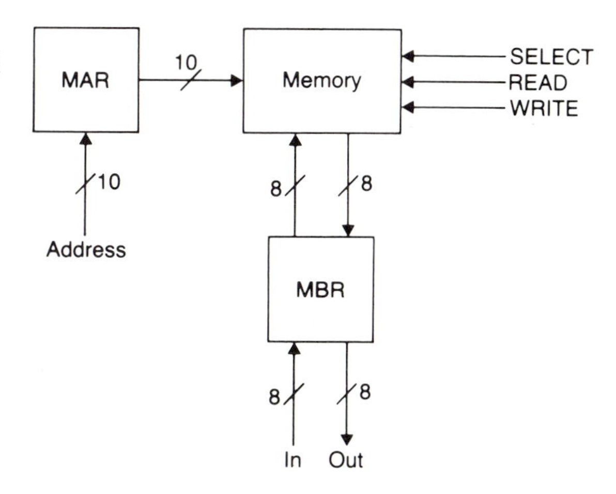

FIGURE 12.2 **Block Diagram of 1K × 8-Bit Memory with External Registers.**

When the WRITE signal is enabled, the contents of the MBR are stored in the location specified by the address as contained in the MAR. Similarly, with the READ enable signal the word from the memory location specified by the MAR is transferred to the MBR. The WRITE mode involves the following operations:

1. The address bits are loaded into the MAR.
2. The data bits of the word are loaded into the MBR.
3. The WRITE signal is enabled, which transfers the contents of the MBR to the selected memory location. This operation destroys the previous content of that location.

The READ mode involves the following steps:

1. The address bits are loaded into the MAR.
2. The READ signal is enabled.
3. The memory word is transferred from the memory to the MBR. The memory word contents are not destroyed or altered by the READ operation.

Depending on the technology used, a memory could be one of three types: magnetic, semiconductor, or optical. The classification of various memory devices are shown in Figure 12.3. Although LSI technology is now able to generate large-capacity semiconductor memory devices, there are many computers that use core memory that utilizes small magnetic cores as the basic memory element. Because of price, power, and speed requirements, solid-state memory has become the memory of choice for most systems. However, magnetic bubble memory presently is being considered for building large memory systems. The optical memory, likewise, is a relatively new idea to the commercial community. Its impact is yet to be seen, but it definitely is showing significant promise in optical laboratories.

FIGURE 12.3 **Memory Types.**

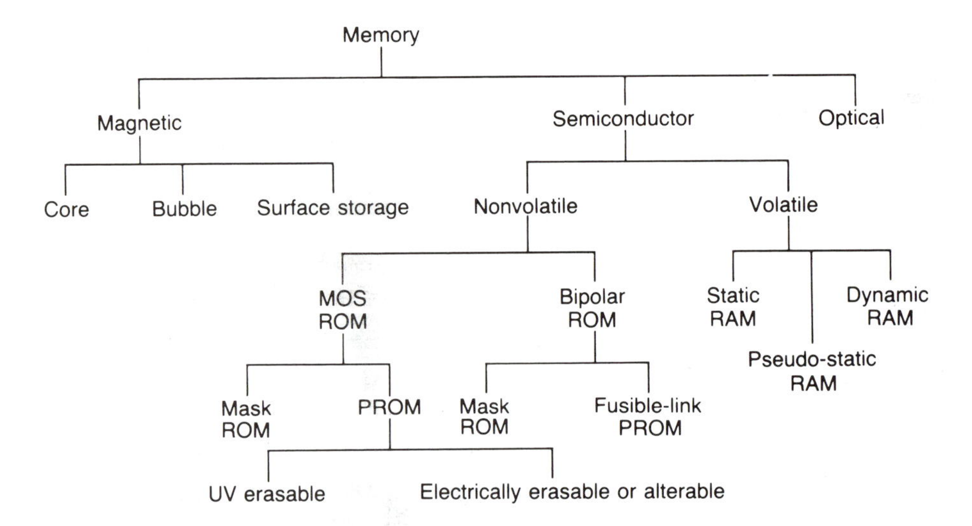

Semiconductor memory systems may be classified into two broad categories: *random access memories* (RAMs) and *read-only memories* (ROMs). RAM, also called a *read/write memory*, is a type of memory where the data may be either written into the memory or read from the memory. A ROM, however, as the name implies, is a memory unit that performs only the READ operation; it does not have the capability to operate in the WRITE mode. RAMs are primarily used for main memories, in smaller memories called scratch-pad memories, and in buffer or cache memories. ROMs are used for permanent storage of programs and data.

RAMs have two storage modes: static and dynamic. *Static RAMs* use FFs to store data and are easier to use than the dynamic RAMs. *Dynamic RAMs* mostly rely on circuit capacitance to store data and must be refreshed periodically to recharge the capacitance. ROMs utilize two semiconductor processes: bipolar and MOS. The *bipolar ROM* is usually faster than the *MOS ROM*. ROMs that are programmed by the manufacturer are known as *mask ROMs*. There is also a class of ROMs that is user-programmable for which various techniques are available for programming and altering stored data. It must be noted that solid-state RAMs are *volatile*, that is, they lose their information once the power is turned off. In comparison, ROMs store information permanently and are termed *nonvolatile*.

The ideal memory would have the following characteristics:

unlimited capacity,

negligible writing and access time,

negligible power consumption,

nonvolatility,

low cost per bit,

high reliability.

With this large number of factors to consider, it is not surprising that memory development has not come to a halt. In the following sections various types of RAMs are discussed. ROMs will not be included here since they were covered once in Chapter 4.

12.3 Magnetic Core Memories

The core RAM was the principal storage device of early computers. The current applications of core RAMs and their historical perspective warrant an introduction to them. Core RAM has the advantage of being nonvolatile; that is, the data are not lost when the power is turned off. One of the more recent evolutions in core memories has been the development of small core memories assembled on printed circuit boards.

Magnetic cores are small doughnut-shaped toroids made of ferromagnetic material. A typical core has an outer diameter of about 0.7 mm and an inner diameter of about 0.5 mm. The data storage is controlled by an axial wire that passes through the center of the core. When a current is passed through the wire, a directional magnetic flux is generated in the core and is retained by the core when the current is withdrawn. If the direction of the current is reversed, the resulting magnetic flux also reverses its direction. Magnetization in one direction is considered to be equivalent to a 1, in the other direction to a 0. The two states of the core are shown respectively in Figures 12.4[*a*] and [*b*]. Figure 12.4[*c*] shows the corresponding hysteresis characteristics of the core material. The *hysteresis loop* is a plot of the magnetic field intensity, or flux density, B, which is proportional to the amount of current flowing along the core axis versus H, the magnetic force. The magnetic force is directly proportional to the amount of current flowing. The best core material has a hysteresis loop that approximately approaches a rectangle.

The core, as illustrated by the hysteresis curve in Figure 12.4[*c*], uses a positive current for the WRITE mode and a negative current for the READ mode. In the WRITE mode the positive current

FIGURE 12.4 **Magnetic Core: [*a*] Logic 1 State, [*b*] Logic 0 State, and [*c*] Hysteresis Characteristics.**

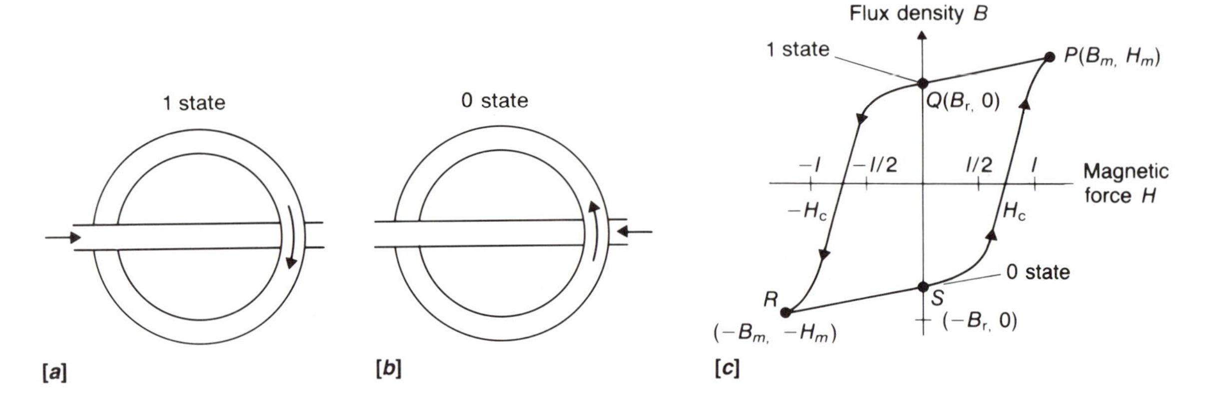

should be large enough to cause the magnetic force to exceed the value of the coercive field strength, H_c. Once this critical value is exceeded, and the current is removed, the flux density will decrease very slightly to the remnant flux density, B_r. This phenomenon is a direct consequence of magnetization. When the core is at this particular coordinate point, *Q*, the core is said to have logic 1 stored in it. For the READ operation a negative current is applied. If sufficient current is supplied, and the magnetic force is just below the critical value, $-H_c$, the core swings through the point *R* and assumes the value at *S*. This state corresponds to a logic 0. A value of current, *I*, is required to change one state of the core to the other. From the characteristics of the hysteresis loop, it can be seen that one-half of the current ($\pm I/2$) is not enough to switch states. The status of the core is not altered as long as the current through it is between $+I/2$ and $-I/2$. A flux equivalent to a current *I* is necessary to alter the core's state.

Cores are usually assembled to form a core plane, as shown in Figure 12.5. Each core in a core plane corresponds to one bit of a word. An *n*-bit word requires *n* planes. There are two wires threading each of the cores, forming a grid structure. These wires are used to address a particular magnetic core in the core matrix. For example, if the vertical wire designated by 10 and the horizontal wire designated by 01 are both activated by passing a current, $I/2$, through each, the core labeled 1001 will be switched to logic 1 due to flux equivalent to that which would occur if a single wire carried a current *I*. Similarly by reversing the direction of both currents, the same core can be switched to 0. Only one of the cores at a time has the equivalent current, *I*, flowing through it. All others have one-half that value, which is insufficient to cause the magnetic field to switch direction.

FIGURE 12.5 **Memory Core Plane.**

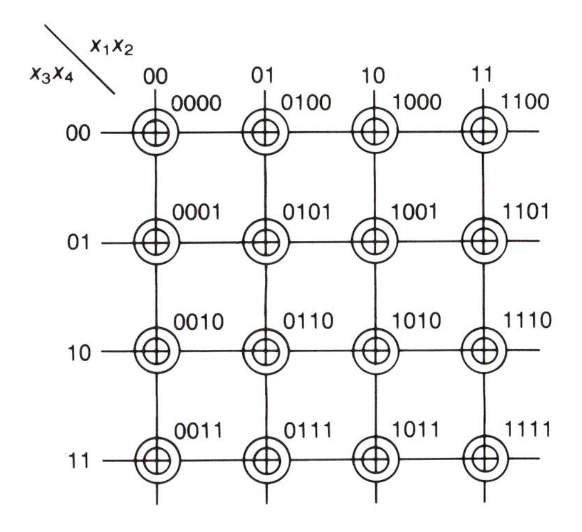

The data are retrieved from a core memory with the help of an additional wire, called a *sense line*, that is threaded through each of the cores and is connected to the input of a *sense amplifier*. Such a configuration is shown in Figure 12.6. The amplifier senses the change of the state. For example, just before the READ pulse occurs, the core is in the logic 1 state. Due to the presence of $-I$, the flux density changes from B_r to the most minimum flux value, $-B_m$, a large flux change. If the READ pulse occurs when the core holds a logic 0, the flux density changes only a little from $-B_r$ to $-B_m$, a small flux change. A large change of flux induces a large voltage at

FIGURE 12.6 **Sense Winding and Sense Amplifier.**

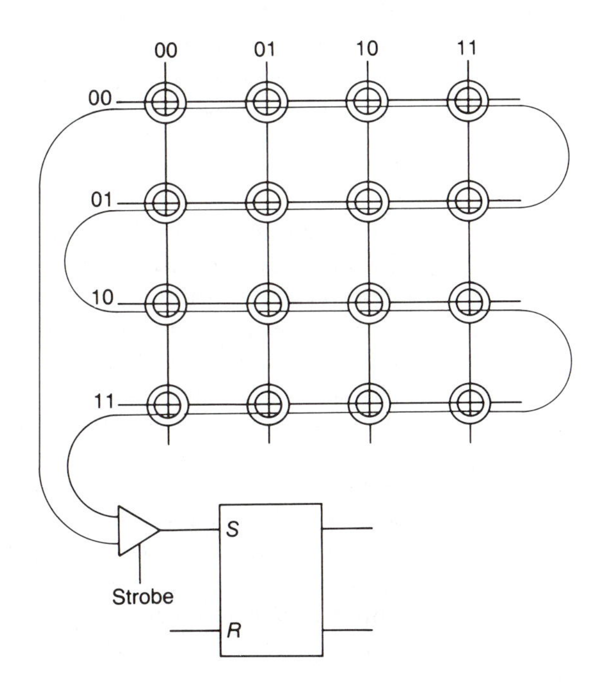

the amplifier. A small flux change induces a small voltage. The amount of voltage induced determines whether a 1 or a 0 was stored in the core. The *strobe* input to the amplifier allows the amplifier to become active only during the READ mode. After a core is read, due to the reading process its content will always be a 0. The READ operation in a magnetic core memory is called a *destructive read-out* since any 1s stored will be destroyed.

Figure 12.7 illustrates the complete timing diagrams for the READ operation. When both vertical (10) and horizontal (01) wires are activated with $-I/2$ current, it implies a READ mode. This mode causes the total current to be $-I$ and switches the core state to 0. If the core is in state 1, then the READ operation detects the flux change, setting the FF shown in Figure 12.6. If the core is in state 0, the change in flux is detected as a 0, which causes the FF to reset. Since the READ process destroys the 1s that are read, each READ cycle must be followed by a WRITE cycle to restore the data to the pre-READ value.

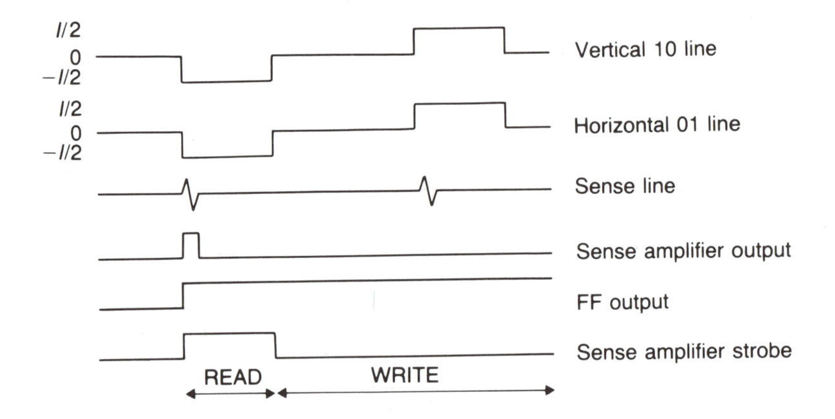

FIGURE 12.7 **READ Timing Diagrams.**

The two lines used for writing 1s and 0s in the cores pass through corresponding cores in each core plane. In an eight-bit memory, if the value 10101010 is to be written into the proper core in each of eight planes, the same two lines pass through each core making up the word. The first action is to clear the word by READing it. Next, current to cause a 1 to be written is applied to the lines that pass through each core in the word. For this process an additional line, called an *inhibit line*, is threaded through the center of each core of the plane, as shown in Figure 12.8. This line is used to inhibit the switching of a core to state 1 whenever a 0 is to be in that core. Presence of $-I/2$ current in the inhibit line of each plane to which a 0 is to be written cancels the flux generated by the positive current flowing through one of the other two enable lines. This cancellation results in the restoration of logic 0. Figure 12.9 shows the corresponding WRITE timing diagrams when the two enable lines are

FIGURE 12.8 **Core Plane with Sense and Inhibit Lines.**

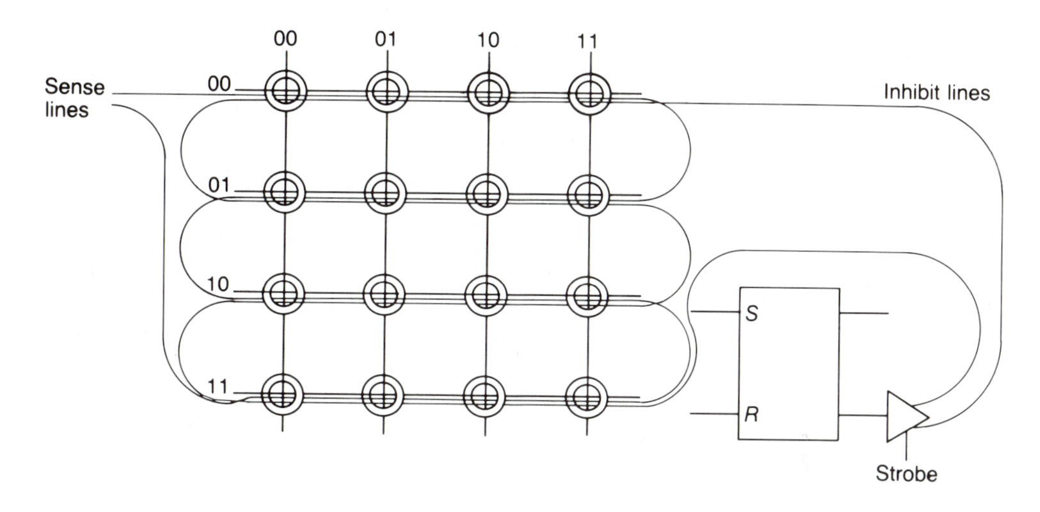

FIGURE 12.9 **WRITE Timing Diagrams.**

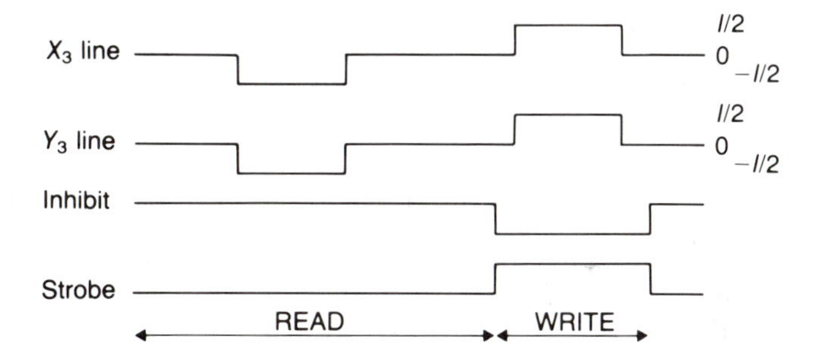

activated with $I/2$ current in each. A 0 on the input line activates the inhibit wire driver on each plane to which a 0 is to be written.

With all the changes in flux that occur in large core memories, voltages can be induced in lines adjacent to cores being switched. This voltage results in electrical noise that must not be interpreted as valid data. To reduce noise, the cores in the memory plane are usually tilted in alternate directions, and wires are routed in such a way that the polarity of one-half of the voltages is opposite to that of the other half. Figure 12.10 shows one such arrangement of a core plane. An $m \times n$ core memory is organized in m core planes. Each core plane contains n number of bits. The usual organization of the core memories involves the inclusion of two address selection wires, one bit-sense wire, and one inhibit wire. A complete three-dimensional configuration of such a memory stack is shown in Figure 12.11.

Core memories have been constructed that combine the functions of the sense and inhibit lines. Since these two lines are never in use at

FIGURE 12.10 **Noise-Free Core Plane.**

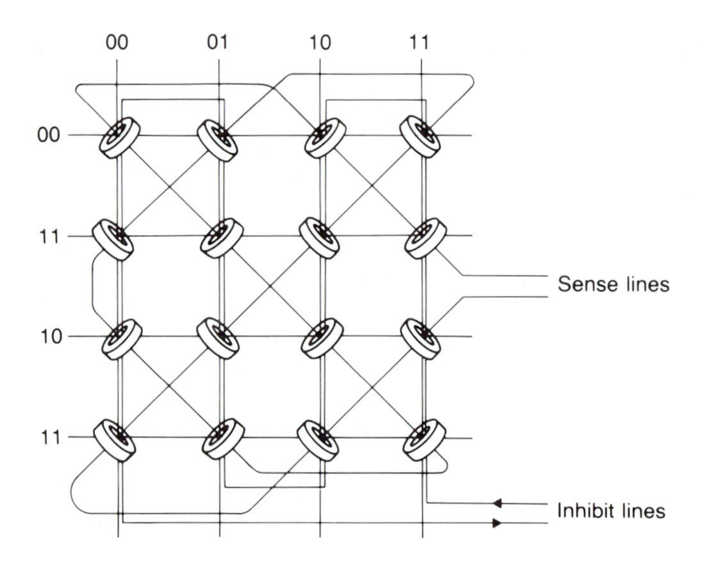

FIGURE 12.11 **Three-Dimensional Core Organization.**

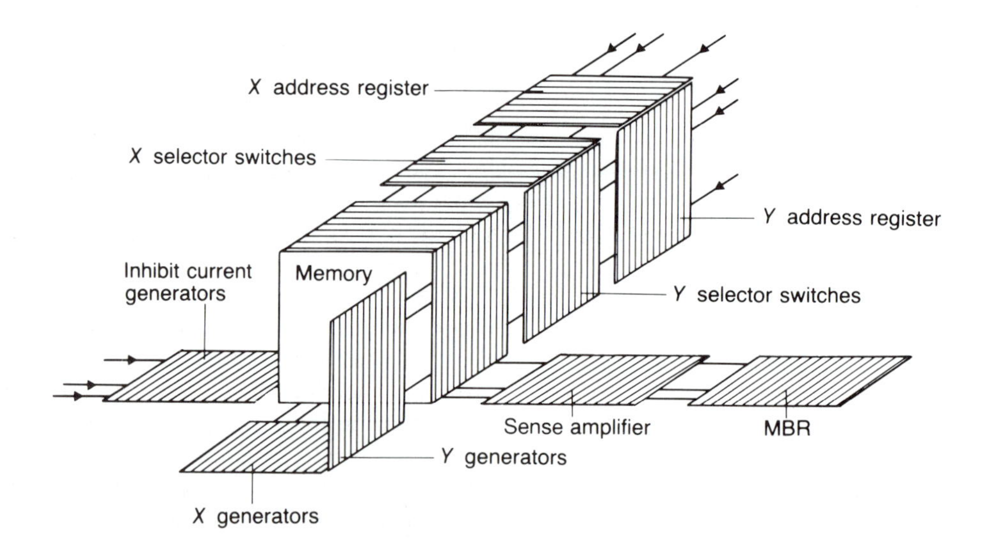

the same time, in three-wire memories their functions are switched electronically. The primary advantages of core memory over semiconductor memory are that it is nonvolatile and can withstand more severe environments, including intense radiation. However, semiconductor memories are smaller and much cheaper, and they consume less power. In addition, the data stored in the semiconductor memories are not destroyed during the READ operation, eliminating the WRITE-after-READ cycle required in core memories.

12.4 Magnetic Bubble Memories

In recent years a number of new technologies have been developed and explored for memory applications. One of these is *magnetic bubble memory*, which has the potential capabilities of having a high bit density and large capacity per chip. It is also less expensive than core memory. The bubble technology has now reached a point where it is destined to play a leading role in the future development of computer hardware. The generation, propagation, and sensing of bubble domains are actually different aspects of the same phenomenon—the interaction of the bubble with its magnetic environment.

In suitably prepared magnetic materials, such as epitaxially grown magnetic garnet films, the directions of magnetization are perpendicular to the film plane. In the absence of external fields, the film plane is divided into two regions of intertwined serpentines, one with upward magnetization and the other with downward magnetization, in order to achieve a minimum-energy magnetostatic equilibrium. As a bias field is applied normal to the film plane, the region with magnetization parallel to the field will grow at the expense of the other region. Finally, the serpentines in the region with magnetization opposite to that of the field will shrink to circular cylindrical domains (*bubbles*). Additional increase of magnetic bias will restore the serpentine domains. The bubble-and-serpentine domain phenomenon has several noteworthy features:

1. Over a stable range of the bias field, stable bubbles exist.
2. A bubble can be annihilated by raising the bias field.
3. A bubble can be elongated by lowering the bias field.

In order to provide an economic alternative to conventional computer memories, bubble memory must have the packing density of one million bubbles per square inch, which restricts the maximum bubble diameters to about 7.5 microns (μm). This particular density requires a uniform film thickness of the same magnitude as the maximum bubble diameter, constant within 1 percent or less.

A digital system using small bubbles requires a method for detecting the presence or absence of bubbles. Since bubbles are tiny, fast-moving domains with very little energy associated with them, detection is a difficult problem. In some of the detection schemes the detector could annihilate the bubble, whereas in other schemes the detector samples the bubbles without destroying them.

It has taken longer than expected to bring this technology to the marketplace. The strongholds of established semiconductor and disk technologies are very much in evidence. However, what we see as the first-generation bubble products are respectable entries. It is predicted that this dual competition with semiconductors and disks will continue into the future. The domain of bubbles indeed appears bright, but significant efforts will be necessary to realize the full potential of magnetic bubble memory.

12.5 RAM Characteristics

Semiconductor RAMs have random-access characteristics similar to those of the magnetic core memories. However, semiconductor RAMs are smaller, faster, and cheaper. The RAM data sheet provides many important timing characteristics that must be met to ensure correct operation. The two basic timing problems that concern designers are the READ cycle timing and the WRITE cycle timing. Of these the READ cycle is less complex.

The *cycle time* of a RAM is the minimum time between successive READ or WRITE operations. The cycle time is usually longer for WRITE operations. There are two particular time durations that are important in the READ cycle: the access time and the READ cycle time. The *access time* indicates the total time that elapses between the application of address inputs and the appearance of data on the data output lines. The *READ cycle time* is the duration between successive READ cycles. Figure 12.12 shows these and other important timing parameters for a READ cycle. The data sheet value for the cycle time for READ, t_{cr}, is usually a minimum value; values greater than this are permissible. This critical value indicates the maximum frequency at which the memory may be read. The parameters t_a and t_{od} are respectively the required access time and the time during which the old data remain valid following the change of an address. The region of the diagram having simultaneous constant values of 0 and 1 corresponds to the constant data bits or address lines. The period when signals are changing is shown by crossed lines. The symbols used in the timing diagram are not necessarily universal and may vary from manufacturer to manufacturer. An additional control input, called the CHIP SELECT, is available for selecting a particular RAM chip. This allows the combination of several chips to create a larger memory.

FIGURE 12.12 **READ Cycle Timing Diagram.**

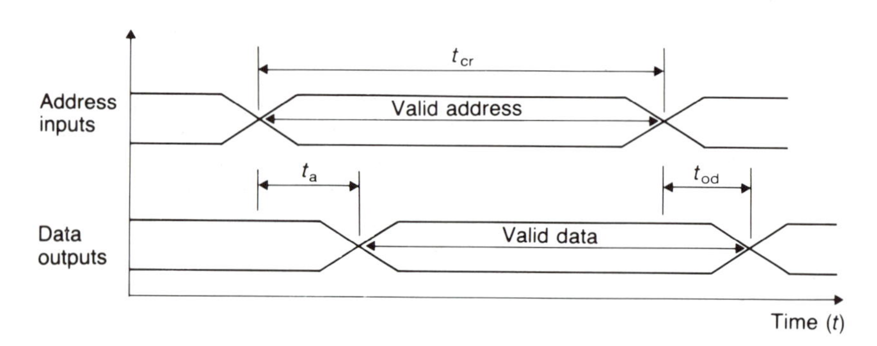

The WRITE cycle timing diagram is somewhat more complicated since it includes the timing for an additional input, $R/\overline{W}$. For a normal READ operation the $R/\overline{W}$ line is held in the high state. During a WRITE cycle the $R/\overline{W}$ line must be held high until the address and data lines are stable; then it is switched low for a specified time to allow the writing of data into the memory. Figure 12.13

FIGURE 12.13 WRITE Cycle Timing Diagram.

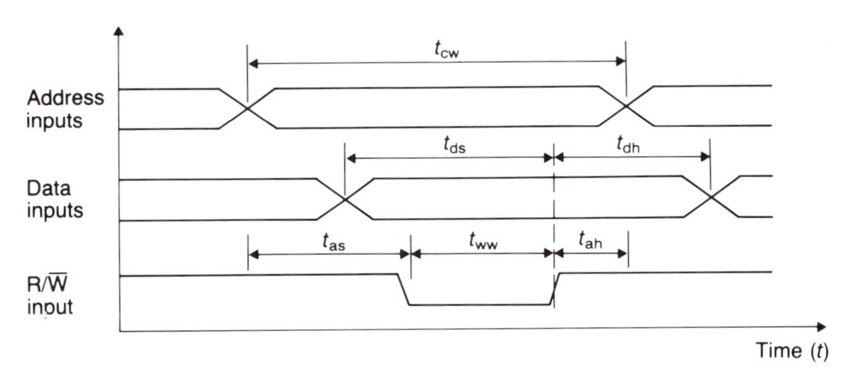

t_{cw} (500 ns): minimum cycle time for WRITE
t_{ds} (350 ns): minimum data setup time
t_{dh} (100 ns): minimum data hold time
t_{as} (200 ns): minimum address setup time
t_{ah} (50 ns): minimum address hold time
t_{ww} (300 ns): minimum WRITE width

demonstrates the WRITE cycle timing and lists important parameters along with their representative values to show their magnitude relationships. These values differ from manufacturer to manufacturer and are often reduced with the implementation of newer fabrication techniques. The next three sections investigate the nature of different RAM technologies.

12.6 Bipolar RAM

In a bipolar RAM the basic memory cell is a flip-flop. This application of FFs was already pointed out in Chapter 6. Special features, however, are added for addressing the cell, reading from it, and writing into it. In addition, one or more control lines may be incorporated to provide a means for combining several units to allow storing a large number of words. These memories are fast, with access times as low as 10 ns.

Figure 12.14 shows the logic diagram of a single-bit bipolar memory cell (U). Even though the unit looks complicated, it is realized from a couple of transistors using the available semiconductor technology. When the READ/$\overline{\text{WRITE}}$ (R/$\overline{\text{W}}$) control is low, the data input is loaded into the *SR* FF selected by the SELECT control. When the R/$\overline{\text{W}}$ control is high, the data bit already stored in the FF appears at the output. Both the input is blocked and the output remains low as long as the SELECT control input is low.

Figure 12.15 shows how 16 bipolar RAM units are connected to form a 16-bit (4 × 4) RAM. The two address lines are decoded by a 2-4 line decoder that enables a particular row of four bits. When

FIGURE 12.14 **Bipolar RAM**

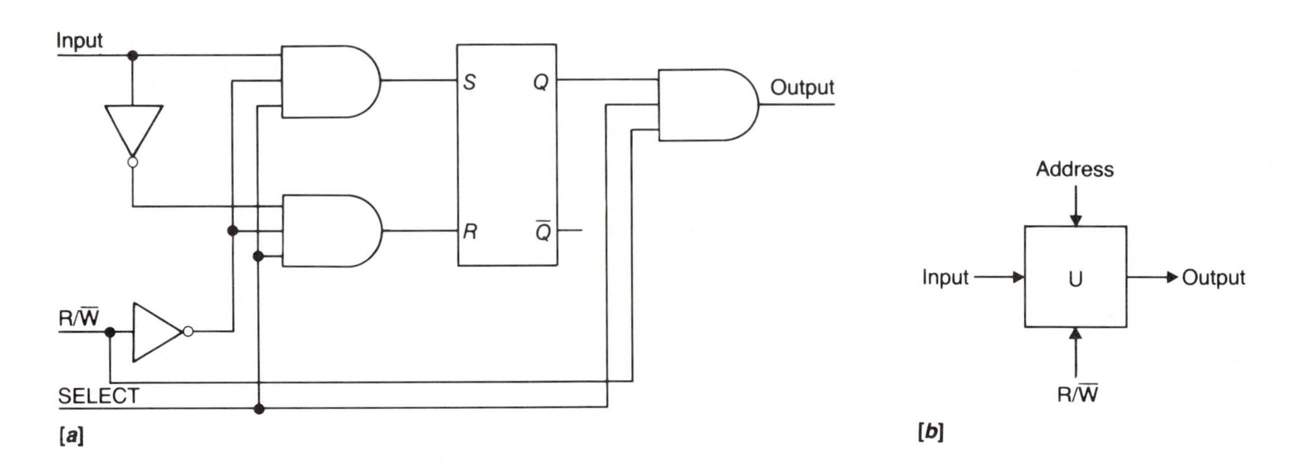

FIGURE 12.15 **16-Bit Bipolar RAM.**

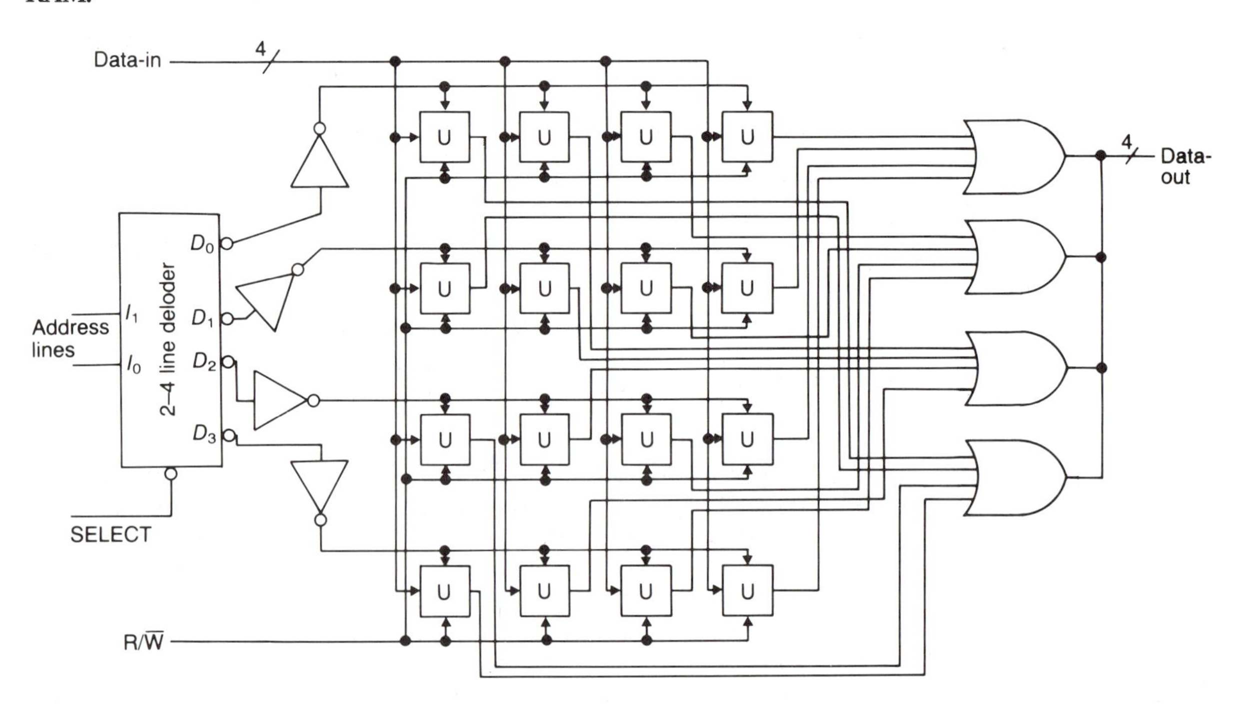

the R/$\overline{\text{W}}$ control is high, the four bits of the selected word go through the four OR gates to the outside. The memory units that are not selected would contribute a logic 0 at the OR inputs. When the R/$\overline{\text{W}}$ control input is low, the four bits of data at the input are loaded into the FFs of a particular row as determined by the address inputs.

Figure 12.16 shows a representation of a 256-bit RAM organized in 16 rows and 16 columns. It can store up to 256 words with one bit each. Each of the 256 ($= 2^8$) FFs can be accessed by activating the appropriate ROW SELECT and COLUMN SELECT lines. The total of eight address lines (A_0–A_7) are divided between the two 4-16 line decoders. Each of the decoders may select up to either 16 columns or 16 rows. The RAM is also provided with a *data-in* and a *data-out* line. In this particular case the row address and the column address can be expressed with a hexadecimal code. For example, an address code of 10101011 $= AB_{10}$ designates the memory address 171_{10} as illustrated in Figure 12.16. This is the 171st cell, counting from left to right, beginning at the top left corner and starting at 0.

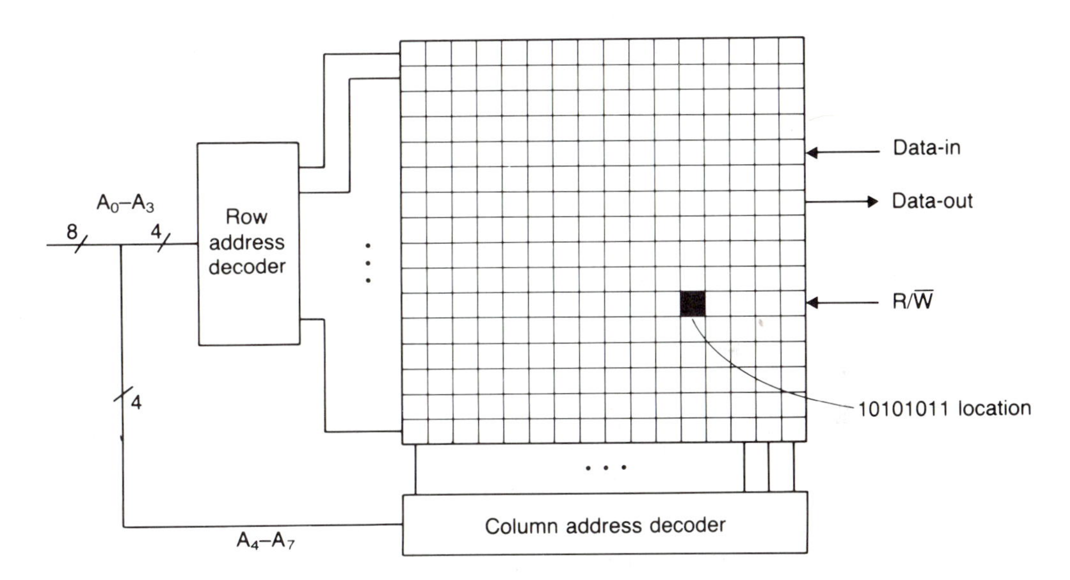

FIGURE 12.16 **Organization of a 256-Bit RAM.**

Using eight 256-bit RAMs, a 256-byte RAM can be made as shown in Figure 12.17. Similarly, Figure 12.18 shows the formation of a 4K-byte memory by using a total of 32 1K-bit RAMs. Each 1K-bit RAM requires 10 address lines since $2^{10} = 1024$. Therefore, 32 of these 1K-bit RAMs are organized in four rows and eight columns in such a way that each row consists of 1K-byte memory. Each of the 1K-bit RAMs has 10 address lines, one data-in, one data-out, one chip select, and one $R/\overline{W}$ line. The remaining two address lines of the 12 needed (since $2^{12} = 4K$) are used to select a particular row of 1K-byte RAM. This selection is accomplished by enabling the corresponding chip selects, CS, via the outputs of a 2-4 line decoder. In order to connect the outputs of each 1K-byte segment of RAM (referred to as one page of RAM) together, the out-

FIGURE 12.17 256-Byte RAM.

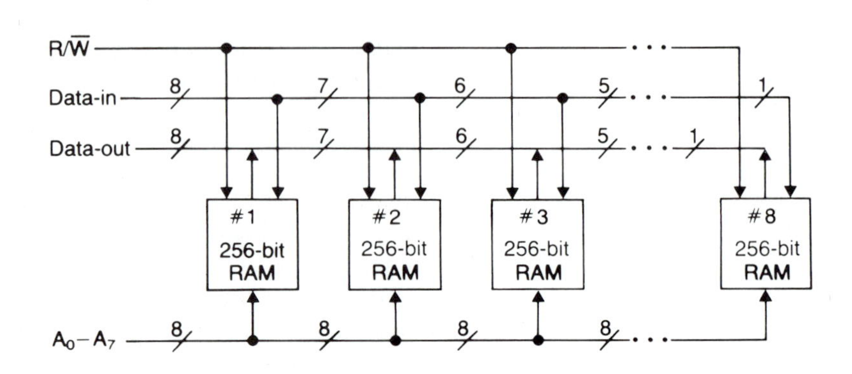

FIGURE 12.18 4K-Byte RAM Using 1K-Bit RAMs.

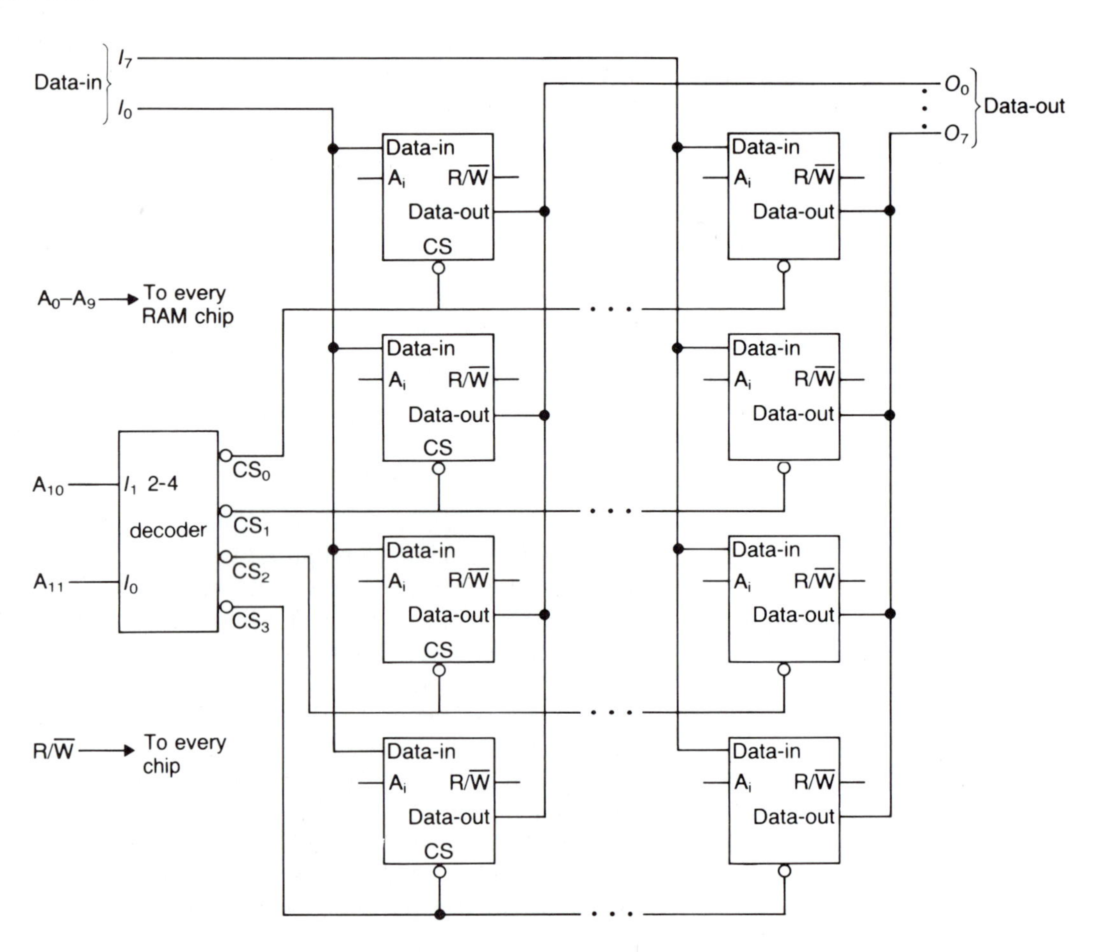

puts must be tri-state outputs that are in the high impedance state unless the chip select is active.

12.7 MOS Static RAM

The MOS (metal-oxide semiconductor) memories are of two types: static and dynamic. In general, the *MOS static RAM* units are composed of MOS field-effect transistor (MOSFET) latches. These memories are comparatively lower in speed than the bipolar RAMs, but cost less, consume less power, and have a high packing density. As long as power is provided the latches retain their binary states. MOS devices have relatively large internal capacitances, and as a consequence their access time is longer than that of bipolar RAMs. In this section we shall restrict ourselves to the discussion of static MOS RAMs.

Figure 12.19 shows one of the typical cross-coupled FFs used in a static MOS RAM. The transistors T_1 and T_2 are the cross-coupled elements of the FF and have relatively low impedances. The two transistors T_3 and T_4 are active loads that behave analogous to resistors, while transistors T_5 and T_6 are used as two-way transmission gates. Either T_1 conducts and T_2 is cut off or vice versa. A static RAM will contain thousands of such FFs, one corresponding to each of the stored bits.

FIGURE 12.19 **Static MOS RAM Cell.**

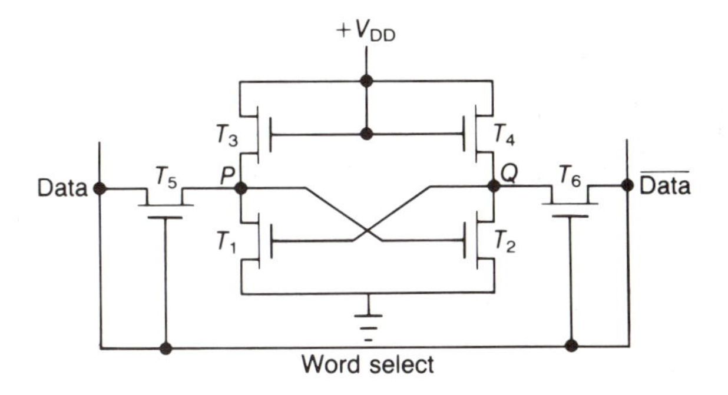

To read the cell, assuming that node P is high and node Q is low, the word line is pulsed, turning on the gate devices T_5 and T_6. The state of the cell is determined by detecting on which digit line the sense current occurs. Writing is performed by forcing the digit lines to the value desired in the cell, thereby over-riding the cell contents. The word line is pulsed, and if the digit line is grounded, then the cell is already in the state to which it was to be written, and no change occurs. If the digit line is grounded, node P is pulled down toward the ground through the gating device T_5. As the level of node P drops below the threshold voltage of T_2, the device turns off and node Q is pulled up through T_6 and T_4. The writing process thus consists of two transients–discharging the initially high node through a gating device operating in common source mode, and charging the initially low node through the other gating device in source-follower operation. The transients overlap to some extent, but it is the charging transient that largely determines cell writing delay in practice.

The major advantages of MOS memories over bipolar memories (e.g., TTL) are that the MOS devices consume less power and that a large memory could be fabricated over a smaller chip area. However, the main disadvantage is that MOS devices are relatively slower than bipolar devices.

12.8 MOS Dynamic RAM

A major drawback of the static cell is that its power dissipation limits the total number of bits per chip due to heat transfer considerations. Complementary MOS (CMOS) devices, however, require little power and allow a higher packing density. These devices store their information as charges on capacitors rather than on FFs. Figure 12.20[*a*] shows one of the memory units in a dynamic RAM. The disadvantage of the dynamic RAM is that it is slower than static RAM. The charge is slowly discharged from the capacitor, causing the stored data bits to be lost unless its charge is refreshed periodically. This refresh requirement complicates the design problem by requiring additional circuitry. Despite the complication of the refreshing process and their slow speed, dynamic RAMs are relatively cheap and are used quite extensively in many applications.

In the dynamic RAM unit the transistors T_1 and T_4 act like resistors to the FF formed by T_2 and T_3. In the READ mode a low is applied at the ROW SELECT, and the currents in the columns (the two columns are respectively designated as 1 and 0) are sensed.

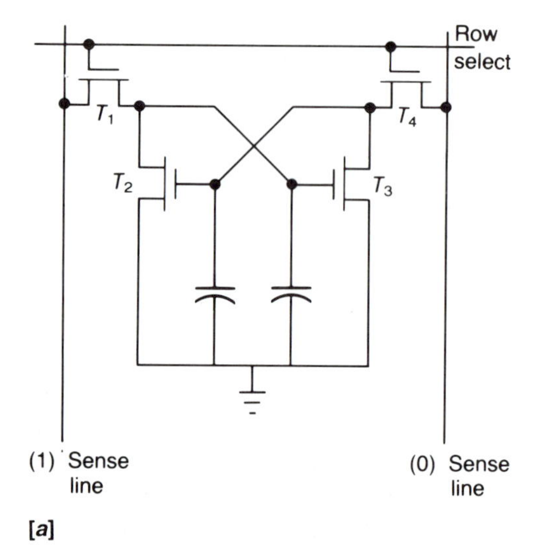

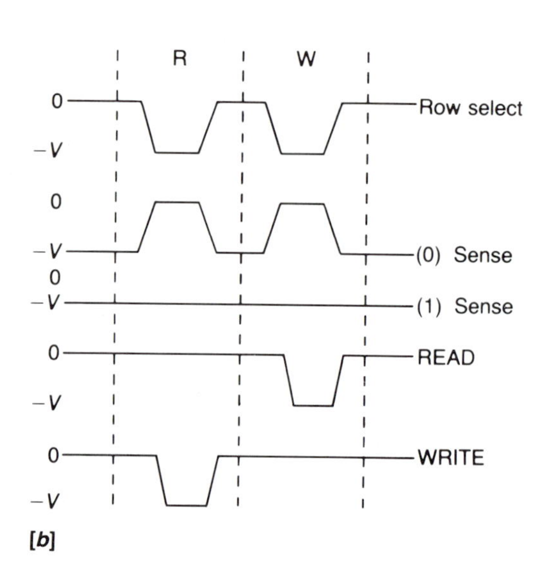

FIGURE 12.20 **Four-Transistor Dynamic MOS RAM Unit: [*a*] Circuit and [*b*] Timing Diagrams.**

The transistors T_1 and T_4 are utilized as row enable transistors. When T_2 is ON, the sense line associated with T_1 carries the current while the other does not. In WRITE mode the ROW SELECT line is enabled and the sense lines are changed to the desired states. The *refreshing* of the unit is accomplished by supplying negative voltage to both row and column lines. The timing diagrams of a representative WRITE operation followed by a READ operation is illustrated in Figure 12.20[*b*], where $-V$ is the most negative voltage.

For the purpose of illustration, Figure 12.21 shows one of the memory units of a three-transistor dynamic MOS RAM and its corresponding timing characteristics. It has four different enable lines: READ DATA (RD), WRITE DATA (WD), READ SELECT (RS), and WRITE SELECT (WS). A negative voltage to WS turns on T_1 and charges the capacitor to a negative voltage. Consequently by applying negative voltage to the WS line and zero voltage to the RS line, a logic 0 may be written. If a negative voltage is applied to the RS line and zero voltage to the WS line, the READ operation can be performed. During the READ mode, the WD line is usually isolated from the capacitor and the RD line is discharged through T_2 and T_3 to zero voltage. The logic state is decoded generally by sensing the voltage of the RD line. The inversion that has taken place is usually counterbalanced by inverting the output.

FIGURE 12.21 **Three-Transistor Dynamic MOS RAM Unit: [*a*] Circuit and [*b*] Timing Diagrams.**

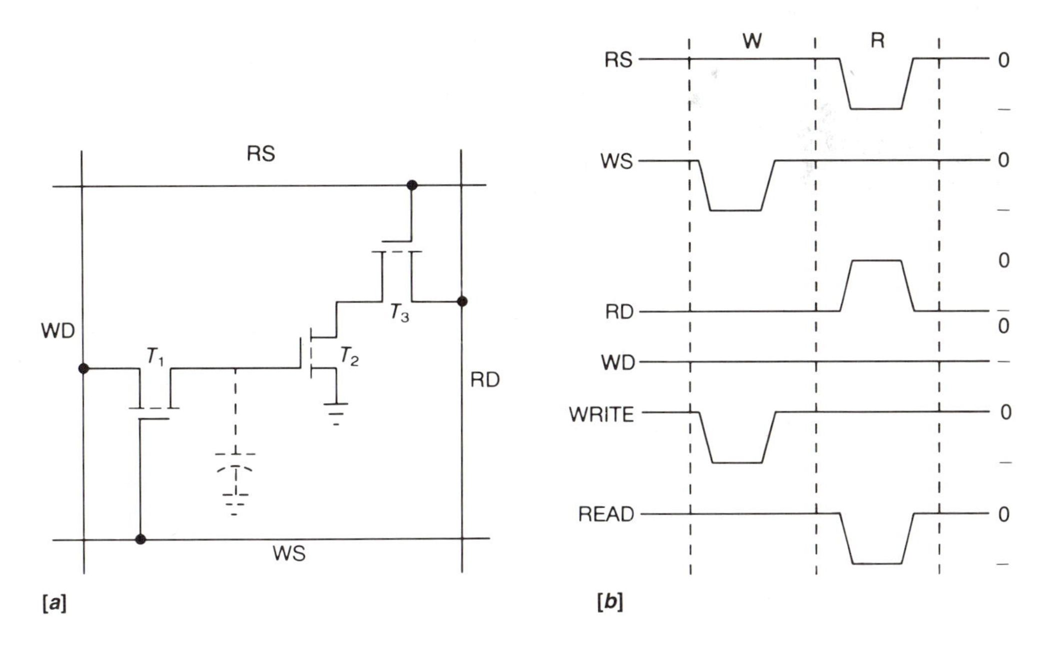

A typical 16K × 1-bit MOS dynamic RAM is shown in Figure 12.22 by the 128 × 128 square matrix of capacitors. It requires 14 address lines (since $2^{14} = 16K$), a data-in line, a data-out line, and an $R/\overline{W}$ line. The seven low-order address bits could be made to

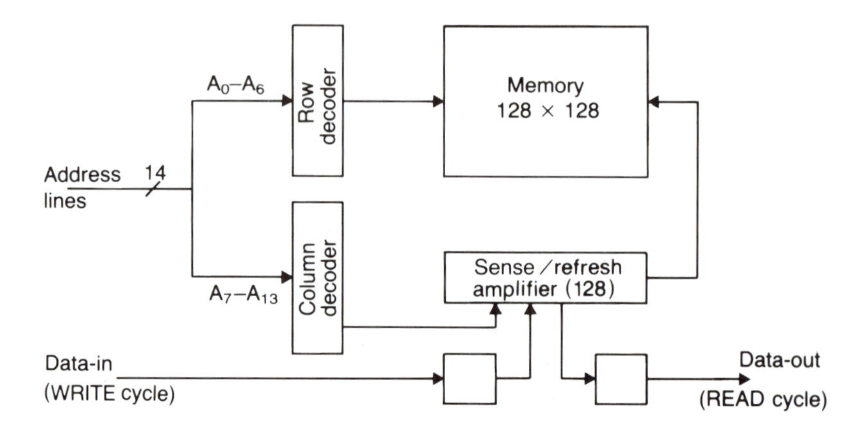

FIGURE 12.22 **Block Diagram of 16K-Bit MOS Dynamic RAM.**

select a particular row, the contents of which are fed into the 128 sense/refresh amplifiers. The seven higher-order address bits select one of the 128 sense amplifiers and place the content on the data-out line. At this time the contents of these 128 bits present in the amplifiers are refreshed by charging the capacitors, that is, by rewriting the data back in the proper memory row. This refreshes all 128 bits in that row.

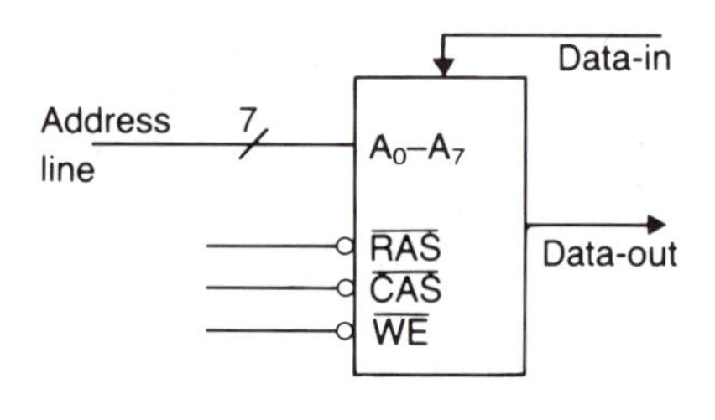

FIGURE 12.23 **Logic Diagram of a 16K-Bit MOS Dynamic RAM.**

Consider a standard 16K-bit MOS dynamic RAM, such as the one shown in Figure 12.23. It usually has seven address inputs instead of 14 due to the pin limitations of the chip. Consequently the 14 address inputs are multiplexed into the row and column latches on the address bus seven bits at a time. The ROW ADDRESS STROBE ($\overline{RAS}$) and the COLUMN ADDRESS STROBE ($\overline{CAS}$) are used respectively to enter the seven low-order address bits and the seven higher-order address bits.

A dynamic RAM requires that each READ cycle be followed by a WRITE cycle so that the data bit is restored to its original location. If new data are to be written into the memory, it is done during the refresh phase. The WRITE ENABLE ($\overline{WE}$) is usually taken low for a READ cycle and high for a WRITE cycle. However, each of the 128 rows of the memory array must be refreshed at a certain special interval (typically 2 ms) to maintain the stored data. This operation is accomplished by repeatedly going through the $\overline{RAS}$ and $\overline{CAS}$ sequences for each of the 128-bit rows.

Dynamic RAMs are important because fewer transistors are needed to store a bit. They are faster than the static RAM and consume relatively less power. However, the refresh cycle often requires additional circuitry and execution time. Consequently a certain minimum memory is required before dynamic RAM becomes economically justified. In a memory array, therefore, smaller memories tend to use static RAMs, whereas in larger memories dynamic RAMs are an economical choice. If each row of memory is accessed within the required refresh time, no additional refresh cycle is necessary. Dynamic RAMs are an obvious choice for applications where it is assured that all rows are repeatedly accessed. Cathode ray tube (CRT) controllers used to display computer data have the information to be displayed stored in dynamic RAM. The memory data are displayed repetitively at a rate that eliminates the refresh requirement altogether.

12.9 PROMs

We discussed ROMs in Chapter 4 for use in designing complex combinational circuits. In a ROM, a semiconductor device (diode, bipolar transistor, or MOSFET) is connected or not connected at the grid intersections based on whether or not the corresponding word bit is a logic 0 or a logic 1. The designer provides the manufacturer with a list of data to be stored in each bit position. A photographic template of the circuit, called a *mask*, is made and used in the fabrication of ROMs. ROM mass production is economically sound only when a large number of these devices are manufactured simultaneously. A disadvantage of the mask-programmed ROMs is that it is impossible to make alterations if a design error is discovered. The anxiety associated with committing a design to ROM has now been eliminated with the availability of a device known as *programmable read-only memory* (*PROM*).

PROMs are nonerasable memories that can be programmed by the user, applying a *PROM programmer*. Once programmed they cannot be altered. However, if an error is found, only one ROM has been ruined. The PROM programmer stores the data by supplying voltages and currents to the solid-state devices connecting the grid points, which is in excess of those encountered in normal operation. This procedure burns open the device, leaving functioning devices at some intersections and none at others, which allows data to be stored. PROMs are available in both bipolar transistor and MOSFET form. A typical PROM organization includes memory arrays, address decoders, and buffers, as shown in Figure 12.24. Since the PROM element varies from device to device, the programming procedure varies from device to device as well.

FIGURE 12.24 **Typical PROM Organization.**

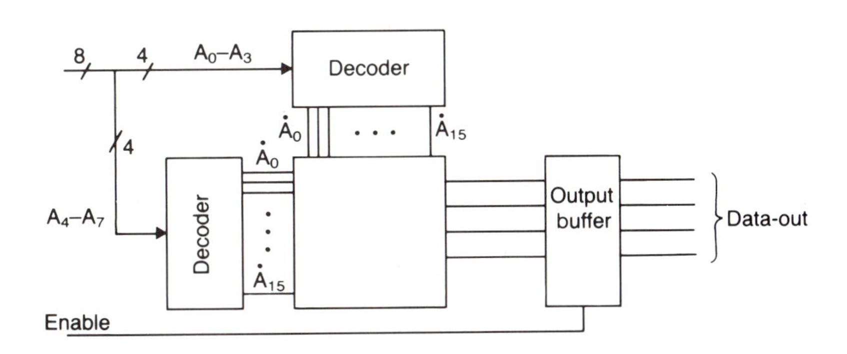

An even better ROM device for system development is an *erasable programmable read-only memory* (*EPROM*). EPROM offers system designers maximum flexibility in changing program instructions and so forth in the development of their systems. In EPROMs, MOS floating gates allow the charge to be stored semipermanently within the gate dielectric or within the varied gate structure. When the programming voltage–induced electric field is removed, a charge remains trapped within the gate. An EPROM is recognizable by the presence of a quartz lid on the top of the IC package. EPROMs are erased by exposing the PROM chip to high-intensity ultraviolet (UV) irradiation. The radiation causes the flow of a photocurrent from the floating gate back to the silicon substrate, which discharges the gate back to its initial status. EPROMs may be programmed and erased as many as a thousand times. Commercially available programmers and UV erasers make EPROMs a valuable development tool. Two other programmable ROMs are the following:

EEPROM. Electrically erasable PROMs (EEPROMs) allow programming of the byte positions and their erasure using voltages only. All values can be destroyed during the erasure procedure.

EAROM. Electrically alterable ROMs are essentially the same as EEPROMs. The altering of a word takes significantly less time than accessing data from a word. An EAROM is a good choice for storing data that must not be lost if power is lost.

EPROMs, EEPROMs, and EAROMs are all nonvolatile storage devices.

12.10 Optical Memories

The growing demand for mass memories has stimulated a variety of memory design techniques. The bit density in magnetic recording on drums, disks, and tapes has no theoretical limit. On the other hand, the conventional magnetic mass memories need mechanical motion of the storage medium, which restricts the bit density to the accuracy of mechanical adjustments, and limits the access time to the millisecond range. This mixing of high-speed circuits with millisecond mechanical systems is a dubious compromise.

One alternative is the use of *optical memories*, which employ either thermal or high field effects of laser power in certain materials to store information. They offer in principle a potential improvement in storage density and access time. There are three broad categories that have emerged in the field of optical processing of information. The first technique, having its origin in spatial filtering studies, is the coherent optical processing technique, which has reached a certain maturity in the sense of being a readily useful method of processing data. The second category includes all the electro-optical effects and devices that are useful in logic, memory, or input/output. The third category includes all optical processing in which signals are carried by light, and logic is performed by interaction between light beams.

The real stimulus to develop an optical information-processing technique is its potential for enormous data-handling capacity. The key to high storage density is the submicron wavelength of light, since the theoretical limit of closeness of stored bits to each other is set by the wavelength of the recording energy. The practical limit is set by particle size in the recording medium. Designers of optical memories take advantage of all the advanced techniques, such as holography, high-resolution photographic emulsions, and photochromism. Most optical memory utilizes parallel storage, that is, all data are written or read out at one time.

An ordinary holographic recording system consists of an optical storage material (e.g., a holographic plate), an object of which a hologram is being recorded, and a laser source that coherently illuminates partly the object and partly the storage material, as shown in Figure 12.25. In the plane of the recording plate, the light beam being reflected or transmitted by the object (object beam) is superimposed on the other part of the laser light guided to the holographic plate (reference beam) directly. The superimposition forms an interference pattern that is used to expose the recording medium. The stored intensity distribution of the interference pattern, in the form of a transmittance function, is the hologram of the object.

FIGURE 12.25 **Ordinary Holographic Recording System.**

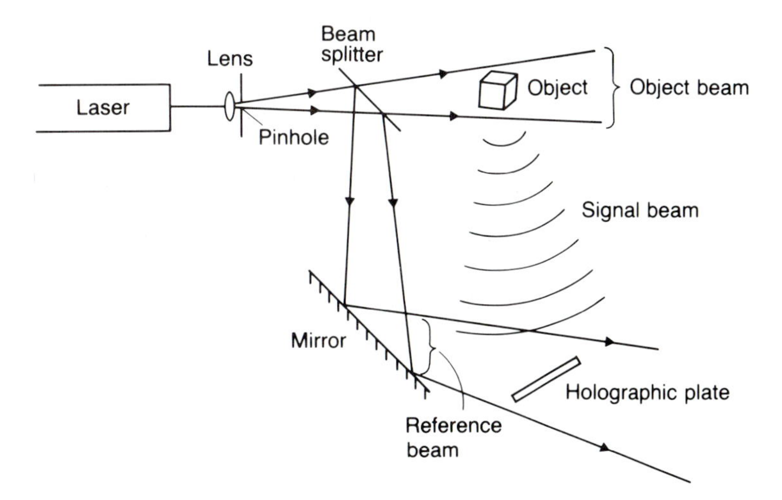

The main ideas of the holographic storage scheme are to create an image of the information to be stored, to use this image as an object in a holographic recording system, and to store the object in a hologram. The information to be stored in a practical memory is being delivered in the form of electronic signals, so a special device known as a *page composer* is necessary to introduce the data in the form of a transparency. This transparency is an array of a binary square matrix, each dot representing either a 1 (transparent dot) or a 0 (opaque dot).

The basic components for a holographic storage system are illustrated in Figure 12.26. The object beam passes through the page composer and combines with the reference beam to form an interference pattern on the storage medium, resulting in a hologram. The page composer is used to convert the incoming data into a one- or two-dimensional optical object that can be recorded holographically. During the read-out operation, the object beam is blocked by a shutter, and the recorded pattern acts like a diffraction grating for the reference beam and projects a reconstructed image at the detector. The detector is a multi-element array with as many elements as there are bits in each stored page. The entire page is immediately available for possible electronic accessing during read-out. The more complex mass memory systems are based on this unit and on additional devices for the addressing of a large number of holograms in the storage plane in which several pages can be stored in parallel. For read-out, one hologram at a time is illuminated by a laser beam and the reconstructed page is imaged onto a common detector array. The addressing of the holograms by a laser beam is controlled by a transverse laser beam deflector. This deflector works as the address unit of the memory and steers a laser beam to each desired hologram position.

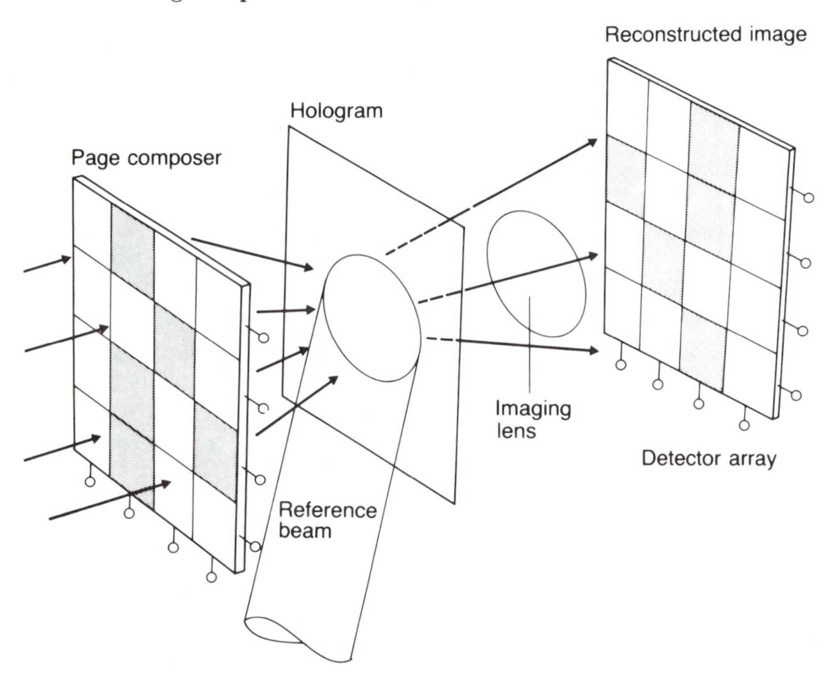

FIGURE 12.26 **Holographic Memory.**

Electro-optic crystals also offer interesting possibilities for the holographic storage of large quantities of data. These recording materials are known to be capable of very high resolution recording without granularity. Unlike "all optical" laser devices whose operation is based on the interaction of an optical signal with laser materials, optoelectronic devices require conversions between optical and electrical energies. Much of the recent interest in optoelectronics stems from its low cost potential, which depends on the fact that relatively large and complex arrays may be fabricated using polycrystalline materials and simple techniques. Moreover, in optoelectronic memories, storage elements are completely independent of their detectors. Read-out from the memory can be serial, parallel, or by block. In addition to this, magneto-optic effects are being utilized now to detect the magnetic state of memory. When the optical electric field causes current carriers to move in the presence of a magnetic field, the carriers experience a magnetic force normal to their motion and the applied magnetic field. This phenomenon is being exploited currently to develop different kinds of magneto-optical memories.

12.11 Summary

This chapter provided an introduction to various kinds of memories. In particular, magnetic, semiconductor, and optical memories were considered. It is imperative that a designer stay current with the rapidly changing memory technology. Selection of the proper storage medium can determine the success of a design.

Problems

1. Determine the RAM contents necessary for generating the sequence $Q_2Q_1Q_0 \rightarrow$ 1, 4, 6, 7, and repeat using the circuit of Figure 12.P1.

FIGURE 12.P1

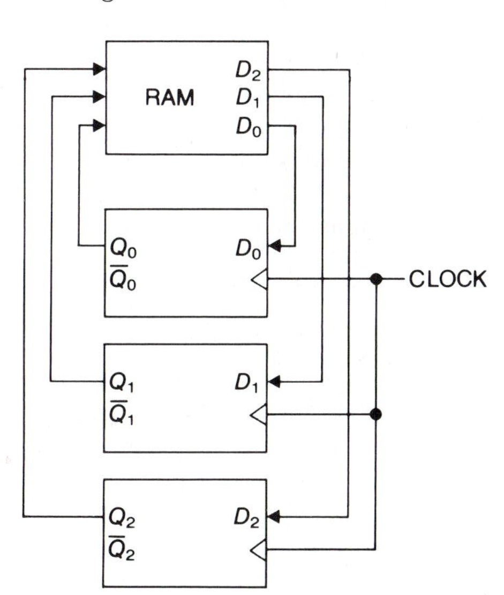

2. Repeat Problem 1 using T FFs.
3. Show the circuit for a 64K × 8 memory made up of 8K × 8 chips.
4. a. Design a 256 × 32-bit RAM from 256 × 8-bit RAMs.
   b. Design a 256 × 8-bit RAM from 32 × 8-bit RAMs.
   c. Design a 4096 × 2-bit RAM from 1K × 1-bit RAMs.
   d. Design a 1K-byte RAM from 256 × 4-bit RAMs.
5. Repeat Problem 4 for ROMs.
6. Design a scale-of-ten counter with a seven-segment display output using D FFs and PLA.
7. Repeat Problem 6 for a scale-of-twelve counter using T FFs.

Suggested Readings

Aagard, R. C.; Lee, T. C.; and Chen, D. "Advanced optical storage techniques for computers." *Appl. Opt.* vol. 11 (1972): 2133.

Ammon, G. J. "Archival optical disc data storage." *Opt. Engn.* vol. 20 (1981): 394.

Bartolini, R. A. "Media for high-density optical recording." *Opt. Engn.* vol. 20 (1981): 382.

Bhandarkar, D. P. "On the performance of magnetic bubble memories in computer systems." *IEEE Trans. Comp.* vol. C-24 (1975): 1125.

Bobeck, A. H.; Bonyhard, P. I.; and Geusic, J. E. "Magnetic bubbles—an emerging new memory technology." *Proc. IEEE.* vol. 63 (1975): 1176.

Cohen, M. S., and Chang, H. "The frontiers of magnetic bubble technology." *Proc. IEEE.* vol. 63 (1975): 1196.

Corsini, P. "*n*-user asynchronous arbiter." *Elect. Lett.* vol. 11 (1975): 1.

Krumme, J. P. "Thermomagnetic recording in thin garnet layers." *Appl. Phys. Lett.* vol. 20 (1972): 451.

Lo, T. C.; Scheuerlein, R. E.; and Tamlyn, R. "A 64K FET dynamic random access memory: design considerations and description." *IBM J. Res. & Dev.* vol. 24 (1980): 318.

Mezrich, R. S. "Magnetic holography." *Appl. Opt.* vol. 9 (1970): 2275.

Majumder, D. D., and Das, J. *Digital Computers' Memory.* Delhi, India: Wiley Eastern Limited, 1980.

Michaelis, P. C., and Bonyhard, P. I. "Magnetic bubble mass memory — module design and operation." *IEEE Trans. Magn.* vol. MAG-9 (1973): 436.

Rajchman, J. A. "Holographic optical memory." *Appl. Opt.* vol. 9 (1970): 2269.

Watkins, J. W.; Boudreaux, N. A.; and Otten, T. H. "Large archival mass memory system using optical diskettes." *Opt. Engn.* vol. 20 (1981): 399.

Weaver, J. E., and Gaylord, T. K. "Evaluation experiments on holographic storage of binary data in electro-optical crystals." *Opt. Engn.* vol. 20 (1981): 404.

CHAPTER THIRTEEN

Processor Design and Microprogramming

13.1 Introduction

In this chapter we will design a simple digital processor with a limited but functional instruction set. The techniques involved in classical hardwired design will be used in the initial design. The classical hardwired design involves deriving the logic equations for each of the many control signals used for control of the processor and implementing them with gate logic. It will be obvious that modifying the initial design requires a tedious rederivation of the control signals. The introduction of another design methodology called microprogrammed control will demonstrate the ease with which the control signals for the same processor can be implemented using ROM. The use of ROMs to implement functions was introduced in Chapter 4.

After studying this chapter and working the exercises at the end of the chapter, you should have a foundation in the design of processors and be prepared for a more in-depth study of:

- Computer architecture;
- Machine language;
- Instruction set implementation in RTL;
- Microprogramming.

13.2 Computer Architecture

A digital processor performs four basic types of primitive operations involving binary data as requested by its set of user-programmable instructions: move operation, unary operation, binary operation, and convert operation. A *move* operation transfers data from one storage location to another through a link (data path). In a *unary* operation a hardware transformation unit (ALU) receives information from a single storage location via a link. The information received is manipulated by the hardware transformation unit before the target storage element receives the result via a link. A *binary*

FIGURE 13.1 **Basic Block Diagram of a Computer.**

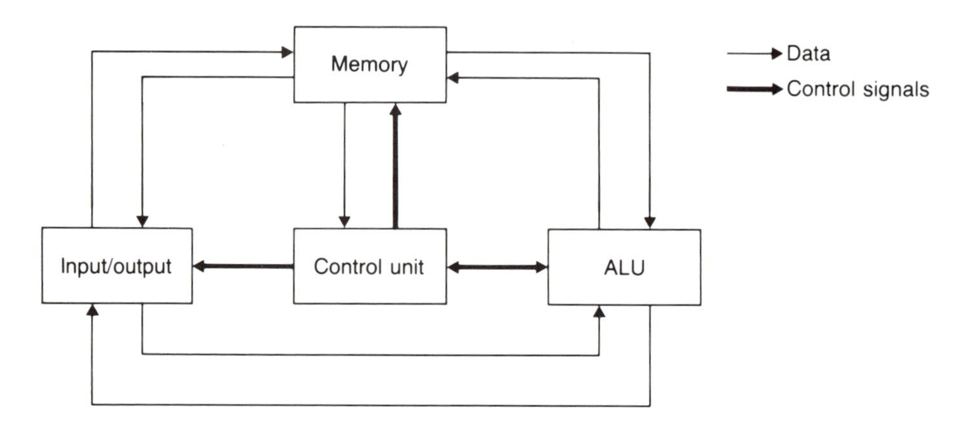

operation involves the hardware transformation unit receiving information from two different storage elements via two separate links, with the resultant operation stored back in a target storage element. A *convert* operation is very much like a unary operation except that the data are transformed into control information.

A computer capable of performing the four basic types of operations consists of the following four basic parts, as shown in Figure 13.1: memory (M), arithmetic and logic unit (ALU), control unit, and input/output (I/O) unit. These four processor components perform the following functions:

1. *Memory.* The memory, made up of devices discussed in Chapter 12, stores both the *operation codes* (*opcodes*) with which the system is programmed and the data on which arithmetic and logic operations will be performed. A memory whose contents are examined without knowledge of the relative placement of opcodes and data is a meaningless table of binary values. To give meaning to the binary information stored, the location of opcodes and data must be ordered in a way that allows the control unit to distinguish between them. In conjunction with the memory, three registers will be defined that have the following functions:
   hold the address of the memory location being written into,
   hold the data to be written into or just read from memory,
   hold the memory address of the next instruction.
2. *Arithmetic logic unit.* In Chapter 5 the ALU was introduced. This unit performs either unary or binary operations. Its function is controlled by the control unit as determined by the opcode of the instruction being executed.

3. *Control unit.* The control unit provides all of the control signals necessary to cause an opcode to be read from memory; the data to be operated on to be read from memory or an external device; the data paths selected; the correct ALU function selected; and the storing of the result in the proper register, memory location, or external device.
4. *Input/output.* To be useful, the computer must have the ability to obtain data or opcodes from external devices and present results to the user or to other devices as control information. Data paths must be provided to which external devices may be connected, and opcodes must be provided that initiate data transfer to and from external devices.

13.3 Machine Language

The control information read from a processor's memory and used by the control unit to supervise the performance of operations on data is called an *instruction.* This information can consist of one or several words of memory. The reader may be familiar with high-level languages such as FORTRAN, PASCAL, or ADA that allow the user to give instructions to the processor in a language very descriptive of the problem for which a solution is sought. The level at which the processor designer operates uses a language that is called *machine language.*

A machine language instruction is made up of a bit pattern that the control unit of the computer decodes to determine the data transfers and operations necessary to perform the intended functions. The instruction is made up of several parts:

1. *Opcode.* The opcode is a group of bits used to indicate the operation to be performed. If n bits of the instruction are reserved for the opcode, 2^n different operations can be coded.
2. *Operand.* The m bits of the instruction used for the operand can perform several functions dependent on the architecture of the processor:
   can be the data to be operated on (immediate data operation);
   can be the address of the data to be operated on (absolute addressing), allowing 2^m locations to be directly addressed;
   can be the address of the address of the data (indirect addressing), allowing 2^m memory locations to be addressed;
   can be an offset from a predetermined address (indexed addressing), allowing addresses 2^{m-1} plus the predetermined address and the predetermined address minus $2^{m-1} + 1$ to be referenced;
   can provide an offset from the address of the instruction

currently being executed (relative addressing), allowing addresses 2^{m-1} plus the current address and the current address minus $2^{m-1} + 1$ to be referenced.

3. *Control bits.* The k bits used as control bits in the instruction identify the addressing mode in use, and the conditions to be checked as part of the instruction operations, such as the sign bit, overflow FF, and other condition FFs.

For machine instructions made up of $n + m + k$ bits, the memory and data paths must be $n + m + k$ bits wide, or more than one word of memory must be used to store an instruction.

In eight-bit microprocessors, opcodes are eight bits wide, allowing the possibility of 256 different operations. The memory used with the processors is also eight bits wide, necessitating multiple words to store instructions. The instructions vary in length from one to three words. The typical forms are as follows:

1. Single-word instructions involve transferring data from one internal register to another with the register code imbedded in the opcode. No operand address is needed:

   XXXXXXXX   Opcode bits

2. Two-word instructions are used to load a register specified by the opcode with data stored in the second word of the two-word instruction, or the second word may be an address, ADR, in the first 256 locations in memory:

   XXXXXXXX   Opcode

   XXXXXXXX   Data or ADR 0–255

3. Three-word instructions allow addressing 2^{16} memory locations:

   XXXXXXXX   Opcode

   XXXXXXXX

   XXXXXXXX   16 bits of address

For an n-bit computer the data paths are most often n bits wide. In general, processors are classified by the bit widths of the data upon which they operate.

In the next section we will begin the design of a limited instruction set four-bit processor, LIS-4.

13.4 LIS-4 Processor Design

The four-bit processor that we will design in the following sections is not necessarily optimal in design or marketable. The more experienced reader will observe ways in which different and perhaps better design decisions could be made. Our intent is to present the design steps and operation of the processor in a way that is not obscured by complex operations.

13.4.1 LIS-4 Processor Architecture

The initial architecture for the LIS-4 processor is given in Figure 13.2. This architecture features two busses. Each of the registers used in the design has a tri-state output connected to the output bus, BO, when its enable output, EO, signal is a logic 1. Each register loads the value on the register input bus, BI, on the falling edge

FIGURE 13.2 **LIS-4 Block Diagram.**

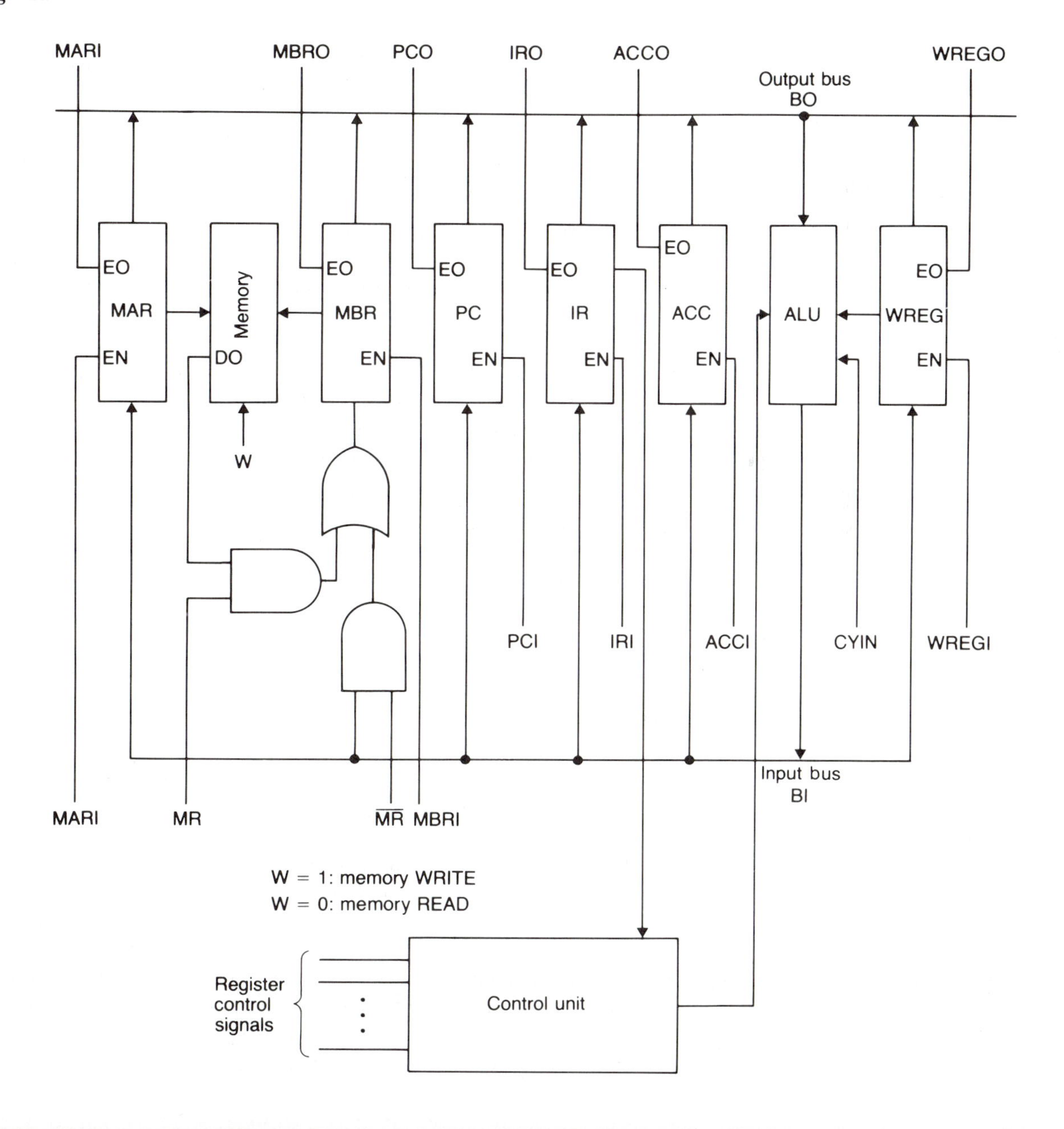

of the system clock when the enable input, EN, is a logic 1. Due to the parallel nature of the operation of the processor, the architecture shown in Figure 13.2 represents the data path for a single bit of an n-bit (in this case four-bit) system.

The functions performed by each four-bit register in Figure 13.2 are as follows:

MAR: The memory address register (MAR) holds the address in memory that is being read from or written to. The size of the MAR determines the number of words in memory that the processor can directly address. In the LIS-4, 16 locations can be addressed directly.

MBR: The memory buffer register (MBR) holds the data read from the memory or the data that are to be written into the memory. The size of the MBR is determined by the number of bits in a data word. The MBR has both a tri-state output connected to BO and a direct connection to the memory input. The source of data for the register is either BI or the memory output. The input source is selected by the memory read signal MR. For memory reads MR should be a 1.

PC: The program counter (PC) stores the address of the next instruction to be executed. It has to be initialized to the address of the first instruction of a program prior to the beginning of execution. During the normal execution of an instruction, the contents of the PC are incremented to point to the next instruction in the program. If an external condition tested by the instruction is met, a different address may be loaded into the PC. Forcing the PC to a nonsequential address is a way of implementing a JUMP instruction.

IR: The instruction register (IR) stores the instruction currently being executed. After the execution of the current instruction, the next instruction is loaded into the IR and the process is repeated.

ACC: The accumulator (ACC) receives the results of two-operand instructions. It can be used as temporary storage in some operations.

ALU: The arithmetic logic unit (ALU), as described in Figure 13.3, is a simple logic unit that operates on one or two binary quantities and places the result on the register input bus, BI. The logic function to be performed is selected by the control unit using signals S1 and S2. The functions to be implemented in the ALU are dictated by the chosen instruction set. The following are ALU operations:

S1,S2 = 0,0 The contents of the bus and the working register, WREG, are ANDed.

S1,S2 = 0,1 The content of the bus is incremented by 1 and

loaded onto BI. The carry-in (CYIN) to the ALU must be a 1 for the increment function.

S1,S2 = 1,0 The contents of the bus and the WREG are added and the result is put on the BI.

S1,S2 = 1,1 The contents of the BO are passed directly to the BI. This movement is referred to as a *fast transfer*.

WREG: The working register stores one of the two operands in a two-operand ALU operation. It can also be used for temporary storage. The WREG has both a tri-state output to the bus and a direct connection to the ALU.

FIGURE 13.3 **Truth Table for the LIS-4 ALU's *i*th Bit.**

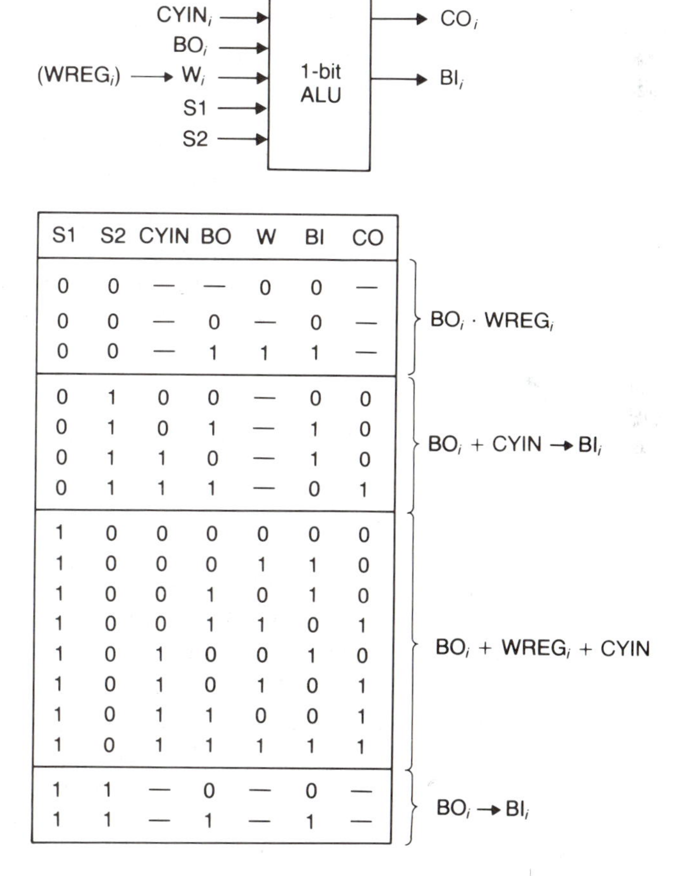

S1	S2	CYIN	BO	W	BI	CO
0	0	—	—	0	0	—
0	0	—	0	—	0	—
0	0	—	1	1	1	—
0	1	0	0	—	0	0
0	1	0	1	—	1	0
0	1	1	0	—	1	0
0	1	1	1	—	0	1
1	0	0	0	0	0	0
1	0	0	0	1	1	0
1	0	0	1	0	1	0
1	0	0	1	1	0	1
1	0	1	0	0	1	0
1	0	1	0	1	0	1
1	0	1	1	0	0	1
1	0	1	1	1	1	1
1	1	—	0	—	0	—
1	1	—	1	—	1	—

13.4.2 LIS-4 Instruction Set

The usefulness of the instructions in a processor is a good indication of how well the designer foresaw the intended application of the machine and how involved the potential users were at the machine conception. Each instruction is given an abbreviation that is called a *mnemonic*. An instruction must convey the operation to be performed (opcode) and the address of the memory location or registers containing the data on which the operation is to be performed (operand). Since the memory of the LIS-4 is only four bits wide, the operand address must be stored in the memory location immediately following the opcode. The initial instructions chosen for implementation on the LIS-4 are as follows:

Mnemonic	Opcode	Function
LA	0000	Load the accumulator
STA	0100	Store the accumulator
ADD	1000	Add the accumulator to the memory
AND	1100	AND the accumulator to the memory
STP	1110	Stop all computation

The descriptions of the five initial instructions are as follows:

LA: The accumulator register is loaded with the contents of the memory location whose address is stored in the memory location immediately following the LA opcode. LA is a two-word instruction.

STA: The content of the accumulator is stored in the memory location whose address is stored in the memory location immediately following the STA opcode. STA is a two-word instruction.

ADD: The accumulator content is added to the content of the memory location whose address is stored in the memory location immediately following the ADD opcode. The result is stored in the accumulator. Add is a two-word instruction.

AND: The contents of the accumulator and the memory whose address is stored in the memory location immediately following the AND opcode are ANDed bit by bit. The result is stored in the accumulator. AND is a two-word instruction.

STP: The execution of this instruction stops all computation and puts the computer in the IDLE state. STP is a one-word instruction.

Even with this limited instruction set and the limitation of 16 memory locations, a program can be written that demonstrates the use of the LIS-4 limited instruction set.

EXAMPLE 13.1

Write a program that adds the contents of locations 13 and 14, multiplies the sum by two, and puts the results in location 15. Show the program in mnemonic form and the complete memory contents before and after the program is run. The data to be operated on, DATA1 and DATA2, are 2 and 3, respectively.

SOLUTION

ADR	Mnemonics	Memory before Run	Memory after Run
0	LA	0000	0000
1	13	1101	1101
2	ADD	1000	1000
3	14	1110	1110
4	STA	0100	0100
5	15	1111	1111
6	ADD	1000	1000
7	15	1111	1111
8	STA	0100	0100
9	15	1111	1111
10	STP	1110	1110
11	—	—	—
12	—	—	—
13	DATA1	0010	0010
14	DATA2	0011	0011
15	—	—	1010

Note that in Example 13.1 the program logic consisted of adding the contents of memory locations 13 and 14, storing the sum in location 15, and then adding the sum, which is still in the accumulator, to the contents of memory location 15, thus doubling the value of the sum. The doubled value in the accumulator is then stored in memory location 15, and the processor is stopped. For the example just considered, the numbers in locations 13 and 14 were chosen to prevent an overflow since the LIS-4 has no overflow detection capability.

After the implementation of the initial five instructions, additional instructions will be added to the processor, and the necessary hardware modifications will be discussed. Initially the processor will be designed using hardwired logic. It will be redesigned using microprogramming techniques after that subject is introduced. The design that follows is based on the assumption that a solid-state static RAM is used and all of the data transfers are synchronous with a system clock. The LIS-4 processor has been built and tested by a senior project group at Wichita State University. Once the computer architecture and the desired instructions are determined, the next step is to determine the necessary register transfers to cause the desired action to take place. Keep in mind that the transfer of data from one register to another occurs at the clock pulse.

13.4.3 Register Transfer Operations

The register transfer operations for the four-bit processor under design are listed as follows:

	ALU Control	
LA Fetch Cycle	S1	S2
MAR ← PC	1	1
MBR ← M	—	—
IR ← MBR	1	1
PC & MAR ← PC + 1	0	1
MBR ← M	—	—
MAR ← MBR	1	1
MBR ← M	—	—
LA Execute Cycle	S1	S2
ACC ← MBR	1	1
PC ← PC + 1	0	1

If these operations were repeated for each instruction, you would find that the portion identified as the *fetch* cycle can be made common to all of the instructions, which simplifies the design and minimizes the number of components required. The following conditions exist at the end of the fetch cycle:

1. The PC holds the address of the memory location that follows the opcode.
2. The IR holds the instruction to be executed during the execute cycle.
3. The MAR holds the value stored in the memory location immediately following the opcode. This value is the operand address.
4. The MBR holds the contents of the memory location whose address is provided with the opcode, that is, the operand.

It is important to remember the contents of each register at the end of the fetch cycle. For some instructions fetch does more than necessary for the execute cycle to perform its function. This does not affect the execute function, however. Since the fetch portion of each instruction cycle is the same, it is necessary only to list the execute cycle register transfers for each of the instructions:

	Execute Cycle	S1	S2
LA:	ACC ← MBR	1	1
	PC ← PC + 1	0	1
STA:	MBR ← ACC	1	1
	M ← MBR	—	—
	PC ← PC + 1	0	1
ADD:	WREG ← MBR	1	1
	ACC ← ACC + WREG	1	0
	PC ← PC + 1	0	1
AND:	WREG ← MBR	1	1
	ACC ← ACC · WREG	0	0
	PC ← PC + 1	0	1
STP:	no execute cycle		

The STP instruction has only a fetch cycle. After the cycle that fetches the STP instruction, the computer goes to an IDLE state with the PC containing the address of the memory location that follows the STP opcode. If the STP opcode is followed immediately by another program, the program can be run by pushing the run switch. The PC contains the address of the instruction.

13.4.4 Timing and Control Characteristics

Both the fetch and execute cycles are sequential processes. The fetch cycle requires seven sequential time intervals to complete the fetch task. Each of the execute cycles for the first four instructions, with the exception of LA, involves three sequential processes. To make all of the execute cycles of identical length and thus simplify the design, a *do-nothing* task is added to the LA instruction. Figure 13.4 shows the timing diagram for the fetch cycle and Figure 13.5

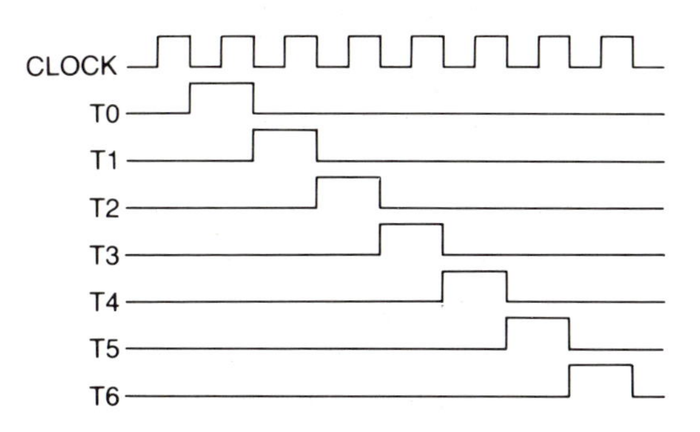

FIGURE 13.4 **Timing Signals for the LIS-4 Processor.**

FIGURE 13.5 **Timing Signal Generator.**

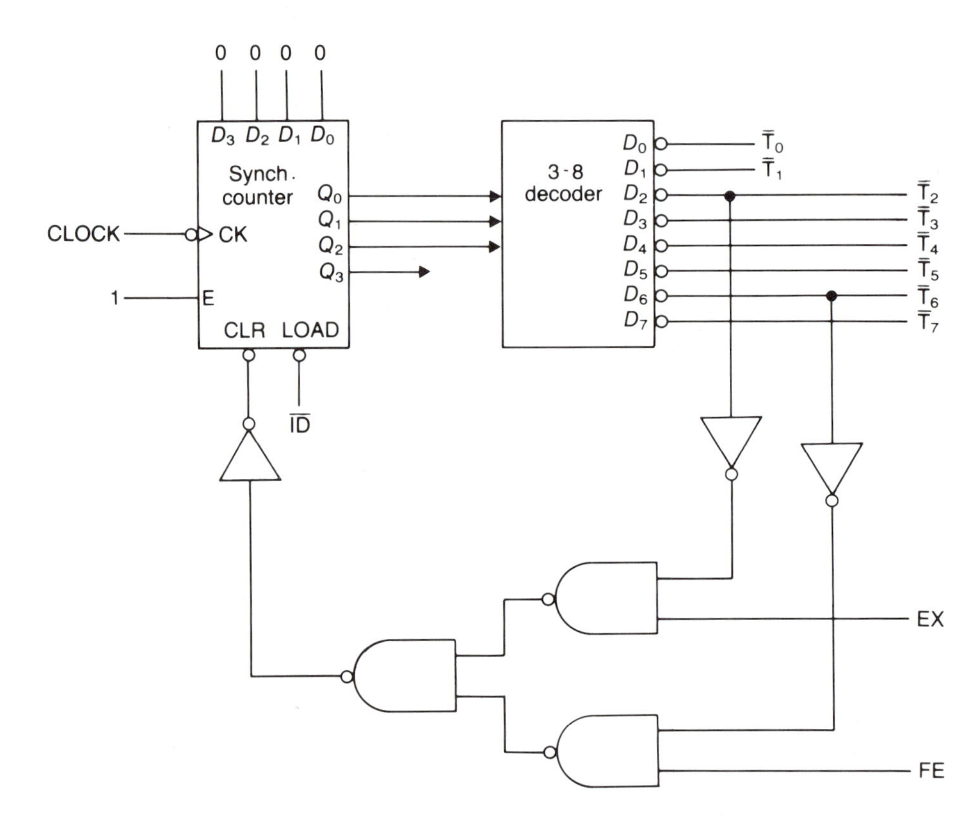

shows a circuit that will generate the timing signals (T) as a function of fetch, FE, execute, EX, and idle, ID. The FE and EX signals are both 1 when the fetch and execute cycles are in progress. The ID signal is a 1 when the processor is not running, which enables the loading of a 0000 into the counter. The circuit allows seven T times in fetch and three T times in execute.

Each of the register transfer operations in the fetch and execute cycles is assigned a distinct time in which the action takes place. Note that it is necessary to distinguish between T*n* in fetch and T*n* in execute. The fetch and execute cycles and their interrelationship with the timing cycles are as follows:

	Fetch Cycle	S1	S2	Time Interval
	MAR ← PC	1	1	FE · T0
	MBR ← M	—	—	FE · T1
	IR ← MBR	1	1	FE · T2
	PC & MAR ← PC + 1	0	1	FE · T3
	MBR ← M	—	—	FE · T4
	MAR ← MBR	1	1	FE · T5
	MBR ← M	—	—	FE · T6

	Execute Cycle	S1	S2	Time Interval
LA:	ACC ← MBR	1	1	EX · T0
		—	—	EX · T1
	PC ← PC + 1	0	1	EX · T2
STA:	MBR ← ACC	1	1	EX · T0
	M ← MBR	—	—	EX · T1
	PC ← PC + 1	0	1	EX · T2
ADD:	WREG ← MBR	1	1	EX · T0
	ACC ← ACC + WREG	1	0	EX · T1
	PC ← PC + 1	0	1	EX · T2
AND:	WREG ← MBR	1	1	EX · T0
	ACC ← ACC · WREG	0	0	EX · T1
	PC ← PC + 1	0	1	EX · T2

13.4.5 LIS-4 System Equations

Assuming that FE and EX signals are available and the opcode is decoded to provide the instruction identification signals, S1 and S2 can be written by examining each T time to determine the corresponding S1 and S2 values. It is easier in this case to write the equations for the complements of S1 and S2. The don't-cares are considered to be 1s.

$$\overline{S1} = FE \cdot T3 + EX \cdot (T2 + AND \cdot T1)$$

$$\overline{S2} = EX \cdot T1$$

Note that for the LA instruction, by including the do-nothing step during the T1 time, the equations are simplified since all cycles end at T2. Next, the equations for PCO (PC output) and PCI (PC input) are written by noting that PCO needs to be a 1 whenever the PC value is to be transferred, and that PCI should be 1 whenever the data are to be transferred from the BI to the PC. Therefore,

$$PCO = FE \cdot (T0 + T3) + EX \cdot T2$$

$$PCI = FE \cdot T3 + EX \cdot T2$$

The equation for MBRO is a little more involved but is determined by examining the register transfer equations in the fetch and execute cycles and determining when the MBR should be connected to BO. Accordingly,

$$\text{MBRO} = \text{FE} \cdot (\text{T2} + \text{T5}) + \text{EX} \cdot \text{T0}(\text{LA} + \text{ADD} + \text{AND})$$

The remaining equations are determined in the same manner and are left as an exercise (see Problem 4).

13.4.6 LIS-4 Instruction Cycle State Control

We now derive the assumed signals FE and EX. Figure 13.6 contains the state diagram of the cycle control circuit. The control signals that cause the transition from one state to another were determined from design decisions made previously. The signals were chosen to be distinct in time and the state occupied.

Idle: During this cycle the machine remains turned on but it does not compute. When the run switch is activated the processor begins operation. A STP instruction returns the processor to the idle state.

Fetch: The processor remains in this state for a total of seven clock periods while the fetch sequential process is being completed. At the end of the T6 pulse, unless a STP instruction has been fetched, the processor enters the execute cycle.

Execute: The execute portion of the instruction is performed during this cycle. At the end of T2, if the run switch is still activated, the fetch cycle is re-entered. If the run switch has been turned off, the idle state is entered.

FIGURE 13.6 **Cycle State Diagram.**

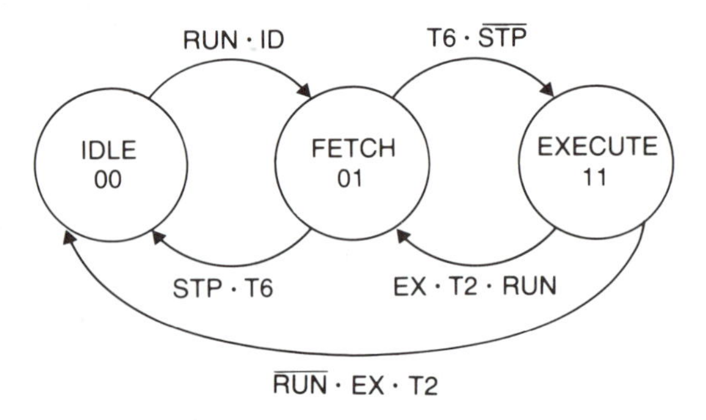

Figure 13.7 shows the necessary circuit for generating the signals FE, EX, and IDLE. The *JK* equations shown in Figure 13.7 were derived by inspection of the state changes desired. Q_1 turns on when a transition is made from FE to EX; thus

$$J_1 = \text{T6} \cdot \overline{\text{STP}}$$

Q_1 turns off when moving from execute to fetch or execute to idle, requiring that

$$K_1 = \overline{\text{RUN}} \cdot \text{EX} \cdot \text{T2} + \text{RUN} \cdot \text{EX} \cdot \text{T2} = \text{EX} \cdot \text{T2}$$

Equations for Q_2 are derived using the same reasoning.

FIGURE 13.7 **Cycle Signal Generator.**

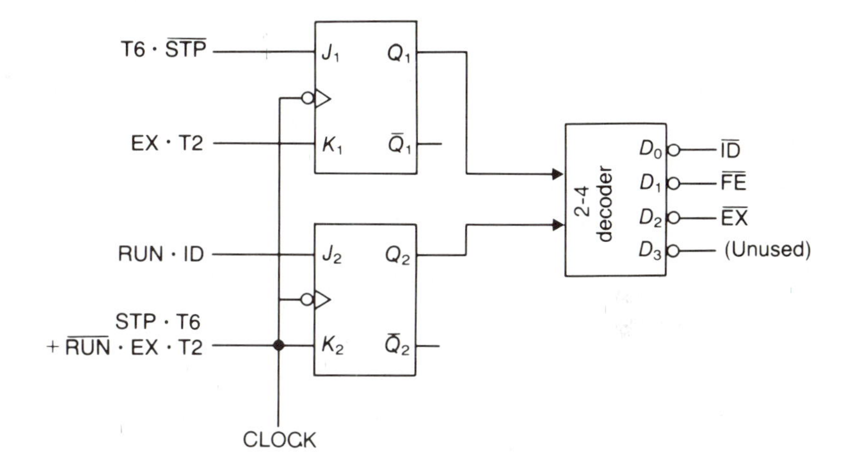

At this point of the design, the only signals that have not been accounted for are the instructions LA, STA, ADD, AND, and STP. These are easily generated as shown in Figure 13.8 by decoding the opcode stored in the IR after the fetch cycle.

Figure 13.9 shows the timing for the processor. The cross-hatched area shows the variations in signals for the indicated instructions. Note in Figure 13.8 that so far only five of the possible 16 opcodes have been used.

FIGURE 13.8 **Instruction Decoding Circuit.**

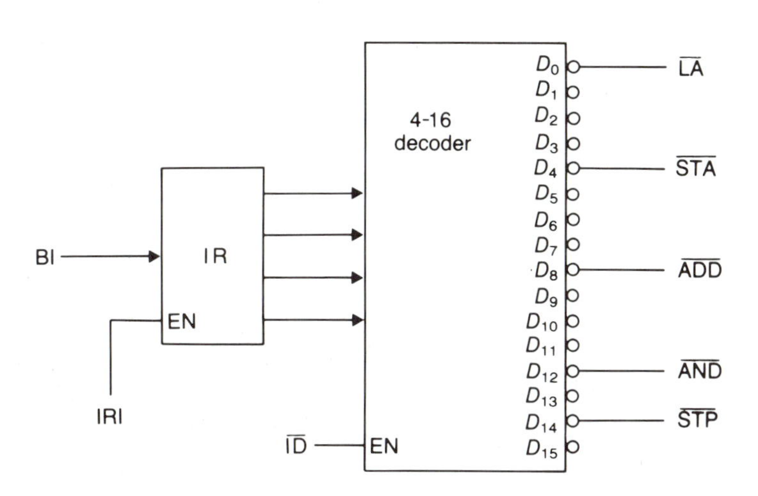

FIGURE 13.9 Processor Timing Diagram.

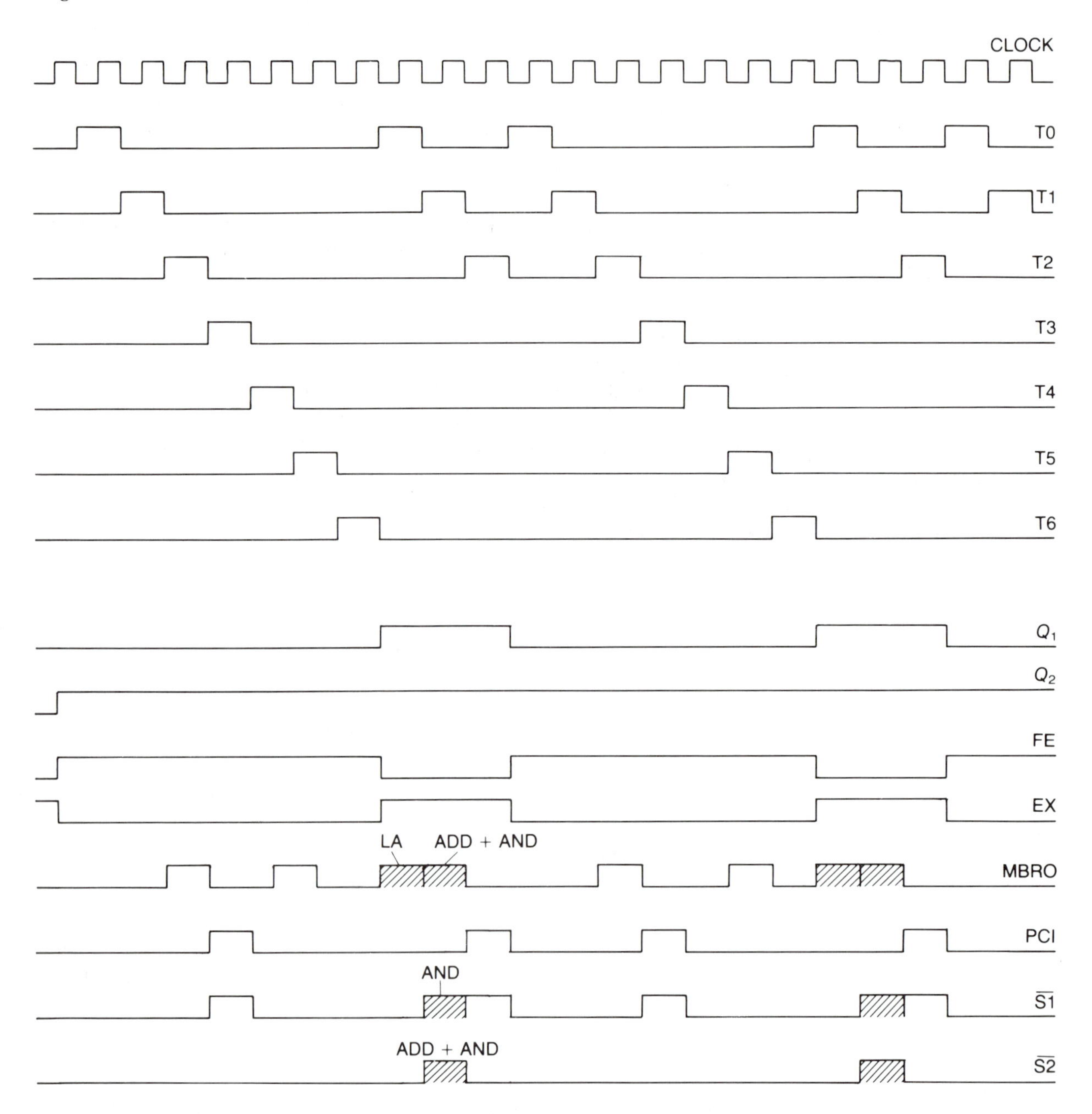

13.4.7 LIS-4 I/O Instructions

Two more instructions will now be added to the initial set of instructions to make the processor more useful. The instructions allow input/output to occur as follows:

Mnemonic	Opcode	Function
OM	0001	Output from memory
IM	0010	Input to memory

OM: The content of the memory location whose address immediately follows the OM opcode is placed on the BI for transfer to the input register of an external device. OM is a two-word instruction.

IM: The value placed on the BI by an external device is stored at the memory location whose address immediately follows the IM opcode. IM is a two-word instruction.

In order to allow the connection of peripherals of any speed, a READY line from the peripheral will first be tested by the processor. If the peripheral register (PR) is not ready, the processor enters a WAIT state until the device turns on its READY signal. This results in a modified sequential cycle state diagram as shown in Figure 13.10. Additions to the LIS-4 processor block diagram are shown in Figure 13.11. Note the new control signals SEND/$\overline{\text{RECEIVE}}$, PRO (peripheral register output), PRI (peripheral register input), and READY.

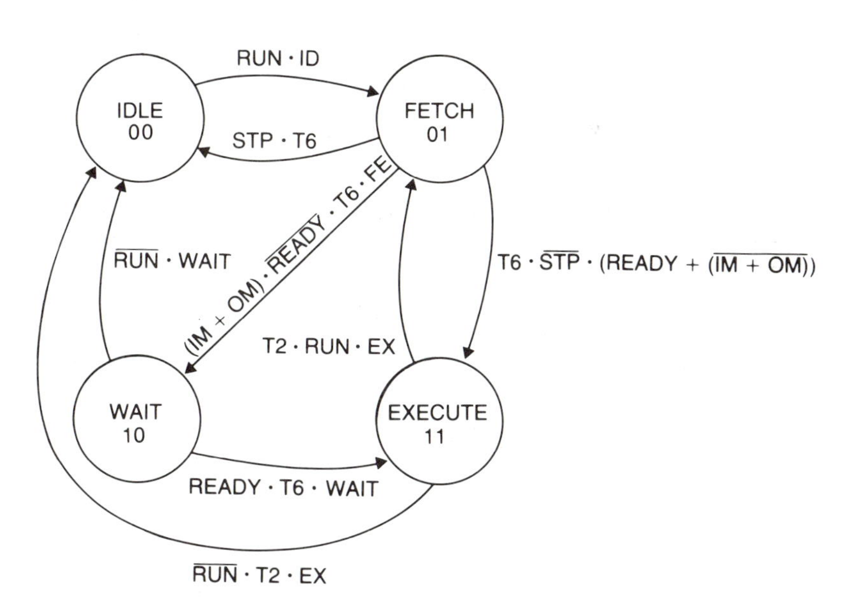

FIGURE 13.10 **Modified Sequential Diagram.**

In the new state diagram a design decision was made to allow the timing pulse to continue to be generated in the WAIT state. The processor will not leave the WAIT state until the READY line is a 1 AND the processor is in T6. The timing circuit of Figure 13.5 will have to be modified to allow T0–T6 to occur in the WAIT

FIGURE 13.11 **Modified Block Diagram of the LIS-4 Processor.**

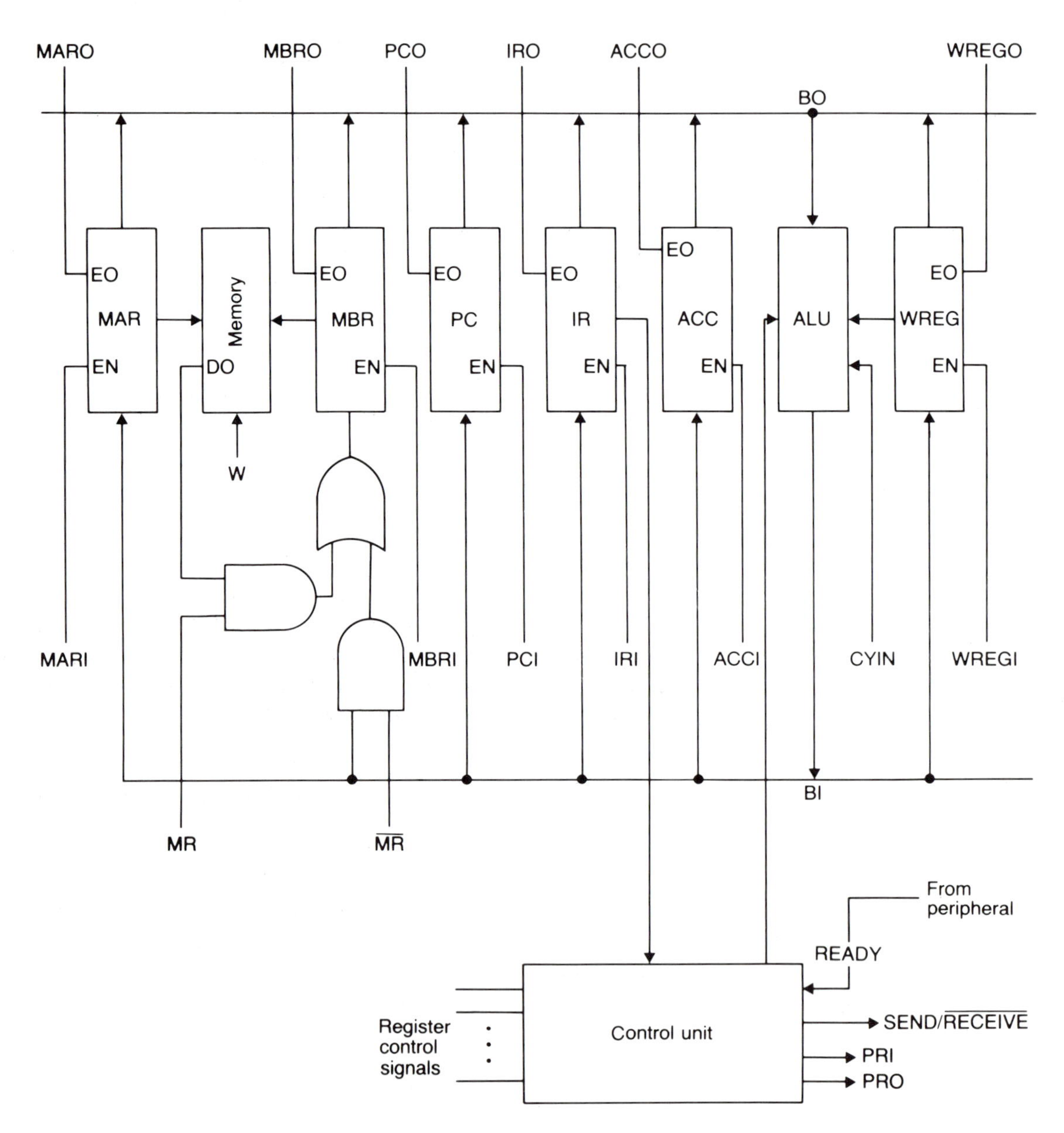

state. With the addition of the WAIT state, the JK inputs for the cycle signal generator circuit are now given by

$$J_1 = \text{T6} \cdot \{\overline{\text{STP}} \cdot \overline{[(\text{IM} + \text{OM})} + \text{READY}] + (\text{IM} + \text{OM}) \cdot \overline{\text{READY}} \cdot \text{FE}\} ;$$

$$K_1 = \text{EX} \cdot \text{T2} + \overline{\text{RUN}} \cdot \text{WAIT} ;$$

$$J_2 = \text{RUN} \cdot \text{ID} + \text{READY} \cdot \text{WAIT} ;$$

$$K_2 = [\text{HALT} + (\text{IM} + \text{OM}) \cdot \overline{\text{READY}} \cdot \text{FE}] \cdot \text{T6} + \overline{\text{RUN}} \cdot \text{T2} \cdot \text{EX}$$

where the execute cycles for the two new instructions are given as follows:

	Execute Cycle	S1	S2	Time Interval
IM:	MBR ← PR	1	1	EX · T0
	M ← MBR	—	—	EX · T1
	PC ← PC + 1	0	1	EX · T2
OM:	PR ← MBR	1	1	EX · T0
	PC ← PC + 1	0	1	EX · T2

Transfer of data between the processor and a peripheral device takes place between the registers in the peripheral and the MBR. Figure 13.12 shows the necessary interface for allowing the transfer of data to take place once the signals from the processor are available. Interfacing processors to external devices will be considered in more detail in Chapter 14.

Consider another programming example making use of the two new instructions.

FIGURE 13.12 **I/O Interface.**

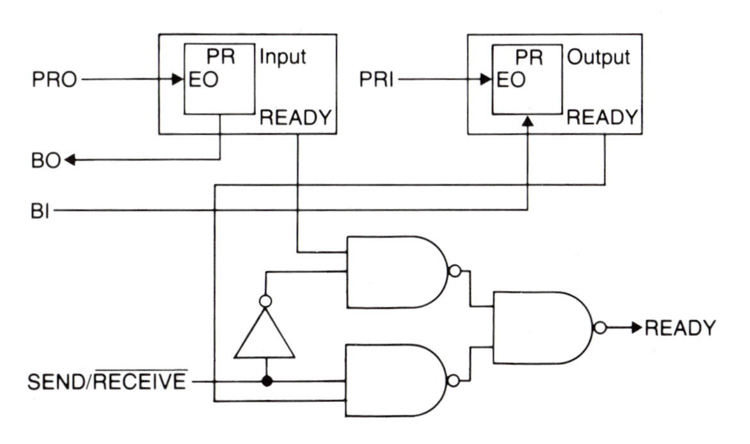

EXAMPLE 13.2

Write a machine language program that inputs a number, adds the number to itself, outputs the result, and stops.

SOLUTION

Address	Instruction	Comments
0000	IM	; INPUT THE VALUE
0001	15	; ADDRESS TO STORE DATA
0010	LA	; PUT THE DATA FROM
0011	15	; LOCATION 15 IN THE ACC
0100	ADD	; ADD ACC DATA
0101	15	; TO DATA IN LOCATION 15
0110	STA	; STORE THE SUM IN
0111	15	; LOCATION 15
1000	OM	; OUTPUT THE CONTENTS
1001	15	; OF LOCATION 15
1010	STP	; STOP THE COMPUTER

To run the program written in Example 13.2, the bit patterns for each instruction, address, and data must be loaded in the indicated addresses, the location of the first instruction loaded in the PC, and the RUN switch actuated.

With the addition of the two new instructions, the control equations for the processor become

$$S1 = FE \cdot T3 + EX \cdot T2 + AND \cdot EX \cdot T1$$

$$S2 = EX \cdot T1$$

$$MBRO = FE \cdot (T2 + T3) + \overline{(STA + IM)} \cdot EX \cdot T0$$

$$PCO = FE \cdot (T0 + T5) + EX \cdot T2$$

$$ACCO = (ADD + AND) \cdot EX \cdot T1 + STA \cdot EX \cdot T0$$

$$MARI = FE \cdot (T0 + T3 + T5)$$

$$MBRI = EX \cdot T0 \cdot (STA + IM) + FE \cdot (T1 + T4 + T6)$$

$$PCI = FE \cdot T3 + EX \cdot T2$$

$$IRI = FE \cdot T2$$

$$ACCI = EX \cdot T0 \cdot LA + EX \cdot T1 \cdot (ADD + AND)$$

$$MR = FE \cdot (T1 + T4 + T6)$$

$$W = EX \cdot T1 \cdot (STA + IM)$$

$$WREGI = EX \cdot T0 \cdot (ADD + AND)$$

$$PRO = EX \cdot T0 \cdot IM$$

$$PRI = EX \cdot T0 \cdot OM$$

$$SEND/\overline{RECEIVE} = OM$$

$$CYIN = FE \cdot T3 + EX \cdot T2$$

WREGO = not used at this point in design

These equations could be implemented to form the control section along with the decode circuitry for the instruction register. This implementation would be of the classical hardwired type. Decreased costs of ROM, RAM, and decoders have made possible, and popular, architectures employing microprogrammed control, the subject of the next section.

13.5 Microprogramming

The concept of *microprogramming* was introduced in 1951 and it has attained significant acceptance as an alternate method of control unit design. Its versatility and structured nature allow a mix of software and hardware so that any required engineering changes can be made more easily.

Computers can be designed to perform very complex operations. Due to the cost involved, such operations are seldom realized in hardware as primitive operations but instead as a sequence of many primitive operations. In Chapter 11 we saw that multiplication and division processes can be performed by a sequence of primitive operations such as addition, shift, scale, rotate, and subtraction. In the first part of this chapter we saw that the design process of the LIS-4 processor involved several different phases: determination of the architecture, formulation of the processor instructions, formulation of control and timing networks, specification of the RTL operations, and design of the control unit.

Prior to using microprogramming techniques, the method of implementing a complex operation in a computer consisted of designing a sequential logic circuit to generate the control signals that activate the primitive operations necessary to perform a complex operation. For a system capable of performing several complex operations, there may be several common subsequences of primitive operations that may be combined to reduce the cost and volume of the hardware. The complex operations appear to the user as *machine language instructions*. When these primitive operations are controlled with conventional logic, the system is called a *hardwired* or *conventional control* system. The sequence of machine language instructions resides in the memory and is operated upon as follows:

Step 1. Under the direction of the control unit, the next machine language instruction is fetched into the instruction register.

Step 2. The contents of the instruction register are decoded by the control unit.

Step 3. The decoded control signal selects the appropriate sequential logic circuit that causes a sequence of primitive operations corresponding to the specific instruction. After completion of the operations making up the instruction, Step 1 is repeated.

13.5.1 Microprogramming Concepts

Microprogramming is an alternative means for implementing a complex operation in the form of a subroutine of primitive operations, called *microoperations*. Microprogramming essentially reduces the complexity and the inflexibility of the control unit. Instead of wired gate logic, a control memory is used where the control information for each primitive operation is stored. The microprogrammed system may be envisioned as a primitive control processor that executes its own microinstructions, causing the inter-register transfer of data. In effect a machine language instruction, also called a *macroinstruction* in microprogramming terminology, is realized by the execution of a sequence of microinstructions or a microprogram. Each microinstruction initiates one or more command signals called *microcommands*. Any operation initiated by a microcommand is a *microoperation*. The memory used to store the microinstructions is known as a *control store* (sometimes also called a *control memory* or *microprogram memory*) that may or may not be separate from the main computer memory. Each of the microinstructions represents a *control word* in the control store. Realizing a complex operation reduces to executing a microprogram as detailed by the following steps:

Step 1. Under the direction of the control unit, the next microinstruction is fetched into a microinstruction register.

Step 2. The control unit translates the microinstruction into the corresponding control signals. The required control signal values are stored either as individual bits in the control word or as groups of bits called *fields* that when decoded provide the control signals.

Step 3. The generated control signals connected to the hardware initiate the intended register transfers.

Step 4. Based on the results of the previous operation, the control unit determines the address of the next microinstruction.

A microprogrammed control unit is illustrated in Figure 13.13. The microinstruction executed by this unit requires a fetch-decode-execute sequence just as the machine language instruction does. The use of microprogrammed control makes the execution of a machine language instruction less complex, and vastly increases the flexibility of the system. Control signals may now be changed by reprogramming bits in the control words. Sequences may be changed by changing the addresses of where control information is stored.

13.5.2 Wilkes's Scheme

The first design implementation of a microprogrammed control unit was by M. V. Wilkes. Figure 13.14 shows the basic organization proposed by Wilkes. The microprogram is stored in a ROM that has two matrices: the control matrix, CM, and the sequencing

FIGURE 13.13 **Microprogrammed Control Configuration.**

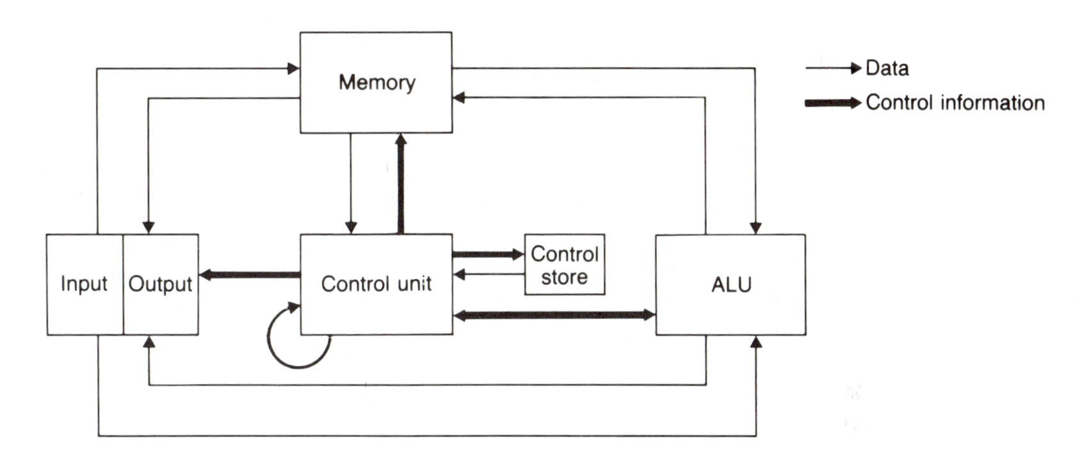

FIGURE 13.14 **Wilkes's Scheme of Microprogrammed Control.**

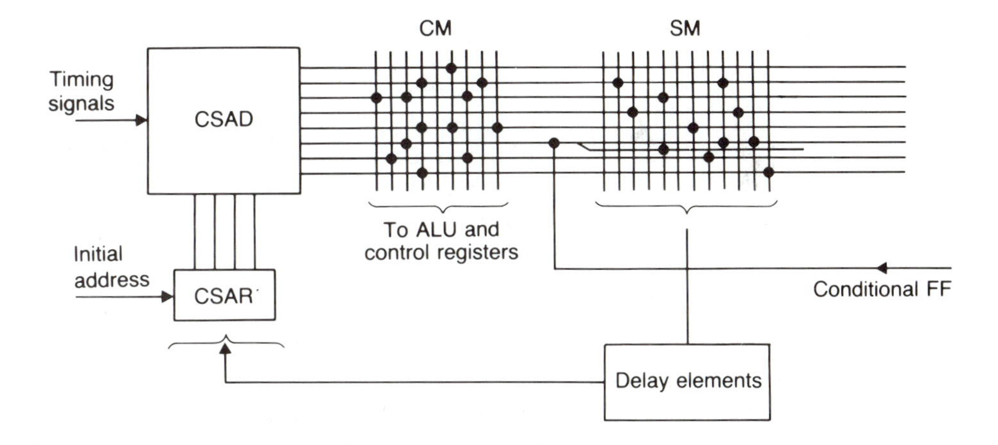

matrix, SM. The output of the CM activates various control signals corresponding to a particular microinstruction, and the output of the SM contains the address of the next microinstruction. The n-bit beginning address is loaded into the control store address register, CSAR, which ultimately is fed into the inputs of the control store address decoder, CSAD. A well-defined timing pulse enables the decoder in selecting one of the two outputs that also serves the ROM matrices. The output of the CM activates a subset of control signals, thus effecting the execution of the corresponding microoperations. The SM outputs are fed into the CSAR via a delay logic network for activating the next microinstruction. In this fashion the microinstructions are executed in a predetermined sequence until the complete microprogram corresponding to the specific machine instruction is over. Like all conventional computers, the microprogram controller provides for the execution of *conditional branching*. The SM consists of an alternate set of horizontal lines, one of which depends on the status of a conditional flip-flop. The conditional FF tests the status of the sign of the accumulator or comparator outputs and selects a microinstruction.

13.5.3 Organization of a Microprogrammed Controller

Wilkes's scheme was limited to using only ROMs. With the advent of fast inexpensive solid-state memories, microprogramming has become more versatile and user-alterable. Computers that allow users to alter the control memory are called *microprogrammable*. The area where user-generated microinstructions are stored is called the *writable control store*, WCS. Changing the control store as the computing machine is running is difficult and dangerous and has been compared to do-it-yourself brain surgery. Figure 13.15 illustrates a very simplified organization of a microprogrammed computer, where CSBR is the control store buffer register and BC is the bus circuitry unit for regulating the flow of information. The execution of a microinstruction begins by selecting the proper microinstruction at the control store location specified by the CSAR and reading the microinstruction into the CSBR (often called the *microinstruction register*, MIR). This is followed by the simultaneous updating of the CSAR for the next microinstruction and execution of the primitive operations as desired by the contents of the CSBR. A sequence of microinstruction steps results in a unique microprogram. A microprogram exists for the implementation of each machine language instruction.

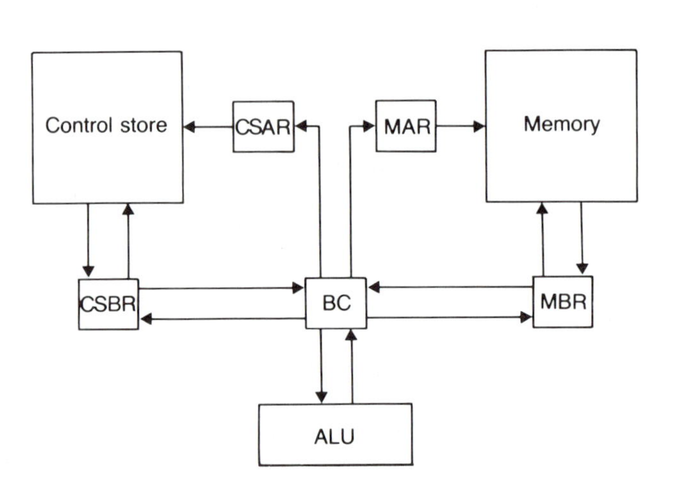

FIGURE 13.15 **Simplified Block Diagram of a Microprogrammed Computer.**

The ALU internal registers need a large number of control signals to enable the necessary register transfer operations. These control signals are realized by programming the control word. An n-bit word could provide up to 2^n different microinstructions. Many control signals may have a bit in the control word dedicated exclusively to them. A field N bits wide could be decoded to generate 2^N mutually exclusive control signals. When each control signal is assigned a bit in the control word, the microprogramming is termed *horizontal.* Horizontal microprogramming allows the most flexibility in utilizing the hardware resources of the machine and provides for the fastest operation. However, horizontal microprogramming results in inefficient utilization of control memory because most of the time only a few microoperations are specified. An alternative technique that can be used in a group of control signals where only one of them is active at a time involves using fewer bits and generating the control signals with a decoder. This type of microprogramming is termed *vertical.* Vertical microprogramming requires fewer bits than horizontal but is slower. Most microprogrammed machines use a format somewhere between the horizontal and vertical format. Such a compromise scheme has a *diagonal* format.

Consider a situation where five bits are reserved for control and eight control signals are required, of which only one is active at a time. Three of the control bits could be used as the input to a 3-8 line decoder for control of the eight mutually exclusive signals, and the remaining two signals could be used for individual control signals. This arrangement provides for the control of a total of 10 microoperations with only five bits.

Microprogramming skills must be developed just as high-level programming skills. The microprogrammer must have a thorough knowledge of the architecture he or she is attempting to control. Now that you have been introduced to microprogramming, we can complete the design of the LIS-4 processor.

13.6 LIS-4 Processor Microprogrammed Control

In order to simplify the addition of instructions to the LIS-4 instruction set and eliminate the necessity of rewriting the control equations with each addition, the microprogrammable control circuit of Figure 13.16 is introduced.

13.6.1 LIS-4 Microprogrammable Control Unit

In Figure 13.16 the control memory consists of a 2^N-word $\times$ M-bit ROM. N is determined by the number of microinstructions required to implement the fetch cycle, and M is determined by the number of control signals required by the LIS-4 architecture. Fig-

FIGURE 13.16 **LIS-4 Microprogrammable Control Section.**

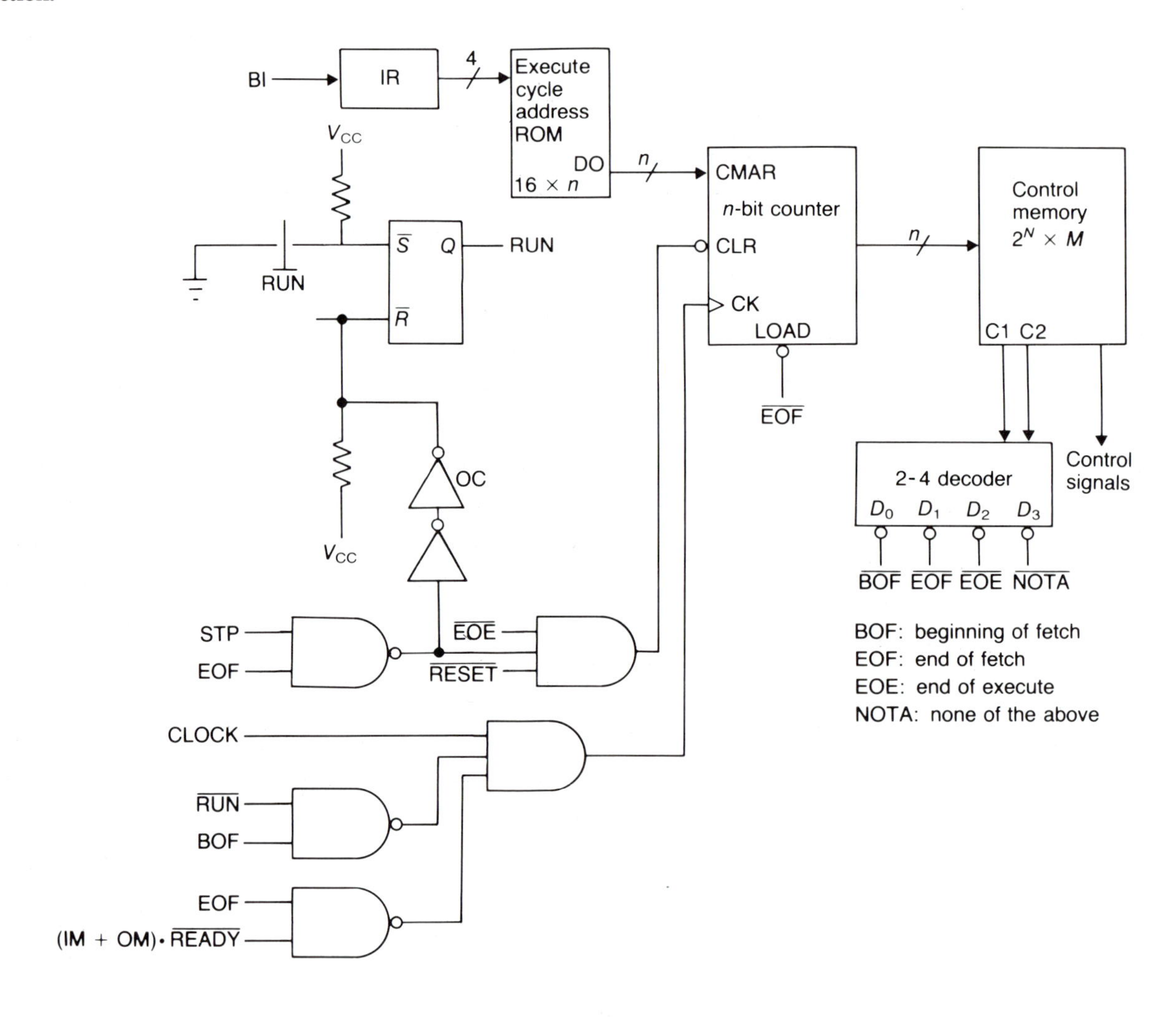

ure 13.17 shows the control ROM contents for the seven instructions implemented in Section 13.4.2. The first seven words correspond to the seven T times required for fetch.

The execute cycle address ROM has as its address input the four opcode bits held in the IR. The opcode then accesses one of the 16 locations in the ROM. Stored in the ROM are the addresses of microprograms in the control ROM that contain the microinstruc-

FIGURE 13.17 **LIS-4 Control ROM Table (Seven Instructions).**

ROM address	0	1	2	3	4	5	6	7	8	9	10	11	12	13	14	15	16	17	18	19	20	21	22
MARI	x			x		x																	
MARO																							
MBRI		x			x		x						x		x								
MBRO			x			x					x						x						
MR		x			x		x																
PRO															x								
W														x		x							
PCI				x								x		x		x		x					
PCO	x			x								x		x		x		x					
IRI			x																				
IRO																							
ACCI											x												
ACCO													x										
WREGI																							
WREGO																							
PRI																	x						
SEND/$\overline{\text{RECEIVE}}$																	x						
CARRY				x								x		x		x		x					
S1	x		x			x					x		x		x		x						
S2	x		x	x		x					x	x	x	x	x	x	x	x					
C2		x	x	x	x	x	x				x		x		x		x						
C1		x	x	x	x	x					x	x	x	x	x	x	x	x					
	T0	T1	T2	T3	T4	T5	T6																
	FETCH										LA		STA		IM		OM						

A cross in a location indicates a 1.

tions for generating the control signals necessary for the execute portion of the opcode held by the IR. The width of the address ROM is determined by the number of words in the control ROM. If the IR held the opcode for the LA instruction, address location 0000 would be accessed. In location 0000 the value 1010 is stored.

The value of 1010 is the location of the two-word microprogram in control ROM that implements the execute portion of the LA instruction. Examine locations 10 and 11 of the control ROM table in Figure 13.17. Note that with the use of microprogrammed control, the number of steps in the execute cycle of a machine instruction can be of any length without increasing the complexity of the system hardware.

The control memory address register, CMAR, is an *n*-bit counter. When the counter is cleared the counter points to the first microinstruction in the fetch cycle. When the end of fetch, EOF, is sensed (control signals C1 and C2 of the control word), the output of the address ROM is loaded into the CMAR, and execution in the control ROM continues from that point. When the end of execute, EOE, is sensed (control signals C1 and C2 of the control word), CMAR is cleared and the fetch microprogram is executed again to fetch the next opcode.

Referring to Figure 13.16, it is assumed that a RUN switch that puts out a momentary low pulse is available. Note that it is still necessary to decode the instructions with a decoder since the STP instruction has no execute and the IM and OM signals are used in the control of the clock and in the I/O circuitry. The microprogrammed LIS-4 differs from the hardwired version in that the clock is turned off in the system when a peripheral is not ready and begins operation on the next clock after the ready line becomes a 1.

13.6.2 LIS-4 Register Additions

Three registers, two decoders, and X-ORs have been added, as shown in Figure 13.18, in order to increase the instruction set and versatility of the LIS-4. Their functions are as follows:

STK: The *stack register* is used to store return addresses when subroutine calling is used. In larger systems the stack register may be a special piece of hardware that allows multiple addresses to be stored in a last in–first out, LIFO, scheme. Each time an address is stored in a LIFO register it is stored on the top of the stack. When an address is read from the stack, it is read off the top of the stack. In our small processor the stack is one deep, that is, it contains one address.

DEV: The *device select register*, DEV, holds the address of the external device with which the processor wishes to communicate. The four-bit register and attached decoder allow the selection of 16 different input and 16 different output devices.

MSWR: The *memory switch register*, MSWR, holds a four-bit quantity that selects one of 16 different 16-word memories. The register and attached decoder allow expansion of the LIS-4

FIGURE 13.18 **LIS-4 Hardware Additions.**

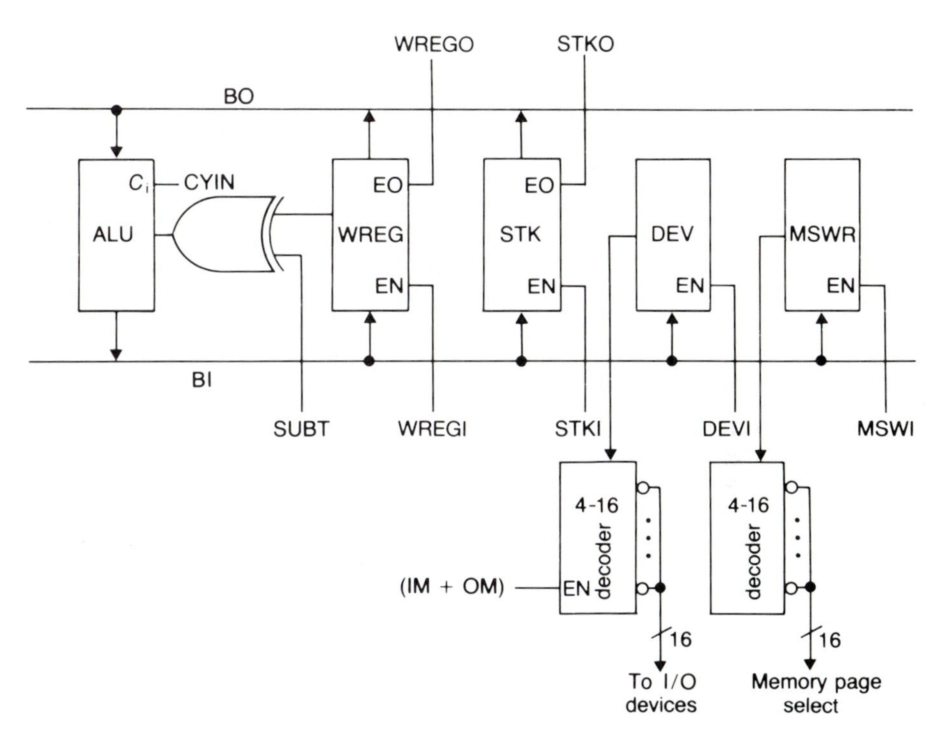

memory size to 256 four-bit words. Each of the 16-word memories will be referred to as a *page* of memory. Using that nomenclature, a 16-page memory is available, and the MSWR and decoder allow the programmer to switch pages.

The addition of the X-OR to each bit of the WREG connection to the ALU will allow taking the 1's and 2's complement of the WREG value and thus the addition of other arithmetic functions.

13.6.3 Expanded LIS-4 Instruction Set

Additional instructions will now be introduced that will make the LIS-4 more functional and give it a more representative set of instructions:

Mnemonic	Opcode	Function
LA	0000	Load accumulator
OM	0001	Output from memory
IM	0010	Input to memory
DS	0011	Device select
STA	0100	Store accumulator
SUBT	0101	Subtract memory from accumulator
RET	0110	Return
MSW	0111	Switch memory page
ADD	1000	Add memory to accumulator
JMSW	1001	Jump to address on other page
SL1	1010	Shift accumulator left one
JSB	1011	Jump to subroutine
AND	1100	AND memory and accumulator
JMP	1101	Jump to specified address
STP	1110	Stop the program
CLR	1111	Clear the accumulator

Most processors have the ability to shift binary quantities right or left one or more positions. With the architecture of the LIS-4, a shift-left instruction can be implemented.

SL1: The SL1 instruction is a single-word instruction that causes the accumulator register value to be shifted left one position.

The RTL description is

SL1: WREG ← ACC
ACC ← WREG + ACC

This action can be programmed by setting

$$ACCO = WREGI = S1 = S2 = 1$$

and

$$ACCI = S1 = C1 = 1$$

respectively in two words of the control ROM and placing the address of the first word of the two-word microprogram for the SL1 execute cycle in the address ROM, in the memory location whose address is the SL1 opcode.

The addition of a jump-to-subroutine, JSB, instruction and a return, RET, instruction gives the programmer the ability to write callable programs.

JSB: This instruction causes the instruction execution sequence to be transferred to the address stored in the memory location immediately following the JSB opcode. The address of the next opcode (the address of the JSB opcode plus two) is stored in the stack register. JSB is a two-word instruction.

RET: This instruction is a single-word instruction that should be the last instruction of a subroutine. The RET instruction transfers the contents of the STK to the IR.

The RTL descriptions of the JSB and RET are

JSB: STK ← PC + 1
PC ← MAR

RET: PC ← STK

The reader may be amused at the thought of a subroutine in a 16-word memory. In order to provide additional memory capability to the LIS-4, the following instructions are added that make use of the memory switch register, MSWR.

MSW: This double-word instruction selects the 16-word *page* of memory to be active. The desired page number specified by the second word of the instruction is placed in the MSW. Program execution, after execution of this instruction, continues at the address of the MSW opcode plus one location but on the newly selected page.

JMSW: This triple-word instruction allows the transfer of the next instruction execution to any address in any of the 16 pages of memory. The second word specifies the page, and the third word specifies the address on the page.

The RTL descriptions are

MSW: MSW ← MAR

JMSW: WREG ← MAR
MAR ← PC + 1
MBR ← M
PC ← MBR
MSWR ← WREG

Note that in the JMSW RTL description the page number that was in the MAR after fetch had to be stored temporarily in the WREG until the current page was read, in order to find the new address to branch to on the new page. If the new page number had been put in the MSWR immediately, the wrong memory address would have been read.

Prior to the hardware additions shown in Figure 13.18, the LIS-4 had limited I/O capability. A new instruction that will utilize the device select register, DEV, has been added:

DS: This double-word instruction allows the selection of one of 16 peripheral devices. The contents of the second word of the instruction, which is the number of the desired peripheral, is placed in the DEV. The device remains selected until a new DS instruction is executed. The RTL description is left as an exercise (Problem 5).

In order to provide the ability to clear the accumulator, a clear instruction is added.

CLR: This single-word instruction causes a value of 0000 to be placed in the accumulator.

The RTL description of the CLR instruction execute cycle is given to emphasize a point:

CLR: $ACC \leftarrow IR + 1$

An opcode of 1111 was chosen for the CLR instruction so that when the IR content is incremented, a 0000 results that is used to clear the ACC. It is imperative that the microprogrammer be aware of the contents and function of each register at each increment of time in order to make maximum use of the hardware. Note also that in the CLR cycle and for other one-word instructions it is not necessary to increment the PC since, at the end of fetch, the PC contains the address of the memory location following the opcode address.

Two additional instructions complete the set of 16 instructions:

SUBT: This double-word instruction causes the contents of the memory location given in the second word to be subtracted from the accumulator. The result is stored in the accumulator. The RTL description is left as an exercise (see Problem 5).

JMP: The second word of this two-word instruction is placed in the program counter, allowing transfer of the program sequence to any position on the currently selected page. The RTL description is left as an exercise (see Problem 5).

The programming of the control ROM in Figure 13.19 to implement the complete LIS-4 instruction set is left as an exercise (see Problem 7). The application of ROM control is not limited to digital processors, but can be applied to any system involving digital control. The ROM is used to store the control signals, and a sequencer is designed to bring the control signals from the ROM in proper sequence.

FIGURE 13.19 **LIS-4 Control ROM Table (16 Instructions).**

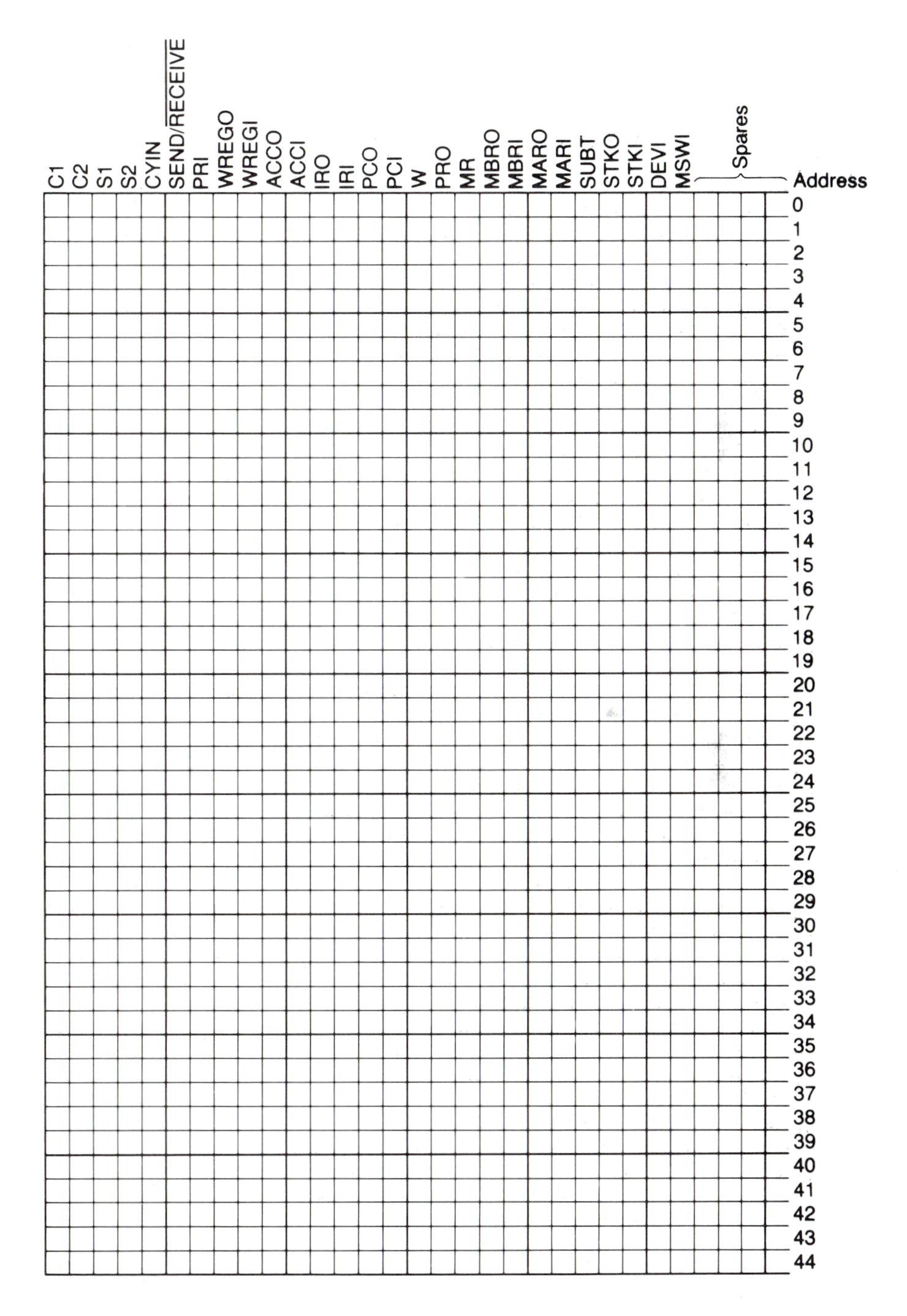

13.7 Microprogrammed Algorithm Control

Microprogramming techniques are applicable to any system that operates under the supervision of a control circuit. As an example we will design a microprogrammed control unit for performing parallel signed magnitude addition/subtraction. Similar designs with hardwired control were considered in Chapter 11.

The algorithm is initiated by supplying a control input E. The process performs addition when $E = 0$ and it performs subtraction when $E = 1$. It is assumed that prior to the start of the algorithm the magnitude of the two n-bit numbers are first stored in the n-bit registers, X and Y, and the corresponding signs are stored in XS and YS FFs. The algorithm process is started when a momentary low pulse is received and used to set a latch that has an output signal called GO.

The algorithm works as follows. In all cases, the X and Y sum or difference will be stored in the X register, and the sign of the result will be stored in XS. If a carry-out occurs, indicating an overflow, a 1 will be stored in an overflow FF, L. The first step in the algorithm involves changing the sign of YS based on the value of E. If XS and YS are now equal, the sum is made and the result stored in L,X with XS providing the correct sign. If L is a 1, an overflow has occurred.

If XS and YS are not the same (after YS is changed according to E), no overflow can occur. To check for the proper sign of the result, Y is subtracted from X and the result stored in L,X. Two conditions must be examined, $X > Y$ and $X < Y$. If $X > Y$, a carry-out will occur for $X - Y$, which indicates that the result of the $X - Y$ operation was the correct one and the result should have the sign in XS; that is, X was larger than Y. If $X <$ Y, then $X - Y$ produces no carry-out and the result stored in X is the 2's complement of the desired result. To correct the result stored in X, the 2's complement of the X content must be taken and XS inverted. The reader is encouraged to try the algorithm on two numbers to verify its correctness.

The RTL description is as follows (BUSY indicates the algorithm is in process):

$S = 0$: BUSY $\leftarrow 0$; IF (GO $= 1$) S $\leftarrow 1$, $L \leftarrow 0$ ELSE
$S \leftarrow 0$;

$S = 1$: BUSY $\leftarrow 1$; IF ($E = 0$) $S \leftarrow 3$;

$S = 2$: $YS \leftarrow \overline{YS}$;

$S = 3$: IF ($XS \oplus YS = 0$) $L,X \leftarrow X + Y$, $S \leftarrow 0$;

$S = 4$: $L,X \leftarrow X + \overline{Y} + 1$;

$S = 5$: IF ($L = 1$) $L \leftarrow 0$; $S \leftarrow 0$;

$S = 6$: $X \leftarrow \overline{X} + 1$; $XS \leftarrow \overline{XS}$; $S \leftarrow 0$;

Closer examination of this algorithm reveals that several of the indicated operations can go on simultaneously, and the RTL description can be reduced to

$S = 0$: BUSY $\leftarrow 0$; IF (GO = 1) $L \leftarrow 0, S \leftarrow 1$ ELSE $S \leftarrow 0$;

$S = 1$: BUSY $\leftarrow 1$; IF $(XS \oplus YS \oplus E)$ $L, X \leftarrow X + Y$, $S \leftarrow 0$;

$S = 2$: $X \leftarrow X + \overline{Y} + 1$; IF (GO = 1) $S \leftarrow 0$;

$S = 3$: $X \leftarrow \overline{X} + 1$; $XS \leftarrow \overline{XS}$; $S \leftarrow 0$;

Figure 13.20 shows the state diagram of the adder/subtracter, Figure 13.21 gives its microprogrammed controlled circuit, and Figure 13.22 the controller ROM table. The control ROM output definitions are

C0, C1	ALU FUNCTION SELECT
C2	ALU CARRY-IN
C3	LOAD X AND L
C4	INVERT XS
C5	BUSY
C6, C7	NEXT ADDRESS
C8, C9	CONDITION MULTIPLEXER CONTROL

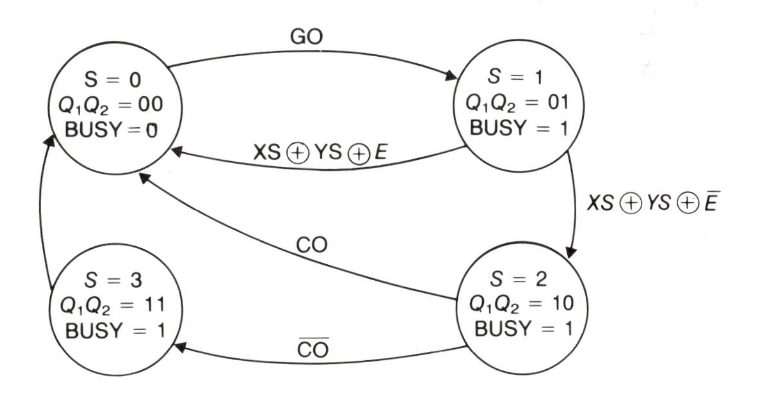

FIGURE 13.20 **Adder/Subtracter State Diagram.**

The ALU functions are as follows:

S1	S2	Function
0	0	TRI-STATE
0	1	$A + B$
1	0	$\overline{A} + B$ + CYIN
1	1	$\overline{B}$ + CYIN

FIGURE 13.21 **Micro-programmed Adder/Subtracter Circuit.**

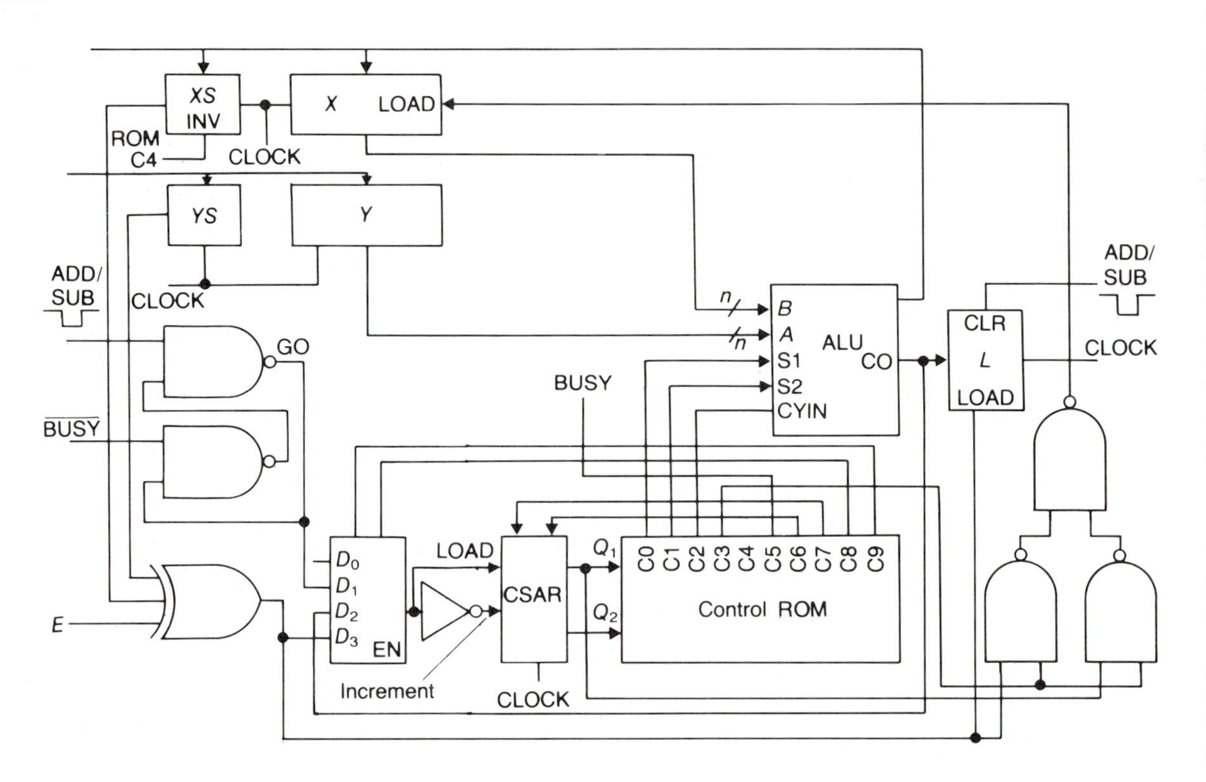

FIGURE 13.22 **Adder/Subtracter Control Table.**

		ROM Outputs									
		Microoperation Control						Next Address		Select	
State	ROM Address	C0	C1	C2	C3	C4	C5	C6	C7	C8	C9
0	00	0	0	0	0	0	0	0	0	0	1
1	01	0	1	0	1	0	1	0	0	1	1
2	10	1	0	1	1	0	1	0	0	1	0
3	11	1	1	1	1	1	1	0	0	0	0

13.8 Summary

The introduction given in this text to processor design is no more than a peek into the vast literature on an ever-expanding subject. Complete texts are devoted to computer architecture, instruction sets, and microprogramming. If the reading and study of this chapter has whetted the reader's appetite for additional information on the subjects presented and made other more specialized texts more understandable, our goal was met.

In the next chapter the subject of connecting devices to computers and some devices typically connected will be introduced.

Problems

1. What are the advantages of using microprogramming instead of the hardwired logic? What are the disadvantages?
2. Design a multiplier circuit to implement the 2's complement multiplication algorithm. List all control signals needed for the algorithm. Develop the microprogrammed control unit and instruction format. Write a microprogram to implement your design.
3. Repeat Problem 2 for 1's complement multiplication.
4. Obtain the remaining processor equations of Section 13.4.5.
5. Give the RTL descriptions not included in the text for the instructions introduced in Section 13.6.3.
6. Complete the control ROM table in Figure 13.17 for the remaining instructions in the original set of seven.
7. Using a control ROM table similar to the one in Figure 13.19, microcode the nine new instructions of the LIS-4 processor as described in Section 13.6.3. Discuss the overall design. Should there be any changes in fetch with the addition of the nine new instructions?
8. Fetch and execute cycles are sometimes overlapped in order to speed up the processor. What additions and changes have to be made in the digital processor in regard to data paths, registers, ALUs, and so on, to allow such an overlap? What changes would have to be made in the fetch and execute cycles to accommodate the change in architecture?
9. Using the instructions ADD, AND, STA, LA, OM, and STP, write a program that adds the values stored in the locations 13, 14, and 15, and outputs the result.
10. Rewrite the fetch cycle for the LIS-4 processor, showing the register transfers only, making use of the fact that a memory read does not use the ALU, BI, and BO.
11. Assume the accumulator register is a shift-right, shift-left register. Also assume a one-of-two data selector is available at each end to supply the shift input. Show the hardware requirements and necessary control signals, and give an RTL description of the following instructions:

    LSL1: Logical shift left 1. A 0 enters at the right end.

    LSR1: Logical shift right 1. A 0 enters at the left end.

ROTR: Rotate right instruction with the LSB entering the MSB and $A_{n-1} \leftarrow A_n$. A_n is the nth bit of the accumulator.

ROTL: Rotate left instruction with the MSB entering the LSB and $A_n \leftarrow A_{n-l}$.

12. Draw the block diagram of an architecture similar to the LIS-4 but with two output busses and two input busses and the necessary control signals to put data on and accept data from either bus. Also add a second ALU connecting the second set of busses. Can you see any advantage in implementing the same set of instructions on this system?
13. Modify the LIS-4 processor as follows. Add a register that stores the ALU output after each operation. The output of this register should attach to a common bus to which each of the other registers has both their inputs and outputs connected. This is called a unibus. Write the RTL description for the fetch and the LA instruction for this architecture.
14. Design a circuit that will perform as an eight-register stack for the LIS-4 processor. Each time the JSB instruction is executed, the return address is stored on the stack. Each time the RET instruction is executed, the last value stored on the stack will be removed and stored in the PC. The register performs as a last in–first out register.
15. Add an FF to the LIS-4 architecture that stores the ALU carry-out after each operation. Show the hardware changes that would be required to accommodate the following instruction:

    JC: Skip the next instruction if the carry is a 1. Execute the next instruction if the carry is a 0.

Suggested Readings

Abraham, E.; Seaton, C. T.; and Smith, S. D. "The optical computer." *Sci. Am.* vol. 248 (1983): 85.

Agerwala, T. "Microprogram optimization: a survey." *IEEE Trans. Comp.* vol. C-25 (1976): 962.

Berndt, H. "Functional microprogramming as a logic design aid." *IEEE Trans. Comp.* vol. C-19 (1970): 902.

Bocker, R. P. "Optical digital RUBIC (rapid unbiased bipolar incoherent calculator) cube processor." *Opt. Engn.* vol. 23 (1984): 26.

Dasgupta, S. "Some aspects of high level microprogramming." *ACM Comp. Surv.* vol. 12 (1980): 295.

Davidson, S., and Shriver, B. D. "An overview of firmware engineering." *Comput.* vol. 11, no. 8 (1978): 21.

Divan, D. M.; Hancock, G. C.; Hope G. S.; and Burton, T. H. "Microprogrammable sequential controller." *IEE. Proc. E., Comp. & Dig. Tech.* vol. 131 (1984): 201.

Grasselli, A., and Montanari, U. "On the minimization of read-only memories in microprogrammed digital computers." *IEEE Trans. Comp.* vol. C-19 (1970): 1111.

Halatsis, C., and Gaitanis, N. "On the minimization of the control store in microprogrammed concepts." *IEEE Trans. Comp.* vol. C-27 (1978): 1189.

Kenny, R. "Microprogramming simplifies control system design." *Comp. Des.* vol. 14 (1975): 96.

Kleir, R. L., and Ramamoorthy, C. V. "Optimization strategies for microprograms." *IEEE Trans. Comp.* vol. C-20 (1971): 783.

Papachristou, C. A. "Method for direct multiway branching in microprogram control." *Elect. Lett.* vol. 17 (1981): 709.

Quatse, J. T., and Keir, R. A. "A parallel accumulator for a general-purpose-computer." *IEEE Trans. Elect. Comp.* vol. EC-16 (1971): 165.

Tokoro, M.; Tamura, E.; and Takizuka, T. "Optimization of microprograms." *IEEE Trans. Comp.* vol. C-30 (1981): 491.

Toong, H. D. "Microprocessors." *Sci. Am.* vol. 237 (1977): 146.

Vandling, G. C., and Waldecker, D. E. "The microprogram technique for digital logic design." *Comp. Des.* vol. 8 (1969): 44.

CHAPTER FOURTEEN

Input/Output and Interfacing Design

14.1 Introduction

In Chapter 13 the LIS-4 processor was designed to demonstrate processor design techniques. In reality, more readers will be involved with the subject of this chapter than with the design of processors. Most engineers and computer scientists encounter the use of processors and/or their application as controllers in electromechanical systems. A proficient application or use of a processor requires an understanding of its I/O (input/output) structure. The flexibility of the I/O structure of a processor will determine the ease with which an application can be made.

Even with careful selection of a processor, a mismatch occurs between the processor and the device(s) to be controlled. The mismatch can occur in data rates, data representation, coding, and the like, or even the physical location of the devices to be controlled. A circuit to reconcile these differences must be designed to reside between the processor and the mismatched device. This circuit is called an *interface circuit*, or simply the *interface*.

In this chapter we will look at small-system I/O, interfacing, LSI interface devices, and A/D and D/A converters. After reading this chapter, studying the examples, and working the exercises at the end of the chapter, you should be able to:

- Understand I/O concepts and terminology;
- Appreciate interfacing problems and solutions;
- Undertake the interfacing of small systems in controller applications with confidence.

14.2 Small-System I/O

All of us fortunate enough not to have some type of communicative disorder communicate continually in one way or another. The rules of communication we have learned are followed with no consideration of their origin or doubt about their effectiveness. A device with which we are all familiar involves aspects of communication between devices that we will discuss later.

If you elect to use the phone, the following action-response sequence occurs:

Action	Response
Pick up the phone.	Dial tone is heard.
Input the address.	Key tones are heard.
Dialing complete.	Phone rings (or busy tone sounds).

At this point in the use of the phone you have initiated communication at a specific location (address) and have been given indicators from the system interface (one of several phone systems) that the process is proceeding as designed.

Now let us consider the action from the perspective of the person being called:

Action	Response
Phone rings (an *interrupt* occurs).	Receiver is lifted.
HELLO.	Message is given.
Phone is hung up and the request is attended to and/or the action occurring prior to the interrupt continues.	

You can imagine the difficulty of using the phone if the only response you received was that the person who was called said, "Hello." The requested action and subsequent response sent back to the requestor to indicate that the request was received is a signal exchange that occurs in many digital system interactions and is sometimes called *handshaking*. In the next section we will examine the ways I/O takes place and the handshaking that is required.

14.2.1 A Typical Microprocessor (μP)

The structure of a small system is shown in Figure 14.1. We will define small systems as eight-bit, μP-based systems. Typical I/O instructions for a small system are as follows:

IN: Data are read from the system data bus. The second word of the instruction, the *port number,* is placed on the μP I/O address bus. The device addressed is one of 255 possible ports (I/O connections).

OUT: Data are output from the μP on the system data bus to one of 255 possible ports or devices. The second word of the instruction, the port number, is placed on the μP I/O bus.

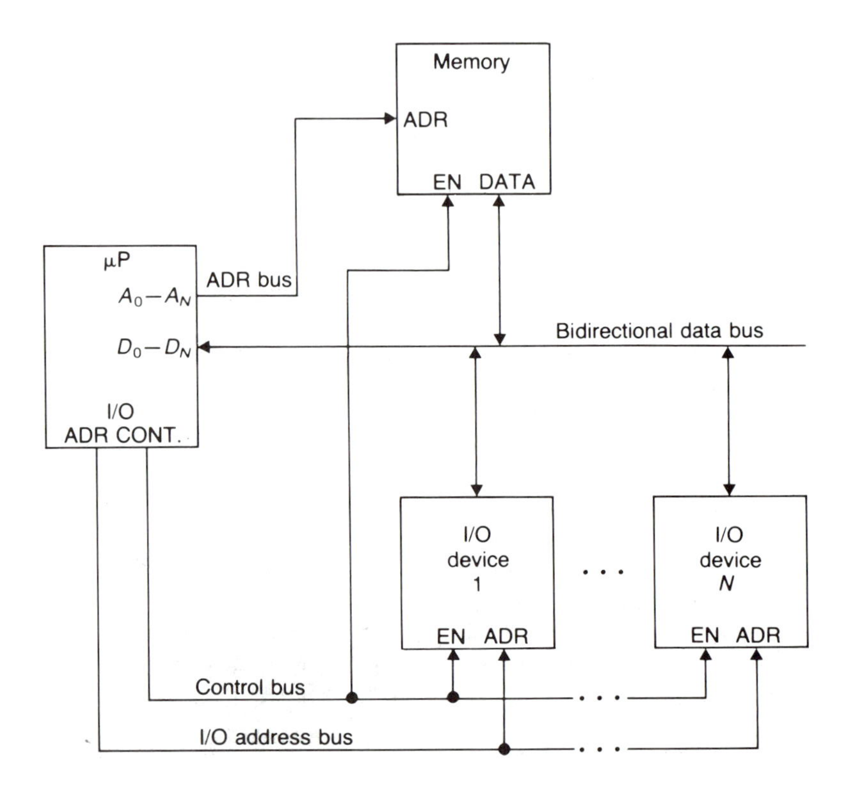

FIGURE 14.1 Small-System Architecture.

In each instruction data are read from or output to a register in the device or the device interface. The accumulator or other register internal to the μP is the source or recipient of the data transferred.

Figure 14.2 shows the data, address, and control signals provided by a μP. The signals have the following functions:

I/OA0–I/OA7: These address signals form the I/O address, which is the second word of the I/O instruction. The I/O address in some μPs appears on the address lines instead of on a separate bus.

A0–A15: These 16 address lines allow the selection of 2^{16} memory locations.

D_0–D_7: These data lines form an eight-bit bidirectional data bus. All parallel data transfers to and from the μP use this bus.

I/$\overline{\text{OMEM}}$: The I/$\overline{\text{OMEM}}$ signal differentiates data transfer between the μP and memory or peripheral devices. This line is low (0) when the memory is to be written to or read from, and is high (1) when the data are to be read from or written to a peripheral device.

RD/$\overline{\text{WR}}$: The RD/$\overline{\text{WR}}$ signal is output by the μP to indicate to the memory and peripheral devices the direction of the data

FIGURE 14.2 Typical μP.

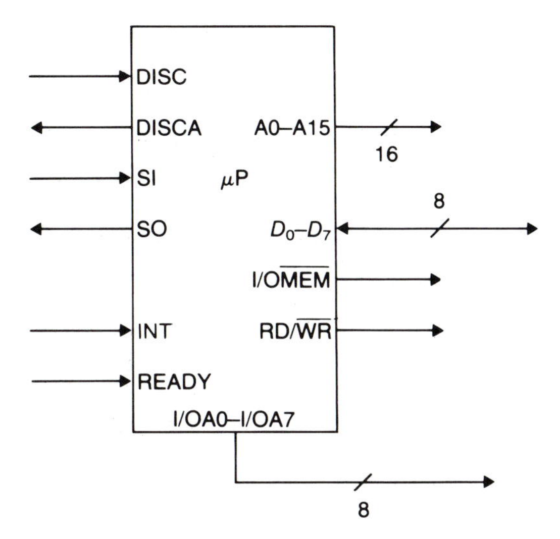

transfer that is to occur. This signal is high when memory or a peripheral is being read, and low when data are being written to memory or a peripheral.

INT: When the interrupt signal, INT, is high, the μP completes the instruction being executed, stores the address of the next instruction to be executed in a stack register, and begins execution in the highest location of memory. Interrupts are discussed further in Section 14.2.4.

DISC: The disconnect signal, DISC, when high, puts all address lines, data bus, and control signals in a tri-state (high impedance) condition.

DISCA: This disconnect accomplished signal (DISCA) remains connected and goes high when the DISC request is accomplished.

SI: The serial input, SI, provides for the serial input of data. A serial input instruction tests the input and if it is high, puts a 1 in the MSB position of the accumulator.

SO: The MSB of the accumulator is loaded into an FF connected to this pin when the serial output instruction, SO, is executed.

READY: If the READY line goes low during a memory or peripheral read or write, the μP will not proceed until the READY line goes high again. Placing the μP in a waiting state by controlling the READY line allows synchronizing the processor with slow memory and peripherals.

The I/O and interrupt discussions that follow will refer to these signals just listed.

14.2.2 I/O-Mapped I/O and Memory-Mapped I/O

If the two I/O instructions introduced in Section 14.2.1 are used to perform I/O, called I/O-mapped I/O, up to 256 external devices can be addressed. The I/O address space consists of 256 locations (I/O *ports*) to which 256 I/O devices may be attached. To allow 256 ports, the eight bits of I/O address would have to be decoded. In systems with eight or fewer ports, the ports, if restricted to port numbers that are powers of two, can be activated by a single address line. For port 4, address line I/OA2 would be used; for port 64, address line I/OA6 would be used; and so on.

A technique called *memory-mapped I/O* increases the ways in which peripheral devices can be addressed. The registers in the peripheral devices to or from which data are transferred are treated as memory locations and given addresses in memory space. This treatment allows the use of all memory reference instructions as I/O instructions. The READY input to the μP can be used to synchronize data transfer with devices that operate at a slower data rate than the μP.

To differentiate between accessing memory and a peripheral using memory-mapped I/O, bit 15 of the address bus typically is set high for all accesses of peripheral devices as memory, and low for actual memory accesses. Figure 14.3 shows the use of both techniques for accessing identical registers.

The left peripheral register in Figure 14.3 can be read by any instruction that accesses memory and uses 8001H (the H signifies hexadecimal) address. Note that NAND gate 1 has as two of its inputs address bits A0 and A15, the two bits that are high in the 8001H address. We have assumed in this example that bits A0 and A15 will both be high only when addressing latch #1. Note also that the page decode for the memory is disabled by A15 when memory-mapped I/O is used. The 15 memory address bits, A0–A14, available for use when using memory-mapped I/O, can be used individually in conjunction with A15 to enable 15 peripheral registers, or decoded to allow addressing up to 32,768 external devices.

Memory-mapped I/O significantly increases the I/O addressing capabilities of a μP; in fact, some μPs have no I/O instructions and use memory-mapped I/O exclusively. The price paid for using memory-mapped I/O is the loss of part of the memory space for use as program storage. When using A15 to indicate memory-mapped I/O, one-half of the memory is unavailable for program or data storage. For most small systems used in control applications, the remaining 32,768 memory locations are sufficient.

The right register in Figure 14.3 is read by an IN instruction and a port number 01H. NAND gate 2 is used to decode the control signals and enable the data on the bus.

FIGURE 14.3 **I/O-Mapped I/O and Memory-Mapped I/O.**

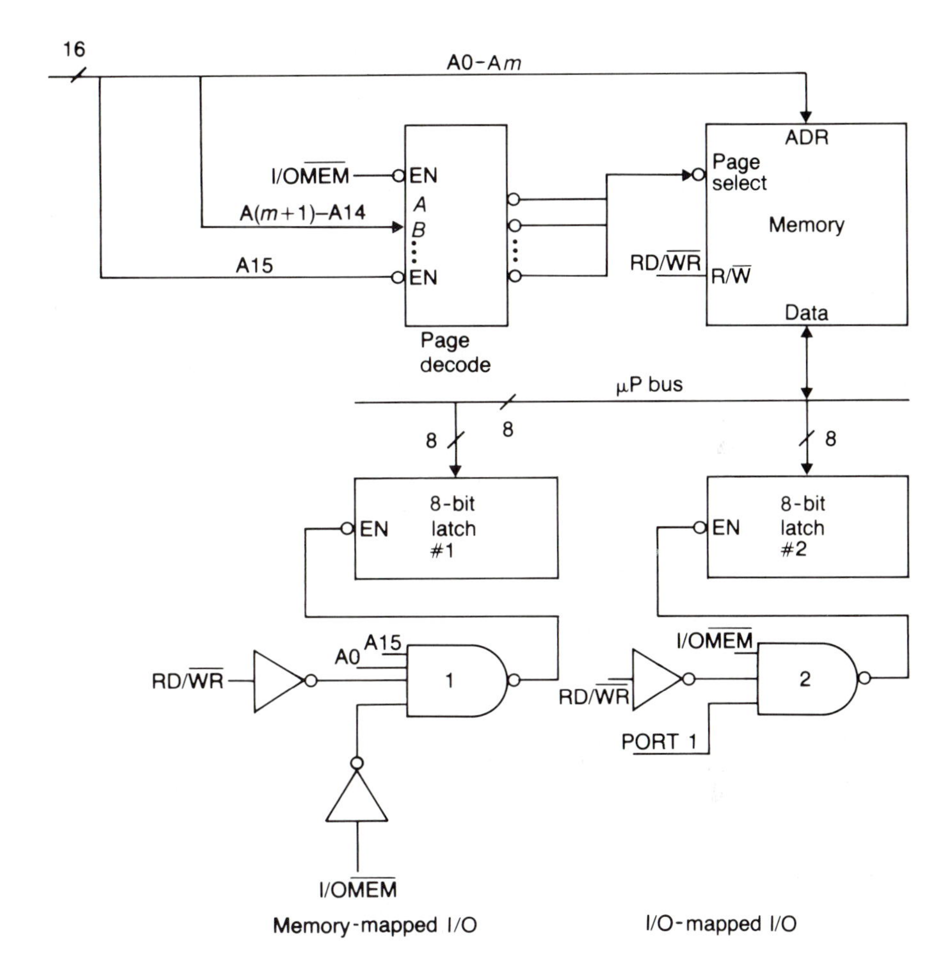

14.2.3 Polling and Interrupts

When a computer executes an input or output instruction, it is performed as part of the instruction sequence under control of the processor's controller. The read or write that results occurs not on demand of the external devices but under processor control. In many applications peripheral devices require output from the processor or have data to be read by the processor asynchronously with respect to internal processor operations. The necessity of a peripheral being serviced when data need to be transferred is handled in two ways: *polling* and *interrupts*.

In order to assure a specified minimum response time in providing processor/peripheral data transfer, the *polling* technique may be used. Figure 14.4 demonstrates this technique. In the example

FIGURE 14.4 **Polling Technique of Peripheral Monitoring.**

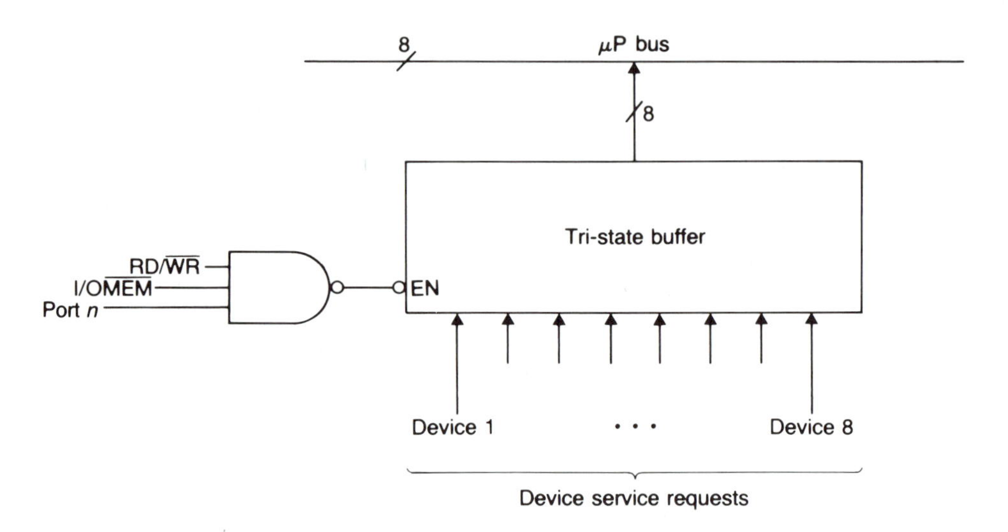

shown, eight devices are connected to a tri-state buffer. A high on any input indicates that the device needs to be serviced; that is, it needs a data transfer. The μP must periodically read the port to which the devices are attached (poll the devices). After reading the status of the devices, the μP performs the necessary data transfers under procedures established by the programmer. If more than one device requires service, the programmer should have established in software the order in which the devices will be serviced.

To work successfully, the polling must occur often enough to allow the data rates required by the devices to be met. The polling process requires a significant amount of overhead in that periodic checks of the peripherals must be made with the possibility that no data transfer is required. The instructions required for the polling operation represent wasted compute time for the processor. A more efficient technique, interrupting, requires an architecture designed to accommodate interrupts.

Interrupts involve external-event-initiated, automatic, subroutine calling. Figure 14.5 demonstrates the interrupt process. During execution of the main program, an interrupt occurs during instruction n. Program control is transferred by hardware to a section of memory where the service routine for the interrupting device is stored. The return point to the main program, instruction $n + 1$, is stored in a last in–first out, LIFO, register. During the execution of the interrupt service routine, another device interrupts (called *nesting* of interrupts), resulting in the following events:

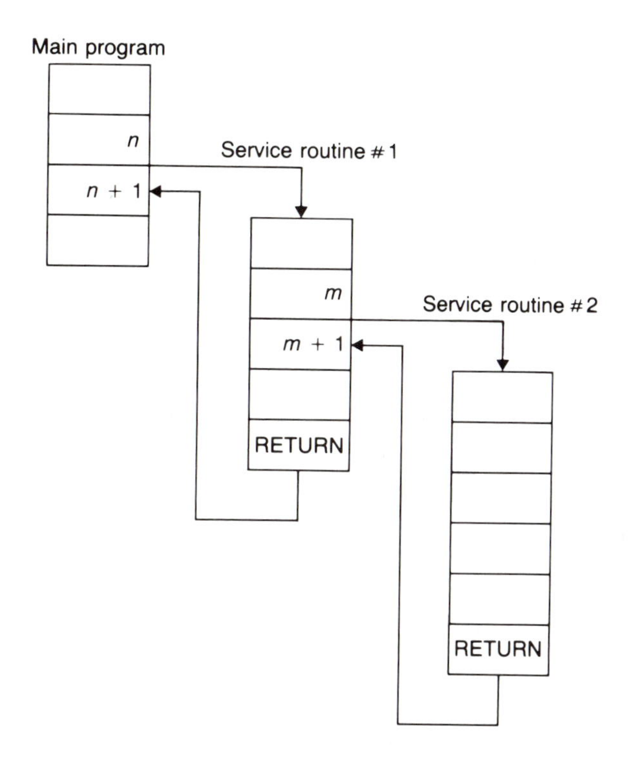

FIGURE 14.5 **Example of an Interrupt System.**

1. The return point to interrupt service routine #1, $m + 1$, is stored on the stack, the LIFO register.
2. The interrupt service routine #2 is completed with the RETURN statement, causing the program control to be transferred to the location at the top of the stack, $m + 1$.
3. The interrupt service routine #1 is completed with the RETURN statement, causing the program control to be transferred to $n + 1$.
4. The main program continues.

Note that with interrupt capability, peripherals are serviced or communicated with only when needed. It is the programmer's responsibility to assure that part of each interrupt routine stores all registers in use when the interrupt occurs and restores the registers to those values prior to returning to the interrupted program. This process is called *saving the state of the machine* and must be accomplished by hardware or software to assure that computation can resume correctly from the point of interruption.

Figure 14.6 pictures a priority interrupt system connected to the μP of Figure 14.2. The system, though not very complex, represents an elegant small-system interrupt structure. The key device is the

FIGURE 14.6 **Priority Interrupt System.**

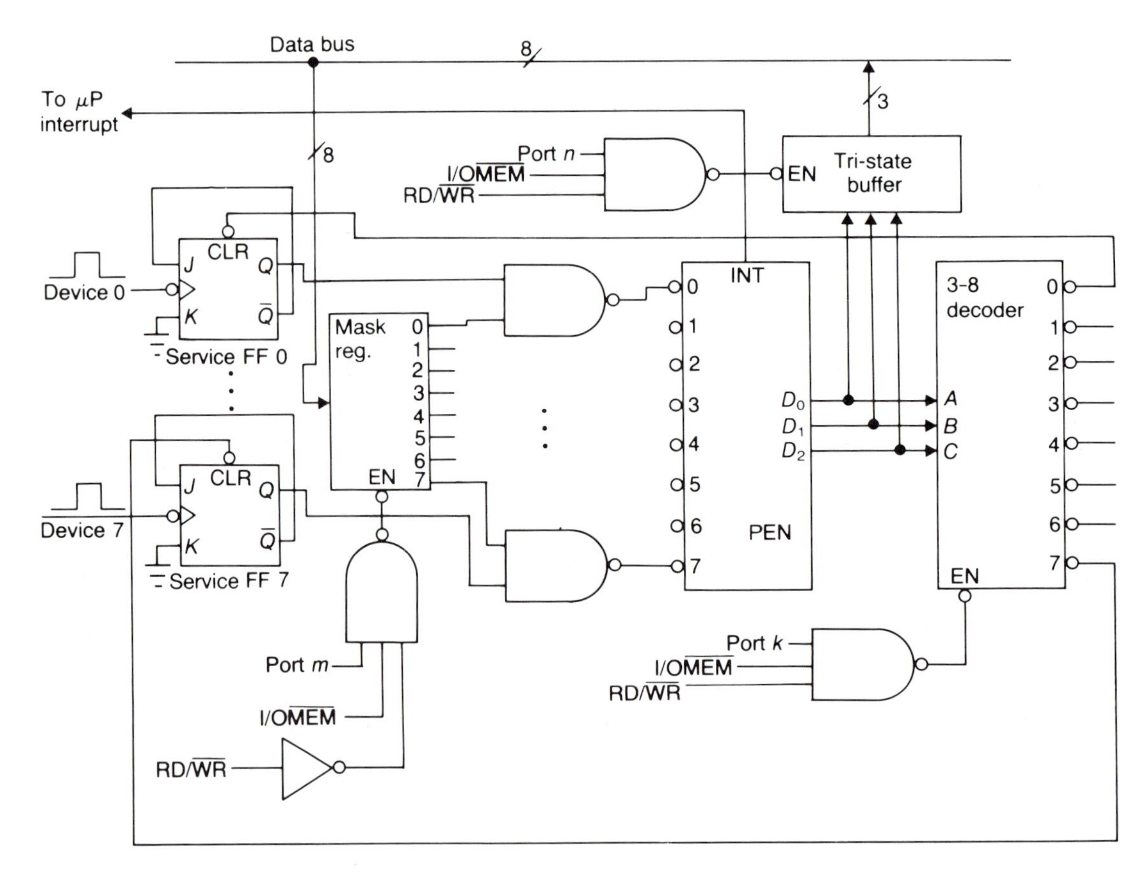

priority encoder, PEN. PENs were introduced in Section 5.8 where a similar encoder was designed. In the PEN used in Figure 14.6 the higher the number of the input, the higher the priority. The output D_2, D_1, D_0 is the binary representation of the number of the highest priority input that is active (high). The INT output goes high when any input is high. This output is connected to the INT input of the μP.

Another device that is included in the priority interrupt system of Figure 14.6 is the *mask register*. A zero in any position in the eight-bit register disables, or masks, the corresponding interrupt. This capability allows the programmer to mask out any interrupt that he or she does not want to interrupt the routine currently being executed by placing a 0 in the corresponding position of the mask register.

The service FF is set by the device to which it is connected by sending a positive pulse to the clock input. The service FF is reset by executing an OUT–to–port *k* instruction. The port *k* decode causes the 3-8 line decoder to be enabled, which resets the interrupt FF that initiated the interrupt sequence.

The interrupt sequence is as follows:

1. A device requiring service pulses its service FF.
2. If the mask bit is 1, the PEN senses an interrupt and INT goes high.
3. The processor program control is forced to the highest location in memory where the programmer has stored a jump-to-an-interrupt handling routine.
4. The program reads D_2, D_1, and D_0 from the PEN, using port *n*.
5. The program outputs a bit pattern to the MASK register using port *m*, which determines the interrupts that can occur during the current service routine.
6. The program executes an OUT–to–port *k* instruction that causes the service FF that initiated the interrupt to be reset.
7. If any unmasked interrupts are active, they now can interrupt the service routine for the previous interrupt.

With the maskable priority interrupt system just discussed, it is apparent that sophisticated control systems can be developed using μPs. In the next section we will investigate connecting devices together that have different data rates and/or different control signal requirements.

14.3 Interfacing

Despite the efforts of processor and peripheral manufacturers to accommodate the features of the other's devices, it is still necessary in many cases to design a circuit that reconciles differences between the control signals and data transfer capabilities of the processor and those of the peripherals. The circuits designed to reconcile differences are called interface circuits or *interfaces*. The goal for the designer is to make the peripheral, as seen by the processor through the interface, and the processor, as seen by the peripheral through the interface, appear as compatible devices. We will look at several of the many devices that are available commercially to use in interface design.

14.3.1 Control Lines and Sense Lines

The simplest external system to interface is a system that has multiple single-line outputs that must be monitored and individual lines to be set high, low, or pulsed for control purposes. The lines to be monitored are called *sense lines* since levels are sensed on individual lines to determine their binary status. The individual lines,

FIGURE 14.7 **Control Lines and Sense Lines.**

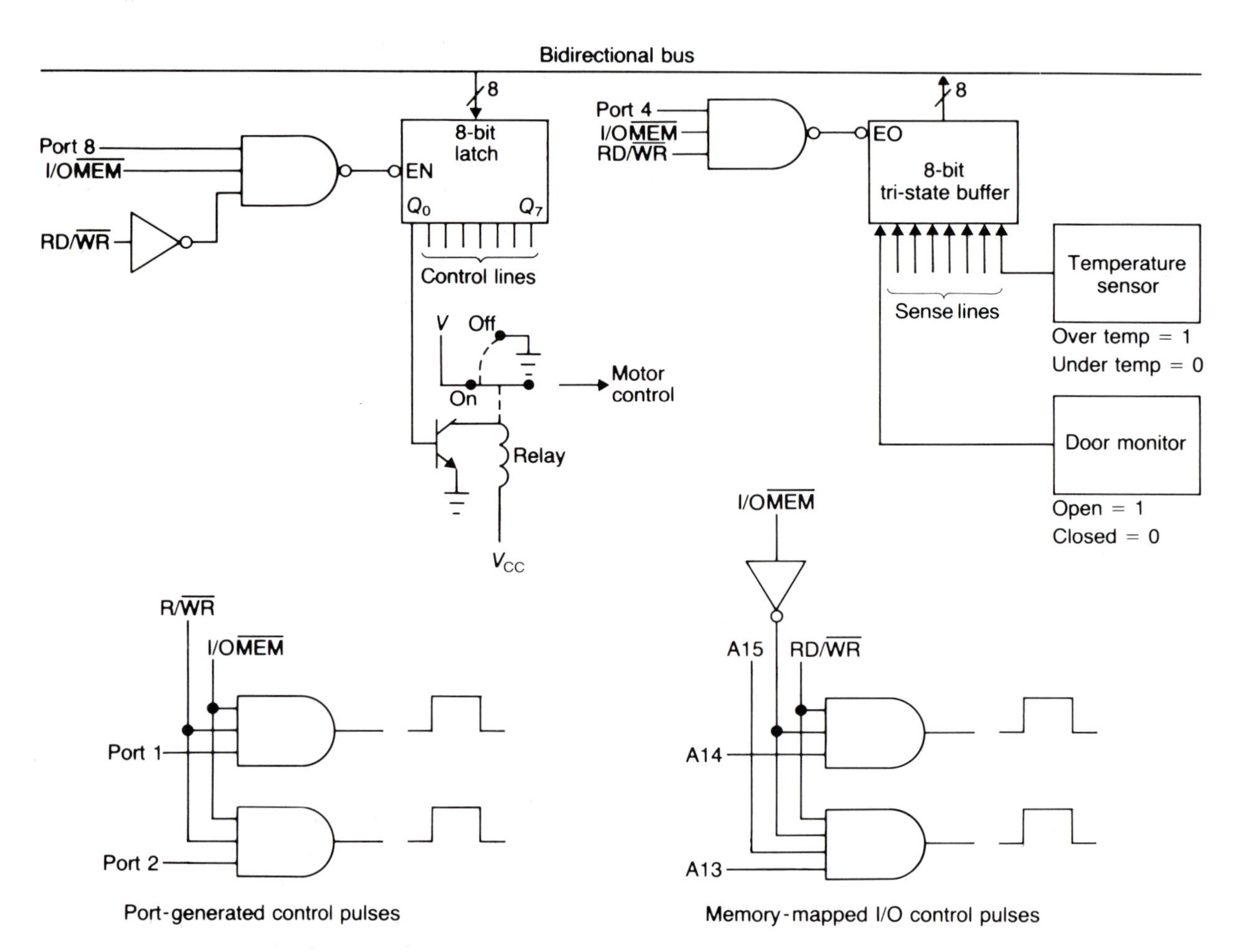

called *control lines*, control single-input, binary-controlled devices. Figure 14.7 gives examples of sense lines and control lines.

The eight-bit latch attached to the bus in Figure 14.7 can be used to set lines high or low for level input devices or can be turned on and off to generate pulses. Assuming that the μP outputs the accumulator contents when the OUT instruction is executed, Figure14.8[*a*] shows the accumulator contents when the OUT instruction is executed and the resultant timing diagram. In Figure 14.7 bit 0 of the latch is shown controlling a relay that turns a motor on and off. The length of pulses output on the latch-controlled control lines can be controlled by software by calling delay loops of known duration.

The section of the circuit in Figure 14.7 labeled *port-generated control pulses* involves decoding a particular port address and using that

FIGURE 14.8 **Generating Control: [*a*] Levels and [*b*] Pulses**

Output Instruction	Accumulator Content
(1) OUT Port 8	00010001
⋮	
(2) OUT Port 8	00010000
⋮	
(3) OUT Port 8	10000001

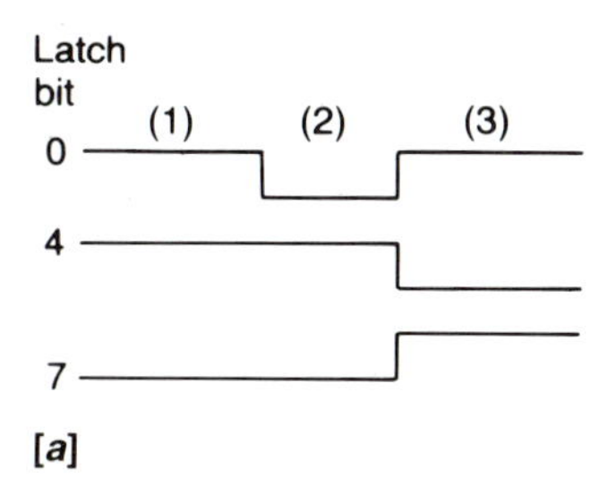

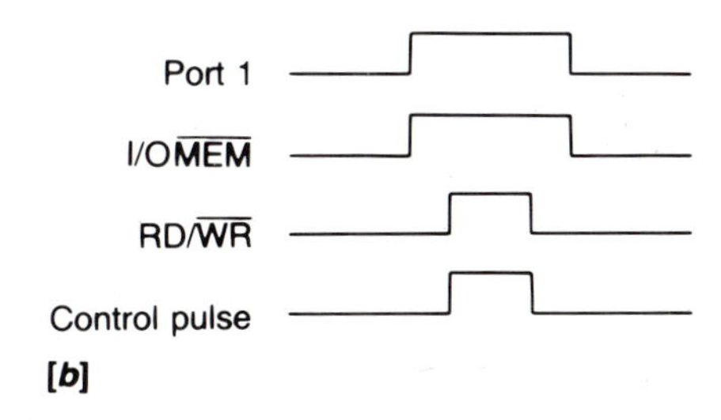

signal with the RD/$\overline{\text{WR}}$ and I/O$\overline{\text{MEM}}$signals to generate pulses, as shown in Figure 14.8[*b*], that can be attached to devices that are controlled with pulse inputs. Each time an OUT–to–port 1 or port 2 is executed, a pulse will be generated.

The memory-mapped I/O control pulse circuit in Figure 14.7 generates a pulse on the decoded lines when any instruction involving a memory read of memory locations C000H or A000H (the H indicates hexadecimal) is executed. The sense line inputs can be from sensors with binary outputs such as window and door sensors in a security system, temperature sensors, and so on. The sense line port would have to be polled or attached to an interrupt system to assure response to an active sensor.

14.3.2 LSI Peripheral Interface Devices

IC manufacturers provide programmable devices that significantly decrease the chip count required for successful connection of a μP to peripherals. The block diagram in Figure 14.9 shows the general structure of a peripheral interface device, PID.

Note that there are two address lines associated with this PID. These two lines select one of the four internal registers to receive

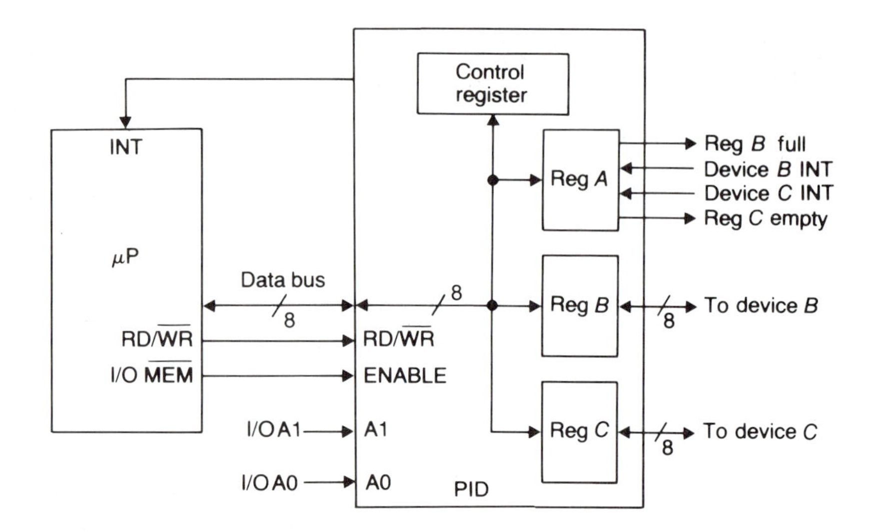

FIGURE 14.9 **Peripheral Interface Device.**

data from the bus or to place data on the bus. Typical functions of the four internal registers are as follows:

1. *Control register*. The control register selects the many modes of operation made possible with the PID. Selectable functions are
   a. data direction of ports *A, B,* and *C,*
   b. structuring of port *A* as control lines for ports *B* and *C,*
   c. selecting specific bits of port *A* to be set or cleared.
2. *Register A*. Register *A* is selectable as either an input or an output port as well as eight control lines usable in conjunction with ports *A* and *B*.
3. *Registers B and C*. These two ports can be selected as latched input ports, latched output ports, or bidirectional ports connecting directly to the system bidirectional bus. Port *A* can be programmed to provide handshake signals for ports *B* and *C*.

The use of the PID involves writing a control word into the control register to select the desired function. Until the control word is altered, the device functions as previously programmed. Figure 14.10 shows a typical function involving register *A* inputs as handshake signals. In this mode two *A* lines are involved in conjunction with port *B*. When port *B* has data transferred to it by the μP, the *data available* line goes high. Whenever the device attached to port *B* needs data, its *data needed* signal goes high. The *data needed* input causes the PID interrupt line, INT, to go high, which when connected to the μP INT causes an interrupt. The μP can determine which port interrupted by reading port *A*.

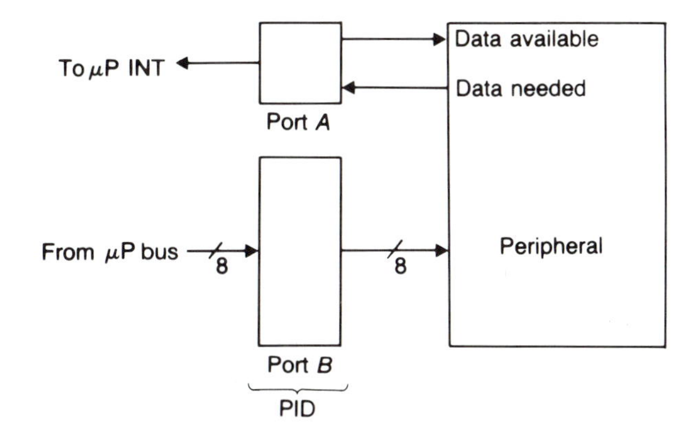

FIGURE 14.10 **Using PID Port *A* for Handshake Signals.**

Commercially available devices vary in function from the description just given but the intent is the same. PIDs can provide the necessary interface between the μP and external devices in many instances. The reader who wishes to use one of these devices will be surprised at the many modes available and the length and detail of the manufacturers' literature explaining the devices. The manufacturers' data must be studied carefully to make the best use of the devices and avoid costly errors.

14.3.3 Serial Communication

Frequently the interface that must be designed involves overcoming the distance separating the processor and the peripheral with which it wishes to communicate. To minimize the wiring cost of linking two distant devices together, serial transmission of data is performed. The μP of Figure 14.2 has SI and SO functions that may be used to perform serial data transmission.

Figure 14.11 shows the timing diagram of the transmission of the first character in a serial transmission. The *MARK* portion of the signal is a high that tells the receiving device that the transmitting device is in standby. When the line goes low (called a *BREAK*) for one clock period, the receiver knows that data follow. Knowing the rate at which data are being sent and synchronizing with the 1 to 0 change when the BREAK occurs, the receiving device can sample

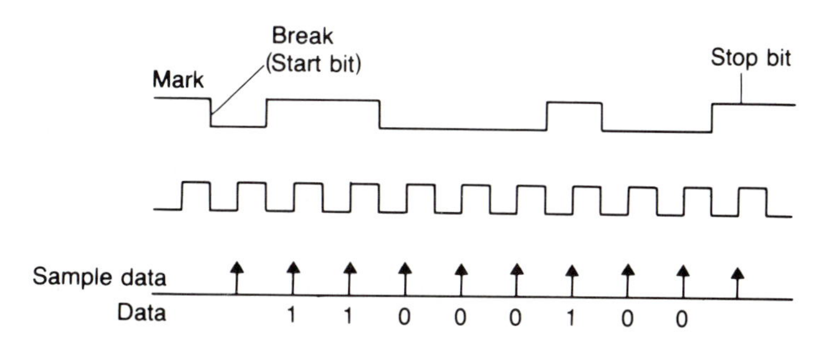

FIGURE 14.11 **Serial Transmission of an Eight-Bit Character.**

the signal during the center of each bit period. After the eight bits have been sent, a high STOP bit is sent. If additional data are to be sent, the process is repeated. The processor can be programmed to perform all of the transmit and receive functions including the timing of the data input samples. Rather than tie up the processor completely with the communication task, the timing and BREAK detection can be handled externally with the read and send data operations requested by the interrupt as shown in Figure 14.12.

The timer in Figure 14.12 detects the MARK to BREAK when the μP is in the receive mode, loads a counter at the input clock rate with one-half the count equivalent to the time between bits of the incoming data, and begins counting. Each time the counter generates a carry (which occurs at the midpoint of the bit period), the processor is interrupted and the data bit read using the SI instruction. In the transmit (Tx) mode the Tx signal is high and the SO output is controlled by the SO instruction. The timer interrupts the μP at the bit rate to cause the SO instruction to output the next bit. The mode, receive or transmit, is sent to the control register in the timer. A more sophisticated communications controller could be designed that would allow the selection of data rates and other functions by the processor. Because of the number of applications requiring serial data transmission, such a circuit is commercially available and is called a *universal synchronous/asynchronous receiver/transmitter, USART.* Figure 14.13 shows a block diagram of a USART and its connection to a μP. The mode, RD/$\overline{\text{WR}}$, and CS (chip select) are used to select the various functions of the USART.

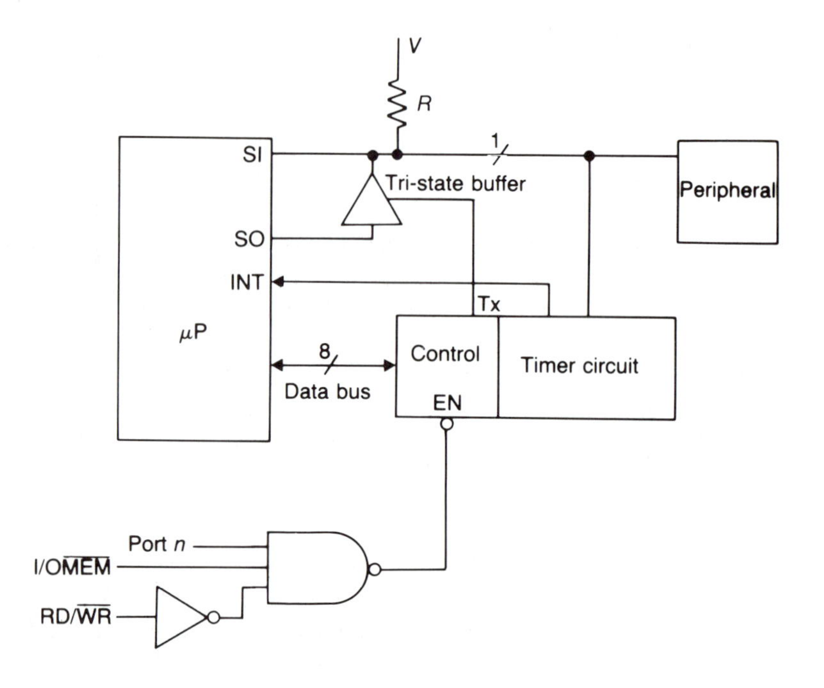

FIGURE 14.12 **Serial Data Transfer.**

FIGURE 14.13 **Block Diagram of a USART.**

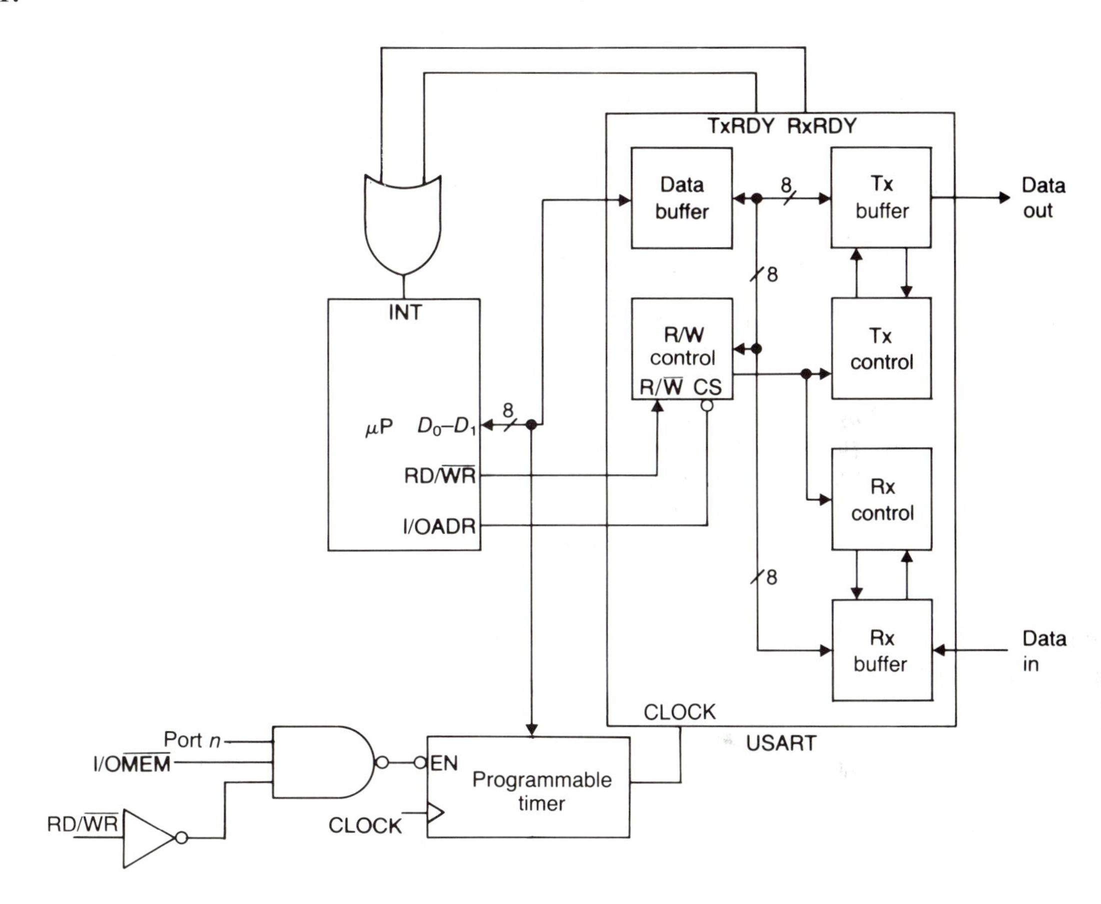

Control information can be written into the R/W control register to select the mode of communication. Eight-bit data are transferred in parallel to the Tx buffer for serial transmission. When the buffer is empty the TxRDY signal is used to interrupt the processor to supply the next eight bits of data. The USART generates the start, stop, and parity bits sent with the data bits. In the receive (Rx) mode when eight bits have been received, the USART makes the RxRDY signal high, which is used to interrupt the processor and cause the data to be read. The Rx control detects the MARK and BREAK with no action required by the processor. The processor is involved only for setup and transferring eight-bit quantities when interrupted.

The programmable timer in Figure 14.13 is another commercially available chip that can be attached to the μP for selecting various bit rates (its function is to output pulses at programmable rates). The application of the USART to different communications modes requires a background in data communications. Be aware that such devices are available in the event you ever have the responsibility for designing a serial communications system.

14.3.4 Direct Memory Access

Direct memory access, DMA, is a process that allows direct transfer of data between a peripheral and memory. Having this capability is important in two situations:

when data rate requirements of the peripheral exceed the I/O capabilities of the processor using I/O-mapped or memory-mapped data transfers,

when programs or data are being read from or written to disk or tape, and the time required using normal I/O input procedures is too time-consuming.

The DMA process involves disconnecting the processor from the control and address busses and allowing the external device to control data transfer between itself and memory. The μP in Figure 14.2 has provisions for accommodating a commercially available DMA controller. Figure 14.14 shows the connections between the μP and a DMA controller.

FIGURE 14.14 **DMA Controller.**

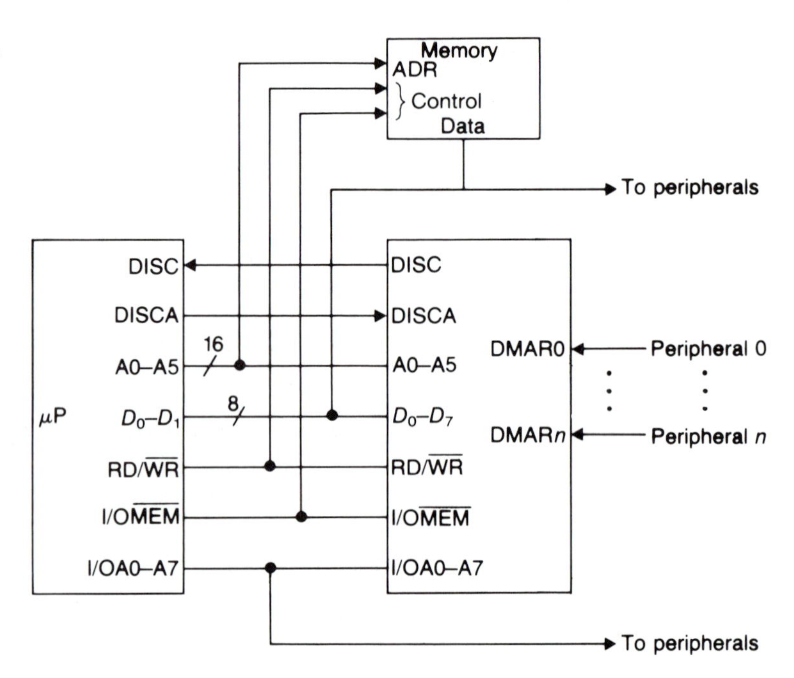

The programmer must cause the initialization of the DMA controller by providing the following information for each device connected to the controller:

the beginning location in memory where data transfer is to begin,

the location where data transfer should end, or the number of words to be transferred,

the priority of the devices connected to the controller.

After the DMA controller is programmed, it monitors the request lines, DMAR*n*. When one or more lines go high, the DMA controller activates the DISC line, which causes the processor to place its address and control busses in the high impedance (DISC) state and output a high on the DISCA line. Using the information stored during the programming of the DMA controller, the data transfer is performed. Upon completion the DMA controller lowers the DISC line, and the processor regains control of the busses.

In systems requiring large data transfers, the addition of a DMA controller significantly increases the system efficiency. Manufacturers' data on the specific controller chosen must be obtained and studied carefully to assure correct implementation.

14.3.5 Other Interface Design Aids

IC manufacturers continually offer new devices to simplify interface design and increase system performance. Designers should be aware of the following devices:

coprocessing units,

floating point processors,

programmable internal timers,

programmable floppy disk controllers,

single/double density floppy disk controllers,

programmable CRT controllers,

programmable keyboard/display interfaces,

programmable voice synthesizers,

bus arbiter chips for interconnection of processors,

programmable communication protocol devices.

Information about these devices and proposed devices can be requested from the manufacturers listed at the end of the chapter. It is the responsibility of a good designer to know what devices exist in order to avoid "redesigning the wheel."

14.4 Analog-to-Digital and Digital-to-Analog Converters

In the application of small processors for the control of electromechanical systems, the digital designer will be required to resolve a serious mismatch in data format when attempting to interface to transducers designed for detecting temperature, pressure, acceleration, velocity, position, and other physical parameters, since most transducers translate physical parameters into a voltage or a current. The interface between voltage and current must convert current or voltage levels to equivalent binary values. The device that performs this function is called appropriately an *analog-to-digital, A/D, converter*. In order to control analog (voltage-controlled) devices, the processor must be interfaced with a *digital-to-analog, D/A, converter*, which converts binary values to equivalent voltage or current. Since an integral part of most A/D converters is a D/A converter, we shall discuss the D/A converter first.

14.4.1 D/A Converters

The function of a D/A converter is to convert a binary value into an equivalent voltage. The number of bits and the reference voltage of the D/A converter determine the significance of a single-bit change in terms of volts per bit (V/bit). The reference voltage, V_{ref}, is determined by the manufacturer of the converter. For some converters there is a range of permissible values. The value of V_{ref} determines the maximum value of output voltage, V_o, that occurs when the binary value being converted has its maximum value. The relationship between the number of bits, V_{ref}, and the resolution of the D/A converter is

$$\text{resolution} = \frac{V_{ref}}{2^n - 1} \text{ V/bit} \qquad [14.1]$$

where n is the number of bits. Figure 14.15 shows the principle behind D/A conversion. The resistor network is called an R-$2R$ bridge. The circuit is unique in that, as connected, the resistance seen in each direction at a node is $2R$, and thus the current divides at each node. The switch in the diagram is replaced by an electronic switch in the actual circuit with a 1 at a bit position connecting the resistance in the leg to V_{ref} and a 0 connecting the resistance to ground. Any number of bits may be added in a similar configuration.

Typically the output voltage, V_o, is amplified using an operational amplifier with a gain that gives a value out of

$$V_o = \frac{V_{ref}(2^n - 1)}{2^n} \qquad [14.2]$$

when all bits are 1 and $V_o = 0$ V for all bits $= 0$. The network in Figure 14.15 would result in a single-polarity D/A converter since no provision is made for connecting a negative V_{ref} to provide the possibility of negative V_o. It is possible to have an output that varies between $\pm V_{ref}$ with the sign of V_{ref} changing with the sign bit of the binary quantity.

FIGURE 14.15 **D/A Circuit Operation.**

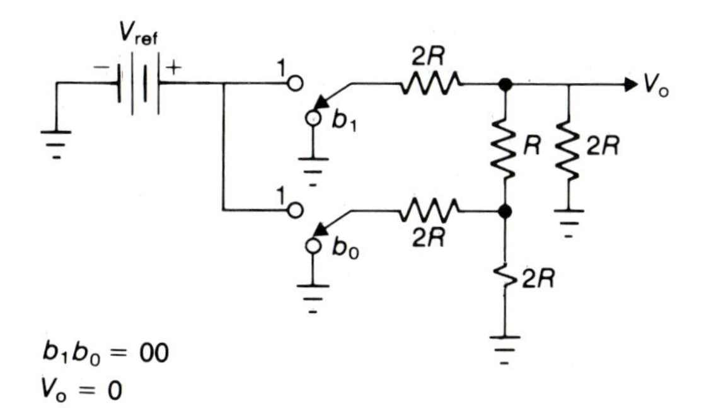

$b_1 b_0 = 00$
$V_o = 0$

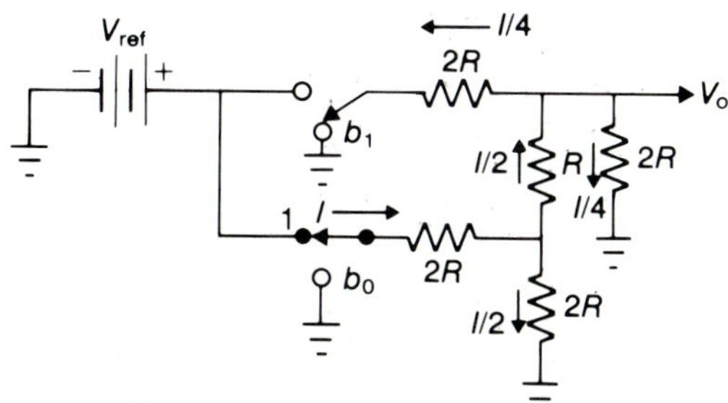

$b_1 b_0 = 01$
$V_o = I/4(2R) = 0.25I(2R)$

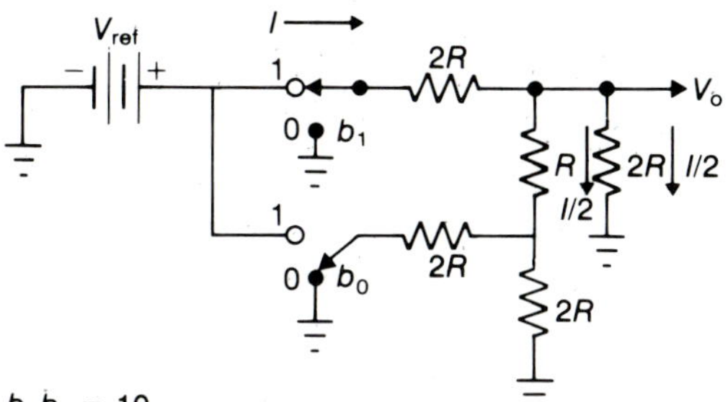

$b_1 b_0 = 10$
$V_o = I/2(2R) = 0.5I(2R)$

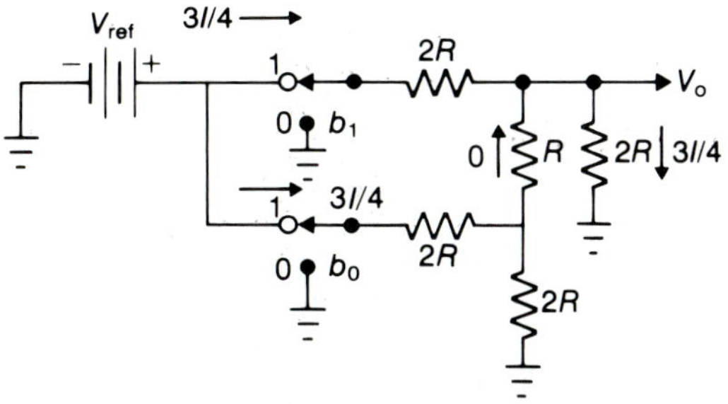

$b_1 b_0 = 11$
$V_o = 3I/4(2R) = 0.75I(2R)$

Figure 14.16 shows a μP connected to a D/A converter and the output of the converter when the processor outputs the contents of an internal register that is incremented by one prior to each output. Some D/A converters have an internal register that would eliminate the register connected to the converter. A register at the input is necessary to hold the binary value being converted. The 5 V reference voltage and eight bits provide a resolution of 0.0196 V/bit. The steps that result from incrementing the binary value one bit at a time are small enough that the D/A converter output appears smooth if viewed on an oscilloscope. Any arithmetic function that could be generated by the processor and scaled to eight bits could be converted to a voltage and displayed on an oscilloscope.

FIGURE 14.16 **D/A connected to a μP.**

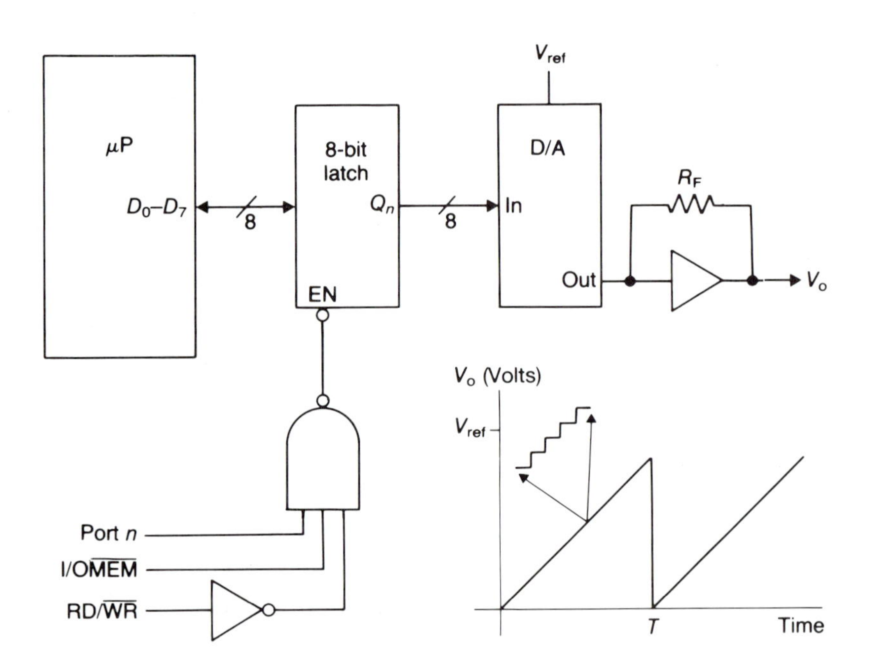

The triangular-shaped device connected to the D/A converter in Figure 14.16 is an *operational amplifier*, *op amp*. The resistorR_F is chosen to provide the desired amplification. The time T, the period of the waveform, is the time required for the processor to generate and output the values from 0 to 255. The manufacturers' data sheets and application notes give suggested configurations for various applications as well as important parameters such as settling time of the output, slew rate, and other parameters that should be covered in an electronics course.

14.4.2 A/D Converters

A/D converters can be designed using several techniques. The most common is called *successive approximation*. A block diagram of a successive approximation converter is shown in Figure 14.17. The triangular device, connected to both the voltage to be converted and the D/A output, is an analog comparator. If the B input to the converter is greater than the A input, the output is a 1. If the A input is greater than B, the output is a 0. The conversion process proceeds as follows:

Step 1. The MSB of the register whose output is connected to the D/A converter is set equal to 1, and all other bits are set equal to 0.

Step 2. If the output of the D/A converter is larger than the voltage being converted, the initial guess was too large and the MSB is turned off. If the D/A converter output is less than the voltage being converted, the bit is left on.

Step 3. The next MSB is now subjected to the process described in Steps 1 and 2. The process is continued for all bits.

FIGURE 14.17 **Block Diagram of a Successive Approximation A/D Converter.**

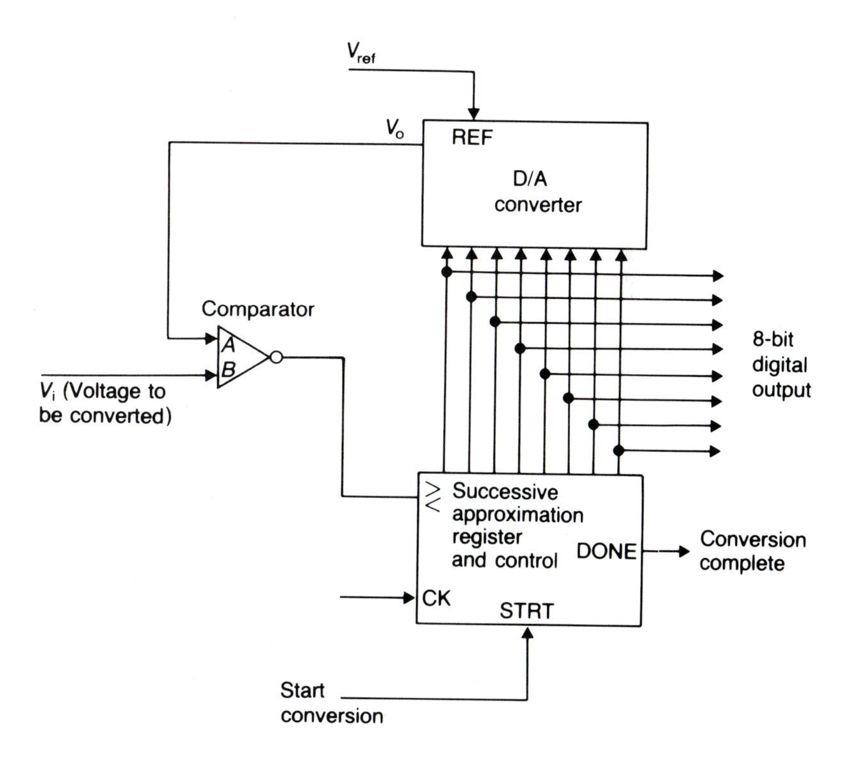

For an eight-bit converter, if a voltage equivalent to 225 bits is input, the register values would be successively as follows:

Binary	Decimal	
10000000	128	< 225
11000000	192	< 225
11100000	224	< 225
11110000	240	> 225
11101000	232	> 225
11100100	228	> 225
11100010	226	> 225
11100001	225	STOP

There are commercially available A/D converters with eight selectable analog inputs, tri-state outputs, and control signals compatible with those provided by typical μPs. Figure 14.18 shows an A/D converter connected to a μP. In Figure 14.18 eight inputs could be connected to the A/D converter, and the input to be converted could be selected by outputting the A/D input number to the latch connected to the data bus and to the address inputs of the converter. The start pulse for the converter is generated using the port-generated control pulses technique discussed in Section 14.3.1. The DONE signal, which indicates the conversion has been completed, is used to interrupt the processor to read the converted value.

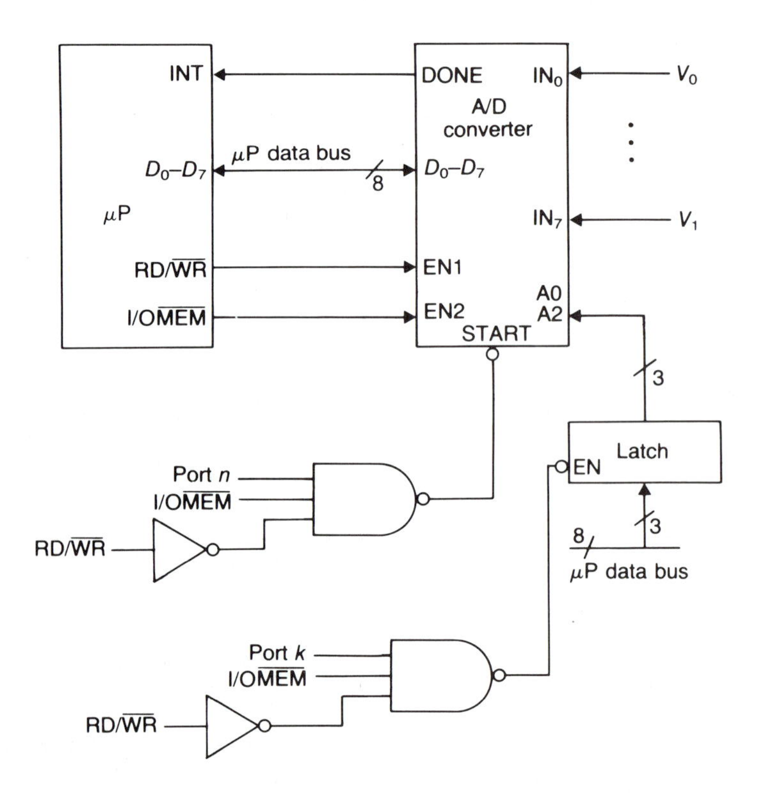

FIGURE 14.18 **A/D Converter Connected to a μP.**

When an A/D converter is used in a system to convert a time-varying voltage, there are signal parameters that must be considered:

Sampling frequency: The minimum *sampling frequency*, or the rate at which conversions are made, must exceed by a factor of two the highest frequency content of the signal being converted to digital form. In most applications the sampling rate is made five to seven times the highest frequency component of the signal being converted in order to recover frequency information. The details of this requirement are beyond the scope of this book. A course in signal analysis or sampled data theory would cover this material.

Conversion time: The *time of conversion* plays a significant role in the accuracy of the resultant conversion value if the signal being converted is time-varying. Figure 14.19 shows an expanded view of a portion of a time-varying signal. A single bit in the converted value is equivalent to

$$\Delta V = \frac{V_{\text{ref}}}{2^n - 1} \text{ V}$$

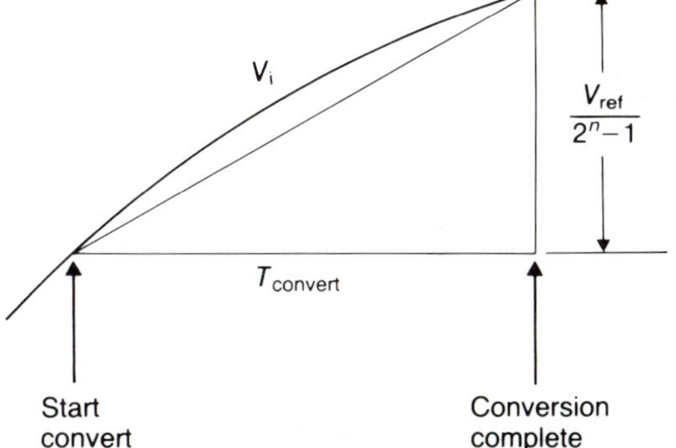

FIGURE 14.19 **Slope Considerations for A/D Conversions.**

If the voltage signal changes by ΔV while in the process of conversion, then we have lost one bit of accuracy in the converted value. To maintain an accuracy of conversion to within the voltage significance of the LSB (i.e., ΔV), the signal cannot change more than ΔV while the conversion is in progress. To assure this condition, the slope shown in Figure 14.19 must meet the following condition:

$$\text{slope}\,|_{\max} < \frac{V_{\text{ref}}}{(2^n - 1)T} \qquad [14.3]$$

T is the convert time of the A/D converter. From differential calculus we know that the slope of a signal at time t is its derivative at time t. Thus,

$$\left.\frac{dV}{dt}\right|_{\max} < \frac{V_{\text{ref}}}{(2^n - 1)T} \qquad [14.4]$$

EXAMPLE 14.1

Assume the input to an A/D converter is $V = 10 \sin(wt)$, the reference voltage is 10 V, a convert time of 1 μs, and eight bits. What frequency signal (f) can we convert to within one bit of accuracy?

SOLUTION

$$\frac{dV}{dt} = 10w \cos(wt)$$

$$\left.\frac{dV}{dt}\right|_{\max} = 10w$$

From Equation [14.4]:

$$10w < \frac{10}{255 \times 10^{-6}}$$

or

$$f = 624 \text{ Hz}$$

A variation of the problem presented in Example 14.1 would be to find the number of bits of accuracy possible for a given signal frequency and convert time. It makes no sense to have a 12-bit A/D converter if, while the conversion is in progress, the voltage changes by the equivalent of the four LSBs.

The problem of converting a varying voltage can be resolved by the use of a sample/hold device. The output of the sample/hold device whose performance is shown in Figure 14.20 follows the input voltage until the HOLD signal becomes a logic 1. The sam-

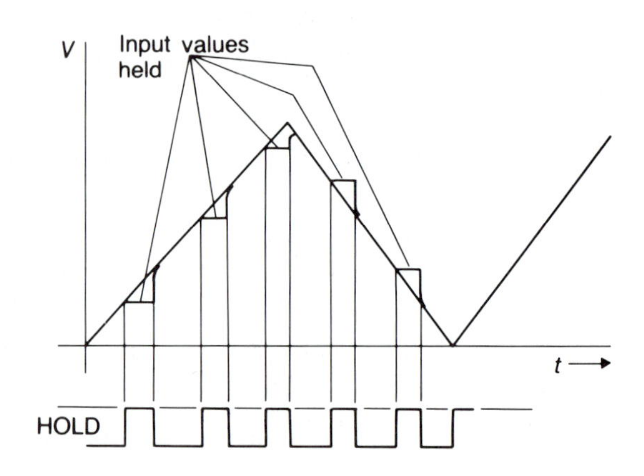

FIGURE 14.20 **Sample/Hold Applied to Varying Voltage.**

ple/hold device then holds the value of the signal until the control signal goes low. This allows an accurate conversion of a signal at a desired time, t.

14.5 Conclusion

This chapter is certainly not all-inclusive in its scope. The reader was introduced to the subject of processor I/O and interfacing to various devices. Becoming proficient in this subject involves continued study of both analog and digital systems, and constant reading of trade magazines and technical journals to keep abreast of new developments and devices. It is the authors' hope that the reader is impressed with the enormity of the task of acquiring design skills and keeping them current. The conclusion of this text is not the end, but a first step.

Problems

1. Design an interface system for interfacing 32 switches such that the processor may determine which switch or switches have been closed after being interrupted by a single switch closure. The switches may be closed for longer than it takes to service the request initiated by the switch. Assume that the switches are bounceless. Circuitry should be added that will allow the processor to respond only once to a switch closure until the switch has been opened and reclosed. Switches 0–15 are of higher priority than switches 16–31. The switches are the only devices that are operating under interrupt control. Use the priority interrupt system illustrated in Figure 14.6.
2. An analog signal $f(t)$ is to be sampled and digitized without the use of a sample/hold circuit. What is the maximum number of bits required in the converter for optimum accuracy (within 1/2 LSB) if the convert time is 1 μs/bit? $f(t) = 5 \sin(100t)$.
3. Design an eight-bit successive approximation D/A converter using the μP of Figure 14.2 as the control circuit. External devices available are an eight-bit register, an eight-bit D/A converter, and an analog comparator. A start conversion signal (pulse) starts the process. The processor should output a pulse when the conversion is complete. Show the device interconnections and a flowchart of the program that would perform the control functions.
4. Select a commercially available 2K $\times$ 8 RAM chip and design a 16K $\times$ 8 memory based on that chip that can be used by the μP in Figure 14.2. Show the connections to the processor.
5. Design a circuit that will take advantage of memory-mapped I/O and provide control signals for transferring data to and from 15 ports.

6. Select any commercially available A/D converter and design the interface to the μP of Figure 14.2.
7. Design the decode circuits to provide control signals for 10 ports using I/O-mapped I/O.
8. Design the interface for a system that upon receipt of a pulse reads a port with address 40H and has provisions to output a number of pulses based on the information read from port 40H. The pulse that initiates the transaction is very short and may not be high long enough for the processor to recognize it. Note that your circuit does not have to know how many pulses to output, but must provide the hardware and data paths for the programmer to use.
9. Find a commercially available peripheral interface device and interface it to the μP of Figure 14.2. The device should have at least two programmable ports.

Suggested Readings

Aaron, M. R., and Mitra, S. K. "Synthesis of resistive digital-to-analog conversion ladders for arbitrary codes with fixed positive weights." *IEEE Trans. Elect. Comp.* vol. EC-16 (1967): 277.

Aasnaes, H. B., and Harrison, T. J. "Triple play speeds A-D conversion." *Electron.* vol. 41 (1968): 69.

Bonivento, C. "On a recursive A/D conversion technique." *Proc. IEEE.* vol. 70 (1982): 1240.

Choudhury, J. K.; Bhattacharyya, S. N.; and Upadhya, R. "Analogue multiplier/divider with digital output." *IEE Proc. G., Elect. Cir. & Syst.* vol. 130 (1983): 101.

Current, K. W. "Quaternary-to-analogy conversion with resistive ladders." *Proc. IEEE.* vol. 70 (1982): 408.

Miyata, T.; Tamagawa, K.; and Watahiki, T. "Ternary-to-analog converters using resistor ladders." *Proc. IEEE.* vol. 67 (1979): 1165.

Morgan, D. R. "A/D conversion using geometric feedback AGC." *IEEE Trans. Comp.* vol. C-24 (1975): 1074.

Shaha, A. R., and Mazumder, B. C. "A flash-type frequency-to-code converter." *Proc. IEEE.* vol. 72 (1984): 530.

Taylor, F. J., and Dirr, W. "A new residue to decimal converter." *Proc. IEEE.* vol. 73 (1985): 378.

Towers, M. S. "Programmable waveform generator using linear interpolation with multiplying D/A converters." *IEE Proc. G., Elect. Cir. & Syst.* vol. 129 (1982): 19.

Woods, J. V., and Zobel, R. N. "Fast synthesized cyclic-parallel analogue-digital converter." *IEE Proc. G., Elect. Cir. & Syst.* vol. 127 (1980): 45.

Yuen, C. K. "Analog-to-Gray code conversion." *IEEE Trans. Comp.* vol. C-27 (1978): 971.

Yuen, C. K. "Another design of the successive approximation register for A/D converters." *Proc. IEEE.* vol. 67. (1979): 873.

Yuen, C. K. "Negabinary A/D conversion." *IEEE Trans. Comp.* vol. C-29 (1980): 740.

Yuen, C. K. "Flexible A/D conversion using a ROM in place of a SAR." *Proc. IEEE.* vol. 71 (1983): 1454.

Zohar, S. "A/D conversion for radix (−2)." *IEEE Trans. Comp.* vol. C-22 (1973): 698.

Zurada, J. M. "Application of multiplying digital-to-analogue converter to digital control of active filter characteristics." *IEE Proc. G., Elect. Cir. & Syst.* vol. 128 (1981): 91.

Zurada, J., and Goodman, K. "Equivalent circuit for multiplying d.a.c. using *R*-2*R* ladder network." *Elect. Lett.* vol. 16 (1980): 925.

Reference Data

Intel Corporation, Literature Department, 3065 Bowers Avenue, Santa Clara, CA 95051:

Microsystem Components Volume I,
Microsystem Components Volume II,
The MCS-80/85 User's Manual,
MCS-48 Family of Single Chip Microcomputers User's Manual,
Microprocessor and Peripheral Handbook.

Texas Instruments, Information Publishing Center, P.O. Box 225012 MS-54, Dallas, TX 75265:

TTL Data Books Volumes 1–4,
Interface Circuits Data Book,
Fundamentals of Microcomputer Design.

Motorola Inc., MOS Integrated Circuits Division, Microprocessor Division, Austin, TX 78721:

Single-Chip Microcomputer Data,
8-Bit Microprocessor & Peripheral Data.

Analog Devices, P.O. Box 796, Norwood, MA 02062-0796:

Microprocessor Systems Handbook,
Analog-Digital Conversion Handbook,
Transducer Interfacing Handbook

INDEX